Lecture Notes in Computer Science 14883

Founding Editors

Gerhard Goos
Juris Hartmanis

AF172632

The series Lecture Notes in Computer Science (LNCS), including its subseries Lecture Notes in Artificial Intelligence (LNAI) and Lecture Notes in Bioinformatics (LNBI), has established itself as a medium for the publication of new developments in computer science and information technology research, teaching, and education.

LNCS enjoys close cooperation with the computer science R & D community, the series counts many renowned academics among its volume editors and paper authors, and collaborates with prestigious societies. Its mission is to serve this international community by providing an invaluable service, mainly focused on the publication of conference and workshop proceedings and postproceedings. LNCS commenced publication in 1973.

Cheqing Jin · Shiyu Yang · Xuequn Shang ·
Haofen Wang · Yong Zhang
Editors

Web Information Systems and Applications

21st International Conference, WISA 2024
Yinchuan, China, August 2–4, 2024
Proceedings

 Springer

Editors
Cheqing Jin
East China Normal University
Shanghai, China

Shiyu Yang
Guangzhou University
Guangzhou, China

Xuequn Shang
Northwestern Polytechnical University
Xi'an Shaanxi, China

Haofen Wang
Tongji University
Shanghai, China

Yong Zhang
Tsinghua University
Beijing, China

ISSN 0302-9743   ISSN 1611-3349 (electronic)
Lecture Notes in Computer Science
ISBN 978-981-97-7706-8   ISBN 978-981-97-7707-5 (eBook)
https://doi.org/10.1007/978-981-97-7707-5

This Springer imprint is published by the registered company Springer Nature Singapore Pte Ltd.
The registered company address is: 152 Beach Road, #21-01/04 Gateway East, Singapore 189721, Singapore

If disposing of this product, please recycle the paper.

Preface

It is our great pleasure to present the proceedings of the 21th CCF Conference on Web Information Systems and Applications in China (WISA 2024). WISA 2024 was organized by the China Computer Federation Technical Committee on Information Systems (CCF TCIS) and Ningxia University. WISA 2024 provided a premium forum for researchers, professionals, practitioners, and officers closely related to information systems and applications to discuss the theme of intelligent information systems, digital transformation, and information system security, focusing on difficult and critical issues in metaverse, knowledge graph, blockchain, and recommendation systems, and the promotion of innovative technology for new application areas of information systems.

WISA 2024 was held in Yinchuan, Ningxia, China, during August 2–4, 2024. WISA 2024 focused on knowledge construction, intelligent computing, large language models, security, and information system applications, emphasizing the technology used to solve the difficult and critical problems in intelligent information systems, digital transformation, and information security.

This year we received 193 submissions, each of which was assigned to at least three Program Committee (PC) members to review. The peer review process was double blind. The thoughtful discussions on each paper by the PC resulted in the selection of 39 full research papers (an acceptance rate of 20.2%) and 11 short papers. The program of WISA 2024 included keynote speeches and topic-specific invited talks by famous experts in various areas of artificial intelligence and information systems to share their cutting-edge technologies and views about the state of the art in academia and industry. Other events included industrial forums and the CCF TCIS salon.

We are grateful to the general chairs, Qingyun Wang (Ningxia University) and Chunxiao Xing (Tsinghua University), as well as to all the PC members and external reviewers who contributed their time and expertise to the paper reviewing process. We would like to thank all the members of the Organizing Committee, and the many volunteers, for their great support in the conference organization. Especially, we would also like to thank the publication chairs, Cheqing Jin (East China Normal University), Shiyu Yang (Guangzhou University), Buyu Wang (Inner Mongolia Agricultural University), Wei Li (Harbin Engineering University), and Minghe Yu (Northeastern University), for their efforts on the publication of the conference proceedings. Last but not least, many thanks to all the authors who submitted their papers to the conference.

July 2024

Xuequn Shang
Haofen Wang
Yong Zhang

Organization

General Co-chairs

Qingyun Wang   Ningxia University, China
Chunxiao Xing   Tsinghua University, China

Program Committee Co-chairs

Xuequn Shang   Northwestern Polytechnical University, China
Haofen Wang   Tongji University, China
Yong Zhang   Tsinghua University, China

Workshop Co-chairs

Xiang Zhao   National University of Defense Technology,
     China
Lemen Chao   Renmin University of China, China

Publication Co-chairs

Cheqing Jin   East China Normal University, China
Shiyu Yang   Guangzhou University, China
Buyu Wang   Inner Mongolia Agricultural University, China
Wei Li   Harbin Engineering University, China
Minghe Yu   Northeastern University, China

Publicity Co-chairs

Xin Wang   Tianjin University, China
Derong Shen   Northeastern University, China
Jianbin Qin   Shenzhen University, China
Li Sun   Ningxia University, China

Challenge Co-chairs

Wei Song   Wuhan University, China
Jiali Mao   East China Normal University, China

Sponsor Co-chairs

Zhenxing Li   Beijing Jieruan Century Information Technology
                            Co., Ltd., China
Shenggen Ju   Sichuan University, China

Website Co-chairs

Li Bi   Ningxia University, China
Bin Xu   Northeastern University, China

CCF Liaison

Yufei Wu   CCF Business Headquarters, China

Organizing Committee Co-chairs

Fang Du   Ningxia University, China
Feng Feng   Ningxia University, China

Program Committee

Baoning Niu   Taiyuan University of Technology, China
Baoyan Song   Liaoning University, China
Bin Li   Yangzhou University, China
Bo Fu   Liaoning Normal University, China
Bohan Li   Nanjing University of Aeronautics and
                            Astronautics, China
Bolin Chen   Northwestern Polytechnical University, China
Buyu Wang   Inner Mongolia Agricultural University, China
Chao Kong   Anhui Polytechnic University, China
Chao Lemen   Renmin University, China

Chen Liu	North China University of Technology, China
Chenchen Sun	Tianjin University of Technology, China
Chengcheng Yu	Shanghai Polytechnic University, China
Cheqing Jin	East China Normal University, China
Chuanqi Tao	Nanjing University of Aeronautics and Astronautics, China
Chunyao Song	Nankai University, China
Dan Yin	Beijing University of Civil Engineering and Architecture, China
Derong Shen	Northeastern University, China
Dian Ouyang	Guangzhou University, China
Dong Li	Liaoning University, China
Erping Zhao	Xizang Minzu University, China
Fan Zhang	Guangzhou University, China
Fang Zhou	East China Normal University, China
Feng Qi	East China Normal University, China
Feng Zhao	Huazhong University of Science and Technology, China
Fengda Zhao	Yanshan University, China
Gansen Zhao	South China Normal University, China
Genggeng Liu	Fuzhou University, China
Guan Yuan	China University of Mining and Technology, China
Guanglai Gao	Inner Mongolia University, China
Guigang Zhang	Institute of Automation, Chinese Academy of Sciences, China
Guojiang Shen	Zhejiang University of Technology, China
Guojun Wang	Guangzhou University, China
Haitao Wang	Zhejiang Lab, China
Haiwei Zhang	Nankai University, China
Haofen Wang	Tongji University, China
Hua Yin	Guangdong University of Finance and Economics, China
Huang Mengixng	Hainan University, China
Hui Li	Guizhou University, China
Jiadong Ren	Yanshan University, China
Jiali Mao	East China Normal University, China
Jianbin Qin	Shenzhen Institute of Computing Science, Shenzhen University, China
Weijin Jiang	Hunan University of Commerce, China
Jianye Yang	Guangzhou University, China
Jiazhen Xi	Huobi Group, China
Jinbao Wang	Harbin Institute of Technology, China

Jinguo You	Kunming University of Science and Technology, China
Jiping Zheng	Nanjing University of Aeronautics and Astronautics, China
Jun Pang	Wuhan University of Science and Technology, China
Jun Wang	iWudao, China
Junying Chen	South China University of Technology, China
Kai Wang	Shanghai Jiao Tong University, China
Kaiqi Zhang	Harbin Institute of Technology, China
Lan You	Hubei University, China
Lei Xu	Nanjing University, China
Li Jiajia	Shenyang Aerospace University, China
Lin Li	Wuhan University of Technology, China
Lin Wang	Wenge Group, China
Lina Chen	Zhejiang Normal University, China
Ling Chen	Yangzhou University, China
Lingyun Song	Northwestern Polytechnical University, China
Linlin Ding	Liaoning University, China
Liping Chen	Tarim University, China
Long Yuan	Nanjing University of Science and Technology, China
Luyi Bai	Northeastern University, China
Lyu Ni	East China Normal University, China
Mei Yu	Tianjin University, China
Meihui Zhang	Beijing Institute of Technology, China
Ming Gao	East China Normal University, China
Minghe Yu	Northeastern University, China
Ningyu Zhang	Zhejiang University, China
Peng Cheng	East China Normal University, China
Qian Zhou	Nanjing University of Posts and Telecommunications, China
Qiaoming Zhu	Soochow University, China
Qingsheng Zhu	Chongqing University, China
Qingzhong Li	Shandong University, China
Qinming He	Zhejiang University, China
Ronghua Li	Beijing Institute of Technology, China
Ruixuan Li	Huazhong University of Science and Technology, China
Shanshan Yao	Shanxi University, China
Shaojie Qiao	Chengdu University of Information Technology, China
Sheng Wang	Wuhan University, China

Shengli Wu	Jiangsu University, China
Shi Lin Huang	Mingyang Digital, China
Shiyu Yang	Guangzhou University, China
Shujuan Jiang	China University of Mining and Technology, China
Shumin Han	Northeastern University, China
Shuo Yu	Dalian University of Technology, China
Shurui Fan	Hebei University of Technology, China
Tianxing Wu	Southeast University, China
Tieke He	Nanjing University, China
Tiezheng Nie	Northeastern University, China
Wei Li	Harbin Engineering University, China
Wei Song	Wuhan University, China
Wei Wang	East China Normal University, China
Wei Yu	Wuhan University, China
Weiguang Qu	Nanjing Normal University, China
Weiyu Guo	Central University of Finance and Economics, China
Weimin Li	Shanghai University, China
Weiwei Ni	Southeast University, China
Xiang Zhao	National University of Defense Technology, China
Xiangfu Meng	Liaoning Technical University, China
Xiangjie Kong	Zhejiang University of Technology, China
Xiangrui Cai	Nankai University, China
Xiaohua Shi	Shanghai Jiao Tong University, China
Xiaojie Yuan	Nankai University, China
Xiaoran Yan	Zhejiang Lab, China
Ximing Li	Jilin University, China
Xinbiao Gan	National University of Defense Technology, China
Xingce Wang	Beijing Normal University, China
Xu Liu	SAP Labs China, China
Xu Lizhen	Southeast University, China
Xueqing Zhao	Xi'an Polytechnic University, China
Xuequn Shang	Northwestern Polytechnical University, China
Xuesong Lu	East China Normal University, China
Yajun Yang	Tianjin University, China
Yanfeng Zhang	Northeastern University, China
Yanhui Ding	Shandong Normal University, China
Yanhui Gu	Nanjing Normal University, China
Yanlong Wen	Nankai University, China

Yanping Chen	Guizhou University, China
Ye Liang	Beijing Foreign Studies University, China
Yi Cai	South China University of Technology, China
Ying Zhang	Nankai University, China
Yinghua Zhou	University of Science and Technology of China, China
Yingxia Shao	Beijing University of Posts and Telecommunications, China
Yong Qi	Xi'an Jiaotong University, China
Yong Tang	South China Normal University, China
Yong Zhang	Tsinghua University, China
Yonggong Ren	Liaoning Normal University, China
Yongquan Dong	Jiangsu Normal University, China
Yu Gu	Northeastern University, China
Yuan Li	North China University of Technology, China
Yuanyuan Zhu	Wuhan University, China
Yue Kou	Northeastern University, China
Yuhua Li	Huazhong University of Science and Technology, China
Yupei Zhang	Northwestern Polytechnical University, China
Yuren Mao	Zhejiang University, China
Zhang Sijia	Dalian Ocean University, China
Zhigang Wang	Ocean University of China, China
Zhiyong Peng	Wuhan University, China
Zhongbin Sun	China University of Mining and Technology, China
Zhongle Xie	Zhejiang University, China
Zhuoming Xu	Hohai University, China
Ziqiang Yu	Yantai University, China
Mingdong Zhu	Henan Institute of Technology, China
Shan Lu	Southeast University, China
Bin Xu	Northeastern University, China

Contents

Knowledge Construction

Intelligent Service

Intelligent Computing

Large Language Model

Security

Information System Applications

Knowledge Construction

Iterative Transfer Knowledge Distillation and Channel Pruning for Unsupervised Cross-Domain Compression

Zhiyuan Wang[1], Long Shi[1(✉)], Zhen Mei[1,2], Xiang Zhao[3], Zhe Wang[4], and Jun Li[1]

[1] School of Electronic and Optical Engineering, Nanjing University of Science and Technology, Nanjing 210094, China
`longshi@njust.edu.cn`
[2] National Mobile Communications Research Laboratory, Southeast University, Nanjing 210096, China
[3] College of Systems Engineering, National University of Defense Technology, Hunan 410073, China
[4] School of Computer Science and Engineering, Nanjing University of Science and Technology, Nanjing 210094, China

Abstract. Practical applications of deep learning are challenged with critical issues that the distributions of training data and testing data are different and the labels of testing data are insufficient. To address these problems, unsupervised domain adaptation (UDA) based transfer learning has gained significant attention. However, advanced deep learning models of UDA are too complex for real-time and resource-constrained applications. In this paper, we propose an iterative transfer model compression (ITMC) method with two key modules, i.e., transfer knowledge distillation (TKD) and adaptive channel pruning (ACP). During each epoch, the TKD module achieves model compression by distilling the knowledge from the teacher model to the student model, while facilitating the transfer of cross-domain knowledge to enhance the performance in the target domain. Concurrently, with the aid of the ACP module, redundant channels in the student model are pruned to reduce the computational cost while retaining the model accuracy. In particular, the alternation of ACP and TKD ensures effective knowledge transfer, balancing the model size and its performance in the target domain. Experimental results demonstrate that ITMC approach achieves higher accuracy under the same compression ratio compared with the state-of-the-art methods.

Keywords: Deep learning · Unsupervised domain adaptation · Adaptive channel pruning · Transfer knowledge distillation

1 Introduction

Deep learning (DL) has ascended to prominence as a foundational technology within the realm of artificial intelligence. As a key tool of DL, convolutional

C. Jin et al. (Eds.): WISA 2024, LNCS 14883, pp. 3–15, 2024.
https://doi.org/10.1007/978-981-97-7707-5_1

neural networks (CNNs) efficiently extract features from data to handle complex tasks such as computer vision and natural language processing [1]. Ideally, the CNN model is trained and tested on a large amount of labeled data and both the training and testing data follow the same distribution. However, in many real-world applications, collecting sufficient labeled testing data is infeasible and the distribution of the testing data is different from that of the source data used for model training. This dissimilarity induces performance deterioration in dataset with unknown distributions (i.e., target domain).

One of the main solutions to this problem is transfer learning [2,3], which transfers the knowledge from the source domain to the target domain. Particularly, as a typical method of transfer learning, unsupervised domain adaptation (UDA) focuses on handling the cases where the target domain is unlabeled. Generally, the UDA methods can be categorized into three types, i.e., sample-based UDA, mapping-based UDA, and adversarial-based UDA. First, sample-based UDA relies on samples with similar distributions across different domains [4]. Second, mapping-based UDA aims to learn a mapping function that projects data from the source and target domains into a shared feature space [5,6]. Third, adversarial-based UDA aims to reduce the domain difference between the source domain and target domain through adversarial training [7].

Another problem is that with the growth of data scale, neural network experiences a remarkable surge in both parameters and complexity. This consumes a large amount of computational resources, which will limit the applicability of UDA in resource-constrained environments. Driven by this issue, network compression has received much attention over recent years and the classical compression methods include quantization [8], pruning [9,10], low-rank decomposition [11], and knowledge distillation (KD) [12,13]. Moreover, researchers have also investigated the combination of multiple compression methods to harness complementary benefits of diverse methods. For example, pruning and KD were combined as a two-stage compression in [14], where the student model was first pruned and then KD was used to improve the performance of the pruned model.

In spite of these benefits, the direct integration of these approaches encounters some limitations. First, the separate optimization of these diverse techniques cannot fully exploit their inherent potential. For example, although [14] employed KD to improve the accuracy of the pruned model, the expressive power of the pruned model deteriorates due to the pruning. Hence, the performance improvement is limited. Second, there are very few attempts that have been made to compress UDA models. In [15], the authors introduced a method called transfer channel pruning (TCP) to compress the deep UDA model, wherein a transfer evaluation criterion based on the maximum mean discrepancy (MMD) loss was applied to identify and prune unimportant channels. However, this technique requires a pre-adapted model for the target domain prior to pruning. Otherwise, the model accuracy may decrease significantly after pruning. In [16], the authors extended the concept of KD to UDA and introduced the knowledge adaptation method. In [17], the authors proposed a progressive KD (PKD) that enables the student model to gradually learn domain-invariant features from the teacher model by dynamically adjusting the importance between UDA and PKD. Since the structure of the student model is randomly selected, there is no guarantee that the student model can be free of redundancy after KD.

To address the aforementioned problems, we propose an iterative transfer model compression (ITMC) method by alternately performing transfer knowledge distillation (TKD) and adaptive channel pruning (ACP). The main objective is to compress deep neural network while simultaneously mitigating the impact of domain shifts. The contributions of this paper are as follows:

- We propose an ITMC algorithm for UDA model compression. In each epoch, TKD and ACP are performed iteratively. In the TKD module, the teacher model and student model are co-trained to make the student model more adaptive to the target domain, while ACP is employed to further simplify the student model by pruning the redundant channels.
- A collaborative loss in the TKD module is designed to help student model training. The objective is to keep the predictions of the student model consistent with those of the teacher model in both the source and target domains. Moreover, a trade-off factor is proposed to balance between fitting labeled data in the source domain and adapting unlabeled data in the target domain.
- To reduce the prediction difference before and after pruning, we develop a flexible pruning strategy for the student model in the ACP module, along with a pruning rate formula to determine whether the pruning should occur in each epoch. As the training progresses, the student model gradually adapts to the pruned model structure with the guidance of the teacher model.
- In our experiments, VGG16 and ResNet50 are employed to evaluate the proposed approach on two datasets, i.e., Office-31 and ImageCLEF-DA. It is shown that ITMC achieves approximately 10% accuracy improvement compared with state-of-the-art model compression methods of UDA.

2   Proposed ITMC Method

2.1   Main Idea

This section presents the proposed ITMC approach for UDA. As shown in Fig. 1, in each epoch, the UDA task is first performed by the teacher model. Subsequently, in the TKD module, we introduce a collaborative loss to help the teacher model $\theta_{T,e}$ distill the learned knowledge from both the source and target domains into the student model $\theta_{S,e}^0$. The student model then calculates the pruning rate and sorts the channels in the ACP module. The pruning is performed through multiple iterations until the target pruning rate of the current epoch is reached. After pruning in each iteration, TKD is used again to make the student model $\theta_{S,e}^1$ adapt to pruned structure and the student model in each iteration is updated as $\theta_{S,e}^2$. The above process is repeated multiple epochs until the student model is converged. In the following subsections, each module of the proposed training process will be introduced in detail.

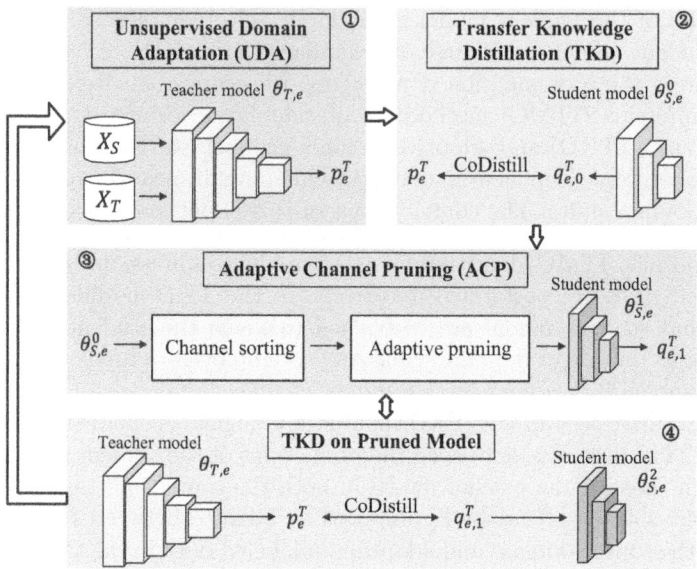

Fig. 1. The training workflow of the proposed ITMC method.

2.2   Transfer Knowledge Distillation (TKD)

In this subsection, we propose a TKD approach to aid the distillation of the source and target domain knowledge from the teacher model into the student model. Additionally, a trade-off factor is introduced to balance the student model between fitting the source domain and adapting to the target domain.

Teacher Model Training of UDA. Prior to TKD, the UDA is performed for the teacher model. The UDA loss of the teacher model consists of the MMD loss and cross-entropy loss [18]. Notably, MMD is commonly utilized in UDA to quantify the distribution discrepancy between the source and target domains [5] and the loss function is defined as

$$
L_{mmd} = \left\| \frac{1}{N_S} \sum_{i=1}^{N_S} \phi(x_i) - \frac{1}{N_T} \sum_{j=1}^{N_T} \phi(x_j) \right\|_{\mathcal{H}}^2, \tag{1}
$$

where the mapping function $\phi(\cdot)$ maps the original data to a reproducing kernel hilbert space denoted by $\mathcal{H}$; x_i and x_j represent the i-th sample of the source domain and the j-th sample of the target domain, respectively; N_S and N_T denote the numbers of samples in source domain and target domain, respectively.

Furthermore, the supervised cross-entropy loss in the source domain can be formulated as $L_{Tce}(\boldsymbol{y}, \hat{\boldsymbol{y}}) = -\frac{1}{N_S} \sum_{i=1}^{N_S} \sum_{c=1}^{C} y_{ic} \log \hat{y}_{ic}$, where C is the number of classes; $\boldsymbol{y}$ and $\hat{\boldsymbol{y}}$ denote the labels and the predictions of training samples respectively; y_{ic} and $\hat{y}_{ic}$ denote the true label and the prediction that the i-th

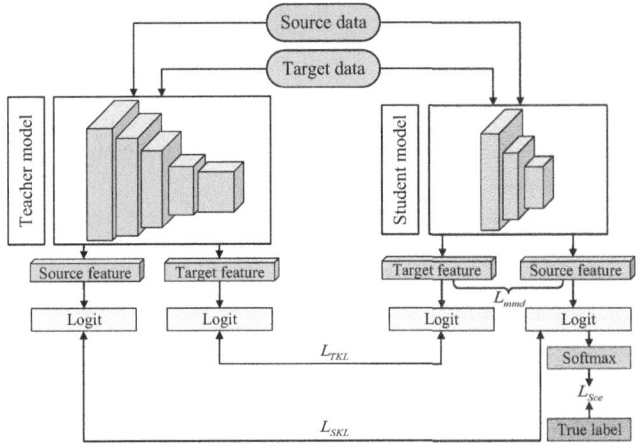

Fig. 2. The training architecture of TKD.

source domain sample belongs to class c respectively. As a result, the total loss function for the teacher model is given by

$$L_T = L_{Tce}(\boldsymbol{y}, \hat{\boldsymbol{y}}) + \alpha L_{mmd}, \qquad (2)$$

where α is a hyperparameter to balance the importance of the cross-entropy loss and the MMD loss [15].

Student Model Training of TKD. In spite of a more complex structure than the student model, the teacher model can effectively extract information from both the source domain and the target domain. Hence, it is preferred to transfer the knowledge learned by the teacher model to the student model, rather than training the student model alone. Unlike the traditional KD method in [12], the proposed TKD prioritizes aligning the predictive distributions of the student model with those of the teacher model for both the source and target domains, with special consideration for scenarios where the target data are unlabeled.

As shown in Fig. 2, we devise a collaborative loss based on Kullback-Leibler (KL) divergence. This loss function quantifies the similarity between the outputs of the student and teacher models in both the source and target domains. The proposed collaborative loss function is given by

$$L_{KL}(\boldsymbol{p}, \boldsymbol{q}) = L_{SKL} + L_{TKL} = -\sum_{i=1}^{N_S}\sum_{c=1}^{C} p_{ic}^T \log \frac{p_{ic}^T}{q_{ic}^T} - \sum_{j=1}^{N_T}\sum_{c=1}^{C} p_{jc}^T \log \frac{p_{jc}^T}{q_{jc}^T}, \qquad (3)$$

where L_{SKL} and L_{TKL} denote the KL loss in the source domain and target domain, respectively; T is the temperature used to soften the model output.

Note that $\boldsymbol{p}$ and $\boldsymbol{q}$ in (3) are the predicted probabilities of the teacher model and the student model respectively, where $p_{ic}^T = \frac{\exp(z_{ic}/T)}{\sum_{k=1}^{C} \exp(z_{ik}/T)}$ $(p_{jc}^T =$

$\frac{\exp(z_{jc}/T)}{\sum_{k=1}^{C}\exp(z_{jk}/T)}$) and $q_{ic}^{T} = \frac{\exp(v_{ic}/T)}{\sum_{k=1}^{C}\exp(v_{ik}/T)}$ ($q_{jc}^{T} = \frac{\exp(v_{jc}/T)}{\sum_{k=1}^{C}\exp(v_{jk}/T)}$) represent the probabilities that the teacher model and the student model predict that the source sample x_i (target sample x_j) belongs to class c at temperature T, respectively. Moreover, z_{ic} (z_{jc}) and v_{ic} (v_{jc}) represent the output logits that the teacher model and the student model predict that source sample x_i (target sample x_j) belongs to class c, respectively.

In pursuit of performance enhancement, the student model is trained by absorbing knowledge from the teacher model and utilizing source domain labeled data. As a result, the overall loss function of the student model is given by

$$L_S = (1 - \beta)L_{Sce} + \beta(L_{KL}(\boldsymbol{p}, \boldsymbol{q}) + L_{mmd}), \tag{4}$$

where L_{Sce} is the cross-entropy loss of the student model on the source domain; L_{mmd} is incorporated to synchronously reduce domain discrepancy, ultimately improving its performance in the target domain; the importance between the different losses is balanced by a dynamic trade-off factor β, given by

$$\beta = 1 - \exp(\frac{-8e}{N_e}). \tag{5}$$

Here, $e \in (0, N_e]$ and N_e is the total number of training epochs.

Since UDA and TKD are performed alternately, the knowledge that can be distilled from the teacher model to the student model is limited at the beginning of the training. In view of this, we suggest that the student model places more importance on learning the source domain knowledge at the beginning of training, and then gradually shifts the importance to the learning of the teacher model's knowledge. Therefore, β gradually increases and converges to 1, as the student model adapts to the target domain.

2.3   Adaptive Channel Pruning (ACP)

In order to further compress the student model, ACP is performed on the student model after TKD. To maintain the performance of the pruned model, we design an adaptive pruning formula that involves the number of training epochs. The process of ACP are shown in Fig. 3.

Channel Sorting Criterion. The goal of channel sorting is to evaluate the importance of channels in each layer in order to prune insignificant channels. To evaluate the contribution of each channel to the model's performance, we apply the channel evaluation criterion in [15] to the i-th channel in the l-th convolutional layer, given by

$$R(\boldsymbol{a}_{l,i}) = \left|(1 - \beta)\frac{\partial L_{Sce}}{\partial \boldsymbol{a}_{l,i}^{s}}\boldsymbol{a}_{l,i}^{s} + \beta(\frac{\partial L_{KL}(\boldsymbol{p}, \boldsymbol{q})}{\partial \boldsymbol{a}_{l,i}^{t}}\boldsymbol{a}_{l,i}^{t} + \frac{\partial L_{mmd}}{\partial \boldsymbol{a}_{l,i}^{t}}\boldsymbol{a}_{l,i}^{t})\right|, \tag{6}$$

where $\boldsymbol{a}_{l,i}^{s}$ and $\boldsymbol{a}_{l,i}^{t}$ represent the activation values with the source domain and target domain data respectively. Notably, a larger value of $R(\boldsymbol{a}_{l,i})$ in (6) indicates a higher importance to the model's performance.

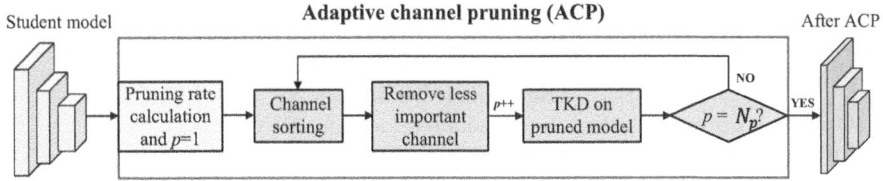

Fig. 3. The process of ACP. First, the pruning rate is calculated based on the adaptive pruning formula. Then, the importance of the channels is ranked and the unimportant channels are iteratively pruned through N_p^e iterations. Finally, the model after each pruning is finetuned by TKD

Adaptive Channel Pruning. According to the criterion in (6), we rank the importance of each channel in the student model. Based on this, we propose an adaptive pruning rule to iteratively prune the least important channels of the model. The pruning rate of the e-th epoch is given by

$$P(e) = \begin{cases} \frac{N_e^f - N_e^s}{60 N_e^2} e, & \text{if } N_e^s \le e < N_e^f, \\ 0, & \text{otherwise,} \end{cases} \qquad (7)$$

where $0 < N_e^s < N_e^f \le N_e$; N_e^s and N_e^f are the epochs that start and complete pruning, respectively.

The motivation of $P(e)$ in (7) is clarified as follows. First, in the early stage of training, the weight parameters of the student model are often unstable. Premature or excessive pruning could result in the loss of vital information, potentially degrading the performance and hindering convergence. Then, as the training progresses, the student model stabilizes, gradually increasing the pruning rate allows the model to become progressively simplified.

Upon determining the pruning rate $P(e)$ in (7), the number of iterations for pruning in the e-th epoch is given by $N_p^e = N_c^e \times P(e)/n$, where N_c^e is the total number of channels of the student model in the e-th epoch; n is the number of channels to prune at each iteration, and there are total nN_p^e channels to be pruned in the e-th epoch.

After pruning n channels for each iteration, the model experiences a reduction in its expressive power. To tackle this issue, we further finetune the pruned student model again through TKD using (4) to adjust the weights and adapt them to the pruned structure. The complete training process of the proposed ITMC approach is outlined in Algorithm 1.

3 Experiments

3.1 Datasets

Office-31. This dataset is commonly used for research on transfer learning. It consists of image data from three different domains: Amazon (A), Webcam (W),

---

Algorithm 1: the ITMC Algorithm for UDA

Input: A teacher model θ_T, a student model θ_S, source-labeled domain
   $X_S = \{(x_i, y_i)\}_{i=1}^{N_S}$, target-unlabeled domain $X_T = \{x_j\}_{j=1}^{N_T}$
Output: A pruned student model adapted for deep UDA

1 **for** $e \leftarrow 1$ **to** N_e **do**
2   Optimize the teacher model θ_T by minimizing L_T in (2) and obtain the optimized model as $\theta_{T,e}$;
3   Optimize the student model θ_S by minimizing L_S in (4) and obtain the optimized model as $\theta_{S,e}^0$;
4   Calculate the pruning rate based $P(e)$ in (7);
5   Calculate the number of pruning iterations N_p^e ;
6   **for** $t \leftarrow 1$ **to** N_p^e **do**
7   **if** the target compression rate of the student model is achieved **then**
8   break;
9   **else**
10   Sort the channels of the student model $\theta_{S,e}^0$ by the channel evaluation criterion in (6);
11   Remove the n least important channels of $\theta_{S,e}^0$ and obtain a student model $\theta_{S,e}^1$;
12   Finetune the pruned student model $\theta_{S,e}^1$ and update it as $\theta_{S,e}^2$;
13   **end**
14   **end**
15 **end**

---

and DSLR (D). Each domain contains 31 classes of target objects commonly found in office environments. However, the number of images varies across the domains. We employ six transfer scenarios to validate our ITMC method, i.e., A → W, A → D, W → D, W → A, D → W, D → A.

ImageCLEF-DA. This dataset serves as the benchmark dataset for the Image-CLEF 2014 domain adaptation challenge, encompassing four distinct domains: Caltech-256 (C), Pascal VOC 2012 (P), ImageNet ILSVRC 2012 (I), and Bing (B). Each domain consists of 12 categories, with each category containing 50 images. Consistent with Office-31, six transfer scenarios are simulated, i.e., C → P, C → I, P → C, P → I, I → C, I → P.

3.2   Baselines

Our experiments are conducted on VGG [19] and ResNet [20] networks, which are widely adopted as backbone networks for UDA methods. The proposed ITMC method is compared with the following baselines: (1) TCP-based UDA model compression [15]; (2) PKD-based UDA model compression [17]; (3) the UDA model is first pruned by TCP, followed by PKD; (4) the UDA model is first compressed by PKD, followed by TCP. Note that the four baselines maintain the same floating point operations (FLOPs) reduction rate as the proposed

ITMC. For the KD module of all the methods, two sets of student-teacher model architectures, i.e., VGG16-VGG19 and ResNet50-ResNet101 are used, which can achieve 21% and 47% FLOPs reduction respectively. For the pruning module, pruning two student models, i.e., VGG16 and ResNet50, achieves 26% and 12% decrease in FLOPs, respectively.

3.3 Implementation Details

All methods are implemented based on PyTorch with an NVIDIA GeForce RTX 4090 GPU. The training starts with pre-trained models on the ImageNet dataset. During the training, all the images are cropped to a fixed size of 224×224 and the training batch size is 32. For TKD, we set the distillation temperature $T = 20$. In the ACP process, n is set to 16, which means that 16 channels are pruned at each iteration. All the networks are trained using stochastic gradient descent (SGD) method with a momentum of 0.9. The learning rate starts from 0.001, and gradually reduces to ensure a stable training.

Table 1. The accuracy (%) with VGG16 as the student model on Office-31 dataset

Training methods	D→W	W→D	W→A	A→W	A→D	D→A	Average
TCP [15]	96.1	99.8	51.2	**76.1**	**76.2**	47.9	74.6
PKD [17]	96.9	100.0	53.3	66.5	70.5	54.2	73.6
PKD→TCP	94.0	98.2	45.3	64.2	66.3	47.3	69.2
TCP→PKD	95.6	99.6	46.8	69.2	73.1	46.9	71.9
ITMC (proposed)	**97.8**	**100.0**	**55.4**	74.0	74.1	**56.3**	**76.3**

Table 2. The accuracy (%) with ResNet50 as the student model on Office-31 dataset

Training methods	D→W	W→D	W→A	A→W	A→D	D→A	Average
TCP [15]	98.2	99.8	55.5	81.8	77.9	50.0	77.2
PKD [17]	96.2	99.4	58.5	78.2	79.5	50.8	77.1
PKD→TCP	96.7	99.2	53.8	79.1	77.7	52.8	76.6
TCP→PKD	93.6	99.6	53.7	79.6	76.5	49.7	75.5
ITMC (proposed)	**98.5**	**100.0**	**62.6**	**84.6**	**82.7**	**60.9**	**81.6**

3.4 Results and Analysis

Office-31 Dataset. The accuracy comparisons between different methods on the Office-31 dataset are shown in Table 1 and Table 2. It is observed that the accuracy can be improved by up to 11.2% compared with all the baselines with ResNet50 in the D→A task. Moreover, it is observed that the proposed method outperforms both TCP and PKD for most transfer scenarios. This is due to the fact that we jointly optimize the two compression methods, where TKD guides the student model to improve the accuracy and ACP further reduces the redundancy of the student model to help create a more compact student model.

Furthermore, we observe that the proposed method is also superior to the straightforward combinations of PKD and TCP, including PKD→TCP and TCP→PKD. This is attributed to the fact that, in TCP→PKD, the pruned student model is not sufficient to learn the teacher model knowledge after TCP. In PKD→TCP, the student model performs TCP without the guidance of the teacher model, making it difficult to capture the complex features in the data.

Table 3. The accuracy (%) with VGG16 as the student model on ImageCLEF-DA dataset

Training methods	C→I	C→P	P→C	P→I	I→P	I→C	Average
TCP [15]	77.8	64.8	87.5	80.5	72.0	90.5	78.9
PKD [17]	75.0	63.8	86.8	82.2	**73.3**	91.3	78.7
PKD→TCP	66.2	55.5	81.8	75.3	67.2	84.7	71.8
TCP→PKD	66.7	56.7	80.0	80.7	67.8	87.8	73.3
ITMC (proposed)	**81.0**	**68.0**	**89.7**	**85.3**	73.0	**92.7**	**81.6**

Table 4. The accuracy (%) with ResNet50 as the student model on ImageCLEF-DA dataset

Training methods	C→I	C→P	P→C	P→I	I→P	I→C	Average
TCP [15]	80.8	66.2	86.5	82.6	75.0	92.5	80.6
PKD [17]	76.5	67.2	85.8	83.3	72.7	92.8	79.7
PKD→TCP	81.8	67.2	87.5	82.7	71.2	91.2	80.3
TCP→PKD	76.2	63.3	87.0	83.0	70.2	93.1	78.8
ITMC (proposed)	**85.8**	**70.2**	**92.3**	**90.7**	**78.2**	**94.7**	**85.3**

ImageCLEF-DA Dataset. The accuracy of different methods on the Image-DA dataset are shown in Table 3 and Table 4. Similar to the Office-31 dataset, we observe that the proposed method achieves higher accuracy than the baselines. For example, in the C→I transfer task with VGG16 as the student model, the proposed method improves the accuracy by about 5% compared with the TCP and PKD, and about 14% compared with the straightforward combinations of PKD and TCP. Additionally, in the P→I transfer task with ResNet50 as the student model, the proposed method improves accuracy by up to 8.1% compared with all the baselines.

Therefore, KD and pruning should be seamlessly orchestrated during the training instead of treating them as two independent phases. Using the proposed ITMC approach, the complementary strengths of the two techniques are maximized to balance the model size and the performance in the target domain.

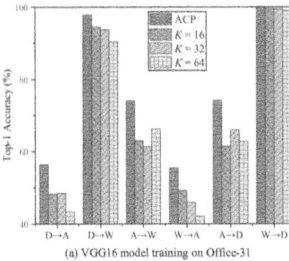
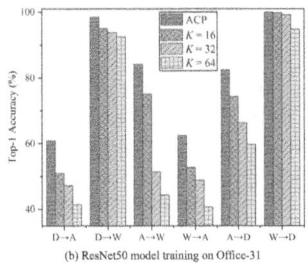

(a) VGG16 model training on Office-31   (b) ResNet50 model training on Office-31

Fig. 4. Top-1 accuracy comparison between the proposed ACP and fixed K-channel pruning under $K = 16, 32, 64$.

Pruning Strategy Analysis. The pruning rate $P(e)$ of (7) plays a crucial role in reducing the model size. To verify the effectiveness of the proposed ACP strategy, we compare the ACP with fixed K-channel pruning in each epoch using the Office-31 dataset as an example. As shown in Fig. 4, in contrast to pruning a fixed number of channels per epoch, the student model achieves higher accuracy under ACP. In addition, the accuracy drops sharply if the fixed number of pruned channels is large, e.g., $K = 64$. This is due to the fact that the weight parameters of the student model are unstable at the beginning of training, and pruning excessive channels may weaken the model's ability to represent critical task-specific features.

4   Conclusion

In this paper, we have proposed ITMC to compress the cross-domain model for UDA. Specifically, a collaborative loss has been developed in the TKD module, which enables the student model to learn feature representations from both the source and target domains. Moreover, a novel ACP module has been designed to iteratively reduce the redundancy of the student model. The iterative TKD and ACP ensure effective knowledge transfer, balancing the model size and its performance in the target domain. Compared with the state-of-the-art methods, the proposed ITMC method has superior performance under different network architectures and datasets with the same compression ratio. It is worth noting that the proposed ITMC is compatible with most of current UDA techniques.

Acknowledgement. This work was supported in part by the National Natural Science Foundation of China (No. 62371239, No. 62201258, No. 62272469, No. 62202232), in part by the Natural Science Foundation of Jiangsu Province under Grant BK20210331, in part by the Jiangsu Specially-Appointed Professor Program 2021, in part by the open research fund of National Mobile Communications Research Laboratory, Southeast University (No. 2023D12), in part by the Fundamental Research Funds for the Central Universities (No. 30923011035), in part by the Science and Technology Innovation Program of Hunan Province (No. 2023RC1007).

References

1. Zhang, Z., et al.: Object relational graph with teacher-recommended learning for video captioning. In: 2020 IEEE/CVF Conference on Computer Vision and Pattern Recognition (CVPR), pp. 13275–13285 (2020). https://doi.org/10.1109/CVPR42600.2020.01329

2. Zhuang, F., et al.: A comprehensive survey on transfer learning. Proc. IEEE **109**(1), 43–76 (2020)

3. Du, J., Nie, T., Dou, W., Shen, D., Kou, Y.: SAREM: semi-supervised active heterogeneous entity matching framework. In: Zhao, X., Yang, S., Wang, X., Li, J. (eds.) WISA 2022. LNCS, vol. 13579, pp. 77–88. Springer, Cham (2022). https://doi.org/10.1007/978-3-031-20309-1_7

4. Dai, W., Yang, Q., Xue, G.R., Yu, Y.: Boosting for transfer learning. In: Proceedings of the 24th International Conference on Machine Learning, pp. 193–200 (2007)

5. Borgwardt, K.M., Gretton, A., Rasch, M.J., Kriegel, H.P., Schölkopf, B., Smola, A.J.: Integrating structured biological data by kernel maximum mean discrepancy. Bioinformatics **22**(14), 49–57 (2006)

6. Mei, Z., Cai, K., Shi, L., Li, J., Chen, L., Immink, K.A.S.: Deep transfer learning-based detection for flash memory channels. IEEE Trans. Commun. (2024)

7. Ganin, Y., et al.: Domain-adversarial training of neural networks. J. Mach. Learn. Res. **17**(1), 1–35 (2016)

8. Yang, J., et al.: Quantization networks. In: Proceedings of the IEEE/CVF Conference on Computer Vision and Pattern Recognition, pp. 7308–7316 (2019)

9. Molchanov, P., Mallya, A., Tyree, S., Frosio, I., Kautz, J.: Importance estimation for neural network pruning. In: Proceedings of the IEEE/CVF Conference on Computer Vision and Pattern Recognition, pp. 11264–11272 (2019)

10. Blalock, D., Gonzalez Ortiz, J.J., Frankle, J., Guttag, J.: What is the state of neural network pruning? Proc. Mach. Learn. Syst. **2**, 129–146 (2020)

11. Idelbayev, Y., Carreira-Perpinán, M.A.: Low-rank compression of neural nets: learning the rank of each layer. In: Proceedings of the IEEE/CVF Conference on Computer Vision and Pattern Recognition, pp. 8049–8059 (2020)

12. Hinton, G., Vinyals, O., Dean, J.: Distilling the knowledge in a neural network. arXiv preprint arXiv:1503.02531 (2015)

13. Gou, J., Yu, B., Maybank, S.J., Tao, D.: Knowledge distillation: a survey. Int. J. Comput. Vision **129**, 1789–1819 (2021)

14. Prakosa, S.W., Leu, J.S., Chen, Z.H.: Improving the accuracy of pruned network using knowledge distillation. Pattern Anal. Appl. **24**, 819–830 (2021)

15. Yu, C., Wang, J., Chen, Y., Wu, Z.: Accelerating deep unsupervised domain adaptation with transfer channel pruning. In: 2019 International Joint Conference on Neural Networks (IJCNN), pp. 1–8. IEEE (2019)

16. Ruder, S., Ghaffari, P., Breslin, J.G.: Knowledge adaptation: teaching to adapt. arXiv preprint arXiv:1702.02052 (2017)

17. Granger, E., Kiran, M., Dolz, J., Blais-Morin, L.A., et al.: Joint progressive knowledge distillation and unsupervised domain adaptation. In: 2020 International Joint Conference on Neural Networks (IJCNN), pp. 1–8. IEEE (2020)

18. Ho, Y., Wookey, S.: The real-world-weight cross-entropy loss function: modeling the costs of mislabeling. IEEE Access **8**, 4806–4813 (2019)
19. Tammina, S.: Transfer learning using VGG-16 with deep convolutional neural network for classifying images. Int. J. Sci. Res. Publ. (IJSRP) **9**(10), 143–150 (2019)
20. Targ, S., Almeida, D., Lyman, K.: Resnet in resnet: generalizing residual architectures. arXiv preprint arXiv:1603.08029 (2016)

Aspect-Based Sentiment Classification Model Based on Multi-view Information Fusion

Yujie Wan, Tianyu Cai, Yilin Li, and Shenggen Ju[✉]

College of Computer Science, Sichuan University, Chegndu 610005, China
jsg@scu.edu.cn

Abstract. Aspect-based sentiment classification is one of the hot tasks in the field of natural language processing. The task aims to judge the sentiment polarity of the target word, also known as the aspect term, specified in the sentence. The current mainstream models aggregate the information of the aspect term neighbor nodes through the graph neural network model to judge the sentiment polarity. Compared with the previous research, this method has achieved obvious results, but it still faces some problems. First of all, the limited scale of the existing public data set constrains the training of the model, and the general knowledge representation ability has certain deficiencies. Secondly, existing methods use single-view information to judge sentiment polarity, but lack multi-view information and corresponding information fusion methods, the complementarity of sentiment feature information from different perspectives has not been studied. To solve the above problems, an aspect-based sentiment classification model based on multi-view information fusion is proposed. By constructing an inference result set from the large language model (LLM), the LLM's results are used to enhance the model's knowledge representation ability. A multi-view information fusion module is proposed to integrate information from two aspects: local fusion and global fusion, and make full use of information from different angles. The experimental results show that the model has higher classification ability than the current mainstream models, and the effectiveness of each module of the model is verified by a variety of experiments.

Keywords: Aspect-based Sentiment Classification · Graph Convolutional Neural Network · Multi-View Information · Information Fusion · Large Language Model

1 Introduction

The Internet provides a platform for people to communicate and express opinions. For instance, new users can assess products or services through comments to increase their purchasing satisfaction. Companies providing goods or services can make improvements based on consumer feedback. Relevant departments can analyze social media posts to evaluate the impact of trending events. Since the 21st century, the Internet has rapidly developed, with data growing exponentially, making sentiment analysis tasks essential in various scenarios, such as those mentioned.

C. Jin et al. (Eds.): WISA 2024, LNCS 14883, pp. 16–28, 2024.
https://doi.org/10.1007/978-981-97-7707-5_2

Current mainstream models utilize syntactic dependency parsing tools to obtain the syntactic dependency tree of text, establishing syntactic relationships between words. They then aggregate information from neighboring nodes using neural network models to determine sentiment polarity. While such methods have shown significant improvement over previous research, they still face several challenges. Firstly, model training is constrained by dataset size, and knowledge representation capabilities need enhancement. Secondly, models rely on a single view as a classification basis, lacking the fusion of multi-view information.

This paper addresses the aforementioned limitations through the following contributions: the paper proposes an aspect-based sentiment classification model based on multi-view information fusion. First, to fully utilize multi-view information, this paper introduces a multi-view information fusion module that performs information fusion at both local and global levels. Second, a LLM inference result set is constructed and used as soft target to guide model training. By fitting the inference results of the LLM, the knowledge representation capability of the model is improved.

2 Related Work

Sentiment polarity classification can be categorized into document-level, sentence-level, and aspect-level based on the granularity of the processing object. The first two belong to coarse-grained sentiment classification, aiming to predict the sentiment tendency of entire documents or sentences [1]. In contrast, aspect-based sentiment classification analyzes at the smallest granularity, aiming to determine the sentiment polarity of specified aspect terms within text. For example, in the sentence "Great food but the service was dreadful." it is inadequate to analyze the entire sentence with just one sentiment polarity. In this text, "Great" expresses positive sentiment towards "food" whereas "dreadful" expresses negative sentiment towards "service" Fine-grained sentiment classification allows for a more specific understanding of various aspects of things.

Early research on aspect-based sentiment classification primarily focused on methods based on sentiment lexicons. With the development of machine learning and the emergence of high-quality datasets, machine learning quickly became the mainstream research method. In recent years, due to advances in hardware computing power and the generation of massive internet data, deep learning-based methods have been widely and effectively applied in aspect-based sentiment classification.

Graph neural networks leverage graph structures for information modeling, propagating information through neighboring nodes and often representing word-level syntactic relationships using syntactic dependency trees. In 2019, Sun et al. [2] proposed a GCN model named CDT based on syntactic dependency trees. This model utilizes syntactic dependency trees to model relationships between words, effectively reducing the distance between aspect terms and opinion words. The syntactic dependency tree is represented as an adjacency matrix, and information propagation is achieved through a graph encoder to enhance aspect term embeddings. Methods based on graph neural networks introduce additional syntactic information into sentiment classification models and provide a basis for information propagation, becoming a recent hot topic in research. Li et al. [3] constructed a syntax GCN module based on syntactic dependency trees and a semantic GCN module based on self-attention mechanism. They integrated syntactic knowledge and semantic information using dual GCN modules and proposed dual regularizers to effectively capture word correlations. Tian et al. [4] argued that graph neural network-based research lacks an effective mechanism to differentiate the importance of dependency relationships. They proposed a type-aware GCN that comprehensively learns syntactic dependency parsing results by combining word relationships and dependency types. To address issues such as neglecting phrase information in models, Wu et al. [5] combined corpus data to build a phrase dependency graph and introduced a phrase dependency GAT. This network models the phrase dependency graph by aggregating node information, edge information, and phrase information. Zhao et al. [6] proposed an aspect term-aware weighting mechanism to control information flow towards aspect terms and introduced a loaded layer focused on aspect terms to mitigate the influence of irrelevant words. Ren et al. [7] enhanced the model's entity representation capability by utilizing a semantic and syntactic dual GCN Module for information extraction and interaction.

3   Model Description

3.1   Model Architecture

The proposed aspect-based sentiment classification model first extracts contextual features from the original text using an embedding layer. It then models the global semantic view of the text using a contrastive learning module, trains syntactic structural view information with a graph convolutional neural module, and incorporates a large language model fitting module to introduce views from LLM. The multi-view information fusion module integrates these three aspects at both local and global levels. The overall structure of the model is illustrated in Fig. 1, and detailed descriptions of each module will be provided in subsequent sections of this chapter.

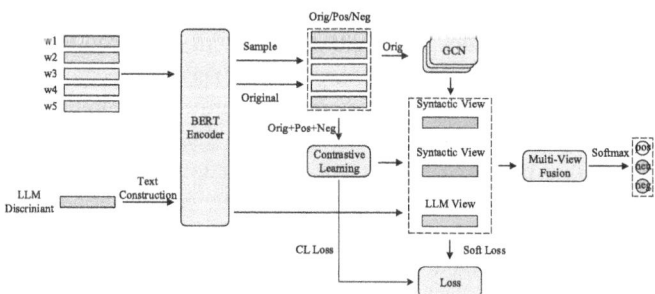

Fig. 1. The Architecture of ABSC Based on Multi-View Information Fusion

3.2   Large Language Model Inference Result Set

In this study, we utilize an interface to invoke a large language model and construct an inference result set based on its outputs. This result set resembles the original dataset and primarily includes the following components for each instance: the original text, the corresponding aspect term, and the inference result from the LLM. If a sentence contains multiple aspect terms, multiple records are created corresponding to each aspect term.

The objective of this paper is to obtain inference results from the LLM using the original dataset. Firstly, we concatenate strings from the dataset to form prompts for querying the LLM. The prompt format is as follows: "Sentence: [text], what is the sentiment polarity of the aspect term [aspect term] in this sentence?", where [text] represents the original text from the dataset, and [aspect term] denotes the aspect term contained in the current text. During the interface invocation, we encountered challenges with inconsistent response formats from the LLM. Some responses consist of complete sentences with redundant information, while others simply provide single words indicating sentiment polarity. These varied response formats complicate the construction of the result set.

To address this, we constrain the format of the LLM's responses using a prompt such as: "A. positive; B. negative; C. neutral. Please answer with option." This format resembles a multiple-choice question and significantly reduces response deviation from the LLM. Standardized responses facilitate string extraction and result set construction. Despite these efforts, certain responses still exhibit format deviations. Therefore, this paper manually corrects these abnormal responses and extracts the inference results to form the final result set.

3.3 Embedding Layer

The embedding layer utilize a BERT Encoder to obtain hidden state vectors of the text. The input sentence is formatted as S_i = [CLS] + SENTENCE + [SEP] + ASPECT + [SEP]. Additionally, to strengthen the connection between aspect terms and sentiment polarity and to enhance the semantic relevance of the LLM inference results, the model's inference results are concatenated into complete text format: "The polarity of [aspect term] is [LLM inference]." Here, [aspect term] represents the aspect term of the current text, and [LLM inference] denotes the inference result of the large language model for the current text.

3.4 Large Language Model Inference Result Fitting

To alleviate the issue of model training being constrained by dataset limitations, we fit the knowledge representation obtained from model training with the inference results from LLM, allowing the model to learn the knowledge representation of the LLM. Through the embedding layer, we obtain the hidden state representation of the LLM inference results H_i^{LLM} = Encoder(L_i), and then perform dimensionality reduction on the hidden state through linear operations, as described in Eq. (1)

$$Hidden_{LLM} = fully\_connected(H_i^{LLM}) \qquad (1)$$

Here, $fully\_connected(\cdot)$ represents a fully connected layer, and $Hidden_{LLM}$ denotes the dimensionally reduced hidden state vector. The dimensionality-reduced result serves as soft target to guide the training of the model in this module. The training loss is defined as shown in Eq. (2).

$$L_{soft} = \frac{\sum_{i=1}^{m} \sum_{j=1}^{n} |f(x_{ij}) - y_{ij}|}{n} \tag{2}$$

The model in this paper utilizes mean absolute error as the soft target loss to fit the inference results from LLM. Here, i represents the index of the sentence involved in computation, j represents the current dimension of the vector being computed, y_{ij} represents the j dimension of the vector for the i text in the LLM's inference result set, and $f(x_{ij})$ represents the computed view vector result through the model.

3.5 Contrastive Learning

This paper aims to learn global semantic view features by contrasting original sentences with semantically similar and dissimilar sentences. We introduced a contrastive learning module [10], where we construct positive samples using dropout and negative samples by replacing words with antonyms.

After computing contextual features using embedding layer, the representation of a sentence vector is obtained by calculating the maximum value at each position of the corresponding feature vectors, as shown in Eq. (3).

$$X_i = f_{\max}(h_1, h_2, h_3, ..., h_m) \tag{3}$$

For sentence $H_i^{ori} = \{h_1, h_2, h_3, ..., h_n\}$, positive samples $H_i^{pos} = \{h_1, h_2, h_3, ..., h_n\}$ and negative samples $H_i^{neg} = \{h_1, h_2, h_3, ..., h_n\}$ are constructed as part of the sample creation process. This paper utilizes the aforementioned data to train sentence vectors for the original sentences. In the aspect-based sentiment classification task, there are three labels: "positive", "negative" and "neutral". The training objective is to minimize the distance between sentence vectors with the same label and maximize the distance between sentence vectors with different labels.

The contrastive learning loss used in this paper is shown in Eq. (4).

$$L_{cl} = -\sum_{i \in Data} \log \frac{f(X_i, X_{i,s})}{f(X_i, X_{i,s}) + f(X_i, X_{i,d})} \tag{4}$$

$$f(X_i, X_{i,s/d}) = \exp(X_i \cdot X_{i,s/d} \cdot \tau) \tag{5}$$

In Eqs. (4), X_i represents the sentence vectors computed from the original dataset, $X_{i,s}$ denotes their corresponding vectors of semantic similarity, and $X_{i,d}$ represents vectors of semantic dissimilarity. Equation (5) is used to calculate the similarity between two sentences, where $X_{i,s/d}$ represents the similarity or dissimilarity of a sentence with respect to its original vector X_i, $X_{i,s}$ (for similarity), or $X_{i,d}$ (for dissimilarity). A larger result indicates greater similarity between the two sentences, with τ as the temperature coefficient.

3.6 Graph Convolutional Module

Using syntactic dependency trees, it is effective to determine whether two words in a sentence have a syntactic relationship. This syntactic relationship shortens the distance between words, facilitating the propagation of information. We introduced a graph convolutional module [10]. This study uses self-attention matrices as adjacency matrices for the graph convolutional layer. The self-attention matrix can directly participate in the graph convolution operation as weights, with Eq. (6) representing the computation process of self-attention.

$$A_k = \frac{QW^Q \times (KW^K)^T}{\sqrt{d}} \tag{6}$$

The output H from the BERT Encoder serves as the query matrix Q and value matrix K, where $W^Q \in \mathbb{R}^{d \times d}$ and $W^K \in \mathbb{R}^{d \times d}$ are learnable weight matrices, and d represents the input node dimension. A^k denotes that this matrix is the k-th attention matrix.

We also use an information masking matrix to mask node information at different distances, as represented by Eq. (7).

$$A_{ij}^k = \begin{cases} A_{ij}^k, d(i,j) < k \\ -\infty, d(i,j) \geq k \end{cases} \tag{7}$$

$d(i,j)$ indicates the shortest syntactic distance between node i and node j.

In this paper, multiple graph convolution operations are applied to aggregate nodes based on various syntactic distances, capturing information from neighboring nodes at different distances. Specifically, the node updates for the l-th layer of graph convolution are defined by Eq. (8).

$$h_i^{l,k} = \sigma(\sum_j^n A_{mask}^k W^{l,k} h_j^{l-1,k} + b^{l,k}) \tag{8}$$

The final output of the graph convolution can be represented as $H_i^k = \{h_1^{l,k}, h_2^{l,k}, h_3^{l,k}, ..., h_n^{l,k}\}$. This paper integrates multiple sets of enhanced representations by averaging them across corresponding sentences.

3.7 Multi-view Information Fusion Module

As mentioned earlier, we obtained the global semantic view through the contrastive learning module, denoted as R_g. We obtained the syntactic structure view vectors through the graph convolution module, which will be referred to as R_s for convenience. Additionally, we obtained the large language model view vectors through the BERT Encoder, denoted as R_l for subsequent representation. To fully leverage the information from the three views, this paper proposes a multi-view information fusion module, as illustrated in Fig. 2.

In the local fusion process, this module first concatenates the three types of view vectors along the 1-st dimension. The concatenated feature vectors are then separately

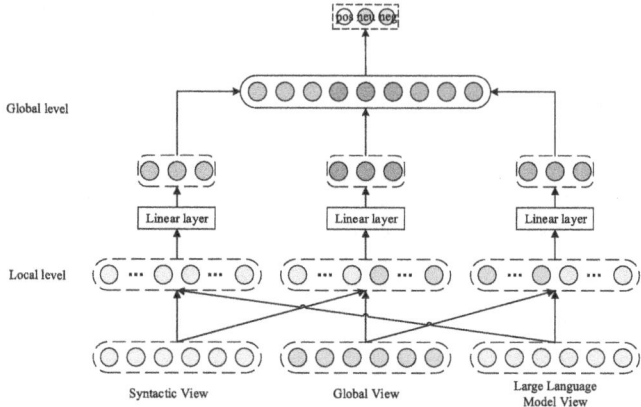

Fig. 2. Module of Multi-View Information Fusion

fed into three independent fully connected layers for dimensionality reduction. Through these steps, we obtain the locally fused syntactic-large language model view vector R_{sl}, the syntactic-global view vector R_{sg}, and the global-large language model view vector R_{gl}. Since the fully connected layers are independent, they have separate parameters. Subsequently, to fully leverage the complementary nature of the three view features, this paper further integrates them at a global level. Specifically, the three feature vectors obtained from the local fusion step are concatenated along the 0-th dimension, reduced in dimensionality using a fully connected layer, and normalized via softmax computation on the resulting vector.

3.8   Loss Function

The loss function of the model in this paper is represented by Eq. (9).

$$L = L_{hard} + L_{soft} + L_{cl} + \lambda||\theta||^2 \tag{9}$$

This loss function mainly consists of four parts. Here, $||\theta||^2$ represents the regularization term, and λ is the regularization coefficient. L_{cl} denotes the contrastive learning loss function. L_{soft} refers to the soft target loss function. The model measures the difference between its inference results and those of the large language model using mean absolute error.

Additionally, during the experimental process, we observed that the large model performs well in classifying data with true positive and negative labels, but shows significant performance degradation when dealing with data labeled as neutral. To avoid fitting errors caused by incorrect soft target, this paper uses cross-entropy loss as a hard target loss to measure the difference between the model's classification results and the true labels. The calculation process is shown in Eq. (10).

$$L_{hard} = -\sum_{i=1}^{n}\sum_{c=1}^{m} y_{ic} \log(\hat{y}_{ic}) \tag{10}$$

Here, c denotes the set of true labels, i represents the index of the sentence currently being computed, y_{ic} indicates the true label of sentence i. When the model's predicted result matches the true label, $y_{ic} = 1$, $\hat{y}_{ic}$ represent the final classification result of the model.

4   Experiment and Analysis

4.1   Dataset

This paper utilized the 14Lap, 14Rest from SemEval [8] and Twitter [9] datasets for the experimental evaluation. The specific information about these three datasets is shown in Table 1. Statistics of Dataset. Additionally, this paper uses the LLM's inference results as soft target to construct a dataset of LLM inference results. The distribution and volume of this dataset are consistent with Table 1.

Table 1. Statistics of Dataset.

Dataset	Positive		Neutral		Negative	
	Train	Test	Train	Test	Train	Test
14Lap	994	341	870	128	464	169
14Rest	2164	728	807	196	637	196
Twitter	1561	173	3127	346	1560	173

4.2   Experiment Environment

The server used for this experiment runs Ubuntu 18.04.1 LTS with an Intel Core i9-12900K CPU, an NVIDIA 3090Ti GPU, and 64GB of memory. The code is written in Python 3.6 and utilizes the PyTorch framework for model development, leveraging the HuggingFace Transformers library version 3.2.0. The model employs a BERT encoder to obtain word embeddings with a dimensionality of 768. The GCN layer depth is set to 1, and the contrastive learning temperature coefficient is set to 0.002. The learning rate is set to 0.002. The regularization coefficient is set to 0.0001. The large language model used in this chapter is OpenAI's ChatGPT, version gpt-3.5-turbo, with a maximum token generation limit of 30. The interface temperature is set to 0.

4.3   Comparison Experiment

To demonstrate the effectiveness of the proposed model, this paper evaluates the model's performance using two metrics: accuracy and macro-F1 score. Table 2 presents a comparison of the experimental results between this paper's model and several baseline models.

Table 2. Experiment Result of Different Models.

Model	14Rest		14Lap		Twitter	
	Acc	Macro-F1	Acc	Macro-F1	Acc	Macro-F1
MGAN [11]	81.25	71.94	75.39	72.47	72.54	70.81
TNet [12]	80.69	71.27	76.54	71.75	74.97	73.60
AEN-Glove [13]	80.98	72.14	73.51	69.04	72.83	69.81
DualGCN [3]	84.27	78.08	78.48	74.74	75.92	74.29
Sentic GCN [14]	84.03	75.38	77.90	74.17	-	-
SSEGCN + BERT [15]	87.31	81.09	81.01	77.96	77.40	76.02
C3DA [16]	86.93	81.23	80.61	77.11	77.08	75.76
Zhao et al. [17]	86.88	81.16	80.56	77.00	76.59	74.67
DMGGAT-Base [18]	87.13	81.19	80.78	77.57	75.99	74.56
DGGCN + BERT [19]	86.89	80.32	81.50	78.51	76.94	75.07
KDGN [20]	87.01	81.94	81.03	78.10	77.64	75.55
SGRPN + BERT [21]	87.21	80.98	81.01	77.99	76.70	74.96
WISI-MIF	**87.94**	**82.97**	**82.91**	**79.62**	**77.70**	**76.36**

The proposed model achieves accuracy scores of 87.94%, 82.91%, and 77.70% on the 14Rest, 14Lap, and Twitter datasets respectively, with corresponding macro-F1 scores of 82.97%, 79.62%, and 76.36%. These metrics outperform other comparative models across all datasets.

Our model demonstrates significant advantages in all dataset metrics, showcasing the ability of the graph convolution to propagate opinion word information to aspect terms based on dependency relationships. Our model addresses the limitation of single-view models by fusing multi-angle information using the multi-view information fusion module.

Compared to these models, our approach effectively learns from large language model views and mitigates overfitting issues caused by small dataset sizes.

4.4   Ablation Experiment

To further validate the effectiveness of the large language model fitting module and the multi-view information fusion module, this study conducted ablation experiments on each module separately. The results are shown in Table 3, where "w/o LF" denotes the model without the large language model fitting module, and "w/o MIF" denotes the model without the multi-view information fusion module.

By introducing the large language model view and employing the large language model fitting module, the LLM can act as a teacher model guiding the student model WISI-MIF in learning, thereby alleviating overfitting caused by dataset limitations. This paper's model can acquire general knowledge, and the experimental results validate the

Table 3. Performance Result of WISI-MIF and Ablation Model

Model	14Rest		14Lap		Twitter	
	Acc.	F1	Acc.	F1	Acc.	F1
WISI-MIF	**87.32**	**81.53**	**80.33**	**77.01**	**78.11**	**76.89**
w/o LF	85.88	80.15	79.75	76.04	74.74	73.05
w/o MIF	86.33	80.29	79.11	75.52	76.22	75.06

effectiveness of the module. Furthermore, the multi-view fusion module proposed in this paper integrates information from three perspectives, which helps the model make more accurate sentiment polarity judgments by leveraging insights obtained from different angles.

4.5 Comparison of Fusion Methods

To further validate the effectiveness of the multi-view information fusion module, this paper conducted comparative experiments with various classical information fusion methods. The accuracy comparison experiment is shown in Fig. 3, and the macro-F1 metric comparison experiment is depicted in Fig. 4.

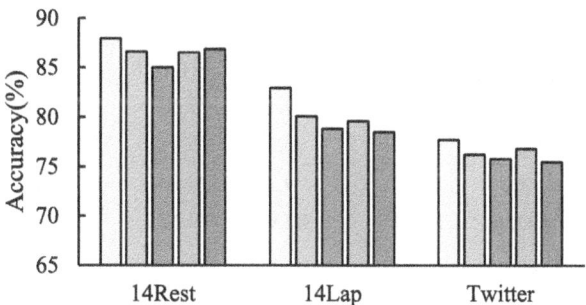

□ Multi-View ▨ Concat ▨ Convolution □ Max ▨ Mean Pooling

Fig. 3. Comparison of Accuracy of Different Fusion Strategies

From the experimental results, it can be seen that the multi-view information fusion module proposed in this paper exhibits better performance compared to other fusion methods.

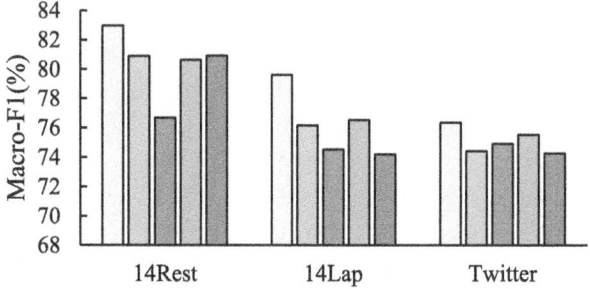

□ Multi-View □ Concat □ Convolution □ Max ▣ Mean Pooling

Fig. 4. Comparison of Macro-F1 of Different Fusion Strategies

4.6 Analysis of Multi-view Ablation Experiment

To validate the effectiveness of the three-view fusion, this paper conducted a multi-view ablation experiment where each experiment removed one of the views, combining the remaining two views for sentiment polarity prediction. The experimental results are shown in Table 4.

Table 4. Result of Multi-View Information Ablation

R_s	R_g	R_l	14Rest		14Lap		Twitter	
			Acc.	Macro-F1	Acc.	Macro-F1	Acc.	Macro-F1
√	√		86.33	80.29	79.11	75.52	76.22	75.06
	√	√	86.86	81.15	80.22	76.41	75.48	74.15
√		√	86.33	80.73	79.75	76.54	75.48	74.69
√	√	√	**87.94**	**82.97**	**82.91**	**79.62**	**77.70**	**76.36**

After removing the LLM view, the model's training was limited by the dataset, indicating a need for improved representation of general knowledge. This highlights the effectiveness of incorporating the large language model view. After removing the syntactic structure view, the model's performance declined. The syntactic dependency tree parsed by the syntactic dependency tool introduces information at the syntactic structure level, which helps to shorten the syntactic distance between aspect terms and opinion words to some extent. Through the GCN module, opinion word information can be transmitted to neighboring nodes, thus improving model performance. After removing the global view, the model's performance declined, indicating that incorporating global view information enhances the model's ability to perceive global information.

5 Conclusion

This paper introduces the large language model fitting module, which uses the inference results of a large language model as soft target to fit the model and learn the knowledge representation of the LLM. To address the limitation of current models lacking multi-view information, this paper introduces the multi-view information fusion module to fully utilize information from different perspectives, including the large language model view, global semantic view, and syntactic structure view.

References

1. Deng, H., Li, Y., Ju, S., et al.: Combines contrastive learning and primary capsule encoder for target sentiment classification. In: Yuan, L., Yang, S., Li, R., Kanoulas, E., Zhao, X. (eds.) WISA 2023. LNCS, vol. 14094, pp. 284–296. Springer, Singapore (2023). https://doi.org/10.1007/978-981-99-6222-8_24
2. Sun, K., Zhang, R., Mensah, S., et al.: Aspect-level sentiment analysis via convolution over dependency tree. In: Proceedings of the 2019 Conference on Empirical Methods in Natural Language Processing and the 9th International Joint Conference on Natural Language Processing, pp. 5679–5688 (2019)
3. Li, R., Chen, H., Feng, F., et al.: Dual graph convolutional networks for aspect-based sentiment analysis. In: Proceedings of the 59th Annual Meeting of the Association for Computational Linguistics and the 11th International Joint Conference on Natural Language Processing, pp. 6319–6329 (2021)
4. Tian, Y., Chen, G., Song, Y.: Aspect-based sentiment analysis with type-aware graph convolutional networks and layer ensemble. In: Proceedings of the 2021 Conference of the North American Chapter of the Association for Computational Linguistics: Human Language Technologies, pp. 2910–2922 (2021)
5. Wu, H., Zhang, Z., Shi, S., et al.: Phrase dependency relational graph attention network for aspect-based sentiment analysis. Knowl.-Based Syst. **236**, 107736 (2022)
6. Zhao, Z., Tang, M., Tang, W., et al.: Graph convolutional network with multiple weight mechanisms for aspect-based sentiment analysis. Neurocomputing **500**, 124–134 (2022)
7. Ren, P., Li, Y., Wang, S., et al.: Sentiment polarity identification method for financial entity based on interactive attention mechanism with two-graph convolutional network. J. Chin. Inf. Process. **37**(12), 129–137+166 (2023)
8. Pontiki, M., Galanis, D., Pavlopoulos, J., et al.: SemEval-2014 task 4: aspect based sentiment analysis. In: Proceedings of the 8th International Workshop on Semantic Evaluation, pp. 27–35 (2014)
9. Dong, L., Wei, F., Tan, C., et al.: Adaptive recursive neural network for target-dependent twitter sentiment classification. In: Proceedings of the 52nd Annual Meeting of the Association for Computational Linguistics, pp. 49–54 (2014)
10. Li, Y., Sun, C., Luo, L., et al.: Aspect-based sentiment classification for word information enhancement based on sentence information. Comput. Sci. 1–12 (2024). http://kns.cnki.net/kcms/detail/50.1075.TP.20231201.1406.014.html
11. Fan, F., Feng, Y., Zhao, D.: Multi-grained attention network for aspect-level sentiment classification. In: Proceedings of the 2018 Conference on Empirical Methods in Natural Language Processing, pp. 3433–3442 (2018)
12. Li, X., Bing, L., Lam, W., et at.: Transformation networks for target-oriented sentiment classification. In: Proceedings of the 56th Annual Meeting of the Association for Computational Linguistics, pp. 946–956 (2018)

13. Song, Y., Wang, J., Jiang, T., et al.: Targeted sentiment classification with attentional encoder network. In: Artificial Neural Networks and Machine Learning–ICANN 2019: Text and Time Series: 28th International Conference on Artificial Neural Networks, pp. 93–103 (2019)
14. Liang, B., Su, H., Gui, L., et al.: Aspect-based sentiment analysis via affective knowledge enhanced graph convolutional networks. Knowl.-Based Syst. **235**, 107643 (2022)
15. Zhang, Z., Zhou, Z., Wang, Y.: SSEGCN: syntactic and semantic enhanced graph convolutional network for aspect-based sentiment analysis. In: Proceedings of the 2022 Conference of the North American Chapter of the Association for Computational Linguistics: Human Language Technologies, pp. 4916–4925 (2022)
16. Wang, B., Ding, L., Zhong, Q., et al.: A contrastive cross-channel data augmentation framework for aspect-based sentiment analysis. In: Proceedings of the 29th International Conference on Computational Linguistics, pp. 6691–6704 (2022)
17. Zhao, G., Luo, Y., Chen, Q., et al.: Aspect-based sentiment analysis via multitask learning for online reviews. Knowl.-Based Syst. **264**, 110326 (2023)
18. Wang, Y., Yang, N., Miao, D., et al.: Dual-channel and multi-granularity gated graph attention network for aspect-based sentiment analysis. Appl. Intell. **53**(11), 13145–13157 (2023)
19. Liu, H., Wu, Y., Li, Q., et al.: Enhancing aspect-based sentiment analysis using a dual-gated graph convolutional network via contextual affective knowledge. Neurocomputing **553**, 126526 (2023)
20. Wu, H., Huang, C., Deng, S.: Improving aspect-based sentiment analysis with knowledge-aware dependency graph network. Inf. Fusion **92**, 289–299 (2023)
21. Huang, W., Cai, S., Li, H., et al.: Structure graph refined information propagate network for aspect-based sentiment analysis. Int. J. Data Warehouse. Min. **19**(1), 1–20 (2023)

GTGNN: Global Graph and Taxonomy Tree for Graph Neural Network Session-Based Recommendation

Zhenhong Wu[1], Yuzheng Liu[1], Xin Shi[1], Xueqing Zhao[1(✉)], Yun Wang[2], and Guigang Zhang[2]

[1] School of Computer Science, Xi'an Polytechnic University, Shaanxi Key Laboratory of Clothing Intelligence, Xi'an, China
{220711004,220721101}@stu.xpu.edu.cn, zhaoxueqing@xpu.edu.cn
[2] Institute of Automation, Chinese Academy of Sciences, Beijing, China
{y.wang,guigang.zhang}@ia.ac.cn

Abstract. The session-based recommendation aims to predict users' short-term decisions by analyzing sequences of anonymous users' historical behavior, often incorporating graph neural networks (GNN) for better performance. However, existing session-based recommendation methods tend to be limited to recommending items that already exist in users' historical sessions, resulting in inadequate ability to recommend new items that users have never interacted with before, which leads to the so-called "information cocoon" phenomenon. To solve this issue, this paper proposes a global graph and taxonomy tree for graph neural network session-based recommendation (GTGNN). First, GTGNN constructs global and session graphs based on all session sequences. Subsequently, the GNN encoder is used to extract the global-level representation and session-level representation of the items. Second, GTGNN uses an item taxonomy tree to learn user intent from the perspective of attention mechanism and historical distribution data respectively, simulating the decision-making process when interacting with new items. Meanwhile, to solve the problem that GNN cannot learn new items, zero-shot learning is introduced to infer potential representations of new items and recommend new items with higher scores to users by calculating the recommendation scores of the corresponding items. Finally, extensive experiments on two real-world datasets (i.e., Amazon G&GF and Yelpsmall) show that the average enhancement of each evaluation metric on these datasets of the GTGNN reaches 3.25% and 3.02%, respectively.

Keywords: GNN · Session-based Recommendation · New Item Recommendation · Global Graph · Taxonomy Tree

1 Introduction

In the Internet era, recommender systems play an important role in areas such as e-commerce, web search, and streaming media, helping them to effectively manage information overload by accurately recognizing user interests. Session-based

C. Jin et al. (Eds.): WISA 2024, LNCS 14883, pp. 29–40, 2024.
https://doi.org/10.1007/978-981-97-7707-5_3

recommendation (SBR) is a commonly used approach in recommendation systems to improve the short-term recommendation efficiency of anonymous users by predicting their next desired items. Traditional SBR methods rely on relationships between items, such as transitions and shared connections, using techniques such as Markov chains [1]. With the development of deep learning technology, neural network-based models are widely adopted in SBR. For instance, GRU4Rec [2] (Hidasi et al.) employs Gated Recurrent Units (GRU) to model sparse session data, to predict the user's next interaction and enhance the effectiveness of SBR. However, RNN-based recommender systems have limitations in dealing with limited user interactions, and it is difficult to accurately estimate user preferences in each session. In recent years, graph neural networks (GNNs) have rapidly emerged. Due to its ability to effectively model the complex topology between users and items, including multilevel relationships and similarity and transformation relationships between items, GNNs perform well in SBR and have become a mainstream method in this field.

In recent years, many GNN-based approaches [4,7] have been introduced to session recommender systems to model complex relationships between neighboring items. For example, SR-GNN [9] is a pioneering approach in SBR systems, which models session data as a directed graph and utilizes GNNs to capture user behavior patterns. TA-GNN [10] introduces a target-aware mechanism based on SR-GNN to better capture the information of predicted target items. GCE-GNN [8] extracts node representations using GNNs by constructing a graph structure that contains both global and local information and combines these representations with item embeddings. However, these methods only recommend items that are already in the session (old items), which tends to reduce the user's interest in interacting with the items and triggers information cocooning. Specifically, these models take the items in the user's history session as inputs and output the probability that these items will be the next recommended items. For example, in Fig. 1(a), the inputs of SR-GNN are V1, V2, V3, and V4, and its outputs are the probabilities of V1, V2, V3, and V4.

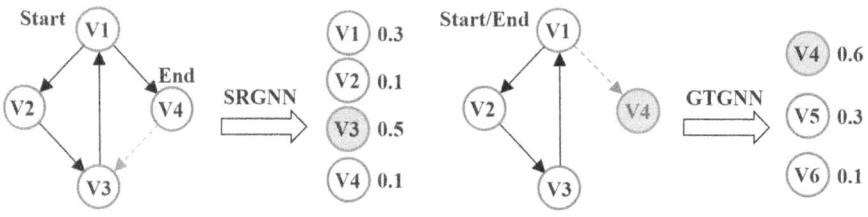

(a) Session-based next item reconmendation   (b) Session-based new item reconmendation

Fig. 1. Difference between session-based next item recommendation and GSNIR

To expand users' choices in session recommendations, this paper introduced the "GNN Session-based New Item Recommendation (GSNIR)" task, whose goal is to predict whether a user will interact with a new item in the future based on the user's previous interactions. Notably, this task is introduced in NirGNN [5]. Figure 1 illustrates the comparison between traditional GNN session-based recommendation methods (e.g., SR-GNN [9]) and the new setup employed in this paper. Although new item recommendations are important for improving user experience, existing GNN-based models perform poorly in this task. As previously mentioned. Therefore, a new GNN model for new item recommendation (GTGNN) is proposed in this paper. Firstly, GTGNN learns item embeddings from the global graph and the session graph, respectively: (i) Learning session-level item representations by relying on pairwise item transformation modeling within the current session; (ii) Learning global-level item representations using pairwise item transformation modeling over all sessions. Second, GTGNN utilizes the taxonomy tree of items to infer whether a user prefers this type of item or the item itself from the attention mechanism and historical distribution data, respectively, to simulate the user's decision-making process when interacting with a new item and predict the user's intention. Meanwhile, to solve the problem of GNN's inability to learn new items, the method introduces zero-shot learning [6] (ZSL) to infer potential representations of new items, thus realizing new item recommendations. Finally, extensive experiments on two real datasets show that GTGNN has good results in handling the GSNIR problem. In summary, the main contributions of the proposed GTGNN are as follows:

1. By introducing the taxonomy tree to construct the global graph learning item representation, the user intention is learned from the user's attention and data distribution, which provides multiple perspectives for the understanding of user preferences and effectively realizes the recommendation of session-based new items.
2. The zero-shot learning is used to effectively solve the problem of the lack of interaction features between new items and users in the traditional recommendations, and effectively improve the performance of session recommendations.
3. Extensive experiments are carried out on two representative real-world datasets, Amazon G&GF, and Yelpsmall, and the experimental results show that the proposed GTGNN outperforms the state-of-the-art methods across multiple evaluation metrics.

The rest of this paper is organized as follows: the problem description is described in Sect. 2; the proposed method is introduced in Sect. 3; Simulation experiments are presented in Sect. 4; and finally, the conclusions of this paper are given in Sect. 5.

2   Problem Description

Let $V = \{v_1, v_2, ..., v_l\}$ denote all the items (old items) that users have interacted with. Each anonymous session is represented as $S = \{v_1^s, v_2^s, ..., v_n^s\}$, where $v_i^s \in V$ denotes the items interacted within session S, and n represents the length of S. Given a set of new items $C = \{c_1, c_2, ..., c_m\}$, the goal of GSNIR is to recommend these new items to users. Specifically, the session-based new item recommendation model outputs the predicted probability ranking for each new item, and the top-k new items with the highest probabilities are recommended to the user. Each item v_i contains a taxonomy tree consisting of three different granularity levels (i.e., t_{i_1}, t_{i_2}, t_{i_3}), such as "beverage", "pop", and "cola". The taxonomy tree provides external information to assist the GNN model in its learning process. Additionally, each item also includes a set of attribute information, such as brand and price, which helps infer the embedding of new items. The embeddings of the three-level taxonomy tree of item v_i are denoted as t_{i_1}, t_{i_2}, t_{i_3}, and the global taxonomy embedding of each item v_i is calculated as $t_i = W(t_{i_1}, t_{i_2}, t_{i_3})$, where $W \in \mathbb{R}^{3d \times d}$ is a multi-layer perceptron (MLP) (with d representing the dimensionality of item embeddings). Word2Vec is used to process the attributes of both old and new items in the session. The attribute embeddings of old and new items are represented as $attr$ and $attr^*$, respectively.

Session Graph Model. Construct the session graph according to the method in SRGNN [9]. Specifically, given a session $S = \{v_1^s, v_2^s, ..., v_n^s\}$, its corresponding session graph $\mathcal{G}_s = (\mathcal{V}_s, \mathcal{E}_s)$ is defined, where $\mathcal{V}_s \subseteq V$ is the set of interacted items in the session graph, and $\mathcal{E}_s$ represents the set of edges, with each edge e_{ij}^s representing adjacent items (v_i^s, v_j^s) in S. Additionally, a self-loop is added to each item in the graph.

Inspired by GEC-GNN [8], the session graph consists of four types of edges, denoted as r_{in}, r_{out}, r_{in-out} and r_{self}. It depends on the relationship between item i and item j. r_{in} indicates transitions only from v_j to v_i, r_{out} indicates transitions only from v_i to v_j, r_{in-out} represents both types of transitions, and r_{self} indicates a self-loop for an item.

Global Graph Model. The session graph effectively captures the complex structure of sessions to learn session-level item embeddings. However, it lacks global information, so it needs to capture item transition information from other sessions for learning item representations. This is referred to as global-level item transition information.

Global-Level Item Transition Modeling. Inspired by GCE-GNN [8], global-level item embedding is learned by constructing a global graph model. It breaks the sequence independence assumption by linking all item pairs through pairwise transformations of all sessions (including the current session). In the global graph, for any item v_i^p in session S_p, its set of neighboring items is defined as the ε-**Neighbor Set** $(\mathcal{N}_\varepsilon(v_i^p))$, and the $\mathcal{N}_\varepsilon(v_i^p)$ can be described by $\mathcal{N}_\varepsilon(v_i^p) = \{v_j^q \mid v_i^p = v_{i'}^q \in S_p \cap S_q; v_j^p \in S_q; j \in [i' - \varepsilon, i' + \varepsilon]; S_p \neq S_q\}$, i' is the order in which item v_i appears in S_q, ε is a hyper-parameter used to control the range of item transitions between v_i^p and the items in S_q.

Global Graph. The global graph aims to capture global-level item transition information to learn embeddings of items across all sessions. It is constructed based on the ε-neighbor set of all items in the sessions, denoted as $\mathcal{G}_g = (\mathcal{V}_g, \mathcal{E}_g)$, where $\mathcal{V}_g \subseteq V$ refers to the set of vertices in the global graph, and $\mathcal{E}_g = \{e_{ij}^g \mid (v_i, v_j) \mid v_i \in V, v_j \in \mathcal{N}_\varepsilon(v_i)\}$ refers to the set of edges in the global graph. Figure 2 illustrates an example of constructing the global graph (with $\varepsilon = 2$).

Additionally, for each adjacent edge $(v_i, v_j)(v_j \in \mathcal{N}_\varepsilon(v_i))$, its frequency across all sessions is used as a weight to distinguish the importance of node neighbors. To enhance efficiency, only the top-N edges with the highest weights are retained for each item v_i on $\mathcal{G}_g$. Thus, $\mathcal{G}_g$ is an undirected weighted graph. For readability, $N_v^g = N_\varepsilon(v)$ is defined in the subsequent text. During testing, for efficiency purposes, the topology structure of the global graph isn't dynamically updated.

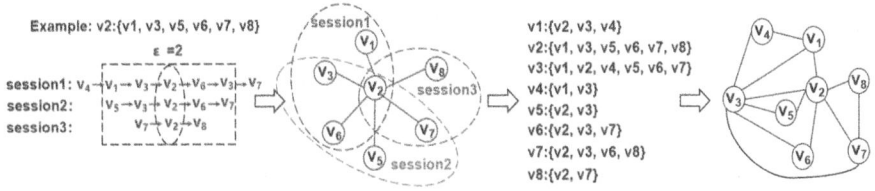

Fig. 2. The construction of a global graph

3 GTGNN

This section provides a detailed explanation of the model's approach. The overall framework of the model is illustrated in Fig. 3.

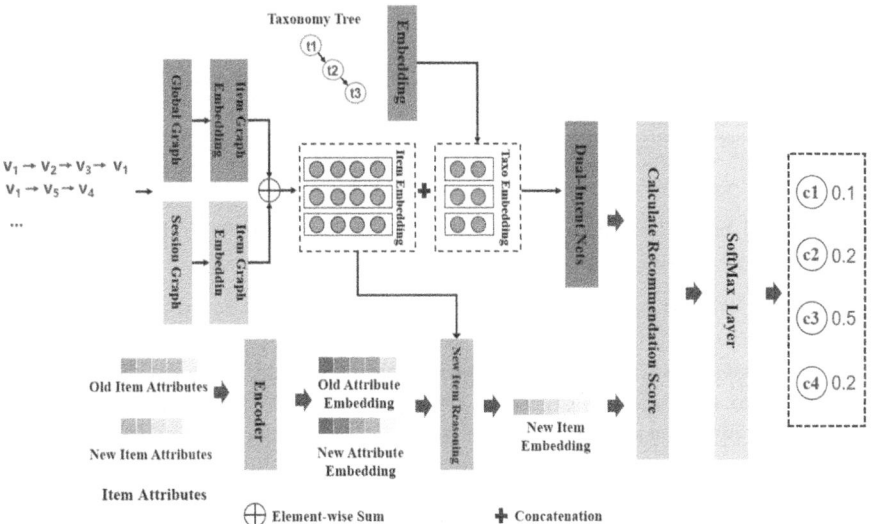

Fig. 3. The overall framework of GTGNN

3.1 Global-Level Item Representation

A project may involve multiple sessions, providing useful project transformation information that aids in current predictions. However, not all ε-neighbor sets of

v are relevant to the user preferences of the current session. Therefore, a graph attention network is used here to aggregate node information. Each network layer comprises two parts: information aggregation, and iterative updating.

Information Aggregation. In fact, the importance weights of different neighbors to a node should vary. The closer a project is to the preferences of the current session, the more important it is for the recommendation. Therefore, the following function is used to calculate the attention coefficients for nodes with different neighbors:

$$\text{Attn}\,(v_i, v_j) = \frac{\exp\left(q_1^T \, \text{LeakyRelu}\,(W_1\,[s \odot v_j]\,\|w_{ij})\right)}{\sum_{v_k \in N_{v_i}^g} \exp\left(q_1^T \, \text{LeakyRelu}\,(W_1\,[s \odot v_k]\,\|w_{ik})\right)}, \tag{1}$$

where LeakyRelu($\cdot$) denotes the activation function, $\odot$ denotes element-wise multiplication, $\|$ denotes vector concatenation, $w_{ij} and w_{ik} \in \mathbb{R}^1$ represent the weights of edge (v_i, v_j) and edge (v_i, v_k) in the global graph, $W_1 \in \mathbb{R}^{d+1 \times d+1}$ and $q_1 \in \mathbb{R}^{d+1}$ are trainable parameters, s represents the features of the current session, which are obtained by computing the average of the representations of items in the current session.

Then, using the attention coefficient calculated above to aggregate the neighbor information $h_{N_{v_i}^g}$,

$$h_{N_{v_i}^g} = \sum_{v_j \in N_{v_i}^g} Attn\,(v_i, v_j)\, v_j. \tag{2}$$

With the neighbor information aggregated using the aforementioned method, the item embedding vectors are updated as follows,

$$v_i^g = \text{agg}\left(v_i, h_{N_{v_i}^g}\right) = \text{relu}\left(W_2\left[v_i \| h_{N_{v_i}^g}\right]\right), \tag{3}$$

where relu($\cdot$) is the activation function, $W_2 \in \mathbb{R}^{d \times 2d}$ represents the transformation weight. v_i^g is the global-level representation of v_i.

Iteration Update. A single aggregation enables node v_i to integrate information from its first-order neighbors. As the iterations proceed, the neighboring nodes of v_i have already aggregated information from their respective neighbors, allowing v_i to aggregate information from its second-order neighbors. Multiple iterations of aggregation can capture information from higher-order neighbors. The representation of the k-th step of the item is as follows:

$$v^{g,(k)} = \text{agg}\left(v^{g,(k-1)}, h_{N_v^g}^{(k-1)}\right). \tag{4}$$

Through the above operations, the global-level item representation $v^{g,(k)}$ after k iterations can be obtained. At the initial iteration, $v^{(0)}$ is set to v.

3.2 Session-Level Item Representation

The session graph comprises pairwise item transitions within the current session. Next, it will discuss how to learn session-level item representations.

Due to the varying importance of neighbors to items in the session graph, an attention mechanism is employed to learn the weights between nodes. Additionally, weight vectors a_{in}, a_{out}, a_{in-out} and a_{self} are set for different relationships. Attention coefficients are computed through element-wise multiplication and nonlinear transformation:

$$attn(v_i, v_j) = \frac{\exp\left(\text{LeakyRelu}\left(a_{r_{ij}}^T (v_i \odot v_j)\right)\right)}{\sum_{v_k \in N_{v_i}^s} \exp\left(\text{LeakyRelu}\left(a_{r_{ik}}^T (v_i \odot v_k)\right)\right)}, \quad (5)$$

where a_{ij} denotes the importance of the features of node v_j to node v_i, LeakyRelu$(\cdot)$ represents the activation function, $a_{r_{ij}}, a_{r_{ik}} \in \mathbb{R}^d$ is the weight vector, r_{ij}, r_{ik} is the relationship between v_i and v_j and between v_i and v_k, and $N_{v_i}^s$ is the first-order neighbors of v_i. Finally, to ensure the comparability of coefficients between different nodes, the softmax function is used to normalize the attention weights.

Next, the output features for each node will be obtained by computing the linear combination of features corresponding to the coefficients,

$$v_i^s = \sum_{v_j \in N_{v_i}^s} attn(v_i, v_j)v_j. \quad (6)$$

Using the methods above, it can obtain the global-level item representation v^g and the session-level item representation v^s. Then, the final item representations h_v can be obtained by merging the two representations, in addition to using dropout on the global-level representation to avoid overfitting,

$$h_v = \text{dropout}\left(v^{g,(k)}\right) + v^s. \quad (7)$$

3.3 Session Intent Learing

Unlike previous works [3,9] that primarily focus on the last item, in this paper, we learn user intent from two aspects: the data distribution and attention mechanism within the session graph. The following is a detailed description.

User α Intent. Firstly, learning user intent from the perspective of the soft-attention mechanism. Recent interactions with items typically reflect the user's short-term preferences more accurately than earlier interactions. Therefore, extracting information from the last interacted item v_n and combining it with each item is crucial. Additionally, the classification tree information offers macro guidance, aiding in understanding user intent. User's α intent I_α can be learned through the following process:

$$\alpha_i = q^T \sigma\left((h_{v_i} \oplus t_i) * W_1 + (h_{v_n} \oplus t_n) * W_2\right), \quad (8)$$

$$I_\alpha = \sum_{i=1}^{n} \alpha_i \left(h_{v_i} \oplus t_i\right), \quad (9)$$

where $\sigma(\cdot)$ is the sigmoid activation function, $\oplus$ is a concatenation function, t_n is the global classification tree embedding of v_n, $W_1, W_2 \in \mathbb{R}^{2d \times d}$ and $q \in \mathbb{R}^d$ control the vector weights. h_{v_i}, h_{v_n} are the final item representations of v_i, v_n.

User β Intent. In recommendation systems, user historical sessions and the information regarding each item's taxonomy or attributes may be incomplete. Relying solely on the user's α-intent is insufficient for effectively mining item intent. Therefore, a Bayesian inference method is introduced, commonly used to make information complete through the inference of new data. The β distribution is a conjugate distribution before the binomial likelihood, hence it can replace the prior distribution to reduce the computational complexity of Bayesian inference. Additionally, the daily behaviors of users (interaction or not) form binomial distributions, which when aggregated over multiple days form β distributions. Thus, GTGNN directly computes the posterior distribution through the item embedding v_i and taxonomy embedding t_i. It considers user preferences from the perspectives of v_i and t_i, understanding whether users are more inclined towards the item itself or its category. v_i and t_i constitute a joint distribution probability. The β distribution represents the probability distribution of users' interest in v_i or t_i under different parameters. For each vertex v_i, the user intent distribution b_i is:

$$b_i = \mathcal{S}\left(\delta\left(h_{v_i}, t_i\right) x^{\phi\left(\varrho\left(h_{v_i}\right)\right)-1}(1-x)^{\phi\left(\varrho\left(t_i\right)\right)-1}\right),$$
$$\delta = \frac{\Gamma\left(\phi\left(\varrho\left(h_{v_i}\right)\right) + \phi\left(\varrho\left(t_i\right)\right)\right)}{\Gamma\left(\phi\left(\varrho\left(h_{v_i}\right)\right)\right)\Gamma\left(\phi\left(\varrho\left(t_i\right)\right)\right)}, \tag{10}$$

where $\mathcal{S}(\cdot)$ is the sample function, x is the variable, $\Gamma(\cdot)$ is the gamma distribution, $\varrho(\cdot)$ is the Softplus activation function, and $\phi(x) = \log\left(1 + e^x\right)$ is the normalization function used to adjust the data distribution range. Specifically, the embeddings of v_i and t_i are processed through the Softplus activation function and normalized to the range (0,1). The focus is primarily on which representation, v_i or t_i, the user is more interested in.

Next, introduce a soft-attention mechanism based on the β distribution to prioritize recently interacted items v_i. The β distribution for item v_i is given by:

$$\beta_i = \frac{h'_{v_i} - \frac{1}{n}\sum_{i=1}^{n}\left(b_i * \left(h_{v_i} \oplus t_i\right)\right)}{\sqrt{\frac{\sum_{i=1}^{n}(b_i - \text{avg}(b))^2}{n}}} W_3,$$
$$h'_{v_i} = b_i * \left(h_{v_i} \oplus t_i\right) + b_n * \left(h_{v_n} \oplus t_n\right), \tag{11}$$

where $\text{avg}(\cdot)$ denotes the average function, and W_3 is an MLP used for dimension reduction. After obtaining the β distribution, it can further derive the user's β intention I_β expressed by items in the session graph, defined as:

$$I_\beta = \sum_{i=1}^{n}\left(h_{v_i} \oplus t_i\right) * \beta_i. \tag{12}$$

By combining the α intent and the β intent, a weighted intent I can be obtained:

$$I = \lambda I_\alpha + (1 - \lambda)I_\beta, \tag{13}$$

where $\lambda \in [0, 1]$ is a parameter controlling the weight of each intent, and the I_α is shown in Eq. 9 above.

3.4   New Item Reasoning Network

Since the embeddings of new items cannot be directly obtained using GNN, the similarity between user intent and new items cannot be computed. Therefore, this paper employs a Zero-Shot Learning (ZSL) based approach to establish a mapping relationship between item semantic attributes and item embeddings and utilizes the semantic information of new item attributes to infer their embeddings. Specifically, given the semantic attributes $attr_i$ of an old item v_i, a transformation function θ is learned to map semantic attribute information to the item embedding space. Specifically, the generated item embedding can be computed as $v_i^* = \theta(attr_i)$, where θ is implemented as an MLP layer. To ensure that the generated item embeddings are in the same space as the original embeddings, an improved Bhattacharyya distance (BCimp) is attempted to narrow the gap between the original embedding v_i and the generated embedding v_i^*. This distance provides an upper limit on the minimum error rate of separability between two discrete probability distributions. The improved Bhattacharyya distance BC_{imp} is given by

$$BC_{imp}\left(v_i, v_i^*\right) = \begin{cases} -\log\left(\sqrt{v_i * v_i^*}\right), \sqrt{v_i * v_i^*} \neq \varnothing. \\ 0 \qquad\qquad\qquad , \text{ otherwise.} \end{cases} \tag{14}$$

To optimize the mapping function θ from the semantic attribute space to the item embedding space, a loss function L_{zero} is designed to minimize the difference between the original item embeddings and the embeddings generated based on the item semantic attributes,

$$L_{\text{zero}}\left(v_i, v_i^*\right) = BC_{imp}\left(v_i, v_i^*\right). \tag{15}$$

3.5   Prediction

By utilizing the optimized mapping function θ, the embedding of a new item c_i is inferred from the embedding of its attribute $attr_i^*$, denoted as $c_i = \theta(attr_i^*)$. Finally, the recommendation score for each new item can be calculated by multiplying the embedding of the new item c_i by the user intent I, defined as follows:

$$\hat{Z}_i = \text{softmax}\left(I^\top * c_i\right). \tag{16}$$

The final loss consists of the CrossEntropy loss and the loss L_{zero} used to optimize the inference ability of the spatial transformation function,

$$L = \gamma(-Z\log(\hat{Z}) - (1-Z)\log(1-\hat{Z})) + (1-\gamma)(\sum_{i=1}^{n} L_{zero}\left(v_i, v_i^*\right), \tag{17}$$

where Z is a one-hot coded vector of ground truth items, $\hat{Z}$ denotes the recommendation score of new items, $\gamma \in [0, 1]$ is a weight controlling the importance of the two losses, and the L_{zero} is shown in Eq. 15 above.

4   Experiments

4.1   Data Description

This experiment runs GTGNN on NVIDIA RTX 4060 8G GPUs. We validated the effectiveness of our proposed method GTGNN on two publicly available datasets, the Amazon G&GF) dataset and the Yelpsmall dataset, the former is a subset of the 2018 Amazon dataset that contains product attributes (e.g., brand and price) and merchant categorization information using hierarchical classification relationships (e.g., "Sports & Outdoors," "Other Sports," "Dance"). These categorization relationships are directly accessible and the latter is randomly selected from the 2021 version of the Yelp dataset, consisting of 6,000 reviews dating from 2016. The taxonomy in the dataset has no hierarchical structure. To generate the classification tree, the following steps were used: pretraining GoogleNews-vectors-negative300 using word2vec, obtaining word vectors for each taxonomy, and then clustering them using K-Means++. The word vectors of the same taxonomy are averaged and clustered again. The 100 taxonomies are first clustered and then the vectors of each taxonomy are summed. The 100-word vectors are clustered into 50 taxonomies, and then into 10 taxonomies, forming a three-level taxonomy tree after three clusters.

In a session, a new item is defined as an item that does not appear in the session, so the last item accessed by the user is selected as the recommended new item based on the timestamp, and the item is deleted from the history session to ensure that it is unique and new to the user. The history session is then reconnected based on the order in which the user interacted with the item.

After processing, the Amazon G&GF dataset contains 18,889 items, 82,251 training sessions, and 20,814 test sessions, while the Yelpsmall dataset contains 14,726 items, 19,035 training sessions, and 2,311 test sessions.

To evaluate the performance of the proposed method, we use other state-of-the-art methods, like SR-GNN [9], GCE-GNN [8], TA-GNN [10] and NirGNN [5], to compare performance with our proposed GTGNN, at the same time, the metrics such as $P@k$ (Precision@k) is used as a measure of prediction accuracy, it can predict the proportion of correctly related results to all returned results, and k denotes the accuracy of the first k items in the list. $MRR@k$(Mean Reciprocal Rank) is the reciprocal rank of correctly recommended items after re-averaging. In our experiment, $k = 10, 20$ are taken as evaluation metrics respectively, and Percentages are used to express these data.

4.2   Experimental Results

Overall Performance. To demonstrate the overall performance of the proposed model, this paper compares it with five session-based GNN recommendation methods. Here, the new results of the GTGNN proposed in this paper with the baseline model on different metrics are shown. Experimental results on the Amazon G&GF and Yelpsmall datasets are presented in Table 1, where the best solution for each metric is highlighted in bold and the second best solution

Table 1. Experiments on Amazon G&GF and Yelpsmall datasets

Method	Amazon				Yelpsmall			
	P@20	MRR@20	P@10	MRR@10	P@20	MRR@20	P@10	MRR@10
SR-GNN	1.4278	0.7434	1.0890	0.7176	1.4799	0.8780	0.5604	0.23 28
TA-GNN	1.6359	0.4924	1.0530	0.4697	1.6155	1.0863	0.5265	0.2101
GCE-GNN	1.7248	1.0190	1.3885	0.9552	1.8203	1.2535	0.6803	0.2581
NirGNN	**2.7610**	0.9314	**1.9762**	0.8110	1.8231	1.2591	0.6513	0.2298
GTGNN(Ours)	1.5855	**1.0389**	1.3404	**0.9987**	**1.9251**	**1.2706**	**0.6923**	**0.2640**

is underlined. It is possible to observe that GTGNN achieves excellent performance gains across different metrics compared to the baseline. On the Yelpsmall dataset, compared to the second-best baseline method, GTGNN improves the performance on the MRR metrics by 4.55% and on the P metrics by 5.59% at $k = 20$, and compared to the second-best baseline method, GTGNN improves the performance on the MRR metrics by 1.76% and on the P metrics by 1.76% at $k = 10$, compared to the second best baseline method, GCE-GNN by 1.76% and 2.29% in the P metric. On the Amazon G&GF dataset, GTGNN improves 1.95% on the $MRR@20$ metric and 4.55% on the $MRR@10$ metric compared to the second-best baseline method. These experimental results show that GTGNN can effectively improve the recommended performance of new items.

Table 2. Ablation study on Yelpsmall

Method	P@20	MRR@20	P@10	MRR@10
GTGNN	**1.9251**	**1.2706**	**0.6923**	**0.2640**
GTGNN w/o Global	1.8231	1.2591	0.6513	0.2298
GTGNN w/o Attr	0.5193	0.1660	0.3462	0.1536

Ablation Study. The ablation study was conducted on the yelpsmall dataset. To elucidate the key components of GTGNN, ablation studies for the global graph and new item inference modules are reported in Table 2. GTGNN w/o Global implies that GTGNN does not have a global graph module. GTGNN w/o Attr implies that GTGNN does not have a new item inference module. The experimental setup is the same as the overall experiment. The results show that the GTGNN model improves after combining all the key components. This reveals the effectiveness of each component.

5 Conclusion

In this paper, a novel GNN-based model GTGNN is proposed to solve the GSNIR problem. To effectively learn user intent, item embeddings are first learned from two perspectives: global graph and session graph. Secondly, the classification tree of each item is utilized to capture the user intent from two aspects of attention and data distribution, respectively. Meanwhile, to solve the problem of the lack of interaction between new items and users, this paper uses a ZSL-based approach to infer the embedding of new item attributes by utilizing their semantic information. Experimental results show that GTGNN can effectively improve the recommendation performance of new items. Future research will build on this study and continue to explore how to further improve the performance of new item recommendations.

Acknowledgment. This work was supported by the National Social Science Foundation of China Art Project(No. 23EH232), the Key Research and Development Program of Shaanxi Province in 2023 (No. 2023-YBGY-404, No. 2023-ZDLGY-48), and Research Center for Culture & Sci-Tech Integration Innovation, Key Research Base of Humanities and Social Sciences of Hubei Province, the Project of Public Digital Cultural Services (GGSZWHFW2024-003), Natural Science Basic Research Program of Shannxi (No. 2021JQ-694) and Shaanxi Province University Young Outstanding Talents Support Program.

References

1. Rendle, S., et al.: Factorizing personalized Markov chains for next-basket recommendation. In: Proceedings of the 19th International Conference on World Wide Web, pp. 811–820 (2010)
2. Hidasi, B., et al.: Session-based recommendations with recurrent neural networks. arXiv preprint arXiv:1511.06939 (2015)
3. Huang, X., et al.: Exploiting item relationships with dual-channel attention networks for session-based recommendation. In: Yuan, L., Yang, S., Li, R., Kanoulas, E., Zhao, X. (eds.) WISA 2023. LNCS, vol. 14094, pp. 198–205. Springer, Cham (2023). https://doi.org/10.1007/978-981-99-6222-8_17
4. Jin, D., et al.: CGMN: a contrastive graph matching network for self-supervised graph similarity learning. arXiv preprint arXiv:2205.15083 (2022)
5. Jin, D., et al.: Dual intent enhanced graph neural network for session-based new item recommendation. In: Proceedings of the ACM Web Conference 2023, pp. 684–693 (2023)
6. Rossi, L., et al.: Generalizability and robustness evaluation of attribute-based zero-shot learning. Neural Netw. **175**, 106278 (2024)
7. Wang, L., et al.: Contrastive graph similarity networks. ACM Trans. Web **18**(2), 1–20 (2024)
8. Wang, Z., et al.: Global context enhanced graph neural networks for session-based recommendation. In: Proceedings of the 43rd International ACM SIGIR Conference on Research and Development in Information Retrieval, pp. 169–178 (2020)
9. Wu, S., et al.: Session-based recommendation with graph neural networks. In: Proceedings of the AAAI Conference on Artificial Intelligence, vol. 33, no. 1, pp. 346–353 (2019)
10. Yu, F., et al.: TAGNN: target attentive graph neural networks for session-based recommendation. In: Proceedings of the 43rd International ACM SIGIR Conference on Research and Development in Information Retrieval, pp. 1921–1924 (2020)

Dual Learning Model of Code Summary and Generation Based on Transformer

Jiaying Wang[1,2], Lijun Cao[1,2], Jing Shan[1,2(✉)], Xiaoxu Song[1,2], and Junyi Jiang[3,4]

[1] Shenyang University of Technology, Shenyang 110870, China
mavis0129@126.com
[2] Shenyang Key Laboratory of Intelligent Technology of Advanced Industrial Equipment Manufacturing, Shenyang 110870, China
[3] Liaoning Technical University, Fuxin 123000, China
[4] Liaoning Economic Vocational and Technical College, Shenyang 110122, China

Abstract. Code summary (CS) produces natural language descriptions based on code snippets, while code generation (CG) produces code snippets based on natural language. Since both tasks are intended to model the relationship between natural language and code snippets, recent research has combined these tasks to improve their performance. The existing approach either relies on LSTM for dual training, which makes it impossible to address the issue of long-distance dependency or has an imbalance in the performance between code generation and code summary. In this paper, an end-to-end model based on Transformer is proposed to handle these problems. We propose two new regularization terms to not only constrain the duality of the two models by explicitly utilizing the probability correlation between CS and CG but also promote alignment between CS and CG models. Based on this, we propose a dual-learning algorithm for CS and CG. Experiments on real Java and Python datasets demonstrated that our model significantly improved the results of CS and CG tasks, surpassing the performance of existing models.

Keywords: Code summary · Code generation · Dual learning · Transformer

1 Introduction

In the field of software engineering, code summary (CS) and code generation (CG) play a crucial role as fundamental components of software development. Since deep learning has made breakthroughs in many fields such as cross-modal generation [12], natural language processing [13] and link prediction [17], researchers are increasingly using neural networks to solve general-purpose CS and CG problems. In an intuitive sense, there exists a significant relationship between CS and CG tasks, where the input of CS corresponds to the output of CG, and vice versa. This interdependence, referred to as "duality", offers valuable constraints that can be utilized to train both tasks effectively. The concept of duality in various AI tasks

© The Author(s), under exclusive license to Springer Nature Singapore Pte Ltd. 2024
C. Jin et al. (Eds.): WISA 2024, LNCS 14883, pp. 41–52, 2024.
https://doi.org/10.1007/978-981-97-7707-5_4

was initially proposed by Xia et al. [16]. Building upon this correlation, they have explored the application of dual models in machine translation, emotion classification, and image recognition, among others.

Wei et al. [15] extended the concept of dual supervised learning to design a dual model based on LSTM and Attention mechanisms, training both CS and CG models simultaneously. To enhance the probabilistic connection between the CS and CG models, they introduced a regularization term in the loss function. It is worth noting that this approach does not address the challenge of long-distance dependencies in code.

CodeT5 [14], a pre-trained model proposed by Wang et al., utilizes the Transformer architecture to perform CS and CG based on pre-training. This approach aims to enhance the alignment between natural language and programming language while simultaneously optimizing the model. However, the methods have not fully capitalized on the inherent connection between these two tasks, i.e. CS and CG are developed and trained independently, albeit with the CS task contributing to the training of the CG task. Thus the performance is not well balanced between code generation and code summary.

To address these challenges, we propose an end-to-end approach that offers ease of training and aims to enhance both CS and CG tasks. Specifically, we explore the potential improvement in code generation performance by leveraging the similarity between summary and code generated by the Transformer model. To achieve a balanced performance between CS and CG tasks, we adopt a dual model architecture that enables the sequential training of both tasks. This approach allows us to capitalize on the synergistic effects of training CS and CG models concurrently, leading to improved performance and a more equitable distribution of outcomes. These findings underscore the effectiveness of our approach in improving the performance of both CS and CG models, highlighting the benefits of explicitly exploiting the probabilistic correlation and enforcing duality between the two tasks.

By leveraging these contributions, our approach achieves a better balance in performance between CS and CG tasks, effectively addresses long dependency issues, and enhances the overall quality of code generation and summarization. The contributions of our work are summarized as follows:

- We propose a dual model architecture that combines Transformer models to address the performance imbalance between CS and CG tasks, as well as overcome the challenge of long-distance dependencies.
- To exploit the probabilistic correlation between CS and CG, we propose dual constraints with two regularization terms. These dual constraints strengthen the duality between the tasks and promote a more consistent and cohesive learning process.
- We conducted extensive experimental studies including the comparisons with the state-of-the-art methods and ablation studies on real-world public datasets. The results demonstrate the effectiveness of our approach.

2 Preliminary

2.1 Code Summary and Code Generation

Code summary can be thought of as a text generation task [2]. Given input code snippets $x = (x_1, x_2, x_3, \ldots, x_m)$, whose purpose is to generate readable natural language (a.k.a. comment) $y = (y_1, y_2, \ldots, y_n)$, which describes the input code snippet x. Let Y be the set of all possible summary sequences. The system tries to find the optimal sequence $y \in Y$ for x, that satisfies:

$$P(y|x) = \prod_{t=1}^{m} P(y_t, x) \tag{1}$$

In contrast, assuming X is the set of all possible code snippets, given a natural language query y, code generation is to generate the best code fragment $x \in X$ for y:

$$P(x|y) = \prod_{t=1}^{n} P(x_t, y) \tag{2}$$

Our objective is to develop two Seq2Seq [7] models, where the input to the CS model comprises the desired outcome of CG model, and vice versa. This arrangement establishes a duality between code summary and code generation, making them dual tasks of each other. By formulating the problem in this manner, we aim to exploit the inherent relationship between the two tasks and leverage their complementary nature to enhance the performance and effectiveness of both CS and CG models.

2.2 Dual Learning

The primary objective of dual learning is to leverage the concept of duality, enabling two tasks to mutually enhance and learn from each other until convergence is achieved. This paradigm involves two tasks: the first task involves mapping samples from input space X to output space Y, while the second task involves mapping samples from input space Y to output space X. Consequently, the initial task aims to model the conditional probability $P(y|x)$, while the latter task focuses on modeling the conditional probability $P(x|y)$. As per the Bayesian theorem [3], if both models operate flawlessly, there should exist the following relationship:

$$P(x, y) = P(x)P(y|x) = P(y)P(x|y) \tag{3}$$

where $P(x, y)$ is a joint probability, $P(x)$ and $P(y)$ are marginal distributions of code x and comment y, and $P(x|y)$ and $P(y|x)$ are conditional probabilities respectively.

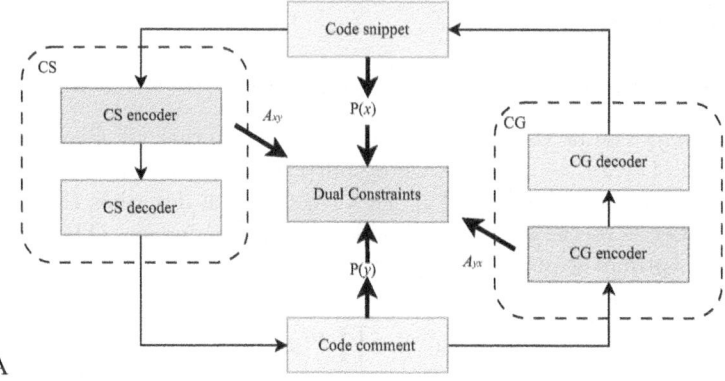

Fig. 1. The overall framework of dual learning of CS and CG

3   Methods

The training framework depicted in Fig. 1 encompasses three main components: the CS (code summary) model, the CG (code generation) model, and the dual constraints. Within the graph, $P(x)$ represents the marginal probability distribution of the variable x, which signifies the likelihood of event x occurring. Similarly, $P(y)$ denotes the marginal probability distribution of the variable y, indicating the probability of event y occurring. $P(x|y)$ represents the conditional probability, denoting the probability of event x occurring given y. On the other hand, $P(y|x)$ represents the conditional probability of event y occurring given x. The attention weights of the CS and CG encoder outputs are denoted as A_{xy} and A_{yx}, respectively. The dual constraints refer to the imposition of regularization terms in the loss function to enforce the duality between the two models.

3.1   Code Summary and Code Generation Module

The Transformer architecture is employed in both the CS and CG models. This architecture comprises stacked layers of multi-head attention and linear transformation layers for both the encoders and decoders. Within each layer, the multi-head attention mechanism is utilized with h attention heads and implements the self-attention mechanism, as illustrated in Eq. 4. This mechanism allows the models to capture and attend to different parts of the input sequence, enabling effective information processing and representation learning.

$$A(Q, K, V) = softmax(\frac{QK^T}{\sqrt{d_k}})V \qquad (4)$$

In the code summary task, the negative logarithmic likelihood serves as the training objective. The corresponding loss function is defined as:

$$L_{xy} = -\frac{1}{m}\Sigma_{t=1}^{m}logP(y_t|y_{<t}, x) \qquad (5)$$

This loss function quantifies the dissimilarity between the predicted output at each time step t (y_t) and the ground truth summary, given the previous generated tokens ($y_{<t}$) and the input code (x). The objective is to minimize this loss function over a batch of training examples, ensuring that the model generates accurate and coherent code summaries.

Similarly, the loss function in the code generation task is defined as:

$$L_{yx} = -\frac{1}{n}\Sigma_{t=1}^{n} log P(x_t|x_{<t}, y) \tag{6}$$

3.2 Dual Constraints

To enforce the duality between the CS and CG models, regularization terms are employed based on the probabilistic correlation between the two models and the symmetry of the weight matrix. We have shown that in the dual model structure, both the CS and CG models are associated with the joint probability distribution $P(x, y)$. According to Eq. 3, the joint probability of event x and event y occurring simultaneously can be expressed as the product of the marginal probability of event x and the conditional probability of event y given event x.

Considering that the CS model is parameterized by θ_{xy}, its conditional probability is denoted as $P(y|x; \theta_{xy})$. Similarly, the CG model is parameterized by θ_{yx}, and its conditional probability is represented as $P(x|y; \theta_{yx})$. By utilizing the Lagrange multiplier method, Eq. 3 is transformed into a regularization term, denoted as L_{dual} as depicted in Eq. 7, which serves as a constraint in the training process.

$$L_{dual} = [\ P(x)P(y|x;\ \theta_{xy}) - P(y)P(x|y;\theta_{yx})]^2 \tag{7}$$

By minimizing the regularization term L_{dual}, we aim to find the optimal solution that satisfies the constraint and enforces similarity between $P(y|x;\theta_{xy})$ and $P(x|y;\theta_{yx})$. This regularization term promotes the alignment and correlation between the two models, enhancing their probabilistic connection. The intention behind incorporating this regularization is to contribute to the training process and improve the overall performance of the dual learning framework.

The CS and CG models exhibit a symmetrical structure, where the output of one model serves as the input for the other, and vice versa. This symmetrical structure establishes an alignment between the tags present in the generated code snippets and the corresponding tags in the generated comments. This alignment can be quantified using a weight matrix. For instance, consider the comment "find the position of a character inside a string" and its corresponding source code "string.find(character)". Irrespective of the generation direction, the word "find" in the comment is consistently aligned with the "find" tag in the source code. Exploiting this alignment, an additional regularization term, denoted as L_{sim}, is introduced. This regularization term leverages the alignment information to further enhance the training process and optimize the performance of the dual learning framework.

In both the CS and CG models, the alignment weight matrix can be obtained using the encoder. Specifically, for the CS model, the weight matrix A_{xy} is computed as $A_{xy} \in (QK^T)^{n \times m}$, while for the CG model, the weight matrix A_{yx} is computed as $A_{yx} \in (QK^T)^{m \times n}$. Here, the element α_{ij} in A_{xy} measures the similarity between the i-th tag in the source code and the j-th tag in the comment. Similarly, the element α_{ji} in A_{yx} measures the similarity between the j-th tag in the comment and the i-th tag in the source code. To ensure consistency, the matrices A_{xy} and A_{yx} are transposed, resulting in A_{yx}^T.

Next, the matrices A_{xy} and A_{yx}^T are expanded column-wise, forming column vectors. These vectors, denoted as A_{xy_i} and $A_{yx_j}^T$, are obtained through dimensionality reduction. Subsequently, softmax is applied to these vectors, producing new vectors $b_i = \mathrm{softmax}(A_{xy_i})$ and $b_j = \mathrm{softmax}(A_{yx_j}^T)$. Finally, the Jensen-Shannon divergence [11], a symmetric method for measuring the similarity between probability distributions, is utilized to calculate the similarity between these two vectors. This similarity measure serves as one of the regularizations, denoted as L_{sim}, which contributes to the training process. Figure 2 provides an overview of this regularization.

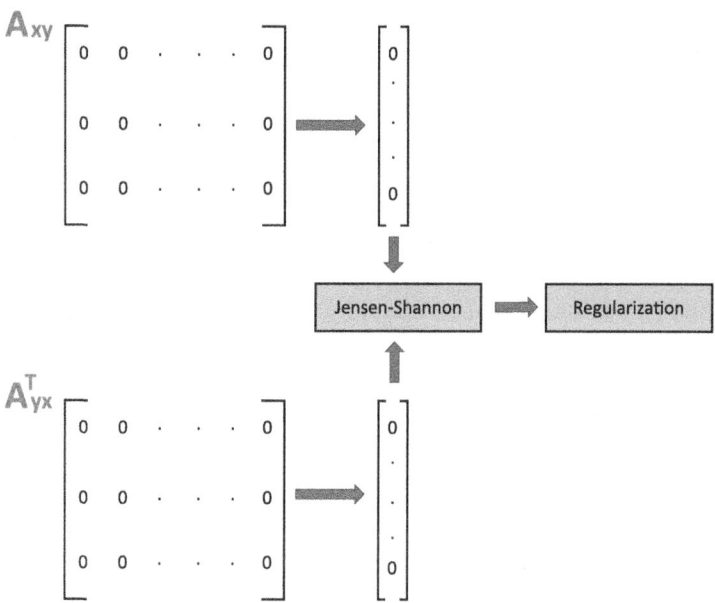

Fig. 2. Jensen-Shannon divergence-based regularization

The algorithm is given in Algorithm 1. It firstly initialize the training process and set the desired number of training iterations. For each iteration, we obtain a minibatch of k pairs of examples, denoted as $\langle (x_i, y_i) \rangle_{i=1}^{k}$. Each pair consists of a code snippet x_i and its corresponding comment y_i. We then calculate the gradients for the parameters θ_{xy} (CS model) and θ_{yx} (CG model). These gradients are computed based on the loss functions and the regularization terms. We update the parameters θ_{xy} and θ_{yx} using the respective optimization methods

opt_1 and opt_2, which use the gradients G_{xy} and G_{yx} to update the parameters. After that, we check for convergence. If the models have converged, the training process is terminated. We repeat the above steps for the specified number of training iterations.

Algorithm 1. Dual learning algorithm of CS and CG

1: **for** number of training iterations **do**
2:      Get a minibatch of k pairs $\langle (x_i, y_i)_{i=1}^k$;
3:      ▷ Calculate the gradients for θ_{xy} and θ_{yx}.
4:      $G_{xy} = \nabla_{\theta_{xy}} \frac{1}{k} \sum_{i=1}^k [L_{xy}(f_{xy}(x_i; \theta_{xy}), y_i) + \lambda_{dual1} L_{dual}(x_i, y_i; \theta_{xy}, \theta_{yx}) + \lambda_{sim1} L_{sim}(x_i, y_i; \theta_{xy}, \theta_{yx})]$;
5:      $G_{yx} = \nabla_{\theta_{yx}} \frac{1}{k} \sum_{i=1}^k [L_{yx}(f_{yx}(y_i; \theta_{yx}), x_i) + \lambda_{dual2} L_{dual}(x_i, y_i; \theta_{xy}, \theta_{yx}) + \lambda_{sim2} L_{sim}(x_i, y_i; \theta_{xy}, \theta_{yx})]$;
6:      ▷ Update θ_{xy} and θ_{yx}
7:      $\theta_{xy} \leftarrow opt_1(\theta_{xy}, G_{xy})$, $\theta_{yx} \leftarrow opt_2(\theta_{yx}, G_{yx})$
8:      **if** models converged **then**
9:          break;
10:     **end if**
11: **end for**

---

4 Experiments

4.1 Datasets

We experimented on a public Java dataset[1] and Python dataset[2]. The statistics for these two data sets are shown in Table 1.

Table 1. Details of the datasets

Datasets	Java	Python
Train	69,708	55,538
Validation	8,714	18,505
Test	8,714	18,505
Avg. tokens in comment	17.7	9.49
Avg. tokens in code	98.8	35.6

4.2 Settings

In the dual model framework, two models are employed: the CS model and the CG model. Both models utilize a transformer architecture. The maximum length for both code and summary is set to 64 in the datasets. To optimize the models, we utilize the RMSprop [5] optimizer with an initial learning rate of 0.0005.

---

[1] https://drive.google.com/open?id=13o4MiELiQoomlly2TCTpbtGee_HdQZxl.
[2] https://drive.google.com/open?id=1XPE1txk9VI0aOT_TdqbAeI58Q8puKVl2.

The batch size is set to 64, and the loss rate is set to 0.1. Additionally, the CS and CG models are initialized using preheated models that have been optimized by RMSprop with an initial learning rate of 0.002. Preheating involves separately pre-training the CS and CG models and then subjecting them to joint training with dual constraints, which accelerates the convergence process.

For the dual learning process, we found that using SGD with an initial learning rate of 0.0005 was appropriate. A learning rate scheduler is employed, where after one cycle, the learning rate is reduced back to its original value multiplied by 0.99. The regularization parameters λ_{dual1} and λ_{dual2} are both set to 0.03, while λ_{sim1} and λ_{sim2} are set to 0.01. The feed-forward network has a dimension of 64, with 2 attention heads and 4 levels. The embedding size and hidden layer dimension are both set to 64.

To construct the vocabulary, we consider all words that have a frequency of at least 3. This approach helps capture meaningful and informative words while filtering out less significant ones. Regarding the choice of optimizer, we opt for the Adam optimizer, which has demonstrated success in various deep learning tasks. We set the initial learning rate to 0.002, allowing the model to efficiently update its parameters during training.

Metrics. To assess the performance of our source code summary generation, we employ three widely used evaluation metrics: BLEU [6], METEOR [9], and ROUGE-L [1] to provide valuable insights into the quality and effectiveness of our generated summaries. Furthermore, to evaluate the source code generation aspect, we primarily rely on the BLEU metric, which has been commonly used in assessing the quality of generated code. For the hyperparameter settings of all benchmark methods, we follow a similar approach outlined in the work of Wei et al. [15].

4.3   Baselines

In our comparative analysis, we benchmarked our model against four state-of-the-art methods in the field. The first method, CODE-NN, is an end-to-end code summary approach proposed by Iyer et al. [8]. This method utilizes LSTM as a decoder and incorporates an attention mechanism at each decoding step.

The second method, DeepCom, is a deep learning-based tool designed for generating source code comments. It employs a neural network model to generate comments given a specific code snippet. Li et al. proposed this approach in their work [10].

Tree2Seq, the third method in our comparison, is a neural network model that operates on a tree structure. It effectively transforms source code into a target sequence by leveraging this tree-based representation. Dong introduced this approach in their study [4].

Lastly, we evaluated our model against LSTMDual, which is a dual model based on the seq2seq architecture. It utilizes recurrent LSTM as both encoders and decoders to generate code and summaries. This approach was proposed by Wei et al. [15].

It is important to acknowledge that our study does not include a comparison with large language models such as ChatGPT. This decision was made due to the significant disparity in the training set sizes between a large language model and our relatively smaller dataset. Conducting a direct comparison under such circumstances would introduce unfairness and bias to the evaluation process.

4.4   Performance of CS and CG

Table 2 presents the performance comparison of code summary methods, where we evaluate our model against four baselines. Our approach consists of two versions: the basic model, which is a Transformer-based source code summary method without the incorporation of our proposed dual constraints, and the dual model, which incorporates the dual constraints. In the table, we highlight the best-performing results in bold and the second-best results with an underline.

Significantly, our model surpasses all four baselines across both languages, exhibiting notable improvements in terms of BLEU, METEOR, and ROUGE-L metrics. Compared to our own basic model, the dual model achieves highly competitive results.

The superior performance of our model can be attributed to the effective utilization of the Transformer model, which effectively addresses the challenge of handling long-distance dependencies. By leveraging the capabilities of the Transformer, our model excels in capturing and modeling intricate relationships between code elements, resulting in more accurate and contextually meaningful code summaries.

The evaluation results revealed that the basic model outperformed the dual model in terms of ROUGE-L metric. This difference in performance can be attributed to the nature of the Python language, which offers a higher degree of flexibility compared to Java. Python code often exhibits greater freedom in its structure and syntax. When dealing with Python code, models need to place greater emphasis on semantic consistency rather than strictly adhering to surface form similarity. The basic model prioritizes semantic correctness and surface form similarity between generated comments and reference comments. However, it does not adequately capture the underlying core semantics between the source code and comments.

In contrast, the dual model incorporates dual constraints, enabling it to focus more on maintaining semantic consistency and relevance between the source code and annotations. By considering the holistic meaning and intent of the code, the dual model aims to generate more contextually appropriate comments. However, this emphasis on semantic consistency and relevance may result in a relatively lower ROUGE-L score compared to the basic model.

Table 2. Performance comparison of code summary

Methods	Java			Python		
	BLEU	METEOR	ROUGE-L	BLEU	METEOR	ROUGE-L
CODE-NN	27.60	12.61	41.00	17.36	9.288	37.81
DeepCom	39.75	23.06	52.67	20.78	9.979	37.35
Tree2seq	37.88	22.55	51.50	20.07	8.957	35.64
LSTMDual	<u>42.39</u>	<u>25.77</u>	<u>53.61</u>	21.80	11.14	39.45
Basic Model	42.01	23.46	51.64	<u>32.52</u>	<u>19.77</u>	**46.73**
Dual Model	**54.12**	**26.51**	**54.56**	**43.25**	**25.43**	<u>45.56</u>

To compare the performance of code generation, we conducted experiments between the basic model, dual model and several other models, including Tree2Seq, and LSTMDual. The results are presented in Table 3. Our models exhibit additional advancements in generating valid code. These results underscore the benefits of leveraging the Transformer architecture, as it enables the model to capture complex relationships and dependencies within the code, leading to more accurate and valid code generation. It is clear from the results that our approach strikes a better balance between CS and CG tasks.

Table 3. Performance comparison of code generation

Methods	Java	Python
	BLEU	BLEU
Tree2seq	13.80	4.472
LSTMDual	17.17	12.09
Basic Model	<u>44.10</u>	**50.21**
Dual Model	**54.43**	<u>41.48</u>

4.5   Ablation Study

To further investigate the impact of the introduced dual constraints, we conducted ablation experiments. In these experiments, we examined the effects of two specific regularization terms: L_{dual}, which enforces the probabilistic connection between the CS and CG models, and L_{sim}, which promotes alignment between the tags in the generated code summary and comment.

From the results in Table 4, it is evident that incorporating these constraints yields improvements in the overall performance of the model. Interestingly, we observed that the application of similar regularization terms yielded a slightly more effective enhancement compared to the probabilistic regularization terms. This may be attributed to the stronger and more explicit nature of the similar constraints, which directly enforce alignment between the generated code and comment.

Naturally, the combination of both types of regularization terms is expected to further enhance the performance. By leveraging both probabilistic and similar constraints, we can effectively capture the probabilistic connection between the CS and CG models, while also promoting alignment and coherence between the generated code and comment.

Table 4. Ablation of CS tasks in different Settings

L_{dual}	L_{sim}	Java			Python		
		BLEU	METEOR	ROUGE-L	BLEU	METEOR	ROUGE-L
-	-	42.01	23.46	51.64	32.52	19.77	46.73
✓	-	42.73	25.54	53.60	35.15	20.57	**47.31**
-	✓	42.96	25.96	53.57	35.25	21.08	47.07
✓	✓	**54.12**	**26.51**	**54.56**	**43.25**	**25.43**	45.56

5 Conclusion

This paper presents a novel approach to constructing a Transformer-based framework for the dual Code Summary (CS) and Code Generation (CG) tasks. Our objective is to develop a joint learning framework that simultaneously trains CS and CG models. In order to strengthen the interdependency between these two tasks during training, we propose the utilization of dual constraints. Specifically, in addition to the constraint applied to the probability, we introduce a constraint that leverages the inherent nature of the similarity mechanism. To evaluate the effectiveness of our proposed model, we conducted extensive experiments on Java and Python datasets. The experimental results demonstrate the superior performance of both the CS model and the CG model, surpassing state-of-the-art methods on these existing datasets.

Acknowledgments. This work is partly supported by the National Natural Science Foundation of China (Nos. 61702346 and 61702345), Basic Scientific Research Project of Liaoning Provincial Department of Education (No. JYTMS20231226), Ministry of Education industry-university cooperative education project (No. 231002108104009) and China Machinery Industry Education Association 2024 annual project on the integration of industry and education (No. ZJJX24CY008).

References

1. Barbella, M., Tortora, G.: Rouge metric evaluation for text summarization techniques. Available at SSRN 4120317 (2022)
2. Brown, T., et al.: Language models are few-shot learners. In: Advances in Neural Information Processing Systems, vol. 33, pp. 1877–1901 (2020)

3. Caprio, M., Sale, Y., Hüllermeier, E., Lee, I.: A novel Bayes' theorem for upper probabilities. In: Cuzzolin, F., Sultana, M. (eds.) Epi UAI 2023. LNCS, vol. 14523, pp. 1–12. Springer, Cham (2023). https://doi.org/10.1007/978-3-031-57963-9_1

4. Dong, L., Lapata, M.: Language to logical form with neural attention. arXiv preprint arXiv:1601.01280 (2016)

5. Elshamy, R., Abu-Elnasr, O., Elhoseny, M., Elmougy, S.: Improving the efficiency of RMSProp optimizer by utilizing Nestrove in deep learning. Sci. Rep. **13**(1), 8814 (2023)

6. Freitag, M., Grangier, D., Caswell, I.: Bleu might be guilty but references are not innocent. arXiv abs/2004.06063 (2020). https://api.semanticscholar.org/CorpusID:215744964

7. Huang, S., Zhou, X., Chin, S.: Application of Seq2Seq models on code correction. Front. Artif. Intell. **4**, 590215 (2021)

8. Iyer, S., Konstas, I., Cheung, A., Zettlemoyer, L.: Summarizing source code using a neural attention model. In: 54th Annual Meeting of the Association for Computational Linguistics 2016, pp. 2073–2083. Association for Computational Linguistics (2016)

9. Kaptchuk, G., Jois, T.M., Green, M., Rubin, A.D.: Meteor: cryptographically secure steganography for realistic distributions. In: Proceedings of the 2021 ACM SIGSAC Conference on Computer and Communications Security, pp. 1529–1548 (2021)

10. Li, B., Yan, M., Xia, X., Hu, X., Li, G., Lo, D.: Deepcommenter: a deep code comment generation tool with hybrid lexical and syntactical information. In: Proceedings of the 28th ACM Joint Meeting on European Software Engineering Conference and Symposium on the Foundations of Software Engineering, pp. 1571–1575 (2020)

11. Sutter, T., Daunhawer, I., Vogt, J.: Multimodal generative learning utilizing jensen-shannon-divergence. In: Advances in Neural Information Processing Systems, vol. 33, pp. 6100–6110 (2020)

12. Wang, J., Hao, S., Shan, J., Song, X.: Visual language–let the product say what you want. In: Proceedings of the AAAI Conference on Artificial Intelligence, pp. 23841–23843 (2024)

13. Wang, J., Shan, J., Santos, O.E., Bao, J.: High quality error-tolerant phrase mining on text corpus. Expert Syst. Appl. **171**, 114557 (2021)

14. Wang, Y., Wang, W., Joty, S., Hoi, S.C.: Codet5: identifier-aware unified pre-trained encoder-decoder models for code understanding and generation. arXiv preprint arXiv:2109.00859 (2021)

15. Wei, B., Li, G., Xia, X., Fu, Z., Jin, Z.: Code generation as a dual task of code summarization. In: Advances in Neural Information Processing Systems, vol. 32 (2019)

16. Xia, Y., Qin, T., Chen, W., Bian, J., Yu, N., Liu, T.Y.: Dual supervised learning. In: International Conference on Machine Learning, pp. 3789–3798. PMLR (2017)

17. Zhai, H., Cao, X., Sun, P., Shen, D., Nie, T., Kou, Y.: Rule-enhanced evolutional dual graph convolutional network for temporal knowledge graph link prediction. In: Yuan, L., Yang, S., Li, R., Kanoulas, E., Zhao, X. (eds.) WISA 2023. LNCS, vol. 14094, pp. 64–75. Springer, Cham (2023). https://doi.org/10.1007/978-981-99-6222-8_6

Relation-Oriented Temporal Knowledge Graphs Completion Based on Recurrent Neural Network

Lin Zhu, Yujing Ke, and Luyi Bai[✉]

School of Computer and Communication Engineering, Northeastern University (Qinhuangdao),
Qinhuangdao 066004, China
baily@neuq.edu.cn

Abstract. Knowledge graphs (KGs) are usually incomplete, and many completion methods have emerged in the field of knowledge graphs. However, most of the current methods learn the nodes on the fixed knowledge graphs by embedding and completing directly, which is computationally complex and difficult to apply concretely. Moreover, most knowledge graphs completion methods ignore the value of temporal information. In this paper, we propose a relational-oriented temporal knowledge graphs completion method based on cyclic neural network named Rnn-Relation. Our model embeds the temporal information into a dynamic space to obtain a new relational representation vector. Then the relational information is trained by cyclic neural network to improve the relevance of relational information and temporal information. Next, the output value of the cyclic neural network layer is recoded to obtain the final value of the relational representation vector. Finally, our model utilizes the negative sample sampling to predict entities, thus significantly improving the performance of the knowledge graphs completion.

Keywords: Deep Learning · Entity Prediction · Knowledge Graphs · Recurrent Neural Network

1 Introduction

In recent years, knowledge graphs technology has gradually been widely used. Knowledge graphs is continuously improved and completed during the creation process [3, 12]. The knowledge graphs completion technology to predict and complete missing data information has become one of the core research issues in the field of knowledge graphs. The current knowledge graphs completion methods are mainly divided into two types of completion methods, namely traditional knowledge graphs completion and temporal knowledge graphs completion.

In the field of traditional knowledge graphs completion, there are currently two mainstream knowledge graphs completion methods: Translation-based Model and Se-mantic Matching Model. One of the most classic knowledge graphs completion methods in traditional knowledge graphs completion is the translation-based completion model TransE [1] proposed by Bordes et al. In addition, many methods have been proposed based on TransE to improve the effect of knowledge graph completion. The semantic matching

C. Jin et al. (Eds.): WISA 2024, LNCS 14883, pp. 53–64, 2024.
https://doi.org/10.1007/978-981-97-7707-5_5

model uses a similarity-based scoring function to construct a knowledge graphs completion model. Representatives of semantic matching models include RESCAL method [14], DistMult method [21] and so on.

Many temporal knowledge graphs data sets have also been created and applied to the real world, such as ICEWS [2] and GDELT [10]. Temporal knowledge graphs have important value in many fields such as recommendation, query and fact prediction systems [18]. Therefore, the task of completing temporal knowledge graphs received more and more attention. However, most of the existing completion methods do not take the temporal information into consideration meaning that only triplets (head, relation, tail) are trained, it can only deal with traditional knowledge graphs but not temporal knowledge graphs. According to the characteristics of the temporal information sequence, people divide the completion task of the temporal knowledge graphs into two specific tasks: interpolation completion method and extrapolation completion method [8]. The form of the interpolation completion task is closer to the general knowledge graphs completion task [9]. The extrapolation completion method of the knowledge graphs is closer to the temporal sequence.

In this paper, we propose a relation-oriented temporal knowledge graphs completion method based on Recurrent Neural Network named Rnn-Relation, which can improve the overall performance of the temporal knowledge graphs completion, and it is also suitable for two key tasks: extrapolation completion and interpolation completion. Our model uses a relationship-oriented recurrent neural network mechanism to model the weight value of the relationship embedded with the timestamp. In addition, through the dynamic space mapping technology, the temporal information is embedded into the representation vector of the relationship. Compared with the existing temporal knowledge graphs completion methods, our method significantly improves the performance of the completion.

The main contributions of this paper:

- Proposing a novel temporal knowledge graphs reasoning model using Recurrent Neural Network, which can model and train complex interactions between relationships.
- Encoding temporal information directly in the learned embedding, which enables time information to be more fully integrated into the knowledge graphs.

The rest of this paper is organized as follows. In Sect. 2, we describe related works briefly. We propose our model and detail its training process in Sect. 3. In Sect. 4, we evaluate our model with some reasoning technologies and experimental results demonstrate that Rnn-Relation outperforms state-of-the-art methods in several aspects. Finally, we conclude our work and propose the future research in Sect. 5.

2 Related Work

In the field of traditional knowledge graphs completion, translation-based models usually use the idea of vector offset to regard the vector distance between two entities as the translation distance, and calculate the authenticity and accuracy of the fact information through the translation distance. As the most typical early translation model method,

Bordes et al. proposed TransE [1], the first knowledge graphs completion model based on translation methods. Next, on the basis of the translation model TransE, a series of method model optimizations are continued, and a series of method models are proposed to improve its performance and adaptability, such as TransH completion model [19], TransR completion model [11], TransD completion model [6] and RotatE completion model [15], etc. In addition, people also use other additional data information of the knowledge graphs to construct a series of joint learning models to further improve the performance of the knowledge graphs completion method [20]. In contrast, the semantic matching model usually measures the authenticity of entity information by combining the underlying semantics between entities and relationship vectors. Match to improve the accuracy of completion. Common models that currently apply this method include the RESCAL method [14], the DistMult method [21], and the HolE method [13]. Similarly, some semantic matching models also introduce other methods to jointly construct the completion model. For example, the ComplEx method [17] adds plural embedding space on the basis of semantic matching, so this model can more efficiently model asymmetric relationships.

To solve the problem of extrapolation completion tasks, García-Durán et al. used recurrent neural networks as a means to learn perceptual representations in temporal texts and proposed TA-TransE and TA-DistMult methods [5] to handle future timestamps. In addition, Trivedi et al. proposed the Know-Evolve method [16] for the task of extrapolation completion of the temporal knowledge graphs. This method learns the non-linear evolution of entity sequence information over time. Another Jin et al. proposed an extrapolation completion method RE-Net [7], which uses a neural network sequence coding model to aggregate the connected entity information of each entity. In contrast, the interpolation completion task of temporal knowledge graph is similar in form to the task of general knowledge graph, so most of the existing interpolation methods try to extend the method of general knowledge graph to adapt to temporal information. It is suitable for interpolation and completion model of temporal knowledge graph. Some interpolation completion models design a scoring function sensitive to temporal information for training, such as TTransE [5] and HyTE [4].

3   Model

3.1   Relation-Oriented Temporal Knowledge Graphs Completion

Interpolation Completion Method. In the task of interpolation completion of the temporal knowledge graphs, the temporal distribution of related fact data exists randomly, that is, the target fact has no linear temporal constraint relative to the existing facts. The goal of the interpolation completion task is to predict the facts that may be missing in the temporal knowledge graphs based on all the known facts to complement the temporal knowledge graphs.

We use the symbol E to represent the data of the entire temporal knowledge graphs. At the same time, the symbol H is used to represent relevant historical facts. For the target fact, there are also two possible forms: task $(s,r,?,t)$ or task $(?,r,o,t)$. Therefore, this section still assumes that the probability distribution of a target fact will be equal to the average of the respective joint probability distributions in these two forms. The calculation method is shown in Eq. (1), where Po(E) corresponds to the task $(s,r,?,t)$ is the joint probability distribution, and Ps(E) is the joint probability distribution of the task $(?,r,o,t)$.

$$P(E) = \frac{(P_o(E) + P_s(E))}{2} \tag{1}$$

For the entire temporal knowledge graphs E, all relevant facts can be used for prediction in the interpolation completion task. Therefore, the joint probability distribution functions of tasks $(s,r,?,t)$ and tasks $(?,r,o,t)$ can be obtained by Eqs. (2) and (3).

$$P_o(E) = \prod_{t=1}^{T} \prod_{(s,r,o,t)E_t} P(o|s, r, t, H(s)) \tag{2}$$

$$P_s(E) = \prod_{t=1}^{T} \prod_{(s,r,o,t)E_t} P(s|o, r, t, H(o)) \tag{3}$$

Extrapolation Completion Method. In the task of extrapolation completion of the temporal knowledge graphs, the relevant fact data will be arranged in the order of the timeline in advance. The goal of the task of extrapolation completion of the temporal knowledge graphs is to predict the facts that may occur in the next adjacent time based on historical facts that have occurred in the past, and then achieve the purpose of completing the overall data.

We use the symbol E to represent the data of the entire temporal knowledge graphs. At the same time, the symbol H is used to represent relevant historical facts. For the target fact, there are two possible forms: $(s,r,?,t)$ or $(?,r,o,t)$. It is assumed that the probability distribution of a target fact will be equal to this. The average value of the respective joint probability distributions in the two forms is calculated as shown in Eqs. (4), where $P_o(E)$ corresponds to the joint probability distribution of the task $(s,r,?,t)$, and Ps(E) is The joint probability distribution of the task $(?,r,o,t)$.

$$P(E) = \frac{(P_o(E) + P_s(E))}{2} \tag{4}$$

In addition, the entire temporal knowledge graphs E can be divided into a series of fact sets $\{E_1,E_2,...,E_t\}$ according to the temporal information, where $E_t = \{(s,r,o,t')E|t' = t\}$ represents the set of facts in the t tense. The task $(s,r,?,t)$ and the task $(?,r,o,t)$ The respective joint probability distribution functions are shown in Eqs. (5) and (6).

$$P_o(E) = \prod_{t=1}^{T} \prod_{(s,r,o,t)E_t} P(o|s, r, t, H_{t-w}(s), ..., H_{t-1}(s)) \tag{5}$$

$$P_s(E) = \prod_{t=1}^{T} \prod_{(s,r,o,t)E_t} P(s|o, r, t, H_{t-w}(o), ..., H_{t-1}(o)) \tag{6}$$

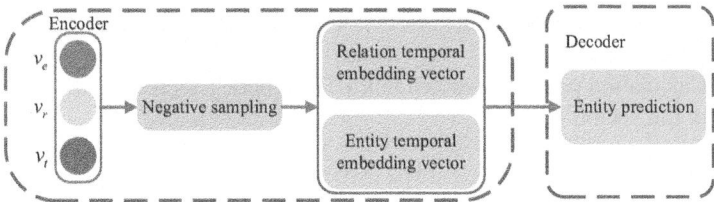

Fig. 1. Framework of our model.

3.2 Training

In this section, the LSTM-based temporal knowledge graphs completion method proposed in this paper is specifically explained. The method can be divided into two modules in the overall framework: encoder module and decoder module (see Fig. 1).

Recurrent Neural Network Coding Module. The entire model is divided into encoder and decoder parts according to the Encoder-Decoder encoding-decoding model framework. LSTM encodes temporal text in the encoder part. The general structure of this part can be divided into mapping layer, recurrent neural network layer, and feedforward layers (see Fig. 2).

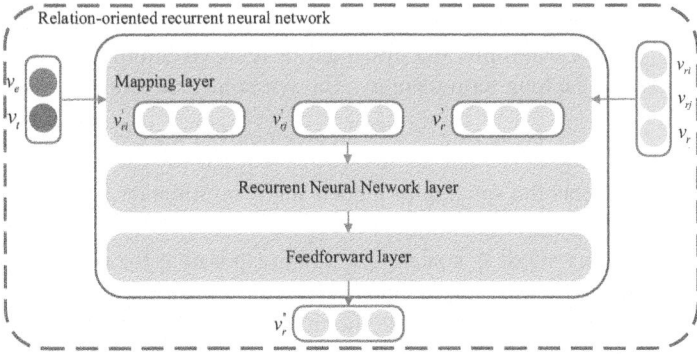

Fig. 2. Schematic diagram of encoding module.

Next, the Mapping layer, Recurrent Neural Network layer and Feedforward layer are explained in order and related formulas are given.

Mapping Layer. In order to reduce the parameters and facilitate the calculation and to adapt the LSTM training model more appropriately, we first preprocess the temporal information. The original timestamp information is (*Years-Months-Days*), which is divided into triples of (*Years, Months, Days*) through array slices, and continues to separate the year, month and date in each timestamp data to obtain the overall embedding of the timestamp v_t: $(d_1, d_2, d_3, d_4, d_5, d_6, d_7, d_8)$, such as dividing (*1999-08-22*)into (*1,9,9,9,0,8,2,2*). In addition, embed the processed temporal information data to obtain low-dimensional vectors $v_y, v_m, v_d \in \mathbb{R}^k$.

After obtaining the embedding vector v_t of the timestamp, first transpose v_t. Then, a dynamic space matrix $W_{et} \in \mathbb{R}^{k \times k}$ is generated by vector multiplication with the entity vector v_s. Calculation process is shown in Eq. (7), where T represents the transposition operation of the vector.

$$W_{et} = v_s v_t^T \tag{7}$$

In addition, considering that in the task $(s,r,?,t)$, only relevant information about the entity s and the relationship r can be obtained, so we choose to extract the relationship information from all the facts with the entity s as the main body, and construct a piece belonging to the entity s The subject relationship chain C_e. Then each relationship $v_{r_i} \in \{r \cup Re\}$ in the relationship chain is projected into the previously constructed spatial matrix $W_{et} \in \mathbb{R}^{k \times k}$. The calculation method is shown in Eq. (8):

$$v_{r_i}' = W_{et} v_{r_i} r_i \in \{r \cup R_e\} \tag{8}$$

Recurrent Neural Network Layer. Through a number of LSTM units to learn, train and store the information contained in the relational sequence, the output data is converted into fixed format data of the input data in each cell output, and then the weight information is updated iteratively. Among them, it can be specifically divided into three processing modules.

1) Long-term memory module: Through the transmitted historical cell information and the specific information of the long-term memory weighting module, the sigmoid function is used to determine the discarded or retained information to obtain the vector content of the long-term memory. The symbolic expression in Eq. (9):

$$memory_{long}^t = c^{t-1} \odot \text{sigmoid}(w_1 * v^t) \tag{9}$$

The v^t represents the splicing of hidden status information of the current input information and the negotiation time, $v^t = [\begin{smallmatrix} x^t \\ h^{t-1} \end{smallmatrix}]$.

2) Short-term memory module: update the state, responsible for passing the current input information and the previous hidden state information into the *tanh* function and the *sigmoid* function to calculate the main content of the short-term memory vector, and its expression is shown in Eq. (10).

$$memory_{short}^t = \tanh(w_3 * v^t) \odot \text{sigmoid}(w_2 * v^t) \tag{10}$$

The $tanh(w_3 * v^t)$ represents the raw data value of short-term memory information. $sigmoid(w_2 * v^t)$ represents gated information.

3) Output module: used to calculate the value and output value of the next hidden state information. The output module comprehensively considers the cell's long-term memory information and short-term memory information, and transfers the previous time step hidden state information and current input information to the *sigmoid* function, and The updated cell state is calculated by the *tanh* function, and finally the hidden state information is obtained by multiplying the output value of the *sigmoid* function and the *tanh* function. Its specific expression is shown in Eq. (11).

$$c^t = memory_{long}^t + memory_{short'}^t \tag{11}$$

The hidden state information is shown in Eq. (12).

$$h^t = \tanh(c^t) \odot sigmoid(w_4 * v^t) \tag{12}$$

The output value is shown in Eq. (13).

$$y^t = sigmoid(h^t) \tag{13}$$

Feedforward Layer. The feedforward layer includes two linear conversion layers and a nonlinear activation function *gelu*. The specific calculation formula is shown in Eq. (14), where $W_f \in \mathbb{R}^{k \times 2k}$ and $W_f \in \mathbb{R}^{k \times 2k}$ are the linear mapping matrices of the two linear conversion layers, and b_f and b_{ff} are the deviations of the mapping matrix.

$$\overline{v_r} = \sigma(v_r'' W_f + b_f)W_{ff} + b_{ff} \tag{14}$$

In addition, in the feedforward layer, this part also uses the residual network to improve the performance of the method. The whole is shown in Eq. (15), where *LN()* represents the Layer Normalization operation.

$$\overline{\overline{v_r}} = LN(\overline{v_r} + v_r'') \tag{15}$$

Finally, the relationship expression vector embedded with temporal information and the initial vector of the relationship are weighted and summed according to the set weights, shown in Eq. (16):

$$v_{rt} = (1 - \beta)v_r + \beta\overline{\overline{v_r}} \tag{16}$$

Through training, a more expressive relationship representation vector embedded with temporal information is obtained, which represents a further improvement of the overall performance of the temporal knowledge graphs completion model method and improves the limitations of existing methods.

We combined the latest translation-based time knowledge graphs interpolation method DE-SimplE, and proposed a new interpolation method. The DE-SimplE's calculation method is shown in Eq. (17), where t_{dim} represents the vector dimension occupied by the temporal information.

$$v_{et} = DE(v_e, t) = v_{e[:t\,dim]}v_{e[t\,dim:]}(w_t t + b_t) \tag{17}$$

In the decoder part, this part uses the original formula of the SimplE method as the scoring function, and the calculation method is shown in Eq. (18).

$$Sco(v_s, v_o', v_r) = sum((v_{st} * v_{rt} * v_{ot}) + (v_{st} * v_{rt} * v_{ot})^{-1}) \tag{18}$$

The overall calculation formula is shown in Eq. (19), where E_{train} refers to the training set of the data.

$$L(s, r, t) = -\sum_{(s,o,r,tE_{train})} \frac{\exp(Sco(v_s, v_o', v_r)_o)}{\sum_{i \in Neg} \exp(Sco(v_s, v_o', v_r)_i)} \tag{19}$$

The global loss function of the relation-oriented temporal knowledge map completion method based on the recurrent neural network is shown in Eq. (20).

$$L = L(s, r, t) + L(o, r, t) \tag{20}$$

Training Process. The training process includes four modules: Input module, output module, hidden unit module, network training module (see Fig. 3). The training process for the completion of relational temporal knowledge graphs based on recurrent neural network is shown in Algorithm 1.

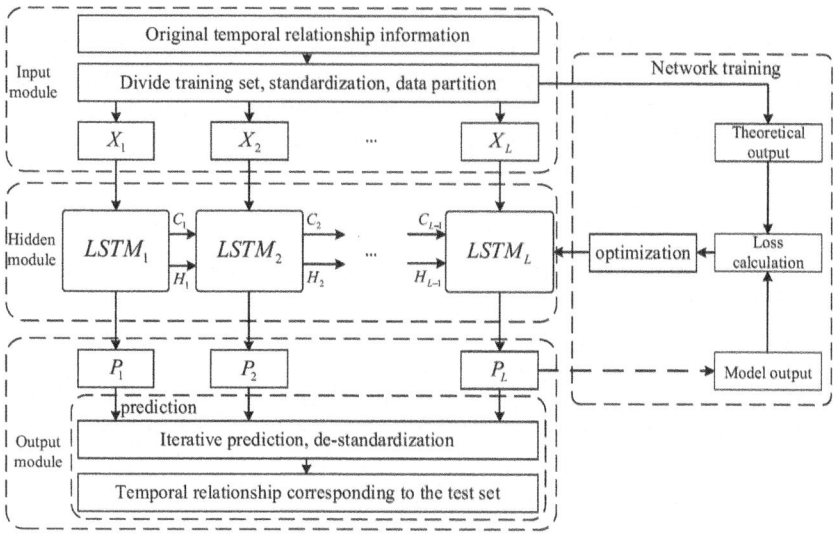

Fig. 3. Schematic diagram based on LSTM.

Algorithm1: Training process

Input: Train set G_{train}, Relational multi-chains R, Window length w, Epoch q, Loss L
Output: The minimum Loss L_{min} on the train set

Initialize: v_e, v_r, v_t ← xavier_uniform(), $L_{min} = MAXINT$

01 for $epoch$ ← 1 to q do

02   L←0.0 // Initialize the Loss

03   for$(s,r,o,t) \in G_{train}$ do
04   split(s,r,o,t)into two tasks $(s,r,?,t)$ and $(?,r,o,t)$
05   for each task in $\{(s,r,?,t),(?,r,o,t)\}$ do
06   map relations to the matrix space by Equation
07   calculate the weight between relations by LSTM
08   encode the final relation vector by Equation

09   calculate $Sco(v_s,v_o{'},v_r)$ and $Sco(v_o,v_s{'},v_r)$ by Equation

10   calculate $L_{(s,r,t)}$ and $L_{(o,r,t)}$ by Equation

11         $L = L_{(s,r,t)} + L_{(o,r,t)}$
12         $L_{min} = \min(L_{min}, L)$
13   Update parameters e_s, e_r, e_o with Adam optimizer
14 return L_{min}

4 Experiments

4.1 Experimental Setup

Datasets. In this experiment, we used the baseline common sub-data sets: ICEWS14 data set, ICEWS05-15 data set and GDELT data set. In the course of this experiment, the standard of Goel et al. was used to divide the three data sets into training set, validation set and test set.

Baseline Methods. We evaluate our model with some reasoning methods: (1) We have summarized some common and reasoning models to train event datasets by ignoring the time points, such as TransE, SimplE, DistMult. (2) We also evaluate with state-of-the-art temporal reasoning models on temporal knowledge graphs, including TTransE, HyTE, TA-DisMult, DE-TransE, DE-DistMult and DE-SimplE.

Experimental Details. The parameter settings of our method are as follows: the candidate value of the method learning rate {0.01, 0.001, 0.0001}, the vector embedding dimension d {50, 100, 200}, the training data batch size B {256, 512, 1024}, and the discard probability γ {0.0, 0.2, 0.4}. Set the weight of the proportion of the relational temporal vector = 0.5, and set the dimension t_{dim} of the entity's temporal information to $d*0.32$.

Following convention, we use Mean Reciprocal Ranks (MRR) and Hits@1/3/10 (proportion of correct entity in top 1/3/10 predictions) as evaluation metrics.

4.2 Experimental Results and Performance Analysis

The performance of this model is tested on three data sets. The test model comes from the best performance model results on the verification set. The experimental results and comparisons of this model and the existing advanced knowledge graphs completion model and temporal knowledge graphs completion model on the three data sets can be clearly seen from the following tables, the Rnn-Relation method has a great improvement in all performance evaluation indicators.

As shown in the Table 1 and Table 2, on the ICEWS14 dataset, the relation-oriented temporal knowledge graphs completion method based on cyclic neural network has improved in performance indicators such as MRR, Hits@1, Hits@3, and Hits@10. In addition, on the ICEWS05-15 data set, the performance improvement of the relation-oriented temporal knowledge graphs completion method based on the cyclic neural network is even more significant. Performance improvements have been achieved in indicators such as MRR, Hits@1, Hits@3, and Hits@10. As shown in the Table 3, for the GDELT data set, the comparison results of the relation-oriented temporal knowledge graphs completion method based on the cyclic neural network and the baseline method are shown in the table. The Rnn-Relation method in this paper has achieved the current best results on the GDELT data set, and has improved in indicators such as MRR, Hits@1, Hits@3, and Hits@10, Especially the performance improvement effect on Hits@10 is the best.

Table 1. Performance comparisons on ICEWS14 dataset.

Model	ICEWS14			
	MRR	Hit@1	Hit@3	Hit@10
TransE	0.280	9.4	-	63.7
DistMult	0.439	32.3	-	67.2
SimplE	0.458	34.1	51.6	68.7
TTransE	0.255	7.4	-	60.1
HyTE	0.297	10.8	41.6	65.5
TA-TransE	0.275	9.5	-	62.5
TA-DistMult	0.477	36.3	-	68.6
DE-TransE	0.326	12.4	46.7	68.6
DE-DistMult	0.501	39.2	56.9	70.8
DE-SimplE	0.526	41.8	59.2	72.5
Rnn-Relation	**0.544**	**43.3**	**60.2**	**73.9**

Table 2. Performance comparisons on ICEWS05-15 dataset.

Model	ICEWS05-15			
	MRR	Hit@1	Hit@3	Hit@10
TransE	0.294	9.0	-	66.3
DistMult	0.456	33.7	-	69.7
SimplE	0.478	35.9	53.9	70.8
TTransE	0.271	8.4	-	61.6
HyTE	0.316	11.6	44.5	68.1
TA-TransE	0.299	9.6	-	66.8
TA-DistMult	0.474	34.6	-	72.8
DE-TransE	0.314	10.8	45.3	68.5
DE-DistMult	0.484	36.6	54.6	71.8
DE-SimplE	0.513	39.2	57.8	74.8
Rnn-Relation	**0.557**	**46.8**	**65.5**	**79.8**

The comparison of the above results proves that the Rnn-Relation method greatly strengthens the relationship representation by using the relationship-oriented cyclic neural network mechanism and the method of embedding the temporal information into the representation vector of the relationship. The expressive ability of vectors makes full use of the temporal information of the temporal knowledge graphs to greatly improve the accuracy of the completion task of the temporal knowledge graphs. Our method

Table 3. Performance comparisons on GDELT dataset.

Model	GDELT			
	MRR	Hit@1	Hit@3	Hit@10
TransE	0.113	0.0	15.8	31.2
DistMult	0.196	11.7	20.8	34.8
SimplE	0.206	12.4	22.0	36.6
TTransE	0.115	0.0	16.0	31.8
HyTE	0.118	0.0	16.5	32.6
TA-DistMult	0.206	12.4	21.9	36.5
DE-TransE	0.126	0.0	18.1	35.0
DE-DistMult	0.213	13.0	22.8	37.6
DE-SimplE	0.230	14.1	24.8	40.3
Rnn-Relation	**0.275**	**17.1**	**28.9**	**46.5**

has significant performance improvement compared with existing advanced knowledge graphs completion model and temporal knowledge graphs completion model on three data sets, which also proves the Rnn-Relation method has strong applicability.

5 Conclusion

In this paper, we propose Rnn-Relation which takes temporal information into consideration in the task of completing the knowledge graphs. First of all, the triple data with temporal information is embedded in the dynamic space, and the time stamp information is embedded in the relation vector to obtain a new relation representation vector. Through the relationship-oriented recurrent network mechanism, the training of the relational data with temporal information is realized. Then the relational information is trained by cyclic neural network to improve the expression ability of the relational vector, which means the accuracy of the completion of the temporal knowledge graphs is improved. Experimental results show that Rnn-Relation outperforms state-of-the-art methods on Hits@1, Hits@3, Hits@10 and MRR over ICEWS14, ICEWS05-15 and GDELT. It can be seen that our method has different degrees of performance improvement compared with the existing advanced knowledge graphs completion model and temporal knowledge graphs completion model.

For the future work, we will extend our model to take the spatial information into account and complete the task of spatial-temporal knowledge graphs completion.

Acknowledgement. The work was supported by the National Natural Science Foundation of China (61402087), the Natural Science Foundation of Hebei Province (F2022501015), and the Fundamental Research Funds for the Central Universities (2023GFYD003).

References

1. Bordes, A., Usunier, N., Garcia-Duran, A., Weston, J., Yakhnenko, O.: Translating embeddings for modeling multi-relational data. In: Advances in Neural Information Processing Systems, pp. 2787–2795. Curran Associates (2013)
2. Boschee, E., Lautenschlager, J., O'Brien, S., et al.: ICEWS coded event data. Harvard Dataverse (2015)
3. Bollacker, K., Evans, C., Paritosh, P., et al.: Freebase: a collaboratively created graph database for structuring human knowledge. In: The Proceedings of the 2008 ACM SIGMOD International Conference on Management of Data, pp. 1247–1250 (2008)
4. Dasgupta, S.S., Ray, S.N., Talukdar, P.: Hyte: hyperplane-based temporally aware knowledge graph embedding. In: Proceedings of the 2018 Conference on Empirical Methods in Natural Language Processing, Brussels, Belgium, pp. 2001–2011. Association for Computational Linguistics (2018)
5. García-Durán, A., Dumančić, S., Niepert, M.: Learning sequence encoders for temporal knowledge graph completion. arXiv preprint arXiv:1809.03202
6. Ji, G., He, S., Xu, L., et al.: Knowledge graph embedding via dynamic mapping matrix. In: The Proceedings of the 53rd Annual Meeting of the Association for Computational Linguistics and the 7th International Joint Conference on Natural Language Processing (2015)
7. Jin, W., Zhang, C., Szekely, P., Ren, X.: Recurrent Event Network for Reasoning over Temporal Knowledge Graphs. arXiv preprint arXiv:1904.05530
8. Kazemi, S.M., Goel, R., Jain, K., et al.: Relational representation learning for dynamic (knowledge) graphs: a survey. arXiv preprint arXiv:1905.11485
9. Leblay, J., Chekol, M.W.: Deriving validity time in knowledge graph. In: Companion Proceedings of the Web Conference 2018 (2018)
10. Leetaru, K., Schrodt, P.A.: GDELT: global data on events, location, and tone, 1979–2012. In: The ISA Annual Convention. Citeseer (2013)
11. Lin, Y., Liu, Z., Sun, M., et al.: Learning entity and relation embeddings for knowledge graph completion. In: The Proceedings of the AAAI Conference on Artificial Intelligence (2015)
12. Miller, G.A.: WordNet: a lexical database for English. Commun. ACM **38**(11), 39–41 (1997)
13. Nickel, M., Rosasco, L., Poggio, T.: Holographic embeddings of knowledge graphs. In: Proceedings of the AAAI Conference on Artificial Intelligence (2016)
14. Nickel, M., Tresp, V., Kriegel, H.P.: A three-way model for collective learning on multi-relational data. In: ICML (2011)
15. Sun, Z., Deng, Z.H., Nie, J.Y., et al.: Rotate: knowledge graph embedding by relational rotation in complex space. arXiv preprint arXiv:1902.10197
16. Trivedi, R., Dai, H., Wang, Y, Song, L.: Know-evolve: deep temporal reasoning for dynamic knowledge graphs. In: Proceedings of the 34th International Conference on Machine Learning, Sydney, NSW, Australia, vol. 70, pp. 3462–3471. JMLR.org (2017)
17. Trouillon, T., Welbl, J., Riedel, S., et al.: Complex embeddings for simple link prediction. In: International Conference on Machine Learning. PMLR: 2071-2080 (2016)
18. Wang, Q., Mao, Z., Wang, B., Guo, L.: Knowledge graph embedding: a survey of approaches and applications. IEEE Trans. Knowl. Data Eng. **29**(12), 2724–2743 (2017)
19. Wang, Z., Zhang, J., Feng, J., Chen, Z.: Knowledge graph embedding by translating on hyperplanes. In: Proceedings of AAAI, pp. 1112–1119 (2014)
20. Xie, R., Liu, Z., Sun, M.: Representation learning of knowledge graphs with hierarchical types. In: IJCAI (2016)
21. Yang, B., Yih, W., He, X., et al.: Embedding entities and relations for learning and inference in knowledge bases. arXiv preprint arXiv:1412.6575

MMPDRec: A Denoising Model for Knowledge Concepts Recommendation Using Metapaths

Bohan Zhang, Mo Chen(✉), Jing Chen, Minghe Yu, Zhenghao Liu, Bin Xu, and Ge Yu

School of Computer Science and Engineering, Northeastern University, Shenyang 110819, China
chenmo@mail.neu.edu.cn

Abstract. Online education has revolutionized knowledge dissemination, providing unprecedented global access to educational resources. The key to improving online learning environments is effective knowledge concept recommendations tailored to the unique preferences and requirements of each student. While existing GCNs-based recommendation systems contribute significantly to the personalization of content, they frequently neglect the intrinsic relationships between knowledge concepts and struggle to contend with the noise inherent in large-scale educational data, potentially weakening the predictive effect of the model. To address these shortcomings, we propose MMPDRec (Multi-MetaPaths Denoising Recommender System)), an innovative framework that integrates the denoising GCN with a multi-head attention mechanism, considering the diverse reasons students engage with specific knowledge concepts. MMPDRec skillfully captures the subtle patterns in student-concept interactions by utilizing the synergistic impacts of these techniques, yielding a more nuanced understanding that drives the recommendation process. Extensive experiments conducted on a real-world MOOC dataset show that MMPDRec outperforms the state-of-the-art models in predicting and recommending knowledge concepts for intricate online learning scenarios.

Keywords: Knowledge Concepts Recommendation · Graph Convolution Networks(GCNs) · Heterogeneous Information Network (HIN)

1 Introduction

With the rapid development of internet technology, online education has become an indispensable part of modern learning environment [1]. Especially in the context of the global epidemic accelerating the digital transformation of education, the frequency and coverage of online education platforms have increased dramatically. Among all online education platforms, MOOC (Massive Open Online Courses) has been popular among learners as a free online course platform [2].

© The Author(s), under exclusive license to Springer Nature Singapore Pte Ltd. 2024
C. Jin et al. (Eds.): WISA 2024, LNCS 14883, pp. 65–77, 2024.
https://doi.org/10.1007/978-981-97-7707-5_6

Under such circumstances, personalized learning recommendation system is particularly important, it can recommend learning resources suitable for learners according to their learning history, interest preferences and learning goals [3].

In the traditional online education model, the system [4–6] often focuses on recommending complete courses to students, which facilitates student learning to a certain extent but may not fully address personal knowledge needs. Courses are typically composed of multiple video lectures, each addressing distinct knowledge concepts. Directly recommending entire courses may not effectively cater to a student's specific interests in particular concepts [7]. Thus, it is crucial to analyze a user's learning preferences from a detailed perspective and predict the specific knowledge concepts that may capture their interest [8]. In this work, we focus on predicting and recommending knowledge concepts on MOOC platforms that students may find interesting. This can greatly enhance learning efficiency and teaching quality, making the learning process more aligned with individual needs and pace.

Traditional collaborative filtering recommendation algorithms [11,12] primarily focus on the interaction data between users and items, often overlooking information about themselves. This approach can significantly impact the quality of recommendations. In contrast, recommendation systems based on GCNs effectively utilize the topological information within the graph and integrate external information [9]. Thus, even in scenarios where user-item interaction data is relatively sparse, GCNs can enhance the accuracy and robustness of recommendations by learning the associative relationships between nodes [10], thus compensating for the shortcomings of traditional collaborative filtering.

Although GCNs have demonstrated significant success in enhancing recommendation accuracy and managing complex data structures [13–15], they still confront several challenges in practical applications, particularly in recommending resources on MOOC platforms, which necessitate further improvements and optimizations. Firstly, the existence of noisy data seriously affects the fitting ability of the model. Although the large amount of interactive data on online education platforms provides rich input for learning, most of these data may contain low-quality information, such as accidental clicks or non-goal-oriented browsing behaviors, which may cause the model training effect to be impaired and fail to accurately capture the real needs of users. Secondly, recommendation systems often ignore the fine-grained factors that influence user choice, such as the teaching style of the teacher and the environment of the school. Failure to consider these individual differences may result in recommendations that do not accurately reflect the actual needs of students. Finally, the existing recommendation systems often ignore the semantic correlation between learning resources. In the field of education, there is often a prerequisite relationship or interdependence relationship between different knowledge concepts, and the neglect of this relationship may lead to the failure of the recommendation system to provide personalized resources that truly conform to the learning process of students.

Facing the above challenges, we propose the MMPDRec framework. Our main contributions can be summarized as follows:

– Firstly, in order to alleviate the negative impact of noisy data on recommendation performance in the convolution stage, a degree-sensitive edge discarding mechanism is introduced. This mechanism can effectively identify and eliminate the edges that may introduce noise in the model training stage, so as to ensure the robustness of the recommendation system.
– Secondly, this work constructs a HIN with comprehensive information, which includes multiple entities and relationship types. Through GCNs and multi-head attention mechanism, we can comprehensively analyze and understand how different factors affect the choice of knowledge concepts, so as to accurately capture individual needs.
– Finally, we conducted extensive experimental validation on real datasets. The results show that MMPDRec not only achieves SOTA in several important performance metrics, but also shows effectiveness in each modules.

2   The Proposed Framework

In this section, we will introduce our proposed model MMPDRec including four main components. Firstly, HIN is constructed by input data. Then, Denoising GCN is used to propagate and aggregate information in the network. Then we use multi-head attention mechanism to weight and aggregate information from different meta-paths. Finally, the optimized users and knowledge concepts vectors are transformed into specific preference scores by matrix factorization. In the following, we will describe each component in detail.

2.1   Task Definition

In this study, our task is to establish the preference relationship between the users ($\mathcal{U}$) and the knowledge concepts ($\mathcal{K}$). In our model, the learning function $\mathcal{F}$ is designed to map the user u and the knowledge concept k to a preference score $s_{u,k}$, which is calculated based on the embedding vectors of the users and the knowledge concepts, which is generated by encoding and transforming the complex interactions and properties of the input data. Such a design not only helps capture the explicit and implicit preferences of users, but also effectively utilizes rich contextual information to improve recommendation performances.

2.2   Construct HIN

In this part, we focus on constructing a comprehensive HIN to make full use of the rich data resources in the online education environment. As is shown in Fig. 1, constructed HIN contains six entities and seven relationship types in the network. In particular, we particularly emphasize the internal connections between knowledge concepts, which are considered to be crucial for understanding the knowledge structure and learning path. Through this design, our network can comprehensively consider multiple factors and provide a comprehensive data view for the model.

Based on this HIN, our model can deeply mine the implicit intention behind user behavior, accurately identify user learning needs, and recommend the most suitable learning resources for users according to the logical relationships and dependencies between knowledge concepts.

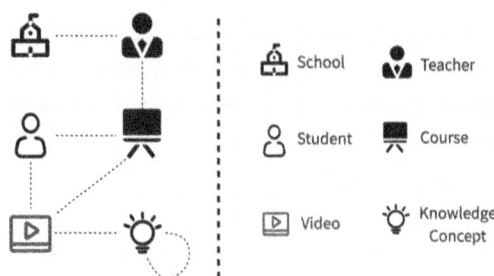

Fig. 1. Different types of entities and relationships in MOOCs

2.3   Denoising GCN

In this study, we detail the design and implementation of a denoising GCN with reference to the previous work [16], especially a recommendation system for users and knowledge concepts in the online education environment. The network effectively deals with noise data and optimizes feature representation through carefully designed steps. The users and knowledge concepts vectors are trained in the same way. The user vectors are used as an example below.

Firstly, the interaction matrix A used in the model is obtained through the meta-path connection, which can reflect the interaction relationship between users or knowledge concepts.

Next, in order to reduce the effect of noise, we adopt Degree-sensitive Edge Dropout to optimize the interaction matrix A. In this process, a retained probability between i and j is calculated according to the Eq. 1:

$$p_{ij} = \frac{1}{\sqrt{d_i}\sqrt{d_j}} \qquad (1)$$

where d_i and d_j represent degrees of i and j respectively. Nodes with high connectivity have a lower retained probability at the margin. Based on these probabilities, the system then selects a number of edges from the total edge set through a random process to discard. This mechanism helps us clean the data input by reducing noise from superfluous connections. The denoised matrix $\hat{A}_p$ is transformed by using the symmetric normalization technique $D^{-1/2}A_p D^{-1/2}$, where D is the diagonal matrix of the node-degree matrix. This normalization helps to reduce the influence of different density regions in the graph, ensure the structural balance of information transmission, and prevent the overbalance phenomenon that may occur in the convolution process of the graph, that is, avoid the excessive influence of some nodes on the learning process. In the convolution operation stage, the Eq. 2 is used for iterative calculation:

$$Y_u^{l+1} = \hat{A}_p(W_u^l X_u^l + b_u^l) \tag{2}$$

where $W_u^l \in \mathbb{R}^{|U| \times d}$ and $b_u^l \in \mathbb{R}^d$ are weight and bias vectors respectively, X_l is the node eigenvector of the l layer and d is the dimension of the vectors. After, the output of each layer is adjusted by calculating the cosine similarity a_{l+1} between the new feature X_{l+1} and the original feature X_0 according to the Eq. 3.

$$a_u^{l+1} = \frac{X^0 \cdot Y_u^{l+1}}{\|X_u^0\| \|Y_u^{l+1}\|} + \epsilon \tag{3}$$

Namely, here ϵ is a small constant, in order to guarantee the numerical stability. Such adjustment helps to maintain the directionality of the feature vector and avoid the degradation of the feature vector in the process of multi-layer propagation.Finally, the output of this layer is expressed as the following Eq. 4

$$X_u^{l+1} = a_u^{l+1} \cdot Y_u^{l+1} \tag{4}$$

Finally, the final users or knowledge concepts representation vector is formed by weighted averaging the outputs X_u^{l+1} of all layers, which is expressed as the following Eq. 5:

$$X_u = \frac{1}{L+1} \sum_{i=0}^{L} X_u^i \tag{5}$$

where L is the number of model layers. This hierarchical and integrated approach not only leverages the depth information of the multi-layer network, but also balances the contributions of each layer to obtain a comprehensive and accurate final representations.

2.4 Multi-head Attention Mechanism

In this part, the multi-head attention mechanism serves as a critical technique for facilitating fine-grained feature fusion. In the construction of recommendation systems, it is crucial to acknowledge that different interaction paths can elucidate distinct user relationships and characteristics from multiple perspectives. By integrating this varied information, we can obtain the more comprehensive understanding of user behavior and preferences.

Specifically, each head in the multi-head attention mechanism independently focuses on a specific meta-path's user vector. For each head, we first calculate the attention weights corresponding to the meta-path using the specific formula as Eq. 6:

$$\alpha_u^{MP_i} = \frac{exp(V_u^T \sigma(W_u e_u^{MP_i}))}{\sum_{j \in |MP|} exp(V_u^T \sigma(W_u e_u^{MP_j}))} \tag{6}$$

where the output $\alpha_u^{MP_i}$ indicates the weight for MP_i^u, and V_u^T and W_u are trainable matrices for users. These weights indicate the importance of different user features. This adaptive approach enables the model to emphasize features that are more critical for tasks. Once the calculation of attention weights for each head is completed, the multi-head attention mechanism combines feature vectors from different meta-paths by weighted summation Eq. 7.

$$e_u = \sum_{j \in |MP|} \alpha_u^{MP_j} X_u^{MP_j} \tag{7}$$

Specifically, the user vectors from each metapath are multiplied by its corresponding attention weight; then all weighted vectors are summed up to form an integrated user vector. The final representation of the user is obtained by simply adding together all weighted user vectors outputted by all heads.

2.5   Matrix Factorization

The matrix decomposition technique can decompose the user-item interaction matrix into low-dimensional potential feature matrices, which represent the potential factors of the user and the item respectively. With this decomposition, previous work has also validated its ability to reveal unexplicitly expressed relationships between users and items, thereby predicting the degree to which users prefer untouched items. Specifically, we compute the potential eigenvectors of users and knowledge concepts in the form of products to obtain their preference scores. The mathematical expression is as Eq. 8:

$$\hat{y}_{u,k} = z_u^T z_c + \gamma \cdot e_u^T M e_k + b_c \tag{8}$$

where z_k and z_k are the underlying vectors of the user and knowledge concepts respectively, and b_c is a bias term. In addition, M is a trainable matrix to let eu in the same space with e_k, and γ is a trainable parameter for the trade-off between the prediction scores from matrix factorization and the user and concept embeddings. The higher this score, the greater the user's preference for that knowledge concept.

2.6   Training Details

The model is trained using a composite loss function that includes Bayesian Personalized Ranking (BPR) [17] and L2 regularization, as shown in the Eq. 9. The BPR loss function is specifically designed to optimize ranking, improving the model's performance in personalized recommendation tasks by training on the relative ranking between known and unknown concepts of each user's preference. In addition, L2 regularization is also introduced to prevent the model from overfitting and improve its generalization ability.

$$\mathcal{L} = \sum_{(u,i,j) \in D_s} -ln(\sigma(\hat{y}_{u,i} - \hat{y}_{u,j})) + \lambda \|\Theta\|^2 \tag{9}$$

Here the λ is designed to balance the two losses and $\sigma(\cdot)$ is *sigmoid* activation function. The parameter optimization of the model uses the Adam optimizer, which is a widely used optimization algorithm for deep learning tasks. The Adam optimizer [18] combines the advantages of momentum and adaptive learning rate to automatically adjust the learning rate during learning of different parameters, accelerating convergence while avoiding falling into local minima.

During the validation process, we adopt a specific strategy to evaluate the recommendation effect of the model. For the validation set, we choose the concept of each user's last interaction as the test target. To fully assess the model's ability to differentiate, each test concept is randomly paired with 99 concepts that the user has not interacted with. Thus, the evaluation process is closer to the real application scenario, and the effectiveness of the model in distinguishing user preferences is tested.

3 Experiment

3.1 Experimental Settings

Dataset. The experiments in this study are mainly based on the Mooccube dataset [19]. Mooccube is a large-scale open online course dataset containing a wide range of courses, user engagement data, and multi-dimensional course characteristics. This makes it ideal for evaluating the effectiveness of recommender systems. The diversity and complexity of this dataset provides a challenging test bed for verifying the validity and applicability of our proposed model. To reduce the experiment workload and keep the comparison fair, we closely follow the settings of the previous work [8] and the statistics are shown in Table 1.

Table 1. Statistics of entities and relations in the Mooccube dataset

Entities	Statistics	Relations	Statistics
users	2,005	user-concept	930,553
concepts	21,037	user-course	13,696
courses	600	course-video	42,117
videos	22,403	teacher-course	1,875
schools	137	video-concept	295,475
teachers	138	course-concept	150,811
		concept-concept	18,354,644

Baselines. In order to fully evaluate the performance of the proposed model, we selected a number of current leading recommendation system algorithms as baselines, including classical matrix decomposition techniques, recently popular deep learning methods, and some algorithms specifically designed for educational recommendation. Each baseline is described below:

- MFBPR [17]: Combining methods of matrix decomposition and personalized ranking, this model focuses on learning implicit feature representations from users' preferences.
- FISM [11]: This model is a collaborative filtering algorithm based on item similarity that captures similarity between items by decomposing the user-item interaction matrix.
- NAIS [12]: Uses neural attention mechanisms to weight the similarity of items in a user's history to provide personalized recommendations.
- ACKRec [7]: Uses content information and information through GCNs to learn the representation of entities from different metapath information to capture the different interests of different students.
- MOOCIR [8]: Designed for Massive Open Online courses (MOOCs) in online education platforms, MOOCIR aims to improve the quality of recommendations by capturing subtle correlations between course content and learner behavior.

Evaluation Metrics. A number of standardized evaluation indicators were used to evaluate the experimental results to comprehensively measure the performance of the recommendation system. The details include:

Hit rate (HR): Measures whether the recommendation list contains items that users actually interact with.

Normalized Discount Accumulation Gain (NDCG): This metric takes into account the position of items in the recommendation list, with correct recommendations higher in the list earning higher points.

Mean Reciprocal Rank (MRR): Evaluates the model's prediction accuracy for the first relevant item in the recommendation list.

3.2 Overall Performance

As shown in Table 2, the overall performance of the MMPDRec model exceeds that of the baseline model. This achievement is mainly due to the fact that the connections between knowledge concepts are more accurately captured. At the same time, the model's unique noise filtering mechanism effectively excluded non-critical information during the training process, reducing the risk of overfitting, and thus improving the generalization ability in the real world recommendation scenario. In short, MMPDRec significantly improves the accuracy and reliability of the recommendation system by optimizing the information extraction process and extracting important features.

Table 2. Comparison of Model Performances

	HR			nDCG			MRR
	k = 5	10	20	k = 5	10	20	
MFBPR	0.668	0.811	0.907	0.481	0.527	0.552	0.448
FISM	0.584	0.701	0.800	0.438	0.476	0.501	0.418
NAIS	0.568	0.691	0.811	0.420	0.461	0.491	0.403
ACKRec	0.659	0.764	0.842	0.503	0.538	0.557	0.475
MOOCIR	0.704	0.836	0.922	0.520	0.562	0.584	0.484
MMPDRec	0.721	0.850	0.933	0.572	0.584	0.593	0.492

3.3 Ablation Study

In this part, we explore the role of key components in the MMPDRec: integration of knowledge concept relationships, denoising GCN, and multi-head attention mechanisms. Three variant models were designed for the ablation experiment as follows:

- w/o kcl:removes of knowledge concept relationships, which causes the model to lose its ability to leverage complex relationships between concepts to enhance recommendations.
- w/o dGCN:removes the denoising GCN, which is designed to refine input features and remove noise information from model learning.
- w/o mha:removes the multi-head attention mechanism, which essentially assigns different attention to different types of information to capture user preferences more finely.

The experimental results in Table 3 show that the performance of these three variant models is lower than that of the complete MMPDRec model. This shows the importance of these components in the original model, each of which has a significant positive impact on the model's recommendation capability.

Table 3. Ablation Study Results To Highlight the Impact of Key Components

	HR		nDCG		MRR
	k = 5	10	k = 5	10	
w/o kcl	0.712	0.847	0.525	0.568	0.489
w/o dGCN	0.701	0.838	0.520	0.562	0.483
w/o mha	0.684	0.821	0.503	0.547	0.470
MMPDRec	0.721	0.850	0.530	0.572	0.492

3.4   Hyper-Parameter Sensitivity

By analyzing the influence of each parameter on the performance, this paper emphasizes the necessity of balanced overparameter setting.

Setting of dropout rate: The results are shown in Table 4. A low drop rate can lead to overfitting because it causes the model to learn noise during training and affects poor model fit. Conversely, a high drop rate eliminates too many connections, hampering the model's ability to capture basic data features. Therefore, choosing a suitable decline rate effectively balances the learning ability of the model and its need for regularization, retaining enough complexity to recognize data patterns.

Table 4. Impact of Dropout Rate on Model Performance

	HR		nDCG		MRR
	k = 5	10	k = 5	10	
rate_0.05	0.700	0.832	0.514	0.557	0.480
rate_0.10	0.721	0.850	0.530	0.572	0.492
rate_0.15	0.695	0.831	0.510	0.554	0.475

Setting the number of GCN layers: The results are shown in Table 5. A limited number of GCN layers may restrict its capacity to capture intricate node relationships in graph-structured data. Conversely, an excessive number of layers can result in over-smoothing and cause node features to become indistinguishable, thereby diminishing the discriminative power of the model. Therefore, selecting an appropriate number of layers is crucial for effective information transmission and feature extraction while maintaining sufficient feature differentiation.

Table 5. Impact of GCN Layer Number on Model Performance

	HR		nDCG		MRR
	k=5	10	k=5	10	
GCN_1	0.708	0.841	0.520	0.562	0.483
GCN_2	0.721	0.850	0.530	0.572	0.492
GCN_3	0.689	0.828	0.499	0.543	0.463

4   Related Work

In order to stimulate students' interest, most recommendation algorithms recommend courses to students. For example, Chen et al. [20] proposed a model utilizing knowledge graph to enhance recommendation systems, thereby addressing the issue of sparse data more effectively. However, since a course often contains multiple videos, making course recommendations directly may overlook a student's interest in a particular knowledge concept. Gong et al. [7] put forward the problem of knowledge concept recommendation, and established the ACKRec framework to naturally integrate rich heterogeneous context-side information into knowledge concept recommendation. Piao [8] used GCNs to learn user and concept representations and combines them into an extended matrix decomposition framework to predict each user's preference for concepts. Jiang et al. [21] proposed a new reinforcement learning framework combined with MOOC knowledge graph to solve the problem of knowledge concept recommendation, which improves the recommendation performance on the basis of interpretability. In order to learn a more accurate representation of nodes, Wang et al. [22] dynamically assign an aspect context to each node by treating aspects as dimensions of student interest.

5   Conclusion and Future Work

In this work, we propose the MMPDRec framework, which is designed to improve the recommendation system by addressing the issue of noise in educational data and the ignored relationship between knowledge concepts. Our experiments show that integrating denoising convolutional networks with multi-head attention systems can enhance the accuracyof recommending knowledge concepts. Future research will explore the integration of more sophisticated graph learning models and their broader application in diverse fields.

Acknowledgments.. This study is funded by National Natural Science Foundation of China (Grant No. 62272093, 62137001), Liaoning Social Science Planning Fund Project (No. L21BSZ075) and the Project of the Association of Fundamental Computing Education in Chinese Universities (2023-AFCEC-184).

References

1. Breslow, L., Pritchard, D.E., DeBoer, J., Stump, G.S., Ho, A.D., Seaton, D.T.: Studying learning in the worldwide classroom: research into edx's first MOOC. Res. Pract. Assess. **8**, 13–25 (2013)
2. Kizilcec, R.F., Piech, C., Schneider, E.: Deconstructing disengagement: analyzing learner subpopulations in massive open online courses. In: Proceedings of the Third International Conference on Learning Analytics and Knowledge, pp. 170-179. ACM, New York (2013)
3. Anderson, A., Huttenlocher, D., Kleinberg, J., Leskovec, J.: Engaging with massive online courses. In: Proceedings of the 23rd International Conference on World Wide Web, pp. 687-698. ACM, New York (2014)
4. Yao, D., Deng, X.: A course teacher recommendation algorithm based on improved latent factor model and personalrank. IEEE Access **9**, 108614–108627 (2021)
5. Zhang, J., Hao, B., Chen, B., Li, C., Chen, H., Sun, J.: Hierarchical reinforcement learning for course recommendation in MOOCs. In: AAAI, vol. 33, pp. 435–442 (2019)

6. Wang, C., Zhu, H., Zhu, C., Zhang, X., Chen, E., Xiong, H.: Personalized employee training course recommendation with career development awareness. In: Proceedings of the Web Conference 2020, pp. 1648–1659. ACM, Taipei (2020). https://doi.org/10.1145/3366423.3380236

7. Gong, J., et al.: Attentional graph convolutional networks for knowledge concept recommendation in MOOCs in a heterogeneous view. In: Proceedings of the 43rd International ACM SIGIR Conference on Research and Development in Information Retrieval, pp. 79-88. ACM, Virtual Event (2020)

8. Piao, G.: Recommending knowledge concepts on MOOC platforms with meta-path-based representation learning. In: Proceedings of the 14th International Conference on Educational Data Mining (EDM 2021), pp. 487–494. International Educational Data Mining Society, Paris (2021)

9. Ying, R., He, R., Chen, K., Eksombatchai, P., Hamilton, W.L., Leskovec, J.: Graph convolutional neural networks for web-scale recommender systems. In: Proceedings of the 24th ACM SIGKDD International Conference on Knowledge Discovery & Data Mining (KDD 2018), pp. 974–983. ACM, London (2018)

10. Wang, X., He, X., Wang, M., Feng, F., Chua, T.-S.: Neural graph collaborative filtering. In: Proceedings of the 42nd International ACM SIGIR Conference on Research and Development in Information Retrieval (SIGIR 2019), pp. 165–174. ACM, Paris (2019)

11. Kabbur, S., Ning, X., Karypis, G.: FISM: factored item similarity models for top-N recommender systems. In: Proceedings of the 19th ACM SIGKDD International Conference on Knowledge Discovery and Data Mining, pp. 659–667. ACM, Chicago (2013)

12. He, X., He, Z., Song, J., Liu, Z., Jiang, Y.-G., Chua, T.-S.: NAIS: neural attentive item similarity model for recommendation. IEEE Trans. Knowl. Data Eng. **30**, 2354–2366 (2018)

13. Wang, X., He, X., Wang, M., Feng, F., Chua, T.-S.: Neural graph collaborative filtering. In: Proceedings of the 42nd International ACM SIGIR Conference on Research and Development in Information Retrieval, pp. 165–174 (2019)

14. He, X., Deng, K., Wang, X., Li, Y., Zhang, Y., Wang, M.: LightGCN: simplifying and powering graph convolution network for recommendation. In: Proceedings of the 43rd International ACM SIGIR Conference on Research and Development in Information Retrieval, pp. 639–648. ACM, Virtual Event (2020)

15. Mao, K., Zhu, J., Xiao, X., Lu, B., Wang, Z., He, X.: UltraGCN: ultra simplification of graph convolutional networks for recommendation. In: Proceedings of the 30th ACM International Conference on Information and Knowledge Management, pp. 1253–1262. ACM, Virtual Event (2021)

16. Zhou, X., Lin, D., Liu, Y., Miao, C.: Layer-refined graph convolutional networks for recommendation. In: 2023 IEEE 39th International Conference on Data Engineering (ICDE), pp. 1247–1259. IEEE, Anaheim (2023)

17. Rendle, S., Freudenthaler, C., Gantner, Z., Schmidt-Thieme, L.: BPR: bayesian personalized ranking from implicit feedback (2012)

18. Kingma, D.P., Ba, J.: ADAM: a method for stochastic optimization. In: Proceedings of the 3rd International Conference on Learning Representations, ICLR 2015, San Diego, CA, USA (2015)

19. Yu, J., et al.: MOOCCube: a large-scale data repository for NLP lMOOCs. In: Proceedings of the 58th Annual Meeting of the Association for Computational Linguistics, pp. 3135–3142. Association for Computational Linguistics, Online (2020)

20. Chen, X., Sun, Y., Zhou, T., Wen, Y., Zhang, F., Zeng, Q.: Recommending Online Course Resources Based on Knowledge Graph. In: Zhao, X., Yang, S., Wang, X., Li, J. (eds.) Web Information Systems and Applications. WISA 2022. Lecture Notes in Computer Science, vol. 13579, pp. 581–588. Springer, Cham (2022). https://doi.org/10.1007/978-3-031-20309-1_51

21. Jiang, L., et al.: Reinforced explainable knowledge concept recommendation in MOOCs. ACM Trans. Intell. Syst. Technol. **14**, 1–20 (2023)

22. Wang, X., Jia, L., Guo, L., Liu, F.: Multi-aspect heterogeneous information network for MOOC knowledge concept recommendation. Appl. Intell. **53**, 11951–11965 (2023)

SPR: A Similar Projection Revisor for Complex Logical Reasoning over Knowledge Graphs

Tingting Wang[1], Yuxuan Tang[1], Ruolin Li[1], Duo Yu[1], Bowen Feng[2], Feng Ding[1], Shuo Yu[1(✉)], and Yanming Shen[1]

[1] Dalian University of Technology, 116024 Dalian, China
{tyx,liruolin}@mail.dlut.edu.cn, {dingfeng,yushuo,shen}@dlut.edu.cn
[2] School of Informatics, University of Edinburgh, Edinburgh, Scotland EH89YL, UK
B.Feng-2@sms.ed.ac.uk

Abstract. Complex logical reasoning over knowledge graphs (KGs) aims to infer target answers by first-order logic queries. Compared to link prediction or multi-hops reasoning tasks, complex logical reasoning is more related to users need of information retrieval. However, existing complex logical reasoning methods only pay attention to modeling entities, relations or operators, and ignore the assistance of auxiliary information in KGs. To address this issue, we propose **S**imilar **P**rojection **R**evisor (SPR), a component based on similar triples projection for complex logical reasoning tasks. Firstly, SPR utilizes auxiliary information in KGs, in terms of relations in triples which are as same as the reasoning step. Secondly, SPR revises the bias projection in the current query via projection revising module. As a pluggable component, SPR can be embedded in complex logical reasoning baselines to improve their performance without changing baselines structure. Experimental results on two benchmark datasets demonstrate that SPR can increase baselines performance by 1.2% and 1.1% respectively.

Keywords: Knowledge graphs · Complex logical reasoning · Auxiliary Information

1 Introduction

Reasoning over knowledge graphs (KGs) can be used to answer user queries based on observed facts, and it has attracted extensive attentions in many domains such as question answering, recommendation system, and drug discovery [1–3]. Compared to link prediction (or single-hop) and multi-hops reasoning tasks, complex logical reasoning over KGs gradually infers target entity from a known entity by answering First-Order Logical (FOL) queries [4,5]. FOL queries include logical operations, such as existential quantifier ($\exists$), conjuction ($\wedge$), and disjunction ($\vee$).

For example, the question *"Who is the coach of Curry's teammate and Thompson's teammate?"* can be expressed as FOL queries, as shown in Fig. 1.

© The Author(s), under exclusive license to Springer Nature Singapore Pte Ltd. 2024
C. Jin et al. (Eds.): WISA 2024, LNCS 14883, pp. 78–90, 2024.
https://doi.org/10.1007/978-981-97-7707-5_7

The query can transform the natural language query into an undirected computation graph which is beneficial for extracting auxiliary information in KGs. However, complex logical reasoning is a challenging task because existing KGs are always incomplete and large-scale, making the queries hard to answer.

To address the aforementioned issues, existing methods can be classified into two research directions [5]. The first direction adapts KGs embedding methods which are successful in knowledge graph completion. They map the queries to low-dimensional continuous spaces by iteratively executing logical operators based on the query computation graphs [4,6–8]. The other direction focuses on pre-trained transformer architectures for obtaining richer semantic patterns such as entities or relations. These methods are mainly incorporated structural knowledge into transformers [9–13].

Despite the success achieved, there are still limitations of existing methods for reasoning over KGs. Because these methods primarily focus on modeling entities, relations, or operators, yet overlook the assistance of auxiliary information within KGs. One type of auxiliary information can refer to the information obtained from the given KGs, such as a triple $h = (entity_{head}, teammate, entity_{tail})$, with a relation as same as the current query $q = (entity_{head}, teammate, ?)$. In KGs, different entities projecting the same relation (e.g., *teammate*) often tend to have similar bias from the target answer. Therefore, paying attention to this type of auxiliary information within KGs can provide assistance for complex logical reasoning.

Inspired by this, we propose a Similar Projection Revisor (SPR), a component emphasizes on triples with the same relations as auxiliary information for complex logical reasoning. A complex logical reasoning process typically involves at least one triple projection operation, and each triple projection operation contains one relation. For example, the complex query illustrated Fig. 1 can be decomposed into two sub-queries: 1)"$WhoisthecoachofCurry'steammate?$", and 2) "$WhoisthecoachofThompson'steammate?$" Thus, the target answer can be reasoned over the overlapped entities from these two sub-queries. Specifically, SPR utilizes these similar triples, which projects the triples into low-dimensional continuous space. If a query contains various triples, multiple projection are performed. Additionally, SPR also employs an advanced revising strategy, which firstly selects one triple from a triple set with the same relations, and then uses the remaining triples to revise it via the bias revising module. Finally, we conduct extensive experiments on two benchmark datasets to demonstrate the effectiveness of SPR. The main contributions are summarized as follows:

- We propose a Similar Projection Revisor (SPR) for complex logical reasoning over KGs, which leverages the assistance of auxiliary information within KGs.
- The proposed SPR uses similar triples projection with the same relations and we design an advanced bias revising module to rectify similar triples projection bias.
- We evaluate our SPR on two widely-used KG bench-marks, i.e., NELL-995-betae and FB15k-237-betae. Experimental results show that SPR significantly and effectively outperforms existing state-of-the-art methods.

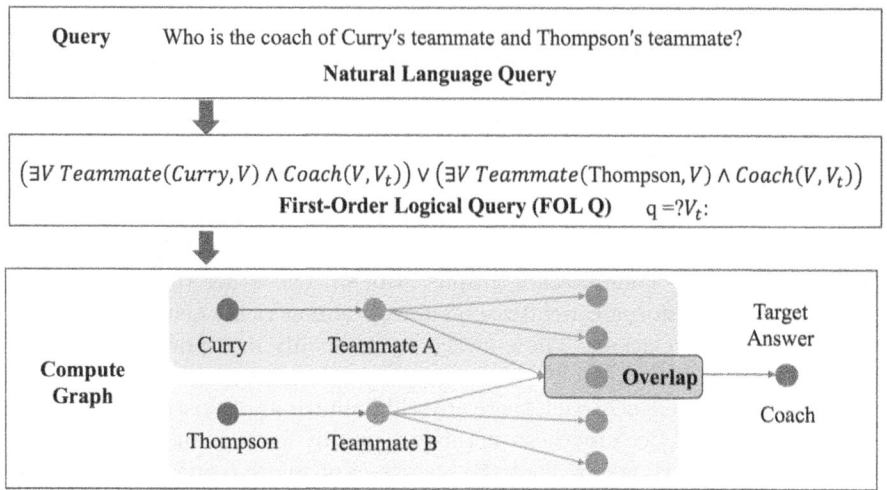

Fig. 1. Natural language query, first-order logical query and computation graph

2 Related Work

2.1 Knowledge Graph Completion

The objective of the knowledge graph completion task is to predict missing links in KGs. Approaches can be broadly categorized into three types: based on distributed representations, based on neural networks, and based on logical rules [14].

Approaches based on distributed representations employ scoring functions to assess the accuracy of triples and subsequently rank all triples. TransE [15] postulates that entities and relationships undergo transformations within the same Euclidean space, with transformations simplified to vector addition. TransH [16], however, argues that TransE underperforms in one-to-many or many-to-many scenarios, suggesting that entity transformations should occur in different spaces dictated by the relationship. RotatE [17] improves on the simple vector addition transformations by employing rotation-based transformations, thereby mathematically satisfying properties such as symmetry and transitivity essential for knowledge graph completion.

Neural network-based methods, known for their robust fitting capabilities, can model more complex features. This category can be subdivided further, for instance, methods based on Convolutional Neural Networks (CNNs) [18] argue that CNNs possess enhanced fitting abilities. ConvE [19], utilizes CNNs to manage entities and relations within graphs and employs two-dimensional convolution to boost interactions between representations. Methods based on Graph Neural Networks (GNNs) [20] suggest that the message-passing attributes of GNNs effectively model entities within graphs. CompGCN [21] introduces an integration of aggregation functions for various entity and relationship types into the GNN model, addressing forward, reverse, and self-loop relations with distinct parameters. RED-GNN [22] extends beyond message passing in graph neural networks by incorporating the relational directed graph structure defined by the authors to capture the local structure of graphs.

Rule-based methods extract rules from KGs using rule mining tools, typically representing these rules through first-order logic and extracting them via statistical methods. AMIE [23] outlines three rule extension steps to expand an initial rule set and derive the final collection of rules. RLvLR [24] proposes a method for rule extraction on large KGs, devising new search and pruning strategies to address the large search spaces inherent in previous methodologies.

2.2   Complex Logical Reasoning

Complex logical reasoning in KGs involves deriving an answer node from a starting node via a specified FOL query, which includes relational transformations as well as operators like conjunction and disjunction. Due to the inclusion of these new operators, the design of models must consider additional factors.

GQE [25] is among the first to employ a model based on distributed representations for this task, utilizing vector addition from TransE [15] for projection operations in GQE and employing attention-based neural networks for conjunction operations. Query2Box [26] adopts a method based on rectangular embedding (Box Embedding) for designing model projections and conjunction operations, also using disjunctive normal form to transform first-order logic queries, placing disjunctive operations at the end of all operations and selecting the highest-scoring sub-query among various sub-queries to simplify the model design [27]. Graph-based modeling is theoretically superior to point-based modeling because a graph can encompass several points in space, aligning more closely with practical scenarios. ConE [28] compensates for Query2Box's inability to model negation operations, using cones to model queries, with negation operations requiring only the complementation of space.

Transformers [29] have demonstrated exceptional performance across multiple deep learning tasks, prompting researchers to incorporate them into complex logical reasoning tasks within KGs. For example, BiQE [30] inputs the directed acyclic graph representing the query into a Transformer, treating intermediate unknown nodes as masks, with the information outputted by the Transformer constituting the final answer.

3   Problem Definition

We firstly introduce the notations used in this paper, and then describe the problem definition of complex logical reasoning over KGs.

Notations. We define natural language problems by FOL queries, including existential quantifiers ($\exists$), conjunction ($\wedge$) and disjunction ($\vee$). The query set is denoted as Q, and the answer set is denoted as A. And the definition of FOL queries is as follows:

$$
\begin{aligned}
Q[A] &:= \exists A \colon \exists V_1, \ldots, V_m \colon q_1 \vee q_2 \vee \ldots \vee q_d, \\
&\text{where } q_i = e_1^i \wedge e_2^i \wedge \ldots \wedge e_{n_i}^i, \\
&\text{where } e_j^i = p(c, V), \text{ with } V \in A, V_1, \ldots, V_m, c \in \mathcal{E}, p \in \mathbb{R}, \\
&\text{or } e_j^i = p(V', V''), \text{ with } V', V'' \in A, V_1, \ldots, V_m, p \in \mathbb{R}.
\end{aligned}
\tag{1}
$$

In some cases, complex logical reasoning tasks also include negation operations ($\neg$). This operator is usually placed in front of the projection operation $p(a, b)$, represented as $\neg p(a, b)$, indicating that node a is projected to node b via relation p.

Problem Definition. Complex logical reasoning over KGs aims to infer the target answer by the given FOL Queries. Specifically, the problem can be defined as follows: Given a query set Q and an answer set A, where Q often involves logical operations such as conjunction ($\wedge$), disjunction ($\vee$), and existential quantifiers ($\exists$), the objective of complex logical reasoning is to predict the target entities (or nodes) from the anchor entity c, which can be reasoned through a series of relations and logical operations in KGs.

4   Method

This section introduces the Similar Projection Revisor (SPR), which is shown in Fig. 2. Particularly, SPR is mainly divided into two parts: Similar Triples Projection, and Bias Projection Revisor.

4.1   Preliminaries

Similar triples refers to triples that have the same relation as the current query q, and this auxiliary information can assist complex logical reasoning over KGs. In order to take running efficiency into account, we use the following method to implement similar triples sampling strategy. Before training, we define a three-dimensional tensor $T \subseteq \mathbb{N}_+^{r_n \times r_m \times 3}$ to store training triples, where r_n is the number of relations, r_m is the number of triples with the highest frequency of relation occurring in KGs, and 3 represents the number of elements in a triple. If the number of triples for a given relation is less than r_m, it will be repetitively filled until r_m is reached.

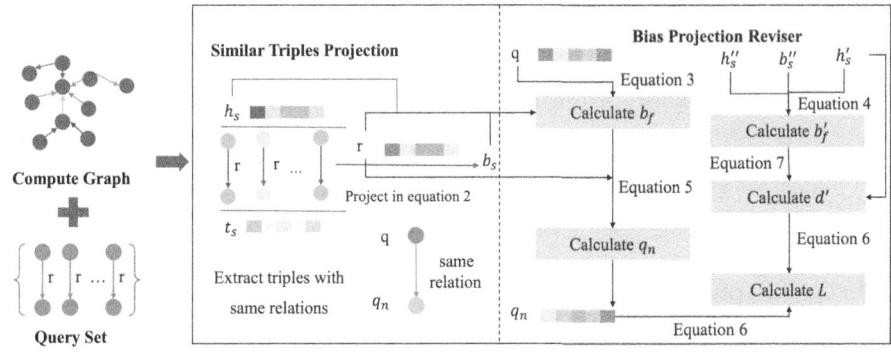

Fig. 2. Overview of the Similar Projection Revisor

During training, this sampling strategy will be quickly realised via the tensor T. In our implementation, each query will sample 25 triples with the same relation in per projection. Since the iteration is performed with a mini-batch data, it would be time-consuming to generate random numbers for each query. So in this paper, we set 25 random numbers generated in $[0, r_m]$ at each projection, which is indexed by the second dimension of the tensor T. After the projection, triples with different relations in tensor T will be disrupted to avoid false correlations among the triples.

4.2 Similar Triples Projection

Considering the assistance of auxiliary information in KGs for complex logical reasoning, we propose a similar triples projection. Similar triples mean that relations in these triples are as the same as the current reasoning step query q. Firstly, we sample a set of similar triples for each query via the tensor T, and then we obtain the bias between the reasoning and the assistance of auxiliary information, by calculating the Euclid distance b_s between the ground-truth answer and the predicted result in the current query. It can be expressed as follows:

$$b_s = t_s - P(h_s, r), \tag{2}$$

where h_s, r, and t_s represent the embedding vector of head entity, relation, and tail entity in similar triples respectively, and $P(,)$ denotes the projection operation. In fact, we can replace calculating the Euclid distance with other methods, such as KL divergence in BetaE [31], according to baselines models.

4.3   Bias Projection Revisor

Bias projection revisor is a revising model, which regards the revised triples as a new query. As mentioned before, complex logical reasoning over KGs usually contains at least one projection operation, and each projection operation contains a relationship. Thus, if a query contains multiple projections, multiple revisions are performed. The strategy of revisions is similar to the Teacher-Student Network [32], which means one entity directs another entity, and can be regarded as a technique of data augmentation.

In the bias projection revisor model, we firstly revise bias in the current query q, and it is expressed as follows,

$$b_f = \text{Linear}_w (h_s - q) \times \text{Linear}_t (b_s \| r), \tag{3}$$

where q represents the current query, $Linear_w$ and $Linear_t$ are linear layers, and $(\|)$ represents concatenating operation. Then, we revise bias in the advanced revised triples, and it is expressed as follows,

$$b'_f = \text{Linear}_w (h''_s - h'_s) \times \text{Linear}_t (b''_s \| r), \tag{4}$$

where h''_s represents the remaining head entities in the similar triples set, h'_s represents the head entities in the advanced revised triples, and b''_s represents the bias in advanced revised triples. The advanced revised triples refer to a set of triples with the same relations as current query, and their projection should be revised in advanced. Finally, we can obtain the query embedding vector q_n through the projection and revising operation, and q_n can be expressed as follows,

$$q_n = P(q, r) + b_f \times \alpha, \tag{5}$$

where α is a hyper-parameter and its value is less than 1. And we define the loss function as:

$$\mathcal{L}_{\text{new}} = L_{pas} (d_{pos}) + L_{ne.g.} (d_{ne.g.}) + L_{pos} (d') \times \beta, \tag{6}$$

$$d' = D (t'_s, P (h'_s, r) + b_f \times \alpha), \tag{7}$$

where d_{pos} denotes the distance between the predicted answer and positive samples, d_{neg} denotes the distance between the predicted answer and negative samples, and β is the hyper-parameter of adjusting the loss of the advanced revisor d'.

5   Experiments

5.1   Experimental Settings

Datasets. We use two datasets from BetaE [31]: FB15k-237-betae and ELL-995-betae. BetaE dataset adds training with negation operations, verification, and test queries, and uses verification and test queries with fewer answers. Training queries include five types: 1p, 2p, 3p, 2i, and 3i. In addition, pi, ip, 2u, and up are also included.

Baselines. The experiments in this paper use four baseline models, including GQE [25], Query2Box [6], BetaE [31], and LogicE [33]. LogicE is a model based on fuzzy logic, using fuzzy logic between [0, 1] for modeling queries, and LogicE has provided multiple variants for modeling queries, such as adopting attention mechanism or not using upper and lower boundary.

Parameter Settings. Hyper-parameters of all models with SPR are consistent with the corresponding baseline models. Some of the hyper-parameters used in baseline models are shown in Table 1:

Table 1. Hyperparameters of baselines for experiments

Model	Batch size	Learning Rate	Embedding Dimension	Number of Iterations
GQE	512	0.0001	800	450000
Query2Box	512	0.0001	400	450000
BetaE	512	0.0001	400	450000
LogicE	512	0.0001	400	450000

5.2   Performance Comparison

Experimental results are shown in Table 2, 3, 4 and Table 5. Table 2 and Table 3 are the MRR results of queries that do not include negation on the two datasets. Q2B in the table is the abbreviation of Query2Box. Table 4 and Table 5 are the MRR results that include negation. The MRR results of the negation operation query only include BetaE and LogicE, because these two models support negation operation. It can be seen that in all types of queries not containing negation operation, SPR can increase performance by 1.2% and 1.1% respectively. The performance of LogicE with SPR on FB15k-237-betae is as the same as the original model. The reason maybe that LogicE uses fuzzy logic to model entities and queries, but SPR only imposes boundary value constraints on fuzzy logic-based models.

Tables 4 and Table 5 show the results of SPR to answer negative queries. Among the four sets of experiments, the results for LogicE increases, but one experimental result for BetaE stays the same and the other decreases. This may be due to the fact that the SPR model does not pay attention to the optimization of the inversion operation. For the inversion operation, SPR simply uses the algorithm in the baseline model without performing more processing. This may result in the failure of the baseline model in inverting operation.

5.3   Ablation Study

In this section, we discuss the effect of adding SPR in two models. The experimental results are shown in Table 6. We conduct experiments on NELL-995-q2b dataset, and we can see that the effect of SPR is significantly higher than not

Table 2. MRR(%) performance of SPR in FB15k-237-betae

Model	1p	2p	3p	2i	3i	pi	ip	2u	up	mean
GQE	35.0	7.2	5.3	23.3	34.6	16.5	10.7	8.2	5.7	16.3
+SPR	**40.4**	**8.6**	**6.6**	**27.0**	**38.3**	**19.0**	**11.7**	**10.4**	**6.6**	**18.7**
Q2B	40.6	9.4	6.8	29.5	42.3	21.2	12.6	11.3	7.6	20.1
+SPR	**40.7**	**9.9**	**7.4**	**30.3**	**43.3**	**21.8**	**13.1**	**12.7**	**8.0**	**20.8**
BetaE	39.0	10.9	10.0	28.8	42.5	22.4	12.6	12.4	9.7	20.9
+SPR	**41.8**	**11.7**	**10.2**	**31.1**	**45.8**	**24.8**	**13.0**	**13.8**	**10.0**	**25.5**
LogicE	41.3	11.8	**10.4**	**31.4**	**43.9**	**23.8**	14.0	13.4	**10.2**	**23.3**
+SPR	**41.7**	**12.1**	10.2	31.2	43.1	**23.8**	**14.1**	**14.8**	10.0	**23.3**
Mean	39.0	9.8	8.1	28.3	40.8	21.0	12.5	11.3	8.3	19.9
+SPR	**41.2**	**10.6**	**8.6**	**29.9**	**42.6**	**22.4**	**13.0**	**12.9**	**8.7**	**21.1**
Diff.	2.2	0.8	0.5	1.6	1.8	1.4	0.5	1.6	0.4	1.2

Table 3. MRR(%) performance of SPR in NELL-995-betae

Model	1p	2p	3p	2i	3i	pi	ip	2u	up	Mean
GQE	32.8	11.9	9.6	27.5	35.2	18.4	14.4	8.5	8.8	18.6
+SPR	**38.2**	**13.2**	**11.0**	**30.3**	**38.6**	**20.6**	**15.9**	**10.6**	**9.9**	**20.9**
Q2B	42.2	14.0	11.2	33.3	44.5	22.4	16.8	11.3	10.3	22.9
+SPR	**43.1**	**14.4**	**11.8**	**34.8**	**46.1**	**23.0**	**17.5**	**12.5**	**10.9**	**23.8**
BetaE	53.0	13.0	11.4	37.6	47.5	24.1	14.3	12.2	8.5	24.6
+SPR	**54.7**	**13.8**	**11.5**	**38.4**	**49.3**	**24.5**	**15.8**	**12.9**	**9.1**	**25.6**
LogicE	58.3	17.7	**15.4**	40.5	**50.4**	**27.3**	19.2	15.9	12.7	28.6
+SPR	**59.0**	**18.5**	15.4	**41.1**	49.6	26.4	**20.2**	**18.2**	**13.4**	**29.1**
Mean	46.6	14.2	11.9	34.7	44.4	23.1	16.2	12.0	10.1	23.7
+SPR	**48.8**	**15.0**	**12.4**	**36.2**	**45.9**	**23.6**	**17.4**	**13.6**	**10.8**	**24.8**
Diff	2.2	0.8	0.5	1.5	1.5	0.5	1.2	1.6	0.7	1.1

Table 4. MRR(%) performance of SPR in FB15k-237-betae while answering queries with negation

Model	2in	3in	inp	pin	pni	Mean
BetaE	**5.1**	**7.9**	7.4	3.6	**3.4**	**5.5**
+SPR	4.9	7.6	**7.6**	**4.0**	3.3	**5.5**
LogicE	4.9	8.2	**7.7**	3.6	3.5	5.6
+SPR	**5.3**	**8.7**	**7.7**	**3.7**	**3.7**	**5.8**

Table 5. MRR(%) performance of SPR in NELL-995-betae while answering queries with negation

Model	2in	3in	inp	pin	pni	Mean
BetaE	**5.1**	**7.8**	10.0	**3.1**	**3.5**	**5.9**
+SPR	4.6	7.4	**10.4**	3.0	3.0	5.7
LogicE	5.3	7.5	11.1	**3.3**	3.8	6.2
+SPR	**5.7**	**7.9**	**11.6**	**3.3**	**4.1**	**6.5**

using SPR. SPR indeed brings gains to the baseline model from different aspects of the knowledge graph. In most cases, SPR can improve the performance of the model because the structure of similar triples projection is simpler and more convenient to use. Taking Query2Box as an example, similar triples projection only need consider one step of reasoning.

Table 6. HITS@3(%) performance of the superposition of SPR in NELL-995-q2b

Model	1p	2p	3p	2i	3i	pi	ip	2u	up	Mean
GQE	40.2	20.3	18.5	23.7	32.4	16.1	8.6	18.6	12.3	21.2
+SPR	**45.3**	**22.0**	**20.3**	**26.5**	**35.9**	**17.3**	**9.9**	**23.2**	**13.7**	**23.8**
Query2Box	52.6	21.8	19.3	30.3	41.7	18.2	8.8	29.9	11.0	26.0
+SPR	**57.0**	**24.6**	**22.7**	**31.6**	**45.0**	**20.6**	9.8	**35.9**	**14.0**	**28.7**

6 Conclusion

In this work, we propose a pluggable component called SPR, which utilizes similar triples projection for complex logical reasoning over KGs. SPR first utilizes similar triples projection, which means that different triples with the same relation have similar bias from the target answer. Then, SPR employs an advanced correction strategy to correct similar triples projection via the bias projection revisor module. In addition, SPR can improve complex logical reasoning performance without changing the baselines models. Experimental results show that SPR significantly and effectively outperforms existing state-of-the-art methods.

Acknowledgments. We would like to thank Yupeng Gao from Dalian University of Technology for his help with this paper. This work is supported by the Fundamental Research Funds for the Central Universities (No. DUT24LAB121).

References

1. Huang, X., Zhang, J., Li, D., Li, P.: Knowledge graph embedding based question answering. In: Proceedings of the Twelfth ACM International Conference on Web Search and Data Mining, WSDM 2019, Melbourne, VIC, Australia, 11–15 February 2019, pp. 105–113. ACM (2019)
2. Xu, F., Liu, J., Lin, Q., Pan, Y., Zhang, L.: Logiformer: a two-branch graph transformer network for interpretable logical reasoning. In: SIGIR 2022: The 45th International ACM SIGIR Conference on Research and Development in Information Retrieval, Madrid, Spain, 11–15 July 2022, pp. 1055–1065. ACM (2022)
3. Ioannidis, V.N., Zheng, D., Karypis, G.: Few-shot link prediction via graph neural networks for covid-19 drug-repurposing. CoRR (2020). arxiv:2007.10261
4. Amayuelas, A., Zhang, S., Rao, S.X., Zhang, C.: Neural methods for logical reasoning over knowledge graphs. In: The Tenth International Conference on Learning Representations, ICLR Virtual Event. OpenReview.net (2022)
5. Liu, Y., Cao, Y., Wang, S., Wang, Q., Bi, G.: Generative models for complex logical reasoning over knowledge graphs. In: Proceedings of the 17th ACM International Conference on Web Search and Data Mining, WSDM 2024, Merida, pp. 492–500. ACM (2024)
6. Ren, H., Hu, W., Leskovec, J.: Query2box: reasoning over knowledge graphs in vector space using box embeddings. In: 8th International Conference on Learning Representations, ICLR 2020, Addis Ababa, Ethiopia, 26–30 April 2020. OpenReview.net (2020)
7. Arakelyan, E., Daza, D., Minervini, P., Cochez, M.: Complex query answering with neural link predictors. In: 9th International Conference on Learning Representations, ICLR 2021, Virtual Event, Austria, 3–7 May 2021. OpenReview.net (2021)
8. Zhang, Z., Wang, J., Chen, J., Ji, S., Wu, F.: Cone: cone embeddings for multi-hop reasoning over knowledge graphs. In: Annual Conference on Neural Information Processing Systems 2021, NeurIPS, Virtual, pp. 19172–19183 (2021)
9. Bi, Z., Cheng, S., Chen, J., Liang, X., Xiong, F., Zhang, N.: Relphormer: relational graph transformer for knowledge graph representations. Neurocomputing **566**, 127044 (2024)
10. Kotnis, B., Lawrence, C., Niepert, M.: Answering complex queries in knowledge graphs with bidirectional sequence encoders. In: Thirty-Third Conference on Innovative Applications of Artificial Intelligence, Virtual Event, pp. 4968–4977. AAAI Press (2021)
11. Liu, X., et al.: Mask and reason: pre-training knowledge graph transformers for complex logical queries. In: The 28th ACM SIGKDD Conference on Knowledge Discovery and Data Mining, Washington, pp. 1120–1130. ACM (2022)
12. Yao, L., Mao, C., Luo, Y.: KG-BERT: BERT for knowledge graph completion. CoRR (2019). arxiv:1909.03193
13. Zhang, W., et al.: Structure pretraining and prompt tuning for knowledge graph transfer. In: Proceedings of the ACM Web Conference, WWW, Austin, pp. 2581–2590. ACM (2023)
14. Chen, X., Jia, S., Xiang, Y.: A review: knowledge reasoning over knowledge graph. Expert Syst. Appl. **141**, 112948 (2020)

15. Bordes, A., Usunier, N., Garcia-Duran, A., Weston, J., Yakhnenko, O.: Translating embeddings for modeling multi-relational data. In: Advances in Neural Information Processing Systems 26: 27th Annual Conference on Neural Information Processing Systems 2013, Lake Tahoe, Nevada, United States, Proceedings of a Meeting Held 5–8 December 2013, pp. 2787–2795 (2013)
16. Wang, Z., Zhang, J., Feng, J., Chen, Z.: Knowledge graph embedding by translating on hyperplanes. In: Proceedings of the AAAI Conference on Artificial Intelligence, vol. 28, pp. 1112–1119. AAAI Press (2014)
17. Sun, Z., Deng, Z.H., Nie, J.Y., Tang, J.: Rotate: knowledge graph embedding by relational rotation in complex space. OpenReview.net (2019)
18. LeCun, Y., et al.: Handwritten digit recognition with a back-propagation network, vol. 2, pp. 396–404. Morgan Kaufmann (1989)
19. Dettmers, T., Minervini, P., Stenetorp, P., Riedel, S.: Convolutional 2D knowledge graph embeddings. In: Proceedings of the AAAI Conference on Artificial Intelligence, vol. 32, pp. 1811–1818. AAAI Press (2018)
20. Kipf, T.N., Welling, M.: Semi-supervised classification with graph convolutional networks (2016). arxiv:1609.02907
21. Vashishth, S., Sanyal, S., Nitin, V., Talukdar, P.: Composition-based multi-relational graph convolutional networks (2019). arxiv:1911.03082
22. Zhang, Y., Yao, Q.: Knowledge graph reasoning with relational digraph. In: Proceedings of the ACM Web Conference 2022, pp. 912–924. ACM (2022)
23. Galárraga, L.A., Teflioudi, C., Hose, K., Suchanek, F.: Amie: association rule mining under incomplete evidence in ontological knowledge bases. In: Proceedings of the 22nd International Conference on World Wide Web, pp. 413–422. International World Wide Web Conferences Steering Committee/ACM (2013)
24. Omran, P.G., Wang, K., Wang, Z.: Scalable rule learning via learning representation. In: IJCAI, pp. 2149–2155. ijcai.org (2018)
25. Hamilton, W., Bajaj, P., Zitnik, M., Jurafsky, D., Leskovec, J.: Embedding logical queries on knowledge graphs. Adv. Neural Inf. Process. Syst. **31**, 2030–2041 (2018)
26. Vilnis, L., Li, X., Murty, S., McCallum, A.: Probabilistic embedding of knowledge graphs with box lattice measures. In: Proceedings of the 56th Annual Meeting of the Association for Computational Linguistics, ACL 2018, Melbourne, Australia, 15–20 July 2018, vol. 1: Long Papers, pp. 263–272. Association for Computational Linguistics (2018)
27. Davey, B.A., Priestley, H.A.: Introduction to Lattices and Order, 2nd edn. Cambridge University Press, Cambridge (2002). https://doi.org/10.1017/CBO9780511809088
28. Zhang, Z., Wang, J., Chen, J., Ji, S., Wu, F.: Cone: cone embeddings for multi-hop reasoning over knowledge graphs. In: Advances in Neural Information Processing Systems 34: Annual Conference on Neural Information Processing Systems 2021, NeurIPS 2021, 6–14 December 2021, virtual, pp. 19172–19183 (2021)
29. Vaswani, A., et al.: Attention is all you need. In: Advances in Neural Information Processing Systems 30: Annual Conference on Neural Information Processing Systems 2017, Long Beach, CA, USA, 4–9 December 2017, pp. 5998–6008 (2017)
30. Kotnis, B., Lawrence, C., Niepert, M.: Answering complex queries in knowledge graphs with bidirectional sequence encoders. In: Thirty-Fifth AAAI Conference on Artificial Intelligence, AAAI 2021, Thirty-Third Conference on Innovative Applications of Artificial Intelligence, IAAI 2021, The Eleventh Symposium on Educational Advances in Artificial Intelligence, EAAI 2021, Virtual Event, 2-9 February 2021, pp. 4968–4977. AAAI Press (2021)

31. Ren, H., Leskovec, J.: Beta embeddings for multi-hop logical reasoning in knowledge graphs. In: Annual Conference on Neural Information Processing Systems, NeurIPS , Virtual (2020)
32. Hinton, G., Vinyals, O., Dean, J.: Distilling the knowledge in a neural network. arXiv preprint arXiv:1503.02531 (2015)
33. Luus, F.P.S., et al.: Logic embeddings for complex query answering. CoRR (2021). arxiv:2103.00418

An Generative Entity Relation Extraction Model Based on UIE for Legal Text

Hua Yin, Shuo Huang, ZhiJian Wang, Yong Ye$^{(\boxtimes)}$, and WenHui Zhu

Guangdong University of Finance and Economics, Guangzhou 510000, China
{yinhua,zjian}@gdufe.edu.cn, 1372327667@qq.com

Abstract. Entity relation extraction is a basic technology in legal text analysis. The complexity of legal text description makes the traditional methods suffer from the problems of error propagation and low efficiency. Although generation based methods provide new solutions, domain-specific design is still necessary. We define ten entity types and three relation types for traffic accident crime cases. And then propose a new generative model UIE-ERNIE-CRF for chinese legal texts based on an universal information extraction model(UIE). Our model is divided into five layers, including input layer, a general semantic representation layer, a task semantic representation layer, a decoding layer, and the output layer. The model identifies specific semantic information required in different tasks through the ERNIE3.0 model. In order to solve the long-distance dependency problem faced by double pointer decoder in the UIE model, sequence labeling decoder CRF is introduced. Compared with the baseline model, UIE-ERNIE-CRF performs well on the precision rate, recall rate and F1 value. And the ablation experiment shows that introducing the ERNIE and CRF is effective for entity relation extraction of legal text.

Keywords: Entity-relation extraction · UIE · ERNIE · CRF

1 Introduction

Automatic structured knowledge extraction can support intelligent judicial applications, such as building judicial knowledge graphs. Entity-relation extraction is one of the important basic tasks. Their methods were categorized into two groups: traditional methods and deep learning-based methods [1]. The traditional rule-based and machine learning based methods require users with domain knowledge, which are suitable for texts with standard structure and predefined relations. Compared with classical extraction methods, the main advantage of deep learning based methods is that the neural network model can automatically learn sentence features without complex feature engineering. They are classified into pipelined and joint extraction. The pipelined methods are limited by the accuracy of the antecedent task and suffers from error propagation [2]. In order to

C. Jin et al. (Eds.): WISA 2024, LNCS 14883, pp. 91–99, 2024.
https://doi.org/10.1007/978-981-97-7707-5_8

effectively integrate the knowledge during entity recognition and relation extraction, the joint learning strategy enhances the relation extraction by sharing the information of upstream texts in the entity recognition task [3,4].

Besides the above model, generation based methods have been proposed one after another, which are based on RNN and adopt the traditional Seq2Seq deep generation framework [5]. However, these generative models do not have significant performances compared to extraction models. The generative pre-training model, such as BART, T5, and GPT, has made it possible to construct effective generative information extraction models. UIE [6], a unified text-to-structure generation architecture, is a generative universal information extraction model with good generalization ability between different tasks in entity recognition and relation extraction.

Although the double pointer decoding adopted by the UIE model has good recognition performance in overlapping entities, due to the long character distance between entities in legal texts and the inclusion of many interactions between irrelevant entities, it will meet noise problems in the entity overlapping scenarios. In contrast, the sequence labeling usually focuses only on the recognition of individual entities, and it is able to better capture the entity structure relations. In addition, CRF [7], as a decoder, enables the sequence labeling approach to capture the dependencies between labels and make joint predictions for each label by the state of the whole sequence, thus improves the performance and generalization ability of the model.

For solving the problem faced in applying UIE on legal text entity-relation extraction, this paper combines UIE with the ERNIE3.0 [8], and designs a new entity-relation extraction model named UIE-ERNIE-CRF, which replaces double pointer decoder with CRF sequence labeling decoder. ERNIE3.0 has a general semantic feature module and a task-specific semantic feature module and introduces the external knowledge during model training to improve efficiency.

2   Related Work

The challenge of AI in Law(CAIL) [9] has been held for six consecutive sessions since 2018. Information extraction tasks have been set up for three consecutive years, involving multiple sub tasks such as named entity recognition(NER) and relation extraction. Due to the professionalism and variety of legal documents, there is currently no unified definition of entity types for NER in the legal field. From the perspective of coarse-grained entity classification, different types of entities such as person names, place names, organizational names, and time can be defined in various legal documents. However, the definition cannot meet the needs of downstream practical tasks which require further entity division [10]. Relation extraction in legal documents is to extract entities such as defendants and victims, as well as their relations (where the relations are pre-defined), and use them to construct triplets (subject, relation, object), where subject represents the main entity, relation represents the relationship, and object represents the guest entity [11].

Information extraction(IE) based on deep learning methods are becoming increasingly mature. The fine-grained information that is difficult to extract directly through rules in legal documents can be identified and extracted, such as identifying judicial entities in case factual descriptions and extracting relations between entities to obtain deep semantic knowledge in legal documents. Gao et al. [12]extracted the relations between judicial entities in key paragraphs of legal documents through improved kernel functions and convolutional neural network methods. Leitner et al. [13] proposed a method for NER of legal documents based on BiLSTM combined with conditional random field models. Wang et al. [14] used character level and word level text representations, combined with self attention mechanisms, to recognize and extract entities such as file numbers and evidence in legal documents. In order to facilitate better interaction between the information learned by the model in the entity recognition and relation extraction stages, and to apply the domain characteristics and knowledge of the Chinese judicial field, Chen [15]proposes a joint entity relation extraction method that incorporates judicial feature representation?which establishes a legal feature dictionary for specific cases and integrates dictionary features into the encoder part of the model through a self attention mechanism, then uses an encoder-decoder model to obtain the vector representation of the corresponding triplet for extracting entities and relations.

3   Research Model

3.1   Legal Entities and Relation Definition

Due to the unique specialization and structural complexity of legal texts, the universal entity and relation definitions do not meet the actual needs of entity relation extraction in legal texts. We choose the traffic accident crime case as the research object. According to the suggestions from domain experts, we define 10 types of entities and 3 types of relationships which can cover the necessary entities and relations in this type of instrument, shown in Table 1 and Table 2.

3.2   Universal Information ExtractionUIE?

UIE is a unified text-to-structure generation architecture that uniformly models different information extraction tasks, adaptively generates target structures, and collaboratively learns common IE capabilities from different knowledge sources. In order to model heterogeneous IE structures, a structure extraction language SEL is used to represent the structure encoding of different information extraction tasks in a unified way, and a structure schema guide using SSI format is used to control what is to be discovered and associated, as well as what is to be generated. The overall framework is shown in the Fig. 1.

3.3   UIE-ERNIE-CRF Model

Entities in legal texts are more likely to be separated by long-distance characters and contain interactions between numerous unrelated entities, which are prone

Table 1. The Entity Label and Description

Label	Description
Defendant	Individuals suspected of violating traffic regulations and causing road traffic accidents due to their actions
Other incident participants	Other individuals or groups involved in the case besides the defendant
Defendant's transportation	The transportation vehicle used by the defendant in the traffic accident
Defendant's driving condition	The specific driving behavior and condition of the defendant at the time of the traffic accident
Participant transportation	Transportation vehicles used by participants other than the defendant
Participant driving condition	The driving or movement of participants other than the defendant at the time of the accident
Place of action	The specific area where traffic accidents occur
Processing unit	The official department or organization responsible for handling, investigating, and ultimately determining responsibility at the accident scene
Determination of defendant responsibility	Official evaluation and conclusion on the legal responsibility of the defendant in a traffic accident
Responsibility	Official legal documents

Table 2. The Relation Type and Description

Type	Description
driving	Details of the defendant's behavior in operating their means of transportation
undertaking	Indicate the types and degrees of responsibilities that the defendant should bear according to legal provisions
issuing	Describe the handling unit's investigation results based on the accident

to introduce additional noise at the encoding stage. The double pointer decoding in UIE has difficulty in capturing dependencies across these long-distance characters. Moreover, searching entity pair in the full legal text has huge computational consumption. In double pointer decoding, the scoring and selecting combinations of each entity pair by start and end points is low efficiency.

For solving the above problem, we propose a UIE-ERNIE-CRF model. It is mainly divided into a general semantic representation layer, a task semantic representation layer and a decoding layer, and the other parts are the input layer and the output layer, as shown in Fig. 2.

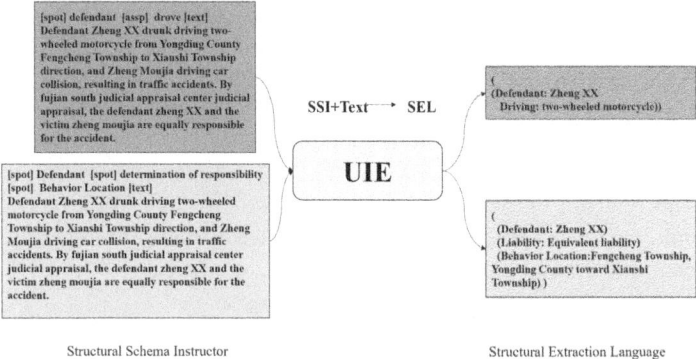

Fig. 1. UIE Framework Diagram

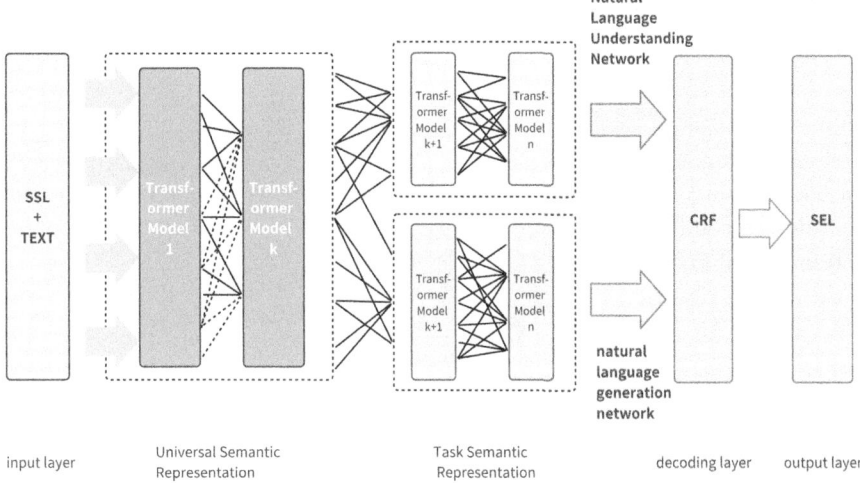

Fig. 2. UIE-ERNIE-CRF Model Structure Diagram

Considering that various natural language processing tasks rely on consistent abstract features at the underlying level, such as lexical and syntactic features, but there are differences in specific requirements at the top level. Therefore, based on the inspiration from the classic model architecture in multitasking learning, where the lower layers share all tasks and the higher layers specialize to specific task, we combine UIE with ERNIE to enable different task paradigms to work together. ERNIE3.0 is a coherent multi-paradigm unified pre-training model. A shared Transformer network is used to accomplish different elaborated completion tasks, while a specific self-attention masking strategy is employed to precisely control the contextual relations in the prediction conditions.

In the decoding stage, Unlike double pointer decoding, sequence labeling is viewed as a structured prediction problem that takes an entire sentence or text fragment as input and produces the same length sequence output. This approach can better capture the structural relations of entities in the text. Also the model can utilize the contextual information of the whole text sequence to improve the accuracy of entity recognition, especially when the distance between entities is very long. Compared with double pointer decoding, the sequence labeling task usually focuses only on the recognition of a single entity without directly involving the relation between two entities. Therefore, it avoids to consider all possible combinations of entity pairs during prediction, simplifying the prediction scope and computational complexity. Therefore, we introduce the CRF layer to the output of the model.

Given the input sequence $X = (x_1, x_2, ..., x_n)$, define P as the n ×k prediction score matrix of the output of the previous layer, where n represents the sentence length, k denotes the total number of label categories. The element P_{ij} in the matrix indicates that the prediction score of the ith word in the sentence assigned as the jth label. The prediction scores sequence $Y = (y_1, y_2, ..., y_n)$ is got, which are calculated as shown in Eq. (1):

$$s(X, Y) = \sum_{i=0}^{n} T_{y_i, y_{i+1}} + \sum_{i=1}^{n} P_{i, y_i} \tag{1}$$

where T denotes the label transfer score matrix, and $T_{i,j}$ means the score transferred from the label with subscript i to the label with subscript j. In order to obtain the conditional probability distribution of Y, normalization is applied to all potential prediction sequences as shown in Eq. (2)

$$p(Y \mid X) = \frac{e^{s(X,Y)}}{\sum_{\tilde{Y} \in Y_x} s(X, \tilde{Y})} \tag{2}$$

where $\widetilde{Y}$ represents the true labeled sequence within the training set and Yx represents all possible labeled sequences. In the training phase of the model, the maximum likelihood principle is used to form the likelihood function by taking the natural logarithm of both sides of the probability p(Y—X), as shown in Eq. (3)

$$\ln(p(Y \mid X)) = s(X, Y) - \ln \left(\sum_{\tilde{Y} \in Y_x} s(X, \tilde{Y}) \right) \tag{3}$$

Subsequently, the most likely sequence of predicted labels is extracted through the decoding process $\overset{*}{Y}$, as shown in Eq. (4)

$$Y^* = \underset{\tilde{Y} \in Y_x}{\mathrm{argmax}}(X, \tilde{Y}) \tag{4}$$

Ultimately, the model transforms the various types of labels into a structured format according to the structured coding paradigm and outputs them as entity relation recognition results.

4   Experiments

4.1   Data Set and Settings

We use CAIL2020 dataset, as well as labeling data from the traffic accident crime judgment. According to the ratio 6:2:2, the dataset was divided into three parts: training set, validation set and test set. The Precision (P), Recall (R) and F1 value are the evaluation metrics. The experimental model parameter settings and their parameter descriptions are shown in Table 3.

Table 3. Configuration of the Experimental Model Parameters

Parameter	Value	Description
per_eval_batch_size	32	Batch size
learning_rate	1×10^{-5}	Maximum learning rate
train_epochs	100	Maximum training period
hidden_act	gelu	hidden layer activation function
attention_dropout	0.3	Stochastic inactivation rate
dropout_prob	0.1	Hidden layer dropout rate
hidden_size	768	Number of neurons in the hidden layer
attention_heads	12	Number of heads of attention mechanisms
hidden_layers	12	Transformer lyers
initializer_range	0.02	standard deviation
weight_decay	0.0	Weight decay value
vocab_size	40000	lexicon
pool_act	tanh	pooling layer activation function

4.2   Experiment Strategies

In order to validate the effectiveness of our model, we have implemented three experiments. UIE is set as the baseline. We set up the ablation experiments on four models: UIE, UIE-ERNIE, UIE-CRF and UIE-ERNIE-CRF. We also consider the situation of overlapping entities, choose the legal texts including the overlapped entities as the test dataset to observe the experimental results. The final experimental results obtained are shown in Table 4.

In the overlapping setting, Our model's Precision, Recall and F1 value is 81.818%, 83.529% and 82.665% respectively.

4.3   Experiment Analysis

From the experimental results, the performance of baseline model UIE is relatively low. Our model achieved the best performance in precision, recall and

Table 4. Experimental Results

experimental model	P (%)	R(%)	F1(%)
UIE	80.899	84.706	82.759
UIE-ERNIE	86.364	89.412	87.861
UIE-CRF	84.091	87.059	85.549
UIE-ERNIE-CRF	89.667	91.765	90.704

F1 value. Compared with UIE, it has improved by 8.768%, 7.059%, and 7.945% respectively. UIE-ERNIE has increased by 5.465%, 4.706%, and 5.102% respectively. Because ERNIE introduces multiple pre-training task types, optimizes the model's learning ability for knowledge at multiple levels such as vocabulary, syntactic structure, and semantic information, the integration of the ERNIE model can improve the performance of baseline when facing the specific field.

Compared with the UIE-ERNIE model, the performance of UIE-CRF model is slightly inferior. The introduction of sequence labeling may improve the model's ability to handle long dependencies and contextual understanding limitations in judicial texts. Sequence labeling assigns a label to each independent element in the sequence and makes decisions based on local features. In some scenarios, different label sequences may be reasonable explanations, so it is difficult to find a unique optimal solution. Therefore,it is necessary to consider introducing other mechanisms to overcome these limitations. In the scenario of overlapping entities, there is promotion space in the performance of UIE-ERNIE-CRF.

5 Conclusion

Entity relation extraction for legal texts is the fundamental task of legal knowledge graph and legal intelligent questions system. The extraction method for general entity relation is not suitable for specific fields. Performance of entity relation extraction methods for Chinese legal text need to be improved. Generative entity relation extraction is a new joint extraction method. We used traffic accident crime case as dataset and collected 1050 expanded data from the Judgment Document Network, then defined ten entity types to describe case facts and three relation types to describe case factual relationships. Based on UIE, We designed an new entity relation extraction model UIE-ERNIE-CRF. To address the issue of long-distance recognition of legal texts, the double pointer decoder in the UIE model is replaced with sequence labeling decoder CRF, and the ERNIE model is introduced to incorporate external knowledge. The effectiveness of the model was demonstrated through experiments.Due to the unbalanced distribution of entities type, it is difficult for the model to recognize the entity and relation with less samples. Subsequently, we will further research unbalanced and overlapping entities problem.

Acknowledgement. This work was supported by the Humanity and Social Science Youth Foundation of Ministry of Education of China (21YJCZH202), the Innovation Team Project of Higher Education of Guangdong Province (2022WCXTD008) and Commission project of Guangdong Province Law Society(GDLS(2024)C12).

References

1. Zhao, X., et al.: A comprehensive survey on deep learning for relation extraction: recent advances and new frontiers. arXiv preprint arXiv:2306.02051 (2023)
2. Liu, Z., et al.: A novel pipelined end-to-end relation extraction framework with entity mentions and contextual semantic representation. Expert Syst. Appl. **228**, 120435 (2023)
3. Shang, Y.M., Huang, H., Mao, X.: Onerel: joint entity and relation extraction with one module in one step. In: Proceedings of the AAAI Conference on Artificial Intelligence, vol. 36, no. 10, pp. 11285–11293 (2022)
4. Chang, H., et al.: JoinER-BART: joint entity and relation extraction with constrained decoding, representation reuse and fusion. IEEE/ACM Trans. Audio Speech Lang. Process. (2023)
5. Han, Z., et al.: SeqViews2SeqLabels: learning 3D global features via aggregating sequential views by RNN with attention. IEEE Trans. Image Process. **28**(2), 658–672 (2018)
6. Lu, Y., et al.: Unified structure generation for universal information extraction, pp. 5755–5772. arXiv preprint arXiv:2203.12277 (2022)
7. Huang, Z., Xu, W., Yu, K.: Bidirectional LSTM-CRF models for sequence tagging. arXiv preprint arXiv:1508.01991 (2015)
8. Yu, S., et al.: ERNIE 3.0: large-scale knowledge enhanced pre-training for language understanding and generation. arXiv preprint arXiv:2107.02137 (2021)
9. Xiao, C., et al.: CAIL2018: a large-scale legal dataset for judgment prediction. arXiv (2018)
10. Gao, H., Hu, Z.: Research progress of named entity recognition in legal documents. J. North China Univ. Technol. **36**(01), 126–135 (2024). issn: 1001-5477
11. Li, X., et al.: Relation enhanced embedding based entities relation extraction from legal documents. J. Chin. Inf. Process. **37**(04), 90–97 (2023). issn: 1003-0077
12. Dan, G., Peng, D., Liu, C.: Entity relation extraction based on CNN in large scale text data. J. Chin. Comput. Syst. **39**(5), 6 (2018)
13. Leitner, E., Rehm, G., Moreno-Schneider, J.: Fine-grained named entity recognition in legal documents. In: Acosta, M., Cudré-Mauroux, P., Maleshkova, M., Pellegrini, T., Sack, H., Sure-Vetter, Y. (eds.) SEMANTiCS 2019. LNCS, vol. 11702, pp. 272–287. Springer, Cham (2019). https://doi.org/10.1007/978-3-030-33220-4_20
14. Wang, D., et al.: Named entity recognition based on JCWA-DLSTM for legal instruments. J. Chin. Inf. Process. **34**(10), 8 (2020)
15. Chen, Y.: Research on entity and relation extraction algorithm forjudgment documents. MA thesis. Dalian University of Technology (2022). https://doi.org/10.26991/d.cnki.gdllu.2021.002661

Uncertain Knowledge Graph Completion with Rule Mining

Yilin Chen[1], Tianxing Wu[1,2(✉)], Yunchang Liu[1], Yuxiang Wang[3], and Guilin Qi[1,2]

[1] Southeast University, Nanjing, China
{cyl,tianxingwu,yunchangliu,gqi}@seu.edu.cn
[2] Key Laboratory of New Generation Artificial Intelligence Technology and its Interdisciplinary Applications (Southeast University), Ministry of Education, Nanjing, China
[3] Hangzhou Dianzi University, Hangzhou, China
lsswyx@hdu.edu.cn

Abstract. To model the uncertainty within knowledge graphs (KGs), existing studies define uncertain knowledge graphs (UKGs), which assign a confidence score to each triple to measure its likelihood of being true and make more precisely downstream tasks such as reasoning and decision making possible. Since KGs usually suffer from the problem of incompleteness, methods of rule mining and reasoning for knowledge graph completion are extensively studied due to their excellent interpretability. However, previous methods are all conducted under deterministic scenarios, neglecting the uncertainty of knowledge, making them unable to be directly applied to UKGs. In this paper, we propose a new framework on uncertain knowledge graph completion with rule mining. The framework is composed of a rule mining model and a confidence prediction model. The rule mining model applies an encoder-decoder network transformer to take rule mining as a sequence-to-sequence task to generate rules. It models the uncertainty in UKGs and infer new triples by differentiable reasoning based on TensorLog with mined rules. The confidence prediction model uses a pre-trained language model to predict the triple confidence given the rules mined. Experiments show that our models significantly outperform various baselines in different evaluation metrics on link prediction and confidence prediction, respectively.

Keywords: Uncertain Knowledge Graph Completion · Rule Mining

1 Introduction

Knowledge graphs (KGs) are structured knowledge bases in the form of graphs to store real-world facts, each of which is represented as a triple (h, r, t) where h and t are head and tail entities respectively, and r is the relation connecting them. Composed of such factual triples, KGs such as YAGO [10], Wikidata [17], and Zhishi.me [19] have been widely used for diverse downstream tasks such as

© The Author(s), under exclusive license to Springer Nature Singapore Pte Ltd. 2024
C. Jin et al. (Eds.): WISA 2024, LNCS 14883, pp. 100–112, 2024.
https://doi.org/10.1007/978-981-97-7707-5_9

question answering [23] and decision making [9]. However, the above KGs neglect the uncertainty of knowledge. The uncertainty arises from two situations. One is the nature of knowledge itself, i.e., some knowledge is inherently uncertain and exists with a certain probability of occurrence, such as the reaction between two proteins. The other one is the inevitable noise introduced with the adoption of automated knowledge graph (KG) construction techniques. To model the uncertainty, uncertain knowledge graphs (UKGs) such as NELL [2] and ConceptNet [13] were proposed, which assign a confidence score to each triple to measure its likelihood of being true.

KGs usually encounter the issue of incompleteness. Although the scale of a KG can be very large, it is impossible to cover all facts in the process of construction. To address this problem, many works focusing on KG completion which can be categorized into two types: embedding-based methods and rule-based ones. Embedding-based methods also known as KG embedding, e.g., TransE [1], RotatE [14], and CompleEx [15]. With the development of pre-trained language models, some embedding models based on pre-trained language models such as KG-BERT [22] and BERTRL [24] have also been proposed. KG embedding is efficient for KG completion because of vector-based computation, but lacks interpretability. Thus, rule-based methods are more preferred due to their symbolic interpretability. The key of such methods (e.g., AMIE [8]) is rule mining by iterating the KG and reasoning with the mined rules. As the scale of KGs grows, rule mining can be considerably time-consuming. To enhance the efficiency, methods like RNNLogic [11] and Ruleformer [20] use neural networks to directly generate rules and also achieve good performance. However, existing rule mining methods **do not model the uncertainty of knowledge and cannot be applied to uncertain KG directly for completion.**

To solve this problem, we propose a new framework for uncertain knowledge graph (UKG) completion with rule mining. The framework is composed of a rule mining model and a confidence prediction model. The rule mining model employs TensorLog [6] to incorporate uncertain information and provide supervision signals to rule generation and reasoning. It is based on an encoder-decoder network transformer [16] and takes rule mining as a sequence-to-sequence generation task. It mines rules effectively because searching in a large relation space is replaced by the generation of a simple sequence. The confidence prediction model takes the mined rules as additional information for predicting the confidence score of the given triple by utilizing a pre-trained language model. To sum up, the contributions of this paper can be summarized as follows:

- We propose a new framework on UKG completion with rule mining, composed of a rule mining model and a confidence prediction model. To the best of our knowledge, it is the first work on rule-based UKG completion.
- We not only design a new rule mining model on UKGs by an encoder-decoder network transformer utilizing TensorLog to model the uncertainty of knowledge, but also propose a new confidence prediction model leveraging a pre-trained language model.

– We conduct experiments on different benchmark datasets for UKG completion tasks including link prediction and confidence prediction. The results show the effectiveness and superiority of our proposed models when compared with the state-of-the-art baselines.

2   Related Work

2.1   Deterministic Knowledge Graph Completion

Previous works on deterministic KG (i.e., all triples are seen as correct without confidence scores) completion can be classified into embedding-based methods and rule-based methods. Embedding-based methods aim to map entities and relations to low-dimensional dense real-value spaces, utilizing vector operations to simulate the relationships between entities in triples. Early embedding-based methods include translation distance methods (e.g., TransE [1]), semantic matching methods (e.g., ComplEx [15]), and neural network methods (e.g., RGCN [12]). With the development of pre-trained language models, researchers applied them in KG embedding learning by using their superior natural language understanding capabilities and rich background textual knowledge. For example, KG-BERT [22] concatenates triples into sentences and takes them as input for BERT [7], transforming the task of KG completion into a sequence classification problem. BERTRL [24] feeds texts of candidate triple instances and their possible reasoning paths to BERT and predicts the existence of the triple.

Rule-based methods aim to mine explicit first-order logic rules from KGs and complete them through rule reasoning. For example, AMIE [8] utilizes search-based methods to mine rules from KGs, employing pruning techniques to enhance algorithm efficiency. RNNLogic [11] and Ruleformer [20] use neural network based generators to produce rules, thereby avoiding the computationally expensive search operations. The use of rules provides good interpretability for KG completion compared with embedding-based methods, and this is also why such methods are increasingly preferred, but these rule-based methods cannot be directly applied to UKG completion since they do not model the uncertainty of knowledge.

2.2   Uncertain Knowledge Graph Completion

Currently, only embedding-based methods [3–5,18,21,25] have been proposed to UKG completion. For example, UKGE [4] is the pioneering work in UKG completion, which simultaneously preserves the structural and uncertainty information of triples in the process of embedding learning. BEUrRE [3] represents entities as boxes (i.e., hypercubes), leveraging the geometric relationships between these cubes to model triples in UKGs. GMUC [25] and GMUC+ [18] further considers the uncertainty of entities and relations themselves, and use gaussian metric learning for few-shot UKG completion. However, these methods lack interpretability and thus UKG completion with rules is worthy to explore.

3   The Proposed Approach

3.1   Problem Definition

Given a head entity h and a relation r, uncertain knowledge graph completion with rule mining refers to mine first-order logical rules and use them to infer the new triple (i.e., predict the tail entity t) with the confidence score s to get the complete new quadruple $\langle (h, r, t), s \rangle$. The first-order logical rules are in the following form:

$$r_h(X, Y) \leftarrow r_1(X, Z_1) \wedge \ldots \wedge r_L(Z_{L-1}, Y) \tag{1}$$

where L denotes the length of rule, $r_1(X, Z_1) \wedge \ldots \wedge r_L(Z_{L-1}, Y)$ is *rule body*, $r_h(X, Y)$ is *rule head*, $\leftarrow$ denotes logical implication, and X, Y, Z_i are variables. We can get grounding rule bodies by replacing variables with entities in KGs, and thus grounding rule heads (i.e., new triples) can be inferred. Besides, the confidence score of each inferred triple needs to be predicted.

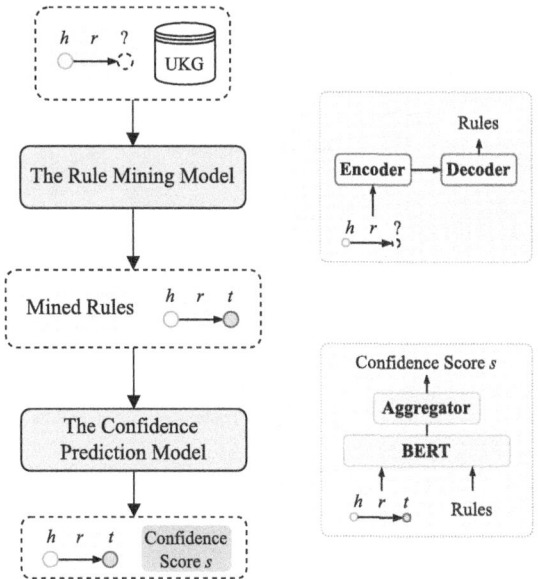

Fig. 1. Our framework on UKG completion with rule mining.

3.2   Framework Overview

The overview of our framework on UKG completion is shown in Fig. 1. It is composed of a rule mining model and a confidence prediction model. The rule mining model, named UKRM (**U**ncertian **K**nowledge graph **R**ule **M**ining), is based on an encoder-decoder network transformer [16]. It treats rule mining as a sequence-to-sequence task that takes the query triple as the input and generates relation distributions for subsequent rule reasoning. We notice that rules are non-differentiable discrete symbols and cannot be utilized for reasoning within end-to-end training, so we transform symbolic reasoning into differentiable vector and matrix operations leveraging TensorLog [6], enabling the entire model to perform end-to-end training. To get discrete rules, we propose a backward rule parsing strategy to sample the best rule from the generated relation distributions. The confidence prediction model, named BCP (**B**ert-based **C**onfidence **P**rediction), encodes the inferred triple and the corresponding rule using BERT to predict the final confidence score. The input sequence consisting of the inferred triple and the rule is pre-processed by the large language model (LLM) GLM-4[1].

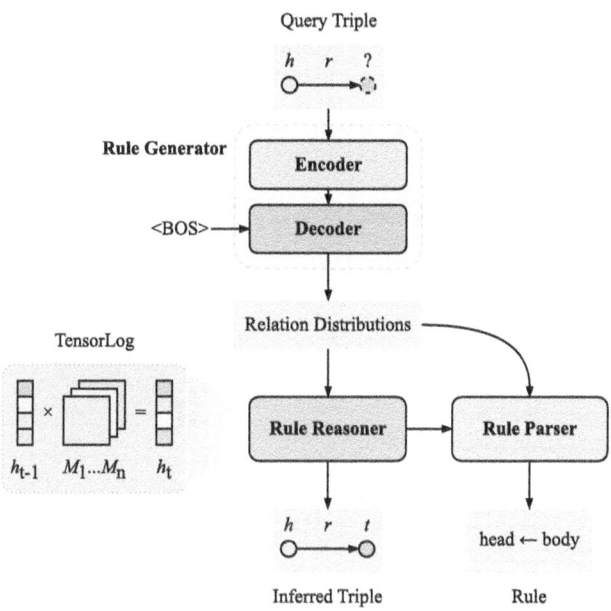

Fig. 2. The overview of our rule mining model UKRM.

3.3 Rule Mining

As shown in Fig. 2, our rule mining model UKRM mainly consists of three components, which are rule generator, rule reasoner, and rule parser. Rule generator takes each query triple as the input and generates a relation distribution for each atom in the rule body. Given the generated relation distributions, rule reasoner infers the tail entity to complete the query triple, and rule parser samples a high-quality interpretable rule for each query triple.

---

[1] https://open.bigmodel.cn/dev/api#glm-4.

Rule Generator. The rule generator is actually an encoder-decoder network transformer. The encoder has N multi-head self-attention layers to encode the string text of the head entity h and relation r in the query triple $(h, r, ?)$. More specifically, h and r are concatenated as the input sequence of the encoder, then we can get the encoding result S'. The decoder which also has N multi-head self-attention layers takes the embedding of special token $<BOS>$ (randomly initialized) as the initial input and performs cross-attention with attention vectors projected from S', generating the distributions of relations in the rule body. Let S_t denote the input sequence to the decoder at the time step t, then the decoding process can be formulated as Eq. 2:

$$R_{t+1} = Softmax(MLP(CrossAttention(S', S_t))) \tag{2}$$

where $CrossAttention$ applies the same strategy in [16] to calculate attention weights by using attention vectors K and V projected by S', and Q projected by S_t, respectively. R_{t+1} denotes the relation probabilistic distribution at time step $t+1$, MLP denotes a multi-layer perception with a fully-connected hidden layer mapping the dimensionality of the output attention vectors to the number of relations, and $Softmax$ is an activation function transforming the input vector into a probability distribution. We repeat this step T times and get the relation distribution of each atom in the rule body with length T. To enable the generated rules with better flexibility, we added inverse relations for the original ones, which means that for the triple (X, r, Y) there will be a new triple (Y, inv_r, X) be added to the KG. Moreover, to facilitate the generation of rules with varying lengths, an additional self-loop relation $<slf>$ is introduced to pad the rule body to the maximum length T.

Rule Reasoner. Rule reasoning is a discrete process and the rule generator is a differentiable model, so the challenge is how to use the relation distributions generated by the rule generator to reason in a differentiable way. To solve this problem, we proposed a rule reasoner based on TensorLog, which can model the uncertainty of knowledge and provide the supervised signal for the rule generator.

In TensorLog, entities are represented as one-hot vectors $\mathbf{e}_i \in \{0, 1\}^{|\mathcal{E}|}$, where $|\mathcal{E}|$ denotes the number of entities in the KG and relations are represented as adjacency matrices $\mathbf{M}_r \in \{0, 1\}^{|\mathcal{E}| \times |\mathcal{E}|}$. $\mathbf{M}_r[i, j] = 1$ means the triple (e_i, r, e_j) is in the KG. Thus, the reasoning result can be calculated by multiplication operations between the entity vector and relation matrix in the following form:

$$\mathbf{e}_t = \mathbf{e}_h \prod_r \mathbf{M}_r \tag{3}$$

where $\mathbf{e}_h$ and $\mathbf{e}_t$ are the head and tail entity vectors respectively, and r is a relation in the rule body. The item in e_t stands for the count of paths from the head entity to the tail entity through the rule body.

To model the uncertainty in UKGs, we re-define the relational matrices $\mathbf{M}_r$ for each relation r, where $\mathbf{M}_r \in [0, 1]^{|\mathcal{E}| \times |\mathcal{E}|}$, and $\mathbf{M}_r[i, j] = s$ indicates that the triple (e_i, r, e_j) exists and the confidence is s. Consequently, each item in $\mathbf{e}_t$ now reflects the weight of an entity as a potential result, incorporating the confidence scores of triples in the reasoning paths (i.e., grounding rule bodies). The weight of the potential result is positively correlated to the number of reasoning paths starting from the head entity to the potential result, and the confidence scores of the triples within these paths. Furthermore, to reason with the relation distributions, we re-written Eq. 3 as follows:

$$\mathbf{Y} = \mathbf{e}_h \prod_{i=1}^{T} \sum_{j=1}^{|\mathcal{R}|} \alpha_{ij} \mathbf{M}_j \tag{4}$$

where α_{ij} represents the probability of the j-th relation in the i-th step generated by the rule generator. T and $|\mathcal{R}|$ denote the length of the rule body and the number of relations in UKG, respectively. $\mathbf{Y}$ is the vector of the reasoning result, denoting the weights of all entities, and it reflects the priority of each entity being the true result. For optimization, the loss function is:

$$loss_{UKRM} = -\log(\max(\gamma, Y_{target})) \tag{5}$$

where Y_{target} denotes the weight of the target tail entity, and γ is a small constant to prevent numerical errors caused by $\log(0)$. In this way, UKRM will assign larger weights to reliable paths, maximizing the weights of target tail entities.

Rule Parser. To sample a high-quality rule body (denoted as the sequence of relations $r_1 \wedge r_2 \wedge \ldots \wedge r_T$) from the generated relation distributions, we introduce a backward rule parser. The key to sampling high-quality rule bodies is to choose the entities and relations that maximize the likelihood of true tail entities. Starting from the tail entity, the reasoning result of entity weight vector can be represented as $\sum_{j=1}^{|\mathcal{R}|} \alpha_{Tj} \mathbf{M}_j \cdot (\mathbf{e}_h \prod_{i=1}^{T-1} \sum_{j=1}^{|\mathcal{R}|} \alpha_{ij} \mathbf{M}_j)$ which is a sum of $|\mathcal{R}|$ vectors (corresponding to $|\mathcal{R}|$ relations respectively) according to Eq. 4. The weight of the tail entity can be calculated as the sum of corresponding entries of the tail entity in these vectors. The relation having the maximum entry for the tail entity in the corresponding vector will be selected as the T-th relation.

When the T-th relation r_T is selected, the reasoning result of entity weight vector can be denoted as $(\mathbf{e}_h \prod_{i=1}^{T-1} \sum_{j=1}^{|\mathcal{R}|} \alpha_{ij} \mathbf{M}_j) \cdot \alpha_{Tr_T} \mathbf{M}_{r_T}$, and we further decompose it as $[w_1, w_2, ..., w_{|\mathcal{E}|}] \cdot [\mathbf{m}_1, \mathbf{m}_2, ..., \mathbf{m}_{|\mathcal{E}|}]$, where w_i is a real number, $\mathbf{m}_i$ is a row vector and $|\mathcal{E}|$ is the number of entities in the UKG, so the entity weight vector can be denoted as $\alpha_{Tr_T} \sum_{i=1}^{|\mathcal{E}|} w_i \mathbf{m}_i$ which is a sum of $|\mathcal{E}|$ vectors (corresponding to $|\mathcal{E}|$ entities respectively). Then the entity having the maximum entry for the tail entity in the corresponding vector will be selected as the entity connecting the $(T-1)$-th and T-th relation. By repeating the above operations, a path from the head entity to the tail entity can be parsed, and we obtain the ground rule for inferring the true tail entity. With this backward greedy parsing strategy, it can be ensured that atoms in the rules are connected in UKGs, and parsed rules contribute maximally to the weight of the true tail entity.

3.4   Confidence Prediction

The success of models such as KG-BERT has demonstrated the potential of pre-trained language models in understanding KGs, ameliorating the shortcomings of previous works that lack semantic alignment with human natural language. Thus, we propose a BERT-based confidence prediction model BCP to predict the confidence score of the triple inferred by the mined rule. The overview of our confidence prediction model BCP is illustrated in Fig. 3. Each inferred triple and the corresponding grounding rule mined by UKRM are taken as the input of the BCP model. We first concatenate the triple and grounding rule as a sequence, and use a template as well as an LLM to preprocess it into the natural language. We then use BERT to encode the natural language sequence to get token embeddings: $E_1, E_2, ..., E_n$, and aggregate them for triple confidence prediction. To illustrate the process of preprocessing more clearly, we give an example in Fig. 3. For the input triple $(charge, relatedto, card)$ and the corresponding grounding rule $(charge, relatedto, card) \leftarrow (charge, relatedto, payment) \wedge (payment, inv\_relatedto, card)$, we concatenate them the template: *[CLS] Question: h r what? Is the correct answer: t? [SEP] Context: grounding rule.* Here, h is a head entity, r is a relation, and t is a tail entity in the input triple. Then, the result is: *[CLS]Question: charge relatedto what? Is the correct answer: card? [SEP] Context: cahrge relatedto payment; card relatedto payment [SEP].* It has become a prompt sequence but there is still a gap between it and natural language due to the irregular expression of entities and relations in KGs, e.g., *'relatedto'* should be *'is related to'*. To address this issue, the texts in the KGs are further processed by the LLM GLM-4, ultimately yielding more natural expressions. Thus, the final form of the example above is: *[CLS]Question: charge*

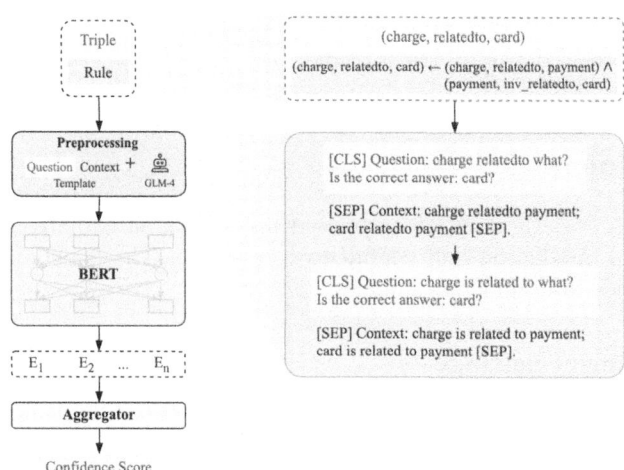

Fig. 3. The overview of our confidence prediction model BCP.

Table 1. Statistical results of datasets.

Dataset	#Entity	#Relation	#Train	#Valid	#Test
CN15K	15,000	36	204,984	16,881	19,293
NL27K	27,221	404	149,001	12,278	14,034

is related to what? Is the correct answer: card? [SEP] Context: charge is related to payment; card is related to payment [SEP].

After preprocessing, the sequence is encoded by BERT to get token embeddings, and different aggregators are applied to calculate the triple confidence. To explore the optimal aggregator, we employ three aggregation strategies, which are the CLS strategy, the average pooling strategy, and the attention strategy. For the CLS strategy, the output embedding of the [CLS] token is directly used to predict the confidence score. For the average pooling strategy, we take the average of all output embeddings of the whole sequence for subsequent prediction. As for the attention strategy, we use the embedding of the first token to calculate the attention scores with all output embeddings and take the weighted sum of them by using these scores as the aggregation result. With the encoding aggregation result, we utilize a feedforward neural network (FNN) to compute the triple confidence, thereby enhancing the generalization ability of BCP. The FNN is denoted as $FNN(\mathbf{x}) = W_1(ReLU(W_2\mathbf{x} + \mathbf{b}_2)) + b_1$, where $\mathbf{x}$ is the aggregation result, $W_2 \in \mathbf{R}^{d \times 4d}$ and $W_1 \in \mathbf{R}^{4d \times 1}$ are weight matrixes, b_1 and $\mathbf{b}_2$ denote bias items, $ReLU(\cdot) = \max(0, \cdot)$ is activation function, and d is the dimension of x. The optimization is ultimately performed using the mean squared error loss function as follows:

$$loss_{BCP} = \frac{1}{n}\sum_{i=1}^{n}(\hat{y}_i - y_i)^2 \tag{6}$$

where y_i denotes the ground truth confidence, and $\hat{y}_i$ represents the confidence predicted by BCP.

4   Experiments

In this section, we evaluated our proposed framework on two UKG completion tasks, i.e., link prediction and confidence prediction. The benchmark datasets and source code are publicly available[2].

4.1   Experiment Settings

Datasets. We evaluated our models UKRM and BCP on two datasets [4] CN15K and NL27K, respectively. CN15K is a subset of the commonsense UKG ConceptNet [13] and NL27K is a subset of the UKG NELL [2]. Confidence scores of triples in both datasets are mapped to [0, 1]. Table 1 summarizes the data statistics.

---

[2] https://github.com/seucoin/UKRM.

Baselines. We compared our models with UKG completion models as well as deterministic KG completion models. Deterministic KG completion baselines include a classic translation-based model TransE and a complex space vector rotation based model RotatE. Both models haven been verified to have the good capability of generalization. UKG completion baselines include UKGE, BEUrRE, PASSLEAF [5], and UKGsE [21], which are the state-of-the-art embedding-based UKG completion models.

Evaluation Protocol. For link prediction, we adopted the mean reciprocal rank (MRR) and Hits@k as the evaluation metrics. MRR is the average of the reciprocal ranks of the correct entities, while Hits@k measures the frequency with which the correct tail entity is found within the top-k ranked entities. For confidence prediction, we used mean squared error (MSE) and mean absolute error (MAE) as the evaluation metrics. MSE is the average of the squared differences between the actual values and the predicted ones. MAE is the average of the absolute differences between the actual values and the predicted ones.

4.2   Link Prediction

The comparison results of the link prediction task are shown in Table 2. The best results is annotated in bold, while the second-best result is annotated with an underscore. The UKRM_de model is the version of the UKRM without uncertainty modeling, and it simply sets all confidence scores to one. From the table, it can be observed that the UKRM model outperforms all the baselines on both datasets. This fully demonstrates the effectiveness of the UKRM. Meanwhile, on the two datasets, UKRM outperforms UKRM_de, indicating that modeling uncertainty can assist the model in obtaining more accurate reasoning results.

To further analyze the performance of UKRM and UKRM_de, we conducted statistical analysis on the distributions of confidence scores in both datasets, as shown in Fig. 4. The figure reveals an excessive concentration of confidence

Table 2. Link prediction results.

Model	CN15K		NL27K	
	MRR	Hits@10	MRR	Hits@10
TransE	0.115	0.277	0.261	0.555
RotatE	0.126	0.297	0.438	0.600
UKGE	0.142	0.209	0.440	0.607
BEUrRE	0.103	0.248	0.358	0.358
PASSLEAF	0.122	0.262	0.463	0.626
UKGsE	0.010	0.020	0.057	0.108
UKRM_de	0.199	0.342	0.793	**0.893**
UKRM	**0.204**	**0.349**	**0.795**	**0.893**

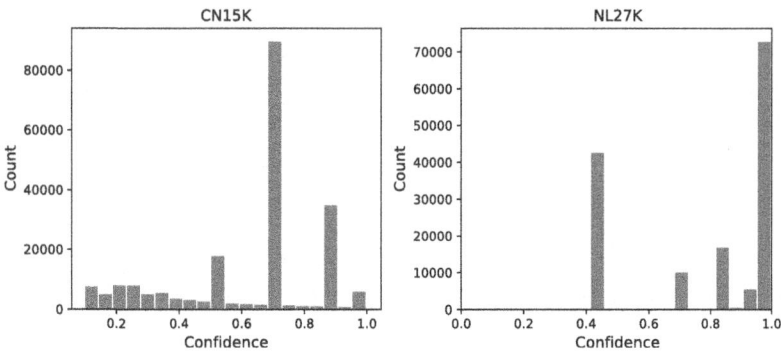

Fig. 4. The distribution of confidence scores on both datasets.

scores in both CN15K and NL27K datasets, particularly evident in NL27K, where confidence scores predominantly fall within $[0.4, 0.5]$ and $[0.9, 1.0]$. This leads to the degradation in the effectiveness of uncertainty modeling, and this is why UKRM and UKRM_de have comparable performance on NL27K.

4.3   Confidence Prediction

The comparison results of the confidence prediction task are shown in Table 3 and the best results are annotated in bold. BCP w/o rule drops the information of rules and takes only triples as the input. BCP_cls, BCP_avg and BCP_attn represent the CLS, average pooling, and attention aggregation strategies, respectively. From the table, it can be seen that BCP model outperforms all the baselines on both datasets. Without rules, the performance of BCP will decrease, particularly evident on NL27K. Among the three different aggregation strategies, the CLS strategy performs the best indicating that the use of the CLS token embedding trained in the pre-training phase of BERT can lead to better performance on confidence prediction.

Table 3. Confidence prediction results.

Model	CN15K		NL27K	
	MSE	MAE	MSE	MAE
UKGE	0.19988	0.36206	0.02861	0.05967
BEUrRE	0.11656	0.28259	0.08920	0.22194
PASSLEAF	0.09375	0.24801	0.01949	0.06253
UKGsE	0.10274	0.25564	0.12202	0.27605
BCP w/o rule	0.02724	0.11109	0.04303	0.15700
BCP_cls	**0.02612**	**0.11037**	**0.01638**	**0.05532**
BCP_avg	0.02650	0.11259	0.01754	0.07112
BCP_attn	0.02654	0.11270	0.01983	0.07187

5   Conclusion

In this paper, we introduce the first framework on UKG completion with rule mining. The framework is composed of two models UKRM (the rule mining model) and BCP (the confidence prediction model). UKRM models the uncertainty of knowledge, enabling the mining of higher quality first-order logical rules with transformer and TensorLog. BCP predicts the accurate confidence score for each inferred triple using BERT. Experiment results validate the advanced performance of our models in the link prediction and confidence prediction tasks.

Acknowledgement. This work is supported by the NSFC (Grant No. 62376058, 52378009, 62276063), the Fundamental Research Funds for the Central Universities (2242022R40045), ZhiShan Young Scholar Program of Southeast University, and the Big Data Computing Center of Southeast University.

References

1. Bordes, A., Usunier, N., Garcia-Duran, A., Weston, J., Yakhnenko, O.: Translating embeddings for modeling multi-relational data. In: Proceedings of NIPS, pp. 2787–2795 (2013)
2. Carlson, A., Betteridge, J., Kisiel, B., Settles, B., Hruschka, E., Mitchell, T.: Toward an architecture for never-ending language learning. In: Proceedings of AAAI, pp. 1306–1313 (2010)
3. Chen, X., Boratko, M., Chen, M., Dasgupta, S.S., Li, X.L., McCallum, A.: Probabilistic box embeddings for uncertain knowledge graph reasoning. In: Proceedings of NAACL, pp. 882–893 (2021)
4. Chen, X., Chen, M., Shi, W., Sun, Y., Zaniolo, C.: Embedding uncertain knowledge graphs. In: Proceedings of AAAI, pp. 3363–3370 (2019)
5. Chen, Z.M., Yeh, M.Y., Kuo, T.W.: PASSLEAF: A pool-based semi-supervised learning framework for uncertain knowledge graph embedding. In: Proceedings of AAAI, pp. 4019–4026 (2021)
6. Cohen, W.W.: TensorLog: a differentiable deductive database. arXiv preprint arXiv:1605.06523 (2016)
7. Devlin, J., Chang, M.W., Lee, K., Toutanova, K.: BERT: pre-training of deep bidirectional transformers for language understanding. arXiv preprint arXiv:1810.04805 (2018)
8. Galárraga, L.A., Teflioudi, C., Hose, K., Suchanek, F.: AMIE: association rule mining under incomplete evidence in ontological knowledge bases. In: Proceedings of WWW, pp. 413–422 (2013)
9. Lv, M., Cao, X., Wu, T., Li, Y.: A civil aviation customer service ontology and its applications. Data Intell. **5**(4), 1063–1081 (2023)
10. Pellissier Tanon, T., Weikum, G., Suchanek, F.: YAGO 4: a reason-able knowledge base. In: Proceedings of ESWC, pp. 583–596 (2020)
11. Qu, M., Chen, J., Xhonneux, L.P., Bengio, Y., Tang, J.: Rnnlogic: learning logic rules for reasoning on knowledge graphs. In: Proceedings of ICLR (2020)
12. Schlichtkrull, M., Kipf, T.N., Bloem, P., Van Den Berg, R., Titov, I., Welling, M.: Modeling relational data with graph convolutional networks. In: Proceedings of ESWC, pp. 593–607 (2018)

13. Speer, R., Chin, J., Havasi, C.: ConceptNet 5.5: an open multilingual graph of general knowledge. In: Proceedings of AAAI, pp. 4444–4451 (2017)
14. Sun, Z., Deng, Z.H., Nie, J.Y., Tang, J.: RotatE: knowledge graph embedding by relational rotation in complex space. arXiv preprint arXiv:1902.10197 (2019)
15. Trouillon, T., Welbl, J., Riedel, S., Gaussier, É., Bouchard, G.: Complex embeddings for simple link prediction. In: Proceedings of ICML, pp. 2071–2080 (2016)
16. Vaswani, A., et al.: Attention is all you need. In: Proceedings of NIPS, pp. 5998–6008 (2017)
17. Vrandečić, D., Krötzsch, M.: Wikidata: a free collaborative knowledgebase. Commun. ACM **57**(10), 78–85 (2014)
18. Wang, J., Wu, T., Zhang, J.: Incorporating uncertainty of entities and relations into few-shot uncertain knowledge graph embedding. In: Proceedings of CCKS, pp. 16–28 (2022)
19. Wu, T., et al.: Knowledge graph construction from multiple online encyclopedias. World Wide Web **23**, 2671–2698 (2020)
20. Xu, Z., Ye, P., Chen, H., Zhao, M., Chen, H., Zhang, W.: Ruleformer: context-aware rule mining over knowledge graph. In: Proceedings of COLING, pp. 2551–2560 (2022)
21. Yang, S., Zhang, W., Tang, R., Zhang, M., Huang, Z.: Approximate inferring with confidence predicting based on uncertain knowledge graph embedding. Inf. Sci. **609**, 679–690 (2022)
22. Yao, L., Mao, C., Luo, Y.: KG-BERT: BERT for knowledge graph completion. arXiv preprint arXiv:1909.03193 (2019)
23. Zeng, J., Chen, T.: Interactive model and application of joint knowledge base question answering and semantic matching. In: Proceedings of WISA, pp. 206–217 (2023)
24. Zha, H., Chen, Z., Yan, X.: Inductive relation prediction by BERT. In: Proceedings of AAAI, pp. 5923–5931 (2022)
25. Zhang, J., Wu, T., Qi, G.: Gaussian metric learning for few-shot uncertain knowledge graph completion. In: Proceedings of DASFAA, pp. 256–271 (2021)

Intelligent Service

MAGAN: Mode Information and Attention-Based GAN for Realistic Time Series Data Synthesis

Yi Wang[1], Yi Luo[1(✉)], Peng Ren[2(✉)], Weifan Wang[3(✉)], Xianbo Liu[1(✉)],
Yuhang Hu[1], Zeming Li[1], Xiangkuan Li[4], Wenyao Li[5], and Chunxiao Xing[2]

[1] School of Computer Science and Technology, Beijing Institute of Technology,
Beijing 100081, China
`yiwang_bit@163.com`, {`luoyi,liuxianbo,huyuhang_21,lizeming`}`@bit.edu.cn`
[2] BNRist, DCST, RIIT, Tsinghua University, Beijing 100084, China
{`renpeng,xingcx`}`@tsinghua.edu.cn`
[3] School of Artificial Intelligence, Henan University, Zhengzhou 450046, China
`henuwwf@henu.edu.cn`
[4] Room 602, Unit 2, Building 19, Yuquanxincheng , Laoshan Street,
Shijingshan District, Beijing 100049, China
[5] School of Software, Henan University, Kaifeng 475004, China

Abstract. The demand for efficient processing and analysis of time series data is growing across multiple fields, yet the diverse acquisition of such data is plagued by issues such as insufficient data volume, poor privacy, and uneven data distribution in related technological research. Time series data generation effectively addresses this issue, with Generative Adversarial Network(GAN) based models showing promising performance among existing methods. Nonetheless, these methods overall performance on fidelity issues(e.g., mode collapse, difficulty capturing long-term dependencies) is not particularly outstanding. In this paper, we propose a GAN framework known as Mode information and Attention-based Generative Adversarial Network(MAGAN) which transforms the metadata and sequential data from real data into mode information and temporal information. We employ a GAN based on a multilayer perceptron (MLP) for mode information, while a hierarchical attention network with attention mechanism for temporal information. In addition to fidelity, we evaluated MAGAN based on usefulness and diversity. Experimental results show that the proposed framework significantly outperforms state-of-the-art benchmarks on three typical real-world datasets.

Keywords: synthetic data generation · time series · GAN

1 Introduction

Due to the crucial role of time series data in scientific research, it is extensively employed in various domains, including meteorology, finance, transportation,

Supported by the National Natural Science Foundation of China under Grant 62076027.
X. Li—Independent author.

and healthcare [7,8,18,32]. However, due to privacy constraints and security concerns, time series data obtained from real-world often suffers from issues such as poor data privacy, insufficient volume, inadequate quality, and uneven distribution [19]. Time series data generation, referring to the process of creating synthetic data that mimic the statistical properties and patterns of real-world data, presents a promising avenue for addressing these issues.

Currently, a variety of methods are proposed to generate time series data, including rule-based methods and simulation model-based methods, as detailed in Section 2. Within the realm of time series data generation, methods based on deep learning have exhibited superior performance [26,27]. However, the existing time series data generation methods still suffer from the issue of mode collapse and difficulty in capturing long-term dependencies. These challenges make tasks performed based on the synthetic time series data challenging to yield results of sufficient reference value [6,10,15,24]. Therefore, we present a Mode information and Attention-based GAN(MAGAN), a framework for synthetic time series data to address these challenges.

Our approach has two innovations:

To alleviate the issue of mode collapse, we record the maximum value max and minimum value min of the sequential data after min-max normalization. Then, a Box-Cox transformation is employed for skewness correction to adjust the distribution shape of the data, recording the transformation parameter λ. Here, we collectively refer to the max, min, transformation parameter λ, and metadata m_i after one-hot encoding during the data transformation process as mode information. The extracted mode information is utilized for learning and generating diverse distribution patterns of time series data using a Multi-Layer Perceptron (MLP)-based GAN.

To address the issue of difficulty in capturing long-term dependencies, we introduce an attention mechanism to better capture the dependencies between sequences. The data is processed through multilayer perceptrons for data discrimination or data generation, achieving the capturing of long-term correlations and the generation of long-term time series data.

Our method utilizes the attention mechanism to address the difficulty in capturing long-term dependencies while considering mode collapse. We demonstrate the advantages in a series of experiments on multiple real-world datasets. Quantitatively and Qualitatively, we find that MAGAN achieves significant improvements over state-of-the-art benchmarks in generating realistic time series.

Roadmap: In the rest of the paper, we begin by discussing related work in Section 2. We provide our problem formulation in Section 3. We describe the design of MAGAN in Section 4 and evaluate it in Section 5, before concluding in Section 6.

2   Related Work

Previous work in time series data generation methods can be categorized into rule-based methods, simulation model-based methods, traditional machine learning-based methods, and deep learning-based methods. Rule-based methods employ predefined rules or constraints derived from data attributes or interrelationships to generate data quickly [5,12] but often lack authenticity due to fixed rules, leading to disparities from real data. Simulation model-based methods use computer simulations to model real-world scenarios [22,23], providing more accurate representations but demanding substantial domain knowledge and resulting in higher computational complexity and potential errors. Traditional machine learning-based methods utilize historical data to generate new data efficiently and quickly [21,29], particularly suitable for small datasets, yet may struggle with large-scale data and complex patterns.

Using deep learning methods to generate time series is a popular idea. Existing deep learning-based methods primarily utilize Generative Adversarial Networks(GANs) and Variational Autoencoders(VAEs) to generate time series data. Sami et al. [26] and Chauhan et al. [27] compared GANs and VAEs in their works, with GANs showing advantages in generating high-quality data. Therefore, our work in this paper chooses to explore along the path of GANs.

In current GAN-based methods, higher-quality time series data is often generated by modifying the GAN architecture and training approach [8,14,30,31]. For instance, RCGAN [8], trained on medical data, substitutes the GAN generator and discriminator with recurrent neural networks. Similarly, C-RNN-GAN [20], trained on classical music data, incorporates bidirectional recurrent neural networks into the GAN discriminator. However, these methods exhibit good performance only on specific datasets. TC-GAN [13] takes data length and size as control inputs but performs poorly in long time series problems. COT-GAN [28] combines classical optimal transport methods with additional time-causal constraints but exhibits excellent performance solely on small datasets. TimeGAN [30] splits time series features into static and dynamic categories and combines unsupervised GAN with supervised training of autoregressive models but fails to address mode collapse issues. DoppelGANger [14] treats the maximum and minimum values of each time series as random variables to be learned and generated, normalizing all time series to alleviate pressure on the generator and enrich data diversity; however, its performance on long time series data still requires improvement. While these methods may perform well individually in certain aspects, their overall performance concerning mode collapse and long-term dependencies remains insufficiently distinguished.

By contrast, our proposed method demonstrates outstanding overall performance in addressing mode collapse and difficulty in capturing long-term dependencies issues. Leveraging mode information generation and an attention-based GAN model, our approach effectively generates high-quality long-term time series data based on real-world data.

3   Problem Formulation

For the convenience of subsequent descriptions, we abstract the scope of our datasets as follows:

A dataset $T = \{(D_1, M_1), (D_2, M_2), \ldots, (D_m, M_m)\}$ is defined as a set of m samples. Each sample consists of sequential data D_i and corresponding metadata M_i. Sequential data $D_i = \{(t_1, x_1), (t_2, x_2), \ldots, (t_n, x_n)\}$ consists of a series of timestamps t_i and measurement values x_i, where x_i is a vector that can contain information from multiple dimensions. The statistical model of time series data can be represented using conditional probability as follows:

$$P(X) = P(x_1) P(x_2|x_1) P(x_3|x_1, x_2) \ldots P(x_n|x_1, \ldots, x_{n-1}) \tag{1}$$

Time series data generation refers to the use of specific models or algorithms to analyze and predict patterns, trends, and correlations in time series data, and to generate new time series data based on these analysis results. Typically, time series data generation methods learn the conditional probability distribution of time series data based on real data and some prior knowledge. The data generation method for time series data is abstracted into a model G, which learns model parameters $\theta(T, P)$ from the real-time series data set T and prior knowledge P. It takes random noise z and control information C transformed from the data demand load as inputs and outputs the generated time series D.

$$D = G^{\theta(T,P)}(z, C) \tag{2}$$

The main focus of our work is to construct an effective time series data generation model G to address the challenges. Through adversarial training, our aim is to equip the generator with the ability to generate high fidelity time series data.

4   Proposed Model: Mode Information and Attention-Based GAN(MAGAN)

MAGAN consists of two modules: mode information generation module and temporal information generation module (see Fig. 1). Mode information generation module is designed to learn and extract information about data patterns from raw time series data, thereby assisting in generating diverse time series data, to alleviate the issue of mode collapse. In temporal information module, we introduce attention mechanism to address the challenge of capturing long-term dependencies.

4.1   Mode Information Generation Module

The mode information generation module involves mode information transformation and mode information generation by a GAN constructed based on a Multilayer Perceptron(MLP).

In mode information transformation, sequential data undergo initial min-max normalization to ensure positivity, with recording the maximum value max and minimum value min after normalization. Subsequently, the Box-Cox transformation is applied to adjust the distribution of data, normalizing the data to a Gaussian distribution, with a parameter λ capturing skewness and kurtosis

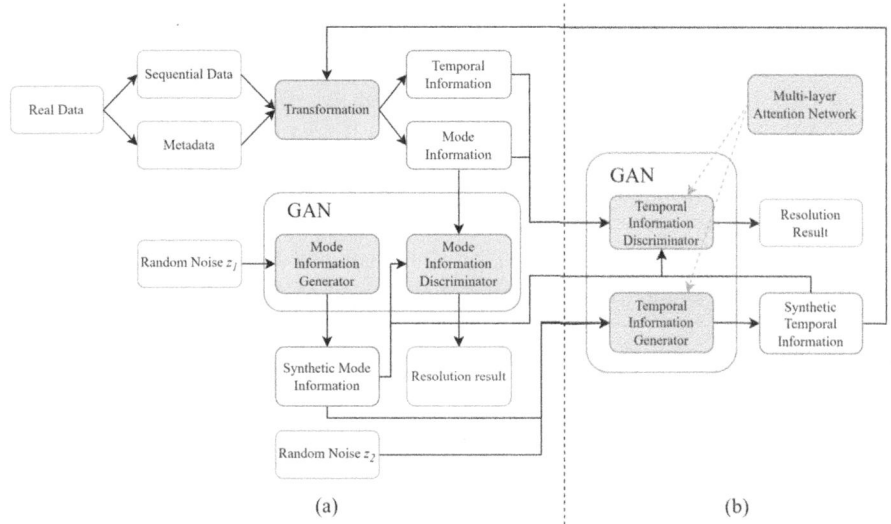

(a)   (b)

Fig. 1. (a) The mode information GAN Network employs an MLP structure to accept input random noise vectors z_1 and mode information. (b) The synthetic mode information, random noise vectors z_2 and real data are utilized as inputs for the temporal information GAN, and an attention mechanism is introduced to address the issue of long-term dependencies.

information. Afterwards, another min-max normalization is conducted to mitigate dimensional disparities among different features. We define the data after transformation as temporal data. For metadata, continuous features are treated similarly to sequential data, while discrete features undergo one-hot encoding. The resultant parameters, denoted as $meta$, combined with min, max, λ, constituting the mode information p, which encapsulates fundamental characteristics of the real data. Generating diverse mode information and using it as input for subsequent temporal data GAN facilitates the generation of diverse temporal data, thereby alleviating the issue of mode collapse.

Given the low-dimensional nature and relatively simple interrelationships of mode information, our study opts for a Multi-Layer Perceptron(MLP) [4,11,25] as the fundamental structure of GAN. However, in different datasets and scenarios, alternative network architectures can be selected based on specific requirements. In mode information GAN, the generator accepts input random noise vectors z_1, generating synthetic mode information (denoted as $G(z; \theta_g)$). The discriminator receives data generated by the generator and real mode information, attempting to differentiate their sources (denoted as $D(x; \theta_d)$). Our objective is to encourage the generator generating high-quality mode information that the discriminator cannot distinguish.

4.2   Temporal Information Generation Module

The main issue faced by capturing and generating long-term sequence is the loss of temporal dependencies between consecutive time points. Methods like TimeGAN [30] and DoppelGANger [14] employ RNN or LSTM methods, which either suffer from short-term memory constraints or partially mitigate long sequence issues but may still lose early information when sequences become excessively long.

In this paper, we employ attention mechanisms to extract correlation information from long sequences, leveraging their ability to dynamically focus on key parts of input data. Attention mechanisms adaptively handle long time series data by dynamically adjusting the focus based on sequence length and time span, capturing both local and global patterns more effectively. Additionally, learned positional encoding is utilized to integrate positional information into the attention computation process, retaining relative distance information within the sequence and enhancing the model's ability to handle sequence data order. Finally, the attention results are aggregated at multiple levels and merged to maximize the capture of long-term sequence correlations for either data discrimination or generation, to alleviate the issue of long-term dependencies.

In temporal information GAN, the generator accepts input random noise vectors z_2 and synthetic mode information p' generated by mode information generator, generating synthetic temporal information (denoted as $G(z, p'; \theta_{g'})$). The discriminator receives real data, synthetic mode information p' and synthetic temporal information generated to distinguish their origins (denoted as $D(x'; \theta_{d'})$). The introduction of the mode information generator enables the generator to better understand the structure and characteristics of the data, facilitating the generation of diverse temporal data. The design aims to enable temporal information GAN to generate time series data with long-term dependencies.

5   Experiment

5.1   Experimental Setup

Dataset. We opt to perform testing and evaluation on three typical real-world time series datasets: the Wikipedia Web Traffic dataset(WWT) [17], the Measurement Lab Broadband Access dataset(MBA) [1], and the StarCraft II game replay dataset(SC2Replay) [2]. Basic statistical information for each time series dataset can be found in Table 1.

WWT dataset tracks the daily page views of different Wikipedia pages from July 1, 2015 to December 31, 2016. Each data sample consists of a sequence of daily page views for a Wikipedia page and three metadata attributes. We access type and agent type as metadata for subsequent training. After removing samples with missing data, the original dataset is reduced to 1177277 samples, with a sequence length of 550 for each sample.

MBA dataset consists of household broadband testing data collected by the Federal Communications Commission in the United States. We selects data from September to October in 2017, comprising hourly traffic measurements from 4378 households. Due to numerous missing values in the original dataset, we follow the data processing method of DoppelGANger [14].

SC2Replay dataset comprises records of high-level StarCraft II game matches. The data records changes in environmental states during the game and all player actions. In this paper, player action data is extracted in chronological order to form sequences with the races of players in the match as metadata, resulting in a one-dimensional time-series dataset consisting of 9196 samples, each with a length of 100.

Table 1. Information about real-world dataset used in experiments

Dataset	Sample size	Sequence length	Sequence information dimension
WWT	117277	550	1
MBA	600	56	2
SC2Replay	9196	100	1

Evaluation Metrics. In our experiment, We draw upon influential evaluation metrics utilized in TimeGAN [30], and access the quality of generated data in three metrics: fidelity, usefulness and diversity.

Fidelity is the core evaluation metric for time series data generation, measuring the similarity between synthetic data and real data, with the goal of making them indistinguishable. To compute fidelity, we train a classifier to distinguish between real and synthetic data, thereby calculating the F1 score as the fidelity metric.

Usefulness refers to the capability of synthetic data to exhibit similar functionality as real data in practical application. We aims for the performance of synthetic data on downstream tasks to be comparable to real data, thereby providing reliable data support for downstream applications. To evaluate usefulness, typical time series classification models are trained on both synthetic and real data separately, and then compares their differences in terms of classification performance.

Diversity refers to the similarity between the distribution of synthetic data and real data, ensuring that the model has good generalizaton ability. Since a time series data typically contains multiple features across multiple time steps, to qualitatively assess diversity, we visualize the data by reducing its dimensionality to observe the coverage of generated data. In this paper, PCA [3] and t-SNE [16] are used to demonstrate the distribution differences between synthetic data and real data.

Baselines. We test our model according to the evaluation metrics and compare MAGAN with Naive GAN [9], RCGAN [8], TimeGAN [30] and Doppel-GANger [14].

5.2   Experimental Results

Fidelity Test Results. As indicates in Table 2. MAGAN achieves the highest fidelity score of 0.913 on the WWT dataset, surpassing all other methods. On the MBA dataset, DoppelGANger achieves the highest fidelity score of 0.824, slightly outperforming other methods, but the proposed MAGAN also reaches a level of 0.797. On the SC2Replay dataset, MAGAN achieves the highest fidelity score of 0.897, surpassing all other methods. Overall, MAGAN demonstrates relatively superior fidelity performance across different datasets, indicating its capability to generate time series data closely resembling real-world data with higher fidelity. While MAGAN does not perform well on the MBA dataset, this may be attributed to the relatively small number of samples in the MBA dataset. Although the attention mechanism of MAGAN aids in focusing on important information, the insufficient number of samples may result in the model excessively focusing on the details and noise of the training data, thereby failing to effectively capture the overall characteristics and distribution of the data.

Table 2. The results on models fidelity and usefulness testing

Metric	Model	WWT	MBA	SC2Replay
Fidelity	Naive GAN	0.286	0.312	0.458
	RCGAN	0.372	0.414	0.583
	TimeGAN	0.501	0.500	0.564
	DoppelGANger	0.867	**0.824**	0.859
	MAGAN	**0.913**	0.797	**0.897**
Usefulness	Naive GAN	0.4396	0.5677	0.4978
	RCGAN	0.3132	0.6109	0.5134
	TimeGAN	0.1151	0.4639	0.3313
	DoppelGANger	0.0427	**0.3131**	0.2131
	MAGAN	**0.0393**	0.4427	**0.0746**

Usefulness Test Results. MAGAN demonstrates superior performance on the WWT dataset and SC2Replay dataset, with usefulness scores of 0.0393 and 0.0746, respectively. On the MBA dataset, MAGAN's usefulness score is only surpassed by DoppelGANger. Overall, MAGAN exhibits excellent performance across different datasets, indicating its high practical usefulness in generating data. This is attributed to the attention mechanism and robust training approach employed by MAGAN, enabling better capture of time series data characteristics and patterns, thus generating time series data with higher practical usefulness.

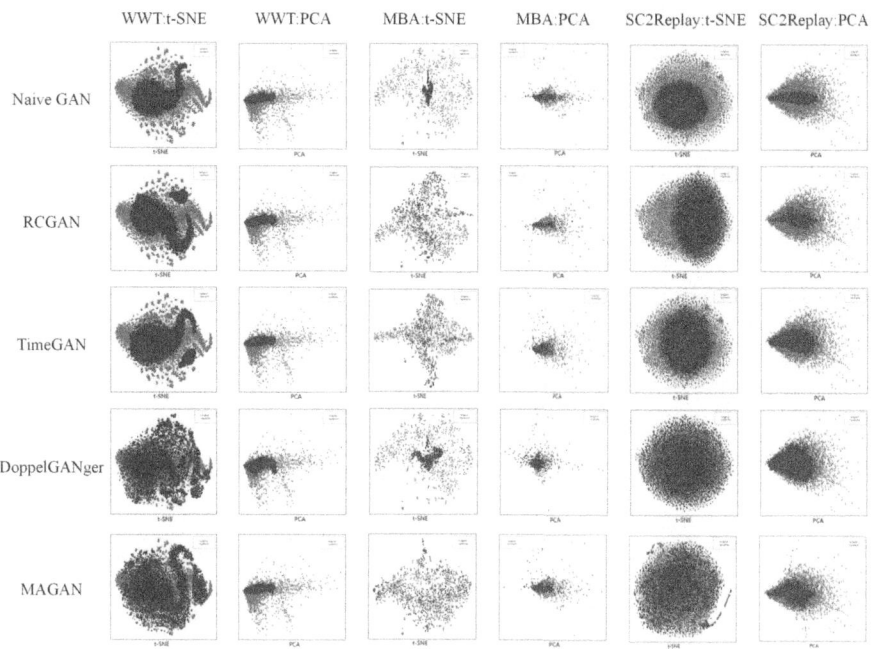

Fig. 2. PCA and t-SNE visualization on WWT, MBA and SC2Replay dataset. Each row provides the visualization for each of the 5 benchmarks. Red denotes original data, and blue denotes synthetic. (Color figure online)

Diversity Test Results. In Fig. 2, MAGAN performs well on three real-world datasets, achieving the best results on two of them. DoppelGANger and TimeGAN also perform relatively well across all three datasets, slightly trailing behind MAGAN. However, Naive GAN and RCGAN show mediocre performance. We attribute the superior performance of MAGAN to its incorporation of kurtosis, skewness, and extremum values as mode information for learning and generating, which together with the metadata from the data source, guiding the generation of sequential information.

6 Conclusion

In this paper, we proposes a Mode information and Attention-based GAN(MAGAN) for time series data generation to address the mode collapse and difficulty in long-term capturing challenges in existing time series data generation methods. A series of experiments were conducted to validate the effectiveness of the proposed method. The results show that this method demonstrates good performance across three key indicators: fidelity, usefulness, and diversity, indicating its broad potential for application. In the future, further work may intend to leverage large language models to facilitate the generation of higher-quality time series data and incorporate multi-condition constraints to enable users to more precisely generate data with specified characteristics.

References

1. Validated data - measuring fixed broadband - eighth report (2018). https://www. fcc.gov/reports-research/reports/measuring-broadband-america/validated-data-measuring-fixed-broadband-eighth
2. Sc2replaystats - starcraft 2 replay replay hosting/training system (2023). https://sc2replaystats.com/
3. Abdi, H., Williams, L.J.: Principal component analysis. Wiley Interdisc. Rev. Comput. Stat. **2**(4), 433–459 (2010)
4. Bishop, C.M.: Neural Networks for Pattern Recognition. Oxford University Press, Oxford (1995)
5. Bruno, N., Chaudhuri, S.: Flexible database generators. In: Proceedings of the 31st International Conference on Very Large Data Bases, pp. 1097–1107 (2005)
6. Dintyala, P., Narechania, A., Arulraj, J.: Sqlcheck: automated detection and diagnosis of SQL anti-patterns. In: Proceedings of the 2020 ACM SIGMOD International Conference on Management of Data, pp. 2331–2345 (2020)
7. Duchon, C., Hale, R.: Time Series Analysis in Meteorology and Climatology: An Introduction. John Wiley & Sons, Hoboken (2012)
8. Esteban, C., Hyland, S.L., Rätsch, G.: Real-valued (medical) time series generation with recurrent conditional gans. arXiv preprint arXiv:1706.02633 (2017)
9. Goodfellow, I., et al.: Generative adversarial nets. Adv. Neural Inf. Process. Syst. **27** (2014)
10. Han, Y., et al.: Cardinality estimation in dbms: a comprehensive benchmark evaluation. arXiv preprint arXiv:2109.05877 (2021)
11. Haykin, S.: Neural Networks and Learning Machines, 3rd edn. Pearson education, New Delhi (2009)
12. Kang, Y., Hyndman, R.J., Li, F.: Gratis: generating time series with diverse and controllable characteristics. Stat. Anal. Data Mining ASA Data Sci. J. **13**(4), 354–376 (2020)
13. Liang, Z., Li, S.: Tc-gan: a transformer-based conditional generative adversarial network for low-dose spect image reconstruction. In: Proceedings of the 2023 15th International Conference on Machine Learning and Computing, pp. 341–347 (2023)
14. Lin, Z., Jain, A., Wang, C., Fanti, G., Sekar, V.: Using gans for sharing networked time series data: challenges, initial promise, and open questions. In: Proceedings of the ACM Internet Measurement Conference, pp. 464–483 (2020)
15. Ma, L., et al.: Mb2: decomposed behavior modeling for self-driving database management systems. In: Proceedings of the 2021 International Conference on Management of Data, pp. 1248–1261 (2021)
16. Van der Maaten, L., Hinton, G.: Visualizing data using t-sne. J. Mach. Learn. Res. **9**(11) (2008)
17. Maggie, Oren Anava, V.K.W.C.: Web traffic time series forecasting (2017). https://kaggle.com/competitions/web-traffic-time-series-forecasting
18. Marazzo, M., Scherre, R., Fernandes, E.: Air transport demand and economic growth in Brazil: a time series analysis. Transport. Res. Part E: Logist. Transport. Rev. **46**(2), 261–269 (2010)

19. McGregor, T., Alcock, S., Karrenberg, D.: The RIPE NCC internet measurement data repository. In: Krishnamurthy, A., Plattner, B. (eds.) PAM 2010. LNCS, vol. 6032, pp. 111–120. Springer, Heidelberg (2010). https://doi.org/10.1007/978-3-642-12334-4_12
20. Mogren, O.: C-RNN-GAN: continuous recurrent neural networks with adversarial training. arXiv preprint arXiv:1611.09904 (2016)
21. Norris, J.R.: Markov Chains, 2nd edn. Cambridge University Press, Cambridge (1998)
22. Odena, A., Olah, C., Shlens, J.: Conditional image synthesis with auxiliary classifier gans. In: International Conference on Machine Learning, pp. 2642–2651. PMLR (2017)
23. Ohm, P.: Broken promises of privacy: responding to the surprising failure of anonymization. UCLA l. Rev. **57**, 1701 (2009)
24. Remil, Y., Bendimerad, A., Mathonat, R., Chaleat, P., Kaytoue, M.: "what makes my queries slow?": subgroup discovery for SQL workload analysis. In: 2021 36th IEEE/ACM International Conference on Automated Software Engineering (ASE), pp. 642–652. IEEE (2021)
25. Rumelhart, D.E., Hinton, G.E., Williams, R.J.: Learning representations by back-propagating errors. Nature **323**(6088), 533–536 (1986)
26. Sami, M., Mobin, I.: A comparative study on variational autoencoders and generative adversarial networks. In: 2019 International Conference of Artificial Intelligence and Information Technology (ICAIIT), pp. 1–5. IEEE (2019)
27. T., J.: Comparative study of gan and vae. Int. J. Comput. Appl. (2018). https://api.semanticscholar.org/CorpusID:53586135
28. Xu, T., Wenliang, L.K., Munn, M., Acciaio, B.: Cot-gan: generating sequential data via causal optimal transport. Adv. Neural. Inf. Process. Syst. **33**, 8798–8809 (2020)
29. Xuan, G., Zhang, W., Chai, P.: Em algorithms of gaussian mixture model and hidden Markov model. In: Proceedings 2001 International Conference on Image Processing (Cat. No. 01CH37205), vol. 1, pp. 145–148. IEEE (2001)
30. Yoon, J., Jarrett, D., Van der Schaar, M.: Time-series generative adversarial networks. Adv. Neural Inf. Process. Syst. **32** (2019)
31. Yu, L., Zhang, W., Wang, J., Yu, Y.: Seqgan: sequence generative adversarial nets with policy gradient. In: Proceedings of the AAAI Conference on Artificial Intelligence, vol. 31 (2017)
32. Zhang, Y., et al.: Hkgb: an inclusive, extensible, intelligent, semi-auto-constructed knowledge graph framework for healthcare with clinicians' expertise incorporated. Inf. Process. Manag. **57**(6), 102324 (2020)

A Study on Context-Matching-Based Joint Training for Chinese Coreference Resolution

Xiangwei Yan[1(✉)], Weiqun Luo[1], Jiabao Wang[1], and Xinyu Shen[2]

[1] Xizang Minzu University, Xianyang 712000, Shaanxi, China
253716275@qq.com
[2] University of Western Australia, Perth Crawley, WA 6009, Australia

Abstract. To address issues such as ineffective control of the number of mentions and neglecting contextual relationships during mention clustering in end-to-end Chinese coreference resolution models, a context-matching-based joint training Chinese coreference resolution model is proposed. The model utilizes RoBERTa(wwm)-large combined with BiLSTM to encode Chinese text, then clusters word embeddings. It uses the results of the word clustering to recognize span mentions, reducing the number of span mentions that need to be processed. Finally, it reclusters the recognized span mentions. During training, the clustering and span mention recognition stages are jointly trained. Furthermore, during clustering, the matching degree of candidate antecedents and mentions within their respective contexts is considered, forming a clustering scoring system that integrates contextual matching.Experiments show that the model achieves an F1 score of 71.11% on the OntoNotes-5.0 Chinese dataset for the CoNLL metric, an improvement of 0.95% compared to the baseline model. On the self-constructed "Tibet News Traffic" dataset, the F1 score reaches 73.62%, an increase of 2.58% compared to the baseline model. These results indicate that the proposed model can effectively enhance the performance of coreference resolution models and demonstrates good generalization capability.

Keywords: coreference resolution · context matching · joint training · Chinese pre-trained models

1 Introduction

Coreference resolution refers to the process of determining which real-world entity a mention in the text refers to [1]. A mention is a word or phrase in the text that refers to an entity, such as a name, pronoun, etc.

In 2017, Lee et al. proposed an end-to-end model for coreference resolution [2]. Innovatively, they treated all spans in the text as potential mentions (a text of length n has $\frac{n(n-1)}{2}$ spans). The core idea is to bypass mention recognition and directly cluster all spans in the text. In 2018, they introduced a coarse-to-fine inference strategy [3], where for all potential antecedents of a given mention, they used a bilinear function to determine the antecedent with the highest coreference probability and then applied more complex antecedent scoring. They also incorporated the English pre-trained model ELMo, which further improved the performance of their coreference resolution model in English. However, the use of an end-to-end architecture in models encounters an issue of exponential growth in the number of mentions when processing long texts, making it computationally impractical to handle all spans directly.

In particular, Dobrovolskii proposed a novel approach based on joint training of tokens (words) and spans [4]. By introducing a span recognition step, spans are recognized before participating in mention clustering, significantly reducing the number of spans that need to be processed and achieving high efficiency. D'Oosterlinck et al. further addressed the challenge of handling conjoined mentions in models that jointly train tokens and spans [5]. This improvement enhances the model's capability to handle conjoined mentions such as "Tom and Mary," thereby further improving the performance of English coreference resolution models.

In Chinese coreference resolution, Kong [6] applied an end-to-end architecture to Chinese text in 2019, integrating structural information by using nodes from parse trees as constraints. This approach filters out text spans that are not feasible, thus reducing computational complexity. Word embeddings were encoded statically. The model did not employ large-scale pretrained models for dynamic context text encoding, which constrained its performance.

In 2021, Huang W [7] introduced the large-scale pretrained Chinese model BERT(wwm)-base for Chinese coreference resolution, adopting Lee's end-to-end architecture. This approach achieved a high performance F1 score in Chinese coreference resolution.

The innovation of this study lies in addressing the lack of contextual information during mention clustering in coreference resolution models. Drawing on the attention mechanism, we constructed contextual vectors for entity mentions and matched the contextual information of each entity mention with that of the others, using this as a crucial criterion for scoring. By deeply analyzing and comparing the contextual relationships between entity mentions, this approach enhances the model's understanding of complex relationships between entities. Due to the significant advantages demonstrated by the joint training approach based on Token and Span, and the gradual development of Chinese pre-trained models, this paper also employed the large-scale Chinese pre-trained model RoBERTa(wwm)-large for text encoding, and adopted the joint training method based on Token and Span.

Based on the above innovations and efforts, this study proposes a context-matching-based joint training Chinese coreference resolution model. Experimental evaluations on the OntoNotes 5.0 Chinese dataset and a custom-built Tibet Traffic News dataset demonstrate superior performance compared to baseline models.

2   Related Work

Coreference resolution algorithms, aimed at understanding the content of texts by identifying chains of coreference relation, have evolved progressively from traditional rule-based and machine learning methods to mainstream deep learning approaches. In 2015, Wiseman et al. introduced the first neural network-based mention-ranking model specifically for coreference resolution [8]. Subsequently, Wiseman et al. employed Recurrent

Neural Networks (RNNs) to better capture global features of coreference resolution [9]. In 2016, Clark and Manning constructed a system consisting of four deep learning modules that defined an entity ranking model [10]. This system comprised mention-pair encoders, entity-pair encoders, mention-ranking models, and entity-ranking models. Clark and Manning also utilized reinforcement learning to enhance the entity ranking model [11], enabling the model to learn to evaluate the importance of various decisions, further improving the model's performance.

Following this, the BERT (Bidirectional Encoder Representations from Transformers) pretrained model emerged, leveraging a bidirectional Transformer architecture to comprehensively capture linguistic features within text [12]. Joshi et al. [13] further applied BERT and its variant SpanBERT [14] to end-to-end coreference resolution. SpanBERT is specifically designed to handle span-based tasks more effectively, enabling better prediction of entity spans and relationships within text. The application of these models not only continues the success of pretrained models in coreference resolution but also significantly enhances resolution performance.

In recent years, new methods for coreference resolution have emerged rapidly. Miculicich et al. [15] proposed a graph-based coreference resolution model where tokens in a document serve as nodes, and their relationships are represented by edges. They utilized an iterative refinement process to optimize predictions of coreference relationships. Toshniwal [16] focused on entity tracking, utilizing fixed-dimensional vector representations derived from pretrained language models to track entities, emphasizing the transient nature of entities. Otmazgin et al. [17] introduced the LINGMESS model, which categorizes coreference decisions into six types and employs an ensemble approach to learn specialized scoring functions for each type, representing the current best span-based method. Liu et al. [18] employed Autoregressive Structured Prediction (ASP) for coreference resolution tasks, transforming the problem into a conditional language modeling task that predicts a series of structured construction actions. Bohnet et al. [19] proposed a seq2seq-based transformation system for coreference resolution. They utilized multilingual T5 as the underlying language model, achieving state-of-the-art results in English coreference resolution.

3   Model

Based on the optimization scheme, the model diagram designed in this paper is shown below. The model includes Token Representation Module, Clustering Module and Span Mention Recognition Module. The overall architecture of the model is shown in Fig. 1:

3.1   Token Representation

The primary step involves employing the RoBERTa(wwm)-large Chinese pre-trained language model as the text encoding layer, aiming to obtain word embeddings that encompass rich contextual semantic features. To accurately predict coreference links, a Bidirectional Long Short-Term Memory network (BiLSTM) was also added. It enhances the construction of word vectors by integrating forward and backward contextual information, thereby improving the modeling of long-distance dependencies and the accurate understanding of word meanings.

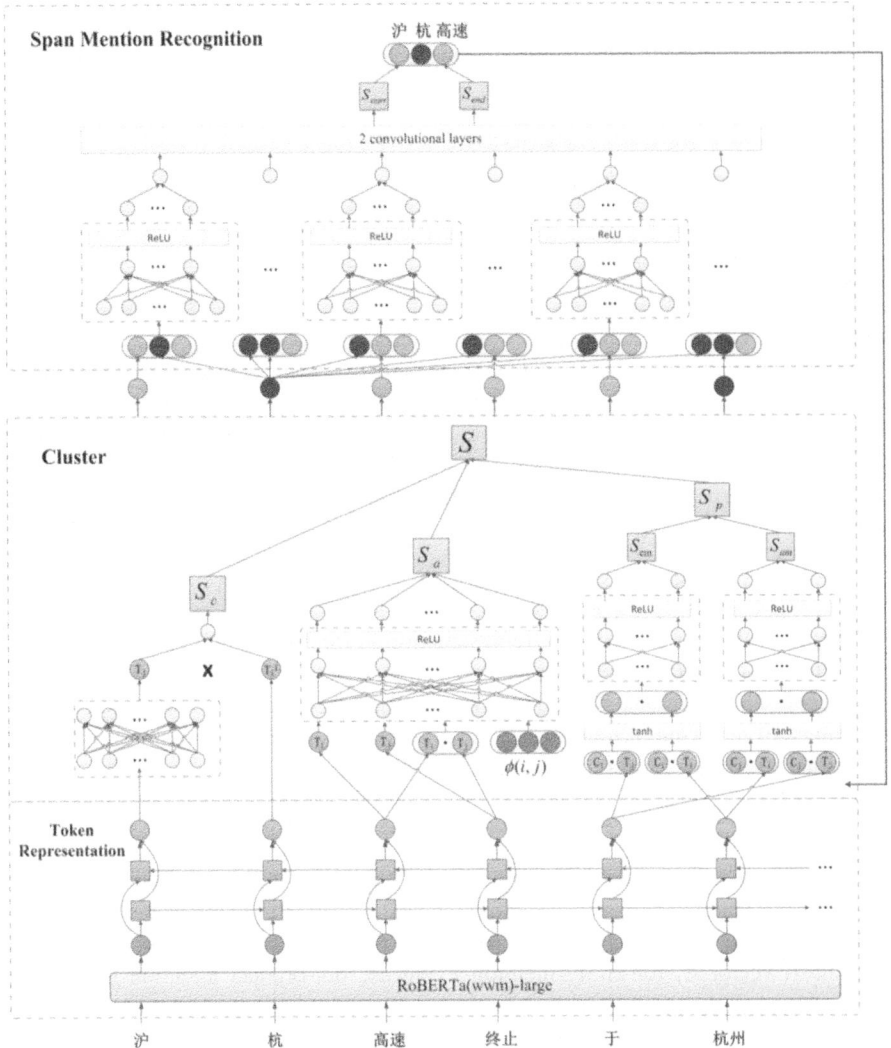

Fig. 1. Model Architecture

3.2 Clusters

The mention clustering aims to match appropriate antecedents for each entity mention. The final coreference score used to evaluate the entity mention and the candidate antecedent consists of three dimensions: coarse pruning score, antecedent score, and context matching score. The sum of these three dimensions yields the ultimate coreference score, forming a scoring system integrated with context matching.

Coarse Pruning. We treat all tokens generated by the Token Representation module as entity mentions. When identifying potential antecedents for each token entity mention, we perform coarse pruning of candidate antecedents. A bilinear function is employed to initially compute a lightweight score $S_c(i, j)$ between each token entity mention and every candidate antecedent, ultimately retaining the most probable candidate antecedents. $S_c(i, j)$ is formulated as Eq. (1):

$$S_c(i, j) = T_i^T \cdot W_c \cdot T_j \tag{1}$$

T_i represents the vector representation of Token i's entity mention. Due to the ease of calculating $S_c(i, j)$ compared to directly computing fine-grained antecedent scores, adopting this coarse-grained preliminary pruning reduces the search scope over the candidate antecedent set, thereby lowering computational complexity.

Antecedent Score. The antecedent score is derived by considering the basic feature vectors of entity mentions and candidate antecedents, along with integrating multiple metadata features such as text category, speaker, and distance between entities. To handle the wide-ranging and unevenly distributed distances between entities, we map the entity distances into predefined intervals: [1, 2, 3, 4, 5, 6, 7, 8–15, 16–31, 32–63, 64 +]. After coarse-grained pruning of candidate antecedents for token entity mentions, the remaining candidates undergo fine-grained scoring $S_a(i, j)$ as Eq. (2).

$$S_a(i, j) = W_a \cdot FFNN_a([T_i, T_j, T_i^\circ T_j, \phi(i, j)]) \tag{2}$$

The input consists of the vector representation T_i for Token i, the vector representation T_j for the candidate antecedent j, the element-wise product $T_i^\circ T_j$, and $\phi(i, j)$. The $\phi(i, j)$ represents additional feature information for Token i and candidate antecedent j, including the type of the current chapter, the speaker information for the current sentence, and distance features. Each type is set as a 20D feature vector.

Context Matching. The context matching score is obtained by computing the matching score between an entity mention and its candidate antecedents within their respective contexts. The key is to establish context vectors for both the entity mention and the candidate antecedents. Given the remarkable success of self-attention mechanisms in natural language processing [20], we adopt a self-attention mechanism to construct the context vector C_i for entity mention i. We use dot-product attention for weight computation and then perform a weighted sum of word vectors in the context to obtain a vector C_i that reflects the relationships within the entity context. The computation process is detailed in Formulas (3) to (5).

$$a_t = \begin{cases} \dfrac{T_i^\circ con_t}{\sqrt{d_{con}}}, & t \notin index(i) \\ 0, & t \in index(i) \end{cases} \tag{3}$$

$$a_{con,t} = \frac{\exp(a_t)}{\sum_{k=START(con)}^{END(con)} \exp(a_k)} \tag{4}$$

$$C_i = \sum_{t=START(con)}^{END(con)} a_{con,t} \cdot con_t \tag{5}$$

where T_i represents the vector representation of entity mention i, con denotes the selected context, con_t represents the t-th word vector in the context, and d_{con} is the dimensionality of the context encoding vector.

After constructing the context vector, the matching score of the candidate antecedent in the context of the entity mention is computed according to Eq. (6).

$$S_{em}(i, j) = W_{em} \cdot FFNN_{em}([\tanh(C_i \circ T_j) \circ \tanh(C_i \circ T_i)]) \tag{6}$$

Here, the symbol $\circ$ denotes inner product operation, tanh is the activation function, FFNN represents feedforward neural network, and W is the weight parameter. $C_i \circ T_i$ represents the relationship between the entity mention and its own context; $C_i \circ T_j$ represents the relationship between the candidate antecedent and the entity mention's context. Computing $C_i \circ T_i$ to provide a alignment criterion for the model.

Similarly, the matching score of the entity mention in the context of the candidate antecedent can be obtained as shown in Eq. (7).

$$S_{am}(i, j) = W_{am} \cdot FFNN_{am}([\tanh(C_j \circ T_i) \circ \tanh(C_j \circ T_j)]) \tag{7}$$

The two types of matching scores are added together to obtain the matching score of the entity mention and the candidate antecedent in their respective contexts, as shown in Eq. (8).

$$S_p(i, j) = S_{em}(i, j) + S_{am}(i, j) \tag{8}$$

3.3 Span Mention Recognition

When the clustering module completes the token-based coreference resolution, the tokens that form the coreference relationships are passed as central words to the span mention recognition module. For each central word, the span mention is predicted by identifying the start and end tokens within the same sentence that form the optimal span with the central word.

Let c be the central word for the span to be predicted, and x be a word in the same sentence as the central word c. For each x, concatenate the embedding of the central word c, the embedding of x, and their distance embedding. Then, pass these concatenated embeddings through a feedforward neural network (FFNN) to extract features, resulting in the intermediate output F(x,c), as shown in Eq. (9).

$$F(x, c) = W_{span} \cdot FFNN_{span}([T_c, T_x, D(x, c)]) \tag{9}$$

where T_c is the word embedding of the center token c, T_x is the word embedding of token x in the same sentence, $D(x,c)$ is the distance embedding between the center token c and token x, and FFNN represents the feedforward neural network.

The output layer consists of two convolutional layers, each with a kernel size of 3. The final two output channels correspond to the scores for each candidate token as the start and end of a span, respectively. The formula (10) is as follows:

$$S_{start}(x, c), S_{end}(x, c) = Conv(F(x, c)) \tag{10}$$

For each center word c, the model predicts the optimal start and end positions of the span by selecting the words with the highest scores from the convolution output, as shown in Eqs. (11) to (12):

$$start(c) = \underset{x \in X}{\operatorname{argmax}} S_{start}(x, c) \tag{11}$$

$$\text{end}(c) = \underset{x \in X}{\text{argmax}} S_{end}(x, c) \tag{12}$$

During inference, the model also considers the position of the center token, ensuring that the start boundary does not fall to the right of the center token and the end boundary does not fall to the left of it. The recognized Span entity mentions are then inputted into the clustering module for Span-level clustering. Together with the clustering results for Token entity mentions, they form the final clustering results.

4 Experiment

4.1 Dataset

OntoNotes 5.0 Chinese Dataset. The OntoNotes 5.0 Chinese dataset is extensive in content and is currently one of the most classic datasets in the field of coreference resolution, as shown in Table 1.

Table 1. OntoNotes 5.0 Chinese dataset

	Train	Dev	Test	All
Document	1810	252	218	2280
Character	750K	110K	90K	950K
Link	74.5K	10.3K	9.2K	94.1K
Mention	102.8K	14.1K	12.8K	129.8K

Tibet Traffic News Dataset. This study constructed the "Tibet Traffic News" dataset sourced from the Tibet Autonomous Region Department of Transportation (https://jtt.xizang.gov.cn/xwzx/). As of January 24, 2024, the dataset comprises 1,060 news reports including sections on "Tibet Traffic Highlights," "Today's Headlines," and "Photo News." The dataset was split into training, validation, and test sets in an 8:1:1 ratio, as shown in Table 2.

Table 2. Tibet Traffic News dataset

	Train	Dev	Test	All
Document	848	106	106	1060
Character	827K	106K	102K	1035K
Link	28K	4K	4K	36K
Mention	78K	10K	10K	98K

4.2 Experimental Results and Analysis

Experimental Results on OntoNotes-5.0 Chinese Dataset. This section conducts experiments on the OntoNotes 5.0 Chinese dataset, evaluating the model performance using the three evaluation metrics: MUC, B-cubed, and CEAF, as well as their average, known as the CoNLL metric. The experimental results are presented in Table 3.

Table 3. Experimental results on OntoNotes 5.0 Chinese dataset

	MUC			B-cubed			CEAF			CoNLL
	Pre	Rec	F1	Pre	Rec	F1	Pre	Rec	F1	Avg.F1
Clark (2016b)	73.85	65.42	69.38	67.53	56.41	61.47	62.84	57.62	60.12	63.66
Clark (2016a)	73.64	65.62	69.40	67.48	56.94	61.76	62.46	58.60	60.47	63.88
Kong (2019)	76.95	64.58	70.21	70.58	54.68	61.60	64.92	55.36	59.75	63.85
bwt-B(2021)	78.97	71.93	75.31	72.05	63.47	67.49	67.51	62.49	64.9	69.23
rbwt-L + BiLSTM (baseline model)	76.79	75.63	76.21	68.89	68.37	68.63	66.66	64.65	65.62	70.16
Our model	78.57	75.47	76.99	70.92	68.13	69.50	69.44	64.43	66.84	71.11

Kong et al. (2019) [6] built upon the model proposed by Lee (2017) by concatenating word vectors, character vectors, and character-level LSTM representations as feature inputs, enabling the model to perform Chinese coreference resolution tasks. The final F1 score achieved was 63.85%, approaching the performance of Clark's method. However, the model architecture followed a more intuitive end-to-end approach, albeit without utilizing pre-trained models for initial text encoding.

bwt-B(2021) [7] utilized the Chinese pre-trained language model BERT(wwm)-base as the text encoder, with its basic architecture inspired by the traditional end-to-end model c2e-coref [3]. After training for 40 epochs, the F1 score reached 68.88%. For comparative purposes, this study extended the training to 100 epochs on the OntoNotes-5.0 Chinese dataset, achieving a final F1 score of 69.23%.

The rbwt + BiLSTM model, built upon the bwt-B model, replaces the text encoding layer BERT(wwm)-base with RoBERTa(wwm)-large + BiLSTM. This configuration forms the baseline model of this study. Achieving an F1 score of 70.16% on the OntoNotes-5.0 Chinese dataset, this improvement can be attributed to the stronger semantic representational capacity of RoBERTa(wwm)-large compared to BERT(wwm)-base. Additionally, BiLSTM effectively integrates contextual information for semantic representation, further enhancing model performance.

Our model employs a text encoding layer using a combination of RoBERTa(wwm)-large and BiLSTM. The model architecture adopts a joint training approach for span mention recognition and mention clustering. During the mention clustering scoring phase, it additionally integrates a context matching score. From the experimental results, we see that our optimal model achieved an F1 score of 71.11% on the OntoNotes-5.0 Chinese dataset, which is 0.95% higher than the baseline model. This improvement is because the joint training approach refines the end-to-end training objectives and includes additional annotations for the dataset to address different training targets, thereby enhancing the model's deep understanding of the text. Additionally, the integrated context matching scoring system further incorporates contextual information from entity mentions and candidate antecedents.

Experimental Results of Tibet Traffic News Dataset. The proposed model in this paper demonstrated similar advantages on our custom-built Tibet Traffic News dataset, indicating that the model has good generalization ability. The experimental results are shown in Table 4.

Table 4. Experimental results of Tibet Traffic News dataset

	MUC			B-cubed			CEAF			CoNLL
	Pre	Rec	F1	Pre	Rec	F1	Pre	Rec	F1	Avg.F1
bwt-B(2021)	73.94	77.84	75.84	66.00	69.87	67.88	63.77	67.41	65.54	69.75
rbwt-L + BiLSTM	71.52	78.29	74.75	67.32	70.68	68.96	66.86	72.14	69.40	71.04
Our model	74.73	78.13	76.39	71.78	73.85	72.80	70.17	73.23	71.67	73.62

We take rbwt-L + BiLSTM as the baseline model of this paper, and apply the scoring system based on context matching and the joint training method to the baseline model to explore the effectiveness of the method in this paper, as shown in Table 5.

Table 5. Ablation experiment

	CoNLL	
	Avg.F1	Δ
rbwt-L + BiLSTM	71.04	
rbwt-L + BiLSTM + Context matching	72.36	+ 1.32
rbwt-L + BiLSTM + Joint training	72.02	+ 0.98
rbwt-L + BiLSTM + Context matching + Joint training	73.62	+ 2.58

In Fig. 2(a), we introduce a scoring system that integrates context matching into the baseline model. The innovation of this system lies in matching the contextual information of entity mentions with each other and using this as one of the key criteria for scoring. By deeply analyzing the contextual relationships between entity mentions, the model is able to better understand the complex relationships between entities, thereby improving the accuracy of coreference resolution. This method effectively addresses the shortcomings of traditional mention clustering by incorporating contextual information. This integration led to a performance improvement of 1.32%, ultimately achieving an F1 score of 72.36%.

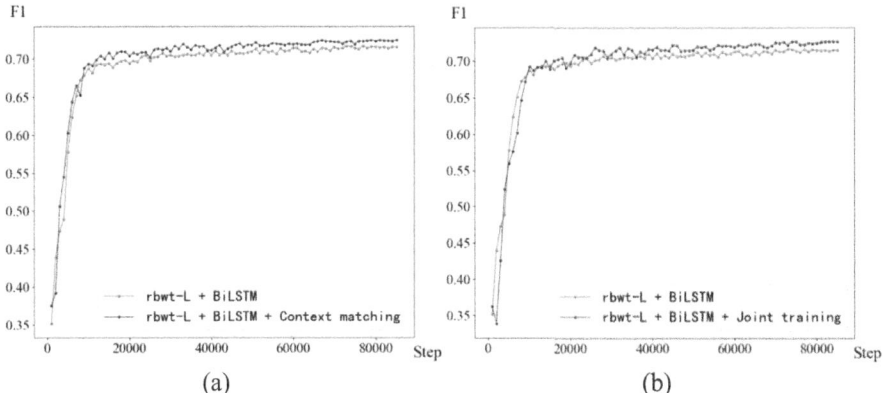

(a)   (b)

Fig. 2. (a) shows the impact of using or not using context matching on the baseline model. (b) illustrates the impact of using or not using the joint training approach on the model.

In Fig. 2(b), based on the baseline model, we transitioned from an end-to-end approach to a token and span joint training approach. We added a span mention recognition module to identify span mentions that include token mentions. Span mention recognition and mention clustering were then jointly trained. This joint training approach involved additional annotations of the dataset, which, compared to the baseline model using an end-to-end architecture, led to a 0.98% performance improvement, achieving an F1 score of 72.02%.

From Fig. 3, it can be seen that the proposed optimal model demonstrates outstanding performance on the custom-built Tibet Traffic News dataset, achieving a final F1 score of 73.62%, which is a 2.58% improvement over the baseline model. This highlights the model's strong generalization capability. The significant enhancement in model performance can be attributed to two key improvements: first, the refinement of training objectives through joint training. We decomposed the end-to-end training objectives into more specific tasks, and additional annotations were made to the dataset for different training goals, which enhanced the model's focus and learning effectiveness for various tasks. Second, the integration of context matching scores: in the process of handling mention clustering, we introduced contextual information of entity mentions and candidate antecedents. This improvement made the model perform better in coreference resolution tasks that require contextual understanding. In summary, these improvements

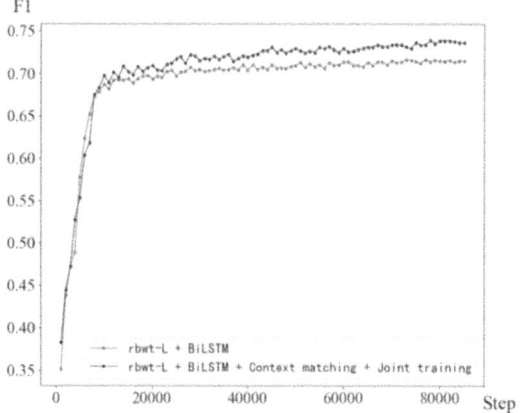

Fig. 3. Comparison of the Baseline Model and the Optimal Model.

collectively enhanced the model's performance in Chinese coreference resolution tasks, demonstrating the effectiveness of advanced encoding methods and refined training strategies in complex language understanding tasks.

5 Conclusion

This paper proposes a context-matching-based joint training Chinese coreference resolution model. Firstly, to tackle the lack of contextual information during the mention clustering stage, an innovative comprehensive scoring system incorporating the contextual relationships of entity mentions was proposed. This system matches the contextual information of each entity mention with that of others and uses this as one of the key criteria for scoring. Secondly, to address the issue of excessive mention quantities when processing long texts using traditional end-to-end architectures, a joint training approach was adopted. This method jointly optimizes the span mention recognition and mention clustering loss. And leveraging the RoBERTa(wwm)-large pre-trained model for processing Chinese. The model was applied to the OntoNotes-5.0 dataset and a custom-built Tibet Traffic News dataset, achieving satisfactory coreference resolution results on both datasets.

References

1. Liu, R., Mao, R., Luu, A.T., Cambria, E.: A brief survey on recent advances in coreference resolution. J. Artif. Intell. Rev. **56**(12), 14439–14481 (2023)
2. Lee, K., He, L., Lewis, M., Zettlemoyer, L.: End-to-end neural coreference resolution. In: 2017 Conference on Empirical Methods in Natural Language Processing, pp. 188–197. ACL (2017)
3. Lee, K., He, L., Zettlemoyer, L.: Higher-order coreference resolution with coarse-to-fine inference. In: 2018 Conference of the North American Chapter of the Association for Computational Linguistics: Human Language Technologies, vol. 2 (Short Papers), pp. 687-692. ACL (2018

4. Dobrovolskii, V.: Word-level coreference resolution. In: 2021 Conference on Empirical Methods in Natural Language Processing, pp. 7670–7675. ACL, Dominican (2021)

5. D'Oosterlinck, K., Bitew, S K., Papineau, B., Potts, C., Demeester, T., Develder, C.: CAW-coref: conjunction-aware word-level coreference resolution. In: 6th Workshop on Computational Models of Reference, Anaphora and Coreference, pp. 8–14. ACL, Singapore (2023)

6. Fang, K., Jian, F.: Incorporating structural information for better coreference resolution. In: 28th International Joint Conference on Artificial Intelligence, pp. 5039–5045. IJCAI (2019)

7. Huang, W.: Research on Chinese Coreference Resolution Methods Based on Pre-trained Language Models. South China University of Technology, Guangzhou (2021)

8. Wiseman, S., Rush, A M., Shieber, S M., Weston J.: Learning anaphoricity and antecedent ranking features for coreference resolution. In: 53rd Annual Meeting of the Association for Computational Linguistics and the 7th International Joint Conference on Natural Language Processing, vol. 1: Long Papers, pp. 1416–1426. ACL, China (2015)

9. Wiseman, S., Rush, A.M., Shieber, S.M.: Learning global features for coreference resolution. In: 2016 Conference of the North American Chapter of the Association for Computational Linguistics: Humanl Language Technologies, pp. 994–1004. ACL, California (2016)

10. Clark, K., Manning, C D.: Improving coreference resolution by learning entity-level distributed representations. In: 54th Annual Meeting of the Association for Computational Linguistics, pp. 643–653, ACL, Germany (2016)

11. Clark, K., Manning, C D.: Deep reinforcement learning for mention-ranking coreference models. In: 2016 Conference on Empirical Methods in Natural Language Processing, pp. 2256–2262, ACL, Texas (2016)

12. Devlin, J., Chang, M.W., Lee, K., Toutanova, K.: Bert: pre-training of deep bidirectional transformers for language understanding. In: 2019 Conference of the North American Chapter of the Association for Computational Linguistics: Human Language Technologies, vol. 1 (Long and Short Papers) on Proceedings, Minneapolis, USA, pp. 4171–4186 (2017)

13. Joshi, M., Levy, O., Weld, D S., Zettlemoyer, L.: BERT for coreference resolution: baselines and analysis. In: 2019 Conference on Empirical Methods in Natural Language Processing and the 9th International Joint Conference on Natural Language Processing, pp. 5803–5808. ACL (2019)

14. Joshi, M., Chen, D., Liu, Y., Weld, D.S., Zettlemoyer, L., Levy, O.: Spanbert: improving pre-training by representing and predicting spans. In: 2020 Transactions of the Association for Computational Linguistics, vol. 8, pp. 64–77. ACL, Cambridge (2020)

15. Miculicich, L., Henderson, J.: Graph refinement for coreference resolution. In: 2022 Findings of the Association for Computational Linguistics, pp. 2732–2742. ACL (2022)

16. Toshniwal S. Efficient and interpretable neural models for entity tracking. arXiv preprint arXiv:2208.14252 (2022)

17. Otmazgin, S., Cattan, A., Goldberg, Y.: Lingmess: linguistically informed multi expert scorers for coreference resolution. In 17th Conference of the European Chapter of the Association for Computational Linguistics, pp. 2752–2760. ACL (2022)

18. Liu, T., Jiang, Y.E., Monath, N., Cotterell, R,. Sachan, M.: Autoregressive structured prediction with language models, pp. 993–1005. ACL (2022)

19. Bohnet, B., Alberti, C., Collins, M.: Coreference resolution through a seq2seq transition-based system. In: 2023 Transactions of the Association for Computational Linguistics, vol. 11, pp. 212–226. MIT Press, Cambridge (2023)

20. Wang, J., Luo, W., Yan, X.: A relation extraction model for enhancing subject features and relational attention. In: 2023 International Conference on Web Information Systems and Applications, pp. 259–270. Springer, Singapore (2023)

DFCDR: Domain-Aware Feature Decoupling and Fusion for Cross-Domain Recommendation

Jinyue Wei[1], Yue Kou[1(✉)], Derong Shen[1], Tiezheng Nie[1], and Dong Li[2]

[1] Northeastern University, Shenyang 110004, China
kouyue@cse.neu.edu.cn
[2] Liaoning University, Shenyang 110036, China

Abstract. Cross-Domain Recommendation (CDR) has indisputably proven its efficacy in alleviating the challenge of data sparsity in Recommender Systems. However, introducing domain-specific preferences from the source domain can introduce irrelevant information to the target domain. Furthermore, directly combining domain-general and domain-specific information may hinder the performance of the target domain. In this paper, we propose a domain-aware feature decoupling and fusion framework for CDR (DFCDR), which enables CDR more trustworthy and accurate. Specifically, we first design a user-level differential privacy method to protect users' privacy within each domain. Then we propose a contrastive learning-based feature decoupling method that achieves two pivotal goals: disentangling users' domain-specific preferences from their domain-general preferences, as well as differentiating between the popular and non-popular features of items. Finally, we present an adaptive feature fusion strategy that leverages a gating network to effectively fuse users' domain-general and domain-specific features in the target domain. We conduct extensive experiments on two real-world datasets. The results demonstrate the effectiveness of our proposed method.

Keywords: Cross-domain recommendation · contrastive learning · feature decoupling · adaptive feature fusion

1 Introduction

With the rapid development of the Internet, the explosive growth of information has led to an increasingly prominent problem of information overload. Recommender systems [8,17] are widely used as important tools for information filtering. However, the problem of data sparsity often impedes a comprehensive understanding of user preferences, while the cold start problem poses difficulties in providing accurate personalized recommendations for new users or items. To address these challenges, researchers have increasingly relied on cross-domain recommendation (CDR) techniques. These techniques leverage information from a source domain to enhance recommendation performance in a target domain,

© The Author(s), under exclusive license to Springer Nature Singapore Pte Ltd. 2024
C. Jin et al. (Eds.): WISA 2024, LNCS 14883, pp. 138–149, 2024.
https://doi.org/10.1007/978-981-97-7707-5_12

effectively leveraging the wealth of data across multiple domains to overcome issues like data sparsity and cold starts.

We study the problem of cross-domain recommendation and address three key issues. First, due to the involvement of information dissemination in cross-domain recommendation, while we strive to make full use of decoupled data, we must also protect users' privacy. For instance, when leveraging user data from the movie domain to enhance book recommendations in another domain, users' personal information in the movie domain may be subject to potential leakage. Therefore, when selecting methods for applying decoupled features, it is crucial to prioritize the protection of user privacy. Second, introducing domain-specific features from the source domain can introduce irrelevant information to the target domain, potentially affecting the accuracy of the recommendations in the target domain. Both users and items can have some domain-specific features. For instance, a user may prioritize performance and brand in the electronics domain, while focusing more on appearance and style in the clothing domain. Furthermore, the features of items, such as popularity, can vary significantly across domains. For example, a movie may gain significant attention due to its strong cast, whereas its original novel counterpart may receive little notice in the book domain. Directly introducing these domain-specific features into the target domain can compromise the accuracy of recommendations. A good model should be able to disentangle users' domain-specific preferences from their domain-general preferences, as well as to differentiate between the popular and non-popular features of items. Finally, decoupled features vary in importance across different domains, and we need to adaptively fuse these features in the target domain. For example, in the movie domain, users may focus more on movie genres and pay less attention to price features. Conversely, in other domains, users may have a higher sensitivity to price. Therefore, it is crucial to adaptively balance these features in the target domain.

Previous works on cross-domain recommendation only decouple user features while ignoring the decoupling of item features [22], and they overlook the importance differences of these features when making final predictions [3,19]. To tackle the aforementioned challenges, we introduce a domain-aware feature decoupling and fusion framework for cross-domain recommendation (DFCDR), aimed at enhancing its trustworthiness and accuracy. First, to rationalize the application of decoupled data, we can employ federated learning [18] and differential privacy [11] prior to decoupling. However, the heterogeneity of data across different domains inevitably affects the overall performance of federated learning. Therefore, we utilize the differential privacy to protect data, introducing mathematically rigorous privacy mechanisms to safeguard data within domains. After decoupling, the aggregation of user information from both the source and target domains has been a prevalent approach in cross-domain recommendation [12]. Nevertheless, the inherent heterogeneity across domains frequently poses challenges in effectively aligning the data. To mitigate this issue, we leverage decoupled data specifically from the target domain for recommendation. Second, in recognition of the inconsistency in user preferences and the influence of item popularity on recommendation performance, we devise a decoupling strategy. In this framework, we decouple user/item features through contrastive learning, employing a self-supervised method on graphs for decoupling. We consider both user and item influences in cross-domain recommendations, differing from prior decoupling methods such as [20]. Third, unlike previous approaches that rely on simple concatenation of decoupled data, we employ gating network to automatically adjust the transmission and processing of information. This allows for better capturing of the complex structures and interactions between features. Moreover, such weighted calculation can provide some interpretability for recommendations. To summarize, our contributions are as follows:

- We design a user-level differential privacy method. Before decoupling, we use the differential privacy method to protect user privacy. After decoupling, we adaptively utilize user features for recommendation in the target domain.
- We propose a contrastive learning-based feature decoupling method that achieves two pivotal goals: disentangling users' domain-specific preferences from their domain-general preferences, as well as differentiating between the popular and non-popular features of items.
- We propose an adaptive feature fusion strategy that leverages a gating network to effectively fuse users' domain-general and domain-specific features in the target domain. This enhances the accuracy and interpretability of cross-domain recommendation.
- We conducted experiments on two real datasets to evaluate the model performance. The results indicate that our DFCDR outperforms state-of-the-art baseline models.

2   Related Work

2.1   Cross-Domain Recommendation

Cross-domain recommendation (CDR) is an important research direction in recommender systems, first introduced by Shlomo Berkovsky et al. [2] in 2007. Its purpose is to enhance the accuracy of recommendations in sparse domains by leveraging information from relatively rich domains. Research in CDR has demonstrated its effectiveness in addressing data sparsity issues [2], overcoming cold start problems [15].

In single-target CDR, the objective is to improve the recommendation performance in the target domain. For example, COAST [23] achieves superior recommendation performance by exploring similarities between users and items across different domains. In dual-target or multi-target CDR, recommendations are made considering two or more target domains simultaneously. Using a three-layer attention framework, TASU [13] creates better user embeddings to improve recommendation performance in two domains. In terms of research methodologies, scholars classify studies on CDR into models based on general entity representations [25], domain mapping [11], heterogeneous graph embedding [9] and multi-domain collaborative training [21].

2.2 Privacy Protection

Currently, privacy protection methods integrated with recommendation systems mainly include homomorphic encryption [6], secure multiparty computation, federated learning [18] and differential privacy [24]. The methods of differential privacy and federated learning are more prevalent. Hussain Ahmad Madni [16] pointed out the issue of gradient leakage in federated learning. Below, we will mainly introduce the method of differential privacy. Hao Zhou et al. [24] proposed a novel lightweight matrix factorization recommendation method. Their approach deploys federated gradient training on users' local Internet of Things (IoT) devices, allowing users to train large-scale data models locally. They designed a two-stage solution to protect the security of user data and reduce the dimensionality of items.

2.3 Information Decoupling

Yoshua Bengio [1] (2013) explored different data representations and the explanatory factors behind the data. In 2019, Jianxin Ma et al. [14] applied it to single-domain recommendation by proposing the MacridVAE method, successfully addressing the problem of learning decoupled representations from user behavior. Ziqiang Cui et al. [4] proposed a BERT model based on Decoupled Attention (DDA), called DDA-BERT, to better utilize multi-information and address issues in sequential recommendation. In recent research, feature decoupling has also been applied to cross-domain recommendations. Ruohan Zhang et al. [19] proposed an innovative approach that decouples domain-invariant and domain-specific representations and introduces contrastive learning to enhance the informativeness of user and item representations in cross-domain recommendation systems.

Overall, research in cross-domain recommendation has made some progress, but challenges remain. Future research may continue to delve into aspects such as feature decoupling and the utilization of multiple information to further enhance the performance and applicability of recommendation systems.

3 Our Proposed Model: DFCDR

Cross-domain recommendation refers to leveraging user-item interaction data R_s from one domain D_s to assist recommendation tasks in another domain D_t. Typically, the source domain D_s has richer data, while the target domain D_t may suffer from data sparsity or lack specific information. The notations used in this paper are listed in Table 1.

As shown in Fig. 1, we propose a domain-aware feature decoupling and fusion framework for cross-domain recommendation (DFCDR), which consists of three parts. The first module is the user-level differential privacy module, primarily tasked with completing the initialization embedding for users/items and the application of differential privacy. The second module is the contrastive learning-based feature decoupling module, it collects features from both domains and

Table 1. Important Notations.

Symbol	Definition
D_s, D_t	Source domain and target domain
e_s, e_t	The user embeddings of source domain and target domain
$\hat{e}_s, \hat{e}_t$	The user embeddings in two domains with privacy protection
i_s, i_t	The item embeddings of source domain and target domain
u_g, u_s	general features and specific features
v_p, v_i	Popularity features and inherent features
R_s, R_t	The interaction matrices of users and items in the two domains

employs contrastive learning methods to design a loss function, separating general and specific features, as well as item popularity and inherent features. The third module is the gating network-based adaptive feature fusion module. It utilizes the decomposed user features for reassignment and predicts ratings using dot product methods to better achieve cross-domain recommendation.

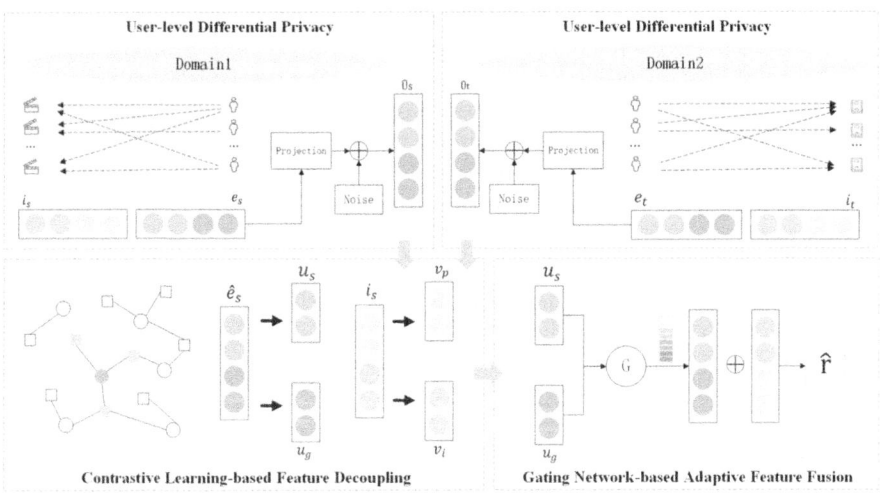

Fig. 1. Overall framework of DFCDR

3.1   User-Level Differential Privacy

Representation Learning. During the representation learning phase, the information used is the user-item rating matrix. We utilize Graph Convolutional Network (GCN) methods to obtain user and item embeddings.

Firstly, the original user-item interaction information is loaded and transformed into an adjacency matrix. User and item embeddings in both the source and target domains are then obtained based on the information from the adjacency matrix.

$$e_i^s = E_i^s x_i^s \tag{1}$$

$$e_i^t = E_i^t x_i^t \tag{2}$$

Where e_i^s and e_i^t denote the initialized embeddings of users in the source and target domains respectively, while E_i^s and E_i^t denote the embedding matrices. x_i^s and x_i^t denote the encodings at initialization.

User-Level Differential Privacy. Building upon the representation learning, we obtain the general embeddings of users in the source and target domains. Subsequently, apply local differential privacy to process the embeddings of users.

Definition 1 (Adjacent Rating Matrices). Given two rating matrices R and R' (both of size $m \times n$), if $R_{k,l} = R'_{k,l}$ for all $(k, l) \neq (i, j)$, where (i, j) is a specific row-column index, then we say that matrices R and R' are adjacent.

Definition 2 (User-level Differential Privacy). Let R and R' be adjacent matrices. Given a random algorithm M and S representing the set of all possible outputs of M, if $Pr[M(D) \in S] \leq e^\varepsilon$ and $Pr[M(D')] \in S] + \delta$, then the algorithm M provides differential privacy protection. Here, ε is the differential privacy budget.

Utilizing the aforementioned method, we perform differential privacy processing on the user data in the source domain and the user embeddings in the target domain, resulting in $\hat{e}_s$ and $\hat{e}_t$. Equation 3 is as follows:

$$\hat{e}_{s/t} = e_{s/t}M + \delta \tag{3}$$

Where e_s and e_t are the original embeddings in the source and target domains, M is the projection matrix, δ is the noise introduced during privacy processing.

3.2 Contrastive Learning-Based Feature Decoupling

To enhance the adaptability of user preference information in recommending items in a certain domain, we decompose the user preference information at the node level, better fulfilling the recommendation.

We use Graph Convolutional Neural Networks (GCN) to obtain multi-layered embedding representations, employ contrastive learning to decompose domain-specific and domain-agnostic preferences. The process of item decoupling is similar to that of user decoupling.

Node-Level Analysis. We utilize Graph Convolutional Networks (GCNs) to analyze user and item information at the node level, enabling the learning of more semantically and structurally meaningful node representations. We consider nodes as users and items, edges as the interaction relationships between users and items. Each convolutional layer processes only first-order neighborhood information and by stacking multiple convolutional layers, we can achieve multi-order neighborhood information propagation. In graph convolution, the representation of each node is a linear combination of itself and its neighboring node representations. This neighbor aggregation is achieved through the adjacency matrix (representing the graph structure), enabling nodes to update their representations using information from adjacent nodes. Equation 4 is summarized as follows:

$$u^k = AGG(u^{(k-1)}, i^k : i \in N_u) \tag{4}$$

Where k denotes the current convolutional layer, $u^{(k-1)}$ denotes the user embeddings from the previous convolutional layer, i^k denotes the item embeddings at the current convolutional layer, N_u denotes all neighboring nodes of user u. $AGG(\cdot)$ denotes the aggregation strategy of the graph convolutional network.

We utilize the results from the last layer of the graph convolutional network to represent the general and specific features of users and the popularity and inherent features of items in each domain. Then, based on the final information embedded by GCN, we decompose the user and item embeddings.

$$u^s_{(g/s)} = M_i Layer^s_l \tag{5}$$

$$v^s_{(p/i)} = M_i Layer^s_l \tag{6}$$

Where $u^s_{(g/s)}$ denotes the universal and specialized features extracted from the source domain, $v^s_{(p/i)}$ represents the popularity and inherent features of items in the source domain; M denotes different mapping matrices; $Layer^s_l$ denotes the final layer embedding of GCN processed in the source domain. The same applies to the target domain.

Feature Decoupling Mechanism. We incorporate mutual information as the basis for preference decomposition.

Mutual Information (MI) measures the degree of mutual dependence between two random variables. The stronger the dependency between the two random variables. Therefore, to achieve domain-general features and domain-specific features, we define the following loss function L_I:

$$L_I = -\log \frac{\exp\left(\frac{I(u^s_g, u^t_g)}{\tau}\right)}{\left(\exp\left(\frac{I(u^s_g, u^t_g)}{\tau}\right) + \exp\left(\frac{I(u^s_g, u^s_s)}{\tau}\right) + \exp\left(\frac{I(u^t_g, u^t_s)}{\tau}\right) + \exp\left(\frac{I(u^s_g, u^t_s)}{\tau}\right)\right)} \tag{7}$$

$$I(A, B) = \cos(AB(\theta)) = \frac{\sum_{k=1}^n x_{1k} x_{2k}}{\sqrt{\sum_{k=1}^n x_{1k}^2} \sqrt{\sum_{k=1}^n x_{2k}^2}} \tag{8}$$

Where τ is a hyperparameter for the Softmax temperature, $I(\cdot)$ is the mutual information function, we use cosine similarity calculation, where x_{1k} and x_{2k} denote the kth value in each embedding vector.

The loss of item decoupling is jointly used with Bayesian Personalized Ranking(BPR) loss for rating prediction in Sect. 3.3.

3.3   Gating Network-Based Adaptive Feature Fusion

We employ a gating mechanism to fuse the user features, enhancing domain adaptability, perform recommendations in the target domain.

Feature Fusion. In the target domain, through feature fusion, cross-domain recommendation systems can more effectively utilize both general and specific information, thereby enhancing recommendation accuracy and personalization.

In many papers, average pooling, max pooling and other methods are commonly used to fuse features. However, there exist differences in embedding distributions between different domains. Therefore, we only utilize information from the target domain to fuse user general and specific features.

Through the gating mechanism, the weights for gate selection are automatically learned and derived through two fully connected layers. Equation 9 is as follows:

$$\omega_i = \text{Sigmoid}\left(w_{s_2}\left(\sigma\left(w_{s_1}\Delta_i + b_{s_1}\right) + b_{s_2}\right)\right) \tag{9}$$

where $w_{s_1}, w_{s_2}, b_{s_1}$ and b_{s_2} are trainable weights and biases, σ denotes the nonlinear activation function $ReLU$. The weights obtained for the target domain user-specific and user-general features are used for fusion calculation, Eq. 10 is as follows:

$$u_{\text{con}} = \omega_1 \hat{u}_g^t + (1 - \omega_1)\hat{u}_s^t \tag{10}$$

where ω_1 denotes the weight of the common features in the target domain calculated by the gating network, $\hat{u}_g^t$ and $\hat{u}_s^t$ are the representations of the user embeddings in the target domain after differential privacy, u_{con} denotes the new user feature used for prediction.

Rating Prediction. We use the dot product method to predict the probability of interaction between users and items in the target domain. Equation 11 is as follows:

$$\hat{r}_{ui}^t = \sigma(u_{\text{con}} \cdot i_t) \tag{11}$$

where σ maps the dot product result into the probability of interaction using the Sigmoid function. For the scoring prediction stage, we utilize the Bayesian Personalized Ranking (BPR) loss to represent the loss function L_B:

$$L_B^S = - \sum_{(u,i\in R^+),(u,j\in R^-)} \ln \sigma(\hat{r}_{ui}^t - r_{uj}^t) \tag{12}$$

Where $u, i \in R^+$ denotes the positive interaction set of user-item pairs, $u, j \in R^-$ denotes the negative interaction set of user-item pairs. Finally, the overall loss function is as follows:

$$L = \alpha L_I + L_B^S + L_B^T + \beta\|\theta\|_2^2 \tag{13}$$

Where α and β denote weights of the loss, controlling the importance of mutual information and preventing overfitting.

Table 2. Statistics on Amazon and Douban datasets.

Datasets	AmazonMovie	AmazonMusic	DoubanMusic	DoubanBook
#users	6995	6995	6529	6529
#items	10000	10000	10000	10000

4   Experiments

In this section, we conduct experiments to answer the following questions: (1) R1: How does our approach perform compared with state-of-the-art CDR methods? (2) R2: Does each module play a role in the model? (3) R3: How does the performance of DFCDR vary with different values of the hyperparameters?

4.1   Experimental Setup

Datasets. We conducted experiments using two real-world datasets: the Amazon dataset and the Douban dataset. The Amazon dataset includes movie and music, while the Douban dataset includes music and book. We retained overlapping users from both domains for experimentation and filtered out users with fewer than 3 interactions. We considered interactions with ratings more significant than 3 as positive examples. We used the same data as described in reference [19]. The basic information of the data is shown in Table 2.

Evaluation Metrics. We selected commonly used evaluation metrics in recommender systems, including HR and NDCG, as the evaluation standards for our experiments.

Baselines. To evaluate the performance of our model, we selected a variety of baseline models for comparison study:

LightGCN [5]: LightGCN proposes a simplified neighborhood aggregation method, using normalized sum to generate representations.

CoNet [7]: CoNet is a collaborative cross-network model facilitating knowledge transfer across domains through cross connections between base networks.

BiTGCF [10]: BiTGCF is a CDR model leveraging a graph collaborative filtering network to integrate users' general features with domain-specific characteristics.

DCCDR [19]: DCCDR introduces an information decoupling approach to enable cross-domain recommendation.

Hyperparameter. In differential privacy, ϵ is a parameter related to the leakage probability. A lower value of ϵ leads to larger introduced noise and higher privacy protection. We select {10e-4, 10e-5, 10e-6, 10e-7, 10e-8} as ϵ respectively. In feature decoupling, the Influence of Decoupling Loss Weight α affects

Table 3. Performance comparison of our method with baselines.

Datasets	AmazonMovie		AmazonMusic		DoubanMusic		DoubanBook	
Metrics	HR@5	NDCG@5	HR@5	NDCG@5	HR@5	NDCG@5	HR@5	NDCG@5
LightGCN	0.232	0.141	0.229	0.171	0.119	0.143	0.160	0.158
CoNet	0.327	0.170	0.235	0.169	0.228	0.170	0.221	0.186
BiTGCF	0.283	0.204	0.315	0.257	0.224	0.167	0.227	0.176
DCCDR	0.376	0.284	0.472	0.359	0.242	0.205	0.240	0.208
DFCDR	0.312	0.293	0.453	0.428	0.251	0.207	0.247	0.210

Table 4. Ablation study results.

Datasets	AmazonMovie			AmazonMusic		
Metrics	HR@10	NDCG@10	MRR@10	HR@10	NDCG@10	MRR@10
DFCDR-P	0.361	0.303	0.286	0.507	0.443	0.424
DFCDR-G	0.352	0.200	0.154	0.526	0.314	0.251
DFCDR-D	0.299	0.245	0.230	0.476	0.412	0.393
DFCDR	0.361	0.309	0.293	0.512	0.445	0.426

the user-specific diagnostic decomposition module. We select $\{0.0001, 0.0005, 0.001, 0.005, 0.01\}$ as α respectively.

By comparing experimental results, we set α as 0.001, we set ϵ as 10e–6.

4.2   Performance Comparison

Baseline Experiment. Table 3 presents the performance comparison between our method and baseline methods. The results indicate that DFCDR performs better on most evaluation metrics across both datasets. Single-domain recommendation methods perform poorly, highlighting the importance of cross-domain information transfer. While DFCDR may experience a slight decrease in performance due to the application of differential privacy methods compared to cross-domain recommendations, overall, it maintains satisfactory performance.

Ablation Study. We conducted an ablation study comparing two variants of DFCRD to evaluate how each module of DFCDR affects performance.

DFCDR-P: This variant removes the part where differential privacy.

DFCDR-G: This variant removes the part where gating network.

DFCDR-D: This variant removes the part where feature decoupling.

The results indicate that without privacy protection, DFCDR performs similarly to DFCDR-P, indicating that the noise effect of differential privacy does not significantly reduce the effectiveness. However, without feature domain adaptive fusion and feature decoupling, the performance decreases, with DFCDR significantly outperforming DFCDR-G and DFCDR-D. The experimental results are displayed in Table 4. The ablation experiment results of the Douban dataset are similar to it.

5   Conclusion

To address the issue of cross-domain recommendation, we propose a novel framework called DFCDR. These modules accomplish three tasks in cross-domain recommendation: privacy preserving information transmission, user/item features decoupling, feature adaptation fusion. Firstly, we employ differential privacy to balance privacy protection and data accuracy. Then, we utilize graph convolutional networks with contrastive learning to achieve feature decomposition. Finally, we use gating network in the target domain to adjust user features and complete cross-domain recommendations. Extensive experiments on two real datasets demonstrate the effectiveness of the model.

Acknowledgements. This work was supported by the National Natural Science Foundation of China under Grant Nos. 62072084, 62172082 and 62072086, the Science and Technology Program Major Project of Liaoning Province of China under Grant No.2022JH1/10400009, the Natural Science Foundation of Liaoning Province of China under Grant No.2022-MS-171, and the Fundamental Research Funds for the Central Universities under Grant No. N2116008.

References

1. Bengio, Y., Courville, A., Vincent,P.: Representation learning: a review and new perspectives. In: IEEE Transactions on Pattern Analysis and Machine Intelligence, vol. 35, pp. 1798–1828 (2013). https://doi.org/10.1109/TPAMI.2013.50
2. Berkovsky, S., Kuflik, T., Ricci, F.: Cross-domain mediation in collaborative filtering. In Proceedings of the 11th International Conference on User Modeling (UM 2007), pp. 355–359 (2007). https://doi.org/10.1007/978-3-540-73078-1_44
3. Cao, J., Lin, X., Cong, X., Ya, J., Liu, T., Wang, B.: DisenCDR: learning disentangled representations for cross-domain recommendation. In: SIGIR, pp. 267–277 (2022)
4. Cui, Z., Su, Y., Lin, F., Yang, C., Zhang,H., Zhang, J.: Dual disentangled attention for multi-information utilization in sequential recommendation. In: IJCNN, pp. 1–8 (2022). https://doi.org/10.1109/IJCNN55064.2022.9892008
5. He, X., Deng, K., Wang, X., Li, Y., Zhang, Y., Wang, M.: LightGCN: simplifying and powering graph convolution network for recommendation. In: SIGIR(2020)
6. Hong, W., Zhang, H., Zhu., J.: FedHD: a privacy-preserving recommendation system with homomorphic encryption and differential privacy. In: ICCSE (2023). https://doi.org/10.1007/978-981-99-2443-1_50
7. Hu, G., Zhang, Y., Yang, Q., CoNet: collaborative cross networks for cross-domain recommendation. In: ACM International Conference on Information and Knowledge Management (2018)
8. Huang, X., Kou, Y., Shen, D., Nie, T.,Li, D.: Exploiting item relationships with dual-channel attention networks for session-based recommendation. In: WISA , pp. 198–205 (2023)

9. Li, J., Peng, Z., Wang, S., Xu, X., Yu, P.S., Hao, Z.: Heterogeneous graph embedding for cross-domain recommendation through adversarial learning. In: DASFAA, vol. 12114 (2020). https://doi.org/10.1007/978-3-030-59419-0_31

10. Liu, M., Li, J., Li, G., Pan, P.: Cross domain recommendation via bi-directional transfer graph collaborative filtering networks. In: CIKM, pp. 885–894 (2020). https://doi.org/10.1145/3340531.3412012

11. Liu, W., et al.: Differentially private sparse mapping for privacy-preserving cross domain recommendation. In: MM, pp. 6243–6252 (2023). https://doi.org/10.1145/3581783.3611924

12. Liu, W., Zheng, X., Su, J., Zheng, L., Chen, C., Hu, M.: Contrastive proxy kernel stein path alignment for cross-domain cold-start recommendation. In: TKDE, vol. 35, pp. 11216–11230 (2023). https://doi.org/10.1109/TKDE.2022.3233789

13. Lu, j., Sun, G., Fang, X., Yang, J., He, W.: A Three-layer attentional framework based on similar users for dual-target cross-domain recommendation. In: DASFAA, pp. 297–313 (2023). https://doi.org/10.1007/978-3-031-30672-3_20

14. Ma, J., Zhou, C., Cui, P., et al.: Learning disentangled representations for recommendation. Adv. Neural Inf. Process. Syst. **32** (2019)

15. Man, T., Shen, H., Jin, X., Cheng, X.: Cross-domain recommendation: an embedding and mapping approach. In: IJCAI, vol. 17. pp. 2464–2470 (2017)

16. Madni, H., Umer, R., Foresti, G.: Blockchain-based swarm learning for the mitigation of gradient leakage in federated learning. In: IEEE Access, pp. 16549–16556 (2023). https://doi.org/10.1109/ACCESS.2023.3246126

17. Resnick, P., Hal, R.: Recommender systems. Commun. ACM **40**(3), 56–58 (1997). https://doi.org/10.1145/245108.245121

18. Wu, Y., Su, L., Wu,L., and Xiong, W., FedDeepFM: a factorization machine-based neural network for recommendation in federated learning. In: IEEE Access, vol. 11, pp. 74182–74190 (2023). https://doi.org/10.1109/ACCESS.2023.3295894

19. Zhang, R., Zang, T., Zhu, Y., Wang, C., Wang, K., Yu, J.: Disentangled contrastive learning for cross-domain recommendation. In: DASFAA, vol. 13944 (2023). https://doi.org/10.1007/978-3-031-30672-3_11

20. Zhang, Y., Shen, Y., Wang, D., Gu, J., Zhang, G.: Connecting unseen domains: cross-domain invariant learning in recommendation. In: SIGIR, pp. 1894–1898 (2023). https://doi.org/10.1145/3539618.3591965

21. Zhang, H., Kong, X., Member, IEEE. et al. Cross-domain collaborative recommendation without overlapping entities based on domain adaptation. Multimedia Syst. **28**, 1621–1637 (2022). https://doi.org/10.1007/s00530-022-00923-9

22. Zhang, H., Zheng, D., Yang, X., Feng, J., Liao, Q.: 2023. Federated Cross-domain Sequential Recommendation via Disentangled Representation Learning, FedDCSR (2023)

23. Zhao, C., Zhao, H., He, M., Zhang, J., Fan, J.: Cross-domain recommendation via user interest alignment. In: WWW, pp. 887–896 (2023). https://doi.org/10.1145/3543507.3583263

24. Zhou, H., Yang, G., Xiang, Y., Bai, Y., Wang, W.: A lightweight matrix factorization for recommendation with local differential privacy in big data In: IEEE Transactions on Big Data, vol. 9, pp. 160–173 (2023). https://doi.org/10.1109/TBDATA.2021.3139125

25. Zhu, F., Chen, C., Wang, Y., Liu, G., Zheng, X.: DTCDR: a framework for dual-target cross-domain recommendation. In: CIKM, pp. 1533–1542 (2019). https://doi.org/10.1145/3357384.3357992

Two-Stage Enhancement for Recommendation Systems Based on Contrastive Learning

Siyu Sun, Tianyu Cai, Fanli Yan, and Shenggen Ju$^{(\boxtimes)}$

Sichuan University, Chengdu 610065, Sichuan, China
sunsiyu@stu.scu.edu.cn

Abstract. Knowledge-aware recommendation systems use knowledge graphs as side information to enhance the performance of recommendation systems. The existing methods often employ graph neural networks to process the relational networks and use contrastive learning to obtain more effective node representations. However, persistent challenges from active users' noisy data and the cold-start problem related to inactive users impact model performance. Recent studies mainly use data augmentation methods to address these issues but only consider the enhancement during the training stage and neglect the pre-training stage. For this reason, we propose a Two-stage Enhancement for Recommendation Systems based on Contrastive Learning (TERSCL). This model primarily integrates two stages of data augmentation methods: one is the pre-training enhancement, where an enhanced adjacency matrix is obtained for subsequent training. The other generates a distinct dropout subgraph during each training round to participate in the subsequent graph encoding and contrastive learning process. Extensive experiments conducted on three benchmark datasets verify that TERSCL can significantly improve recommendation performance and mitigate the cold-start problem.

Keywords: Recommendation system · Graph neural network · Knowledge graph · Contrastive learning · Data augmentation

1 Introduction

Knowledge-aware recommendation is a trending area of recommendation systems with extensive practical applications [1]. However, the knowledge-aware systems face multiple problems, including noise [2], sparsity, bias [3], and long-tail problems [4]. Neighborhood information serves as an integral component of recommendation systems, playing a crucial role in inferring user/item embeddings and extracting higher-order collaborative signals. However, the effectiveness of neighborhood information is constrained by the definition of the adjacency matrix.

The adjacency matrix is typically built upon a bipartite graph of user-item interactions. In some cases, models perform worse for active users, likely due to noisy interactions, which can impact user preference modeling [5]. And additional graph convolution layers introduce more noise. [6]. These issues indicate that the current interaction adjacency matrix design needs improvement. Existing methods use data augmentation such as structural disturbance or node representation disturbance [7]. However, there are two pressing issues remain: (1)The disturbed graphs are used for contrastive learning to enhance data, but the data augmentation before training is not taken into account. (2)The use of random drop operations permanently discards some data, and solely employing dropout subgraphs undervalues the original interaction data's importance.

To address these two issues, we propose a Two-stage Enhancement for Recommendation Systems based on Contrastive Learning (TERSCL). In the pre-training stage, the model generates an enhanced adjacency matrix, providing accurate information on neighborhood relations between users and items, while avoiding noise and sparsity in the adjacency information. In the training stage, the model creates new dropout subgraphs in each round of training through random drop operations, which are used for subsequent graph neural encoding and contrastive learning. In this way, the two-stage enhancement provides superior node representations, effectively improving recommendation accuracy.

The contributions of our work are summarized as follows:

- In the pre-training stage, the top K neighbor nodes for each node are identified, and generate a new enhanced adjacency matrix.
- In the training stage, each training round generates a random dropout subgraph, used as a contrastive view for contrastive learning and as the global views for graph neural encoding.
- Extensive experiments on three datasets confirm TERSCL's superiority, highlighting the effectiveness of enhanced adjacency matrix and dynamic dropout subgraphs.

2   Related Work

2.1   Knowledge-Aware Recommendation Based on GNN

The knowledge-aware recommendation (KGR) treats users and items as nodes in the graph connected by interaction relations. GNN-based methods aggregate information from multi-hop neighbors using GNN's mechanism to model remote connections. Recent studies introduced KGCN [19], label smoothing [20], KGAT [21], and intent modeling [22], utilizing GCN for iterative item embedding aggregation and attention network end-to-end connection for embedding propagation. Despite strong performance in supervised learning, these methods are limited by sparse interaction datasets, hindering recommendation performance improvement.

2.2   Contrastive Learning for Recommendation

Contrastive learning methods minimize distances between the anchor and positive samples while maximizing distances with negative samples to learn node representations. Google's self-supervised learning framework [11] transforms items into two inputs via data augmentation, improving model feature distinguishability. Wu et al. [7] extend user-item-entity graphs into views for contrastive learning, while Shuai et al. constructed review-aware user-item graphs for edge-enhanced contrastive learning. These methods leverage contrastive learning for extracting deeper information but lack integration of fine-grained information.

3   Methodology

We propose a model named TERSCL, including five stages: 1)the Pre-training Enhancement Stage, 2)the Training Enhancement Stage, 3)the Graph Neural Network Encoding Stage, 4)the Cross-view Contrastive Learning Stage, 5)the Model Prediction and Joint Loss Optimization Stage. The model is shown in Fig. 1.

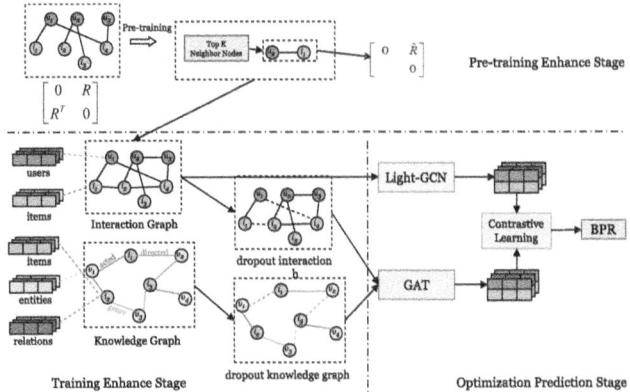

Fig. 1. Illustration of TERSCL

3.1   Pre-training Enhancement Stage

Pre-training Optimization. Before encoding with a graph neural network, TERSCL optimizes the user-item pair adjacency matrix through pre-training. Initially, TERSCL pre-trains using a basic graph convolution method, specifically employing Light-GCN. Light-GCN can capture remote connection information within user-item interactions through recursive aggregation executed T times. The aggregation process of Light-GCN [12] in the t-th layer is shown as follows:

$$
\begin{aligned}
e_u^{(t+1)} &= \sum_{i \in N_u} \frac{1}{\sqrt{|N_u||N_i|}} e_i^{(t)} \\
e_i^{(t+1)} &= \sum_{u \in N_i} \frac{1}{\sqrt{|N_u||N_i|}} e_u^{(t)}
\end{aligned}
\tag{1}
$$

where $e_u^{(t)}$ and $e_i^{(t)}$ are the embeddings of user u and item i in the t-th layer, and N_u and N_i represent the neighbors of user u and item i, respectively. Finally, sum the embeddings of different layers to obtain the final embeddings e_u and e_i, as shown below:

$$e_u = e_u^{(0)} + ... + e_u^{(t)}, e_i = e_i^{(0)} + ... + e_i^{(t)} \tag{2}$$

The BPR loss ensures the prediction scores of items with historical interactions are higher than those of items without interactions. This method enhances the model's understanding of the user's historical behavior, leading to more accurate recommendations of relevant items, as specifically shown below [13]:

$$L = - \sum_{u,i+,i- \in R} \ln(\sigma(e_u^T e_{i+} - e_u^T e_{i-})) \tag{3}$$

where $i+$ represents items that user u has positively interacted with, and $i-$ stands for the negatively sampled items that user u has not interacted with. e_u and e_i are the embeddings obtained from Eq. 2.

The Top K Optimal Neighbor Nodes. To reduce the interference caused by excessive neighbor nodes, our model selects and retains the top K neighbor nodes that contribute the most to the recommendation effect during the pre-training stage.

The model uses the embeddings e_u and e_i obtained from pre-training to generate the top K neighbors for users and items, respectively. A hyperparameter U_k is defined to control the number of neighbor nodes for all users. With the same number of neighbors, the noise from active users with rich neighbors can be reduced. Specifically, for user u, the top K neighbors are obtained by selecting the top U_k elements from the scores of $e_u^T e_I$, as shown in Eq. 4.

$$argmax_{i_1,i_2,...,i_k \in I} e_u^T e_I \tag{4}$$

where e_I represents the set of embeddings for neighbor items of user u; $argmax$ is a function to select the maximum value. Similarly, for items, a hyperparameter I_k is defined through a similar process. The enhanced adjacency matrix $\tilde{R}$ is obtained from the union of user-item interactions generated from both the users and items.

Enhanced Adjacency Matrix. Through the pre-training stage, we obtain an enhanced component for users and items and use this information to generate an enhanced adjacency matrix $\begin{bmatrix} 0 & \tilde{R} \\ \tilde{R}^T & 0 \end{bmatrix}$. This enhanced adjacency matrix is used to replace the original adjacency matrix. The enhanced adjacency matrix will then be used to generate embeddings for user-item interaction prediction.

3.2   Training Enhancement Stage

To enhance contrastive learning with more available views, we propose a novel method that dynamically generates dropout subgraphs. In previous contrastive learning methods [14,15], dropout subgraphs are generated by randomly dropping edges from the original view. However, these methods will lose the information of the original graph. We propose a method that continually generates new subgraphs as auxiliary views for contrastive learning during the training process, which can emphasize the importance of the original views.

Specifically, in each training round, the model dynamically generates a dropout subgraph, while the original view remains unchanged. The edges are randomly deleted according to a fixed ratio ρ to directly disturb the views, obtaining the dropout subgraph $\tilde{\varepsilon}$ of the original view G:

$$\tilde{\varepsilon} = \text{random\_drop}(G, \rho) \tag{5}$$

3.3   Graph Neural Network Coding

The views are divided into the local views and the global views. At this stage, two different graph neural networks are used as graph encoders to capture higher-order structural context for node representations.

For the local views, Light-GCN can effectively encode the high-order context for each node representation of the local views. This process is as shown in Eq. 6.

$$e_u^{(k+1)} = \sum_{u' \in N_u} \frac{1}{\sqrt{|N_u \| N_{u'}|}} e_{u'}^{(k)} \tag{6}$$

where u' is one of the neighbor nodes of user u. N_u represents all neighbor nodes adjacent to user u, $e_u^{(k)}$ is the embedding of user u at the k-th layer. Similarly, the representation of item nodes under the local views $e_i^{(l)}$ is generated.

We use a graph attention network to process the global views and obtain the global views embedding representations e_i^g. This process is shown in Eq. 7 and Eq. 8.

$$e_i^g = e_i^{(0)} + \sum_{v \in N_i} \zeta(v, r_{i,v}) e_v \tag{7}$$

$$\zeta(v, r_{i,v}) = \frac{\exp\left(\text{LeakyReLU}\left(r_{i,v}^T \left[e_v \| e_i^{(0)}\right]\right)\right)}{\sum\limits_{v' \in N_i} \exp\left(\text{LeakyReLU}\left(r_{i,v'}^T \left[e_{v'} \| e_i^{(0)}\right]\right)\right)} \tag{8}$$

where e_v is the initial embedding of entity v; $r_{i,v}$ is the embedding of item i about different relations of entity v; N_i is the set of neighbor entities of item i; $\zeta(v, r_{i,v})$ is the relevance of entity v and specific relation $r_{i,v}$ in the knowledge aggregation process; $e_i^{(0)}$ is the initial embedding of item i; e_i^g is the final representation of item i under the global views; *LeakyReLU* is an activation function used to handle non-linear transformations.

3.4 Cross-view Contrastive Learning

The representation spaces of the global views and the local views are different. To further explore the feature information from both views, TERSCL uses contrastive learning for the nodes of the two views. For each user u, the same user in the two views will be regarded as a positive sample $\{e_u^g, e_u^l\}$; on the other hand, the representations of different users should be different, and their differences need to be emphasized. Therefore, one user is randomly paired with another user and is treated as a negative sample $\{e_u^g, e_{u'}^g\}$. These positive and negative samples are used in contrastive learning, as shown in Eq. 9 below:

$$\mathcal{L}_u = -\ln \frac{\exp(s(e_u^g, e_u^l)/\tau_{cl})}{\exp(s(e_u^g, e_u^l)/\tau_{cl}) + \sum_{u' \neq u}(\exp(s(e_u^g, e_{u'}^l)/\tau_{cl}) + \exp(s(e_u^l, e_{u'}^g)/\tau_{cl}))} \tag{9}$$

where $s(\cdot)$ is the calculation of cosine similarity; τ_{cl} is the temperature parameter of contrastive learning; e_u^g refers to the embedding representation of user u under the global views; $e_u'^g$ is the embedding representation of different user u' from user u under the global views. Similarly, TERSCL employs a similar method for cross-view contrastive learning for item nodes, as shown in Eq. 10.

$$\mathcal{L}_i = -\ln \frac{\exp(s(e_i^g, e_i^l)/\tau_{cl})}{\exp(s(e_i^g, e_i^l)/\tau_{cl}) + \sum_{i' \neq i}(\exp(s(e_i^g, e_{i'}^l)/\tau_{cl}) + \exp(s(e_i^l, e_{i'}^g)/\tau_{cl}))} \tag{10}$$

After obtaining the contrastive learning losses of users and items, respectively, the final contrastive learning loss is shown in Eq. 11.

$$\mathcal{L}_{cl} = \mathcal{L}_i + \mathcal{L}_u \tag{11}$$

3.5 Model Prediction and Joint Loss Optimization

TERSCL obtains the global user representations e_u^g and the global item representations e_i^g through the relational graph attention network, while the local user representations e_u^l and the local item representations e_i^l are obtained through the light-GCN.

The final representations of users e_u^* are obtained by concatenating the user nodes under the two views as shown in Eq. 12. Similarly, the final representations of the items e_i^* are obtained. The matching score $y(u, i)$ between the user and the item is predicted using the dot product of these final representations, as shown in Eq. 12.

$$e_u^* = e_u^g \parallel e_u^l, e_i^* = e_i^g \parallel e_i^l, y(u, i) = e_u^{*T} e_i^* \tag{12}$$

To integrate tasks such as recommendation and contrastive learning, TERSCL uses a multi-task training strategy to optimize the entire model. For the recommendation task, the BPR loss is used to optimize historical data, which can enhance the model's understanding of the user's historical behavior. The BPR loss is shown in Eq. 13. The model parameters are learned by minimizing the loss function shown in Eq. 13, which combines the contrastive loss on the global and local views, the BPR loss, and the intent independence loss.

$$\mathcal{L} = \alpha\mathcal{L}_{BPR} + (1 - \alpha)\mathcal{L}_{cl} + \lambda \parallel \Theta \parallel_2^2 \qquad (13)$$

where Θ is the set of model parameters and α is the hyperparameter that balances the contrastive loss and BPR loss. It should be noted that the BPR loss used here is the BPR loss used in the training stage but not the pre-training stage.

4   Experiments

4.1   Experiment Settings

Dataset. The experiment used three public datasets from different domains: Book-Crossing, MovieLens-1M, and Last.FM. These datasets are from the fields of books, movies, and music, respectively, and they have different scales and sparsity. The dataset statistics are shown as follows. Book-Crossing contains ratings for a variety of books (ranging from 0 to 10). The MovieLens-1M contains 6036 users, 2445 items, and approximately 1 million explicit ratings. Last.FM is collected from the Last.FM online music system (Table 1).

Table 1. Statistics of Datasets

statistical indicators		BookCrossing	MovieLens1M	Last.FM
User-item Interactions	users	17,860	6,036	1,872
	items	14,967	2,445	3,846
	interactions	139,746	753,772	42,346
	sparsity	0.000,523	0.051,075	0.005,882
Knowledge Graph	entities	77,903	182,011	9,366
	relations	25	12	60
	triples	151,500	1 241,996	15,518
	sparsity	0.000,130	0.002,791	0.000,431

Baselines. To demonstrate the effectiveness of the proposed TERSCL, we compared TERSCL with four types of recommendation system models, including CF-based models, embedding-based models, path-based models, and GNN-based models. The brief introductions to these models are as follows:

BPRMF [13] is a typical CF-based model that uses pairwise matrix decomposition to optimize BPR loss based on implicit feedback.

CKE [17] is an embedding-based model that combines structural, text, and visual knowledge in one framework.

PER [18] is a typical path-based model that extracts metapath features to represent the connectivity between users and items.

KGCN [19] is a GNN-based model that enriches item embeddings by iteratively integrating neighborhood information.

KGNN-LS [20] is a GNN-based model that enriches item embeddings through GNN and label smoothing regularization.

KGAT [21] is a GNN-based model that integrates neighbors through an attention mechanism to obtain user and item representations.

KGIN [22] is a GNN-based model that performs GNNs on the proposed user-intent-entity graph at the granularity of user intent.

CG-KGR [23] is a GNN-based model that integrates collaborative signals into knowledge aggregation using GNNs.

RMCEN [24] is a GNN-based model that conducts contrastive learning by adding noise to neighborhood information.

KRSCCL [25] is the latest GNN-based model that introduces the concept of global and local views, and performs cross-view contrastive learning on the two views.

4.2 Experiment Results Analysis

Here we displays the experiment results of TERSCL and all comparison methods. Comparing TERSCL with the best-performing experiments from other methods (where the best result among all is highlighted with an underscore, except TERSCL's best result marked with *), the following analysis can be made (Table 2).

Table 2. Results of AUC and F1 in CTR Prediction

Model	Book-Crossing		MovieLens-1M		Last.FM	
	AUC	F1	AUC	F1	AUC	F1
BPRMF	0.6583	0.6117	0.8920	0.7921	0.7563	0.7010
PER	0.6048	0.5726	0.7124	0.6670	0.6414	0.6033
CKE	0.6759	0.6235	0.9065	0.8024	0.7471	0.6740
RippleNet	0.7211	0.6472	0.9190	0.8422	0.7762	0.7025
KGCN	0.6841	0.6313	0.9090	0.8366	0.8027	0.7086
KGNNLS	0.6762	0.6314	0.9140	0.8410	0.8052	0.7224
KGAT	0.7314	0.6544	0.9140	0.8440	0.8293	0.7424
KGIN	0.7273	0.6614	0.9190	0.8441	0.8486	0.7602
CG-KGR	0.7498	0.6689	0.9110	0.8359	0.8336	0.7433
RMCEN	0.7743*	0.6885*	0.9280	**0.8650**	0.8599	0.7817
KRSCCL	0.7263	0.6710	**0.9329***	0.8610*	0.8686*	0.7991*
TERSCL	**0.7791**	**0.7169**	0.9312	0.8559	**0.9023**	**0.8295**

(1) On three datasets, TERSCL significantly outperforms all state-of-the-art baselines in most metrics. Through the two-stage enhancement, the TER-SCL filters out irrelevant noise from the original data and effectively utilizes structural information. By encoding both the global graph and the local graph, TERSCL captures user preferences and item knowledge information better.

(2) It can be found that improvements on LastFM and Book-crossing are more significant than those on Movielens-1M. However, it couldn't surpass KRSCCL on the movie dataset. This could be attributed to dataset features: the interactions and knowledge graph data on the music and book datasets are sparser than those provided on movies. This indicates that the model has a stronger ability to extract information and can more effectively use sparse labels to achieve a greater impact.

(3) The significant improvement of TERSCL on sparse datasets is attributed to the two innovative modules involved in the two-stage enhancement. These modules effectively utilize existing information, emphasize the most useful parts of information for recommendation, and successfully discard noise.

4.3   Ablation Study

To examine the contributions of the primary components in the model to the final performance, TERSCL and its three variants are compared on three datasets using two evaluation metrics. The experimental results are shown as follows (Table 3).

Table 3. Results of ablation experiments

Model	Book-Crossing		MovieLens-1M		Last.FM	
	AUC	F1	AUC	F1	AUC	F1
TERSCL	0.7791	0.7169	0.9312	0.8557	0.9023	0.8295
w/o enhanced matrix	0.7576	0.6887	0.9306	0.8555	0.8755	0.8095
w/o spanning subgraph	0.7282	0.6809	0.9255	0.8534	0.8600	0.7829
w/o both	0.5106	0.5188	0.9206	0.8482	0.7663	0.7120

The *w/o enhanced matrix* variant refers to the variant of TERSCL where the enhanced matrix module is removed. In this variant, a regular adjacency matrix replaces the enhanced adjacency matrix. The movie dataset shows the least decline, while the book dataset exhibits the most significant decrease. This trend corresponds to the density of the datasets, which reflects the role of the enhanced matrix module in sparse datasets.

The *w/o spanning subgraph* variant refers to the variant that removes the generation of the subgraph. The performance of this variant has decreased, especially in the book dataset, where it drops by 5%. This variant uses the complete and unchanged global view for subsequent training. The ablation results reflect the critical impact of the subgraph generation module and further validate the particular contributions of the two innovative modules on sparse datasets.

The *w/o both* variant refers to the variant that removes both innovative modules. This variant showed the worst ablation results, as expected, confirming the roles of both modules in the experimental outcomes. Particularly on the book dataset, there was a sharp decline. It is believed that the significant changes in the model led to the inability to escape the local optimal point in this dataset, resulting in overfitting.

4.4 Hyperparameter Experiments

Figure 2 illustrates the impact of controlling hyperparameters U_k and I_k, which adjust the number of semantic neighbors, on AUC and F1 scores across three datasets. For the music dataset, an optimal number of neighbors enhances overall performance, with increasing U_k and I_k improving outcomes. However, excessively high numbers of item neighbors can degrade performance due to potential noise introduced by item nodes. In the movie dataset, the number of user neighbors (U_k) significantly influences recommendation performance.

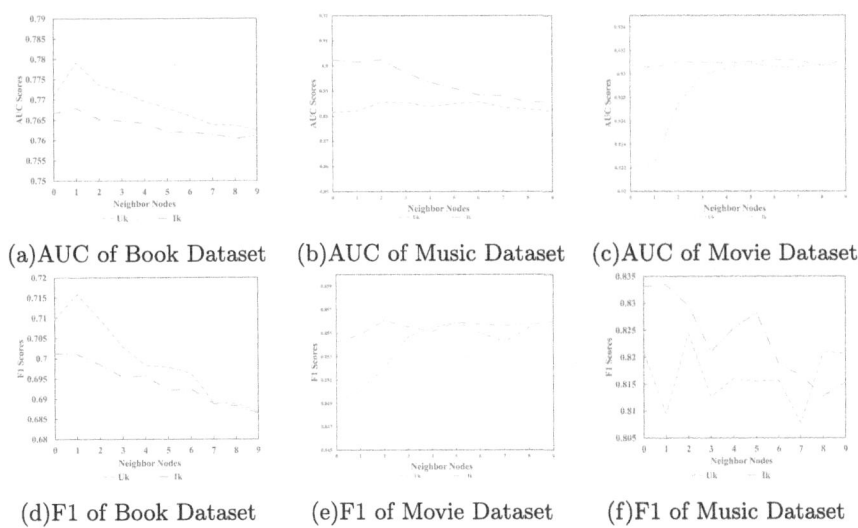

(a)AUC of Book Dataset   (b)AUC of Music Dataset   (c)AUC of Movie Dataset

(d)F1 of Book Dataset   (e)F1 of Movie Dataset   (f)F1 of Music Dataset

Fig. 2. Effect of the Number of Different Semantic Neighbors

4.5 Cold Start Experiments

The cold start problem is a challenge in recommendation systems. Sparse datasets, resulting from reduced historical interactions, impact recommendation effectiveness. To assess model performance on sparse datasets, cold start experiments were conducted on the Last.FM dataset, where 10%, 20%, 30%, 40%, and 50% percentages of data were randomly removed. The experiments utilized the KGIN baseline model for comparison, with AUC scores shown in Fig. 3. From Fig. 3, we can find that despite TERSCL's noticeable performance decline, it consistently outperforms the comparative model, showcasing its robustness in handling sparse datasets and practical challenges. Moreover, when data reduction exceeds 30%, TERSCL maintains steady performance while KGIN sharply declines. This demonstrates TERSCL's ability to extract information effectively from increasingly sparse datasets, emphasizing its robustness and adaptability.

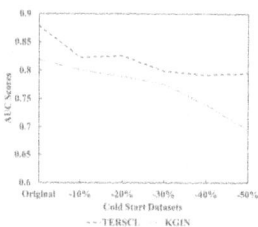

Fig. 3. Cold Start Experiment Results

5   Conclusion

In this paper, we propose a novel knowledge-aware recommendation system. These two-stage enhancement methods effectively alleviate the noise and data sparsity issues during the recommendation process. In the future, further improvements are needed to address the shortcomings in the explainability and diversity of the recommendation results in this study.

Acknowledgement. This work was supported by the Key Programs of the National Natural Science Foundation of China (Grant No.62137001). This work was jointly supported by the Key Technologies R&D Programs of Sichuan Province of China (Grant No.2023YFG0265).

References

1. Chen, X., Sun, Y., Zhou, T., Wen, Y., Zhang, F., Zeng, Q.: Recommending Online Course Resources Based on Knowledge Graph. In: Zhao, X., Yang, S., Wang, X., Li, J. (eds.) WISA 2022. LNCS, vol 13579. Springer, Cham. (2022). https://doi.org/10.1007/978-3-031-20309-1_51
2. Abdollahpouri, H., Mansoury, M., Burke, R., Mobasher, B.: The unfairness of popularity bias in recommendation. ArXiv Preprint ArXiv:1907.13286. (2019)
3. Christakopoulou, E., Karypis, G.: Local item-item models for top-n recommendation. In: Proceedings Of The 10th ACM Conference On Recommender Systems, pp. 67-74 (2016)

4. Clauset, A., Shalizi, C., Newman, M.: Power-law distributions in empirical data. SIAM Rev. **51**, 661–703 (2009)
5. Wang, W., Feng, F., He, X., Nie, L., Chua, T.: Denoising implicit feedback for recommendation. In: Proceedings Of The 14th ACM International Conference On Web Search And Data Mining, pp. 373–381 (2021)
6. Fan, W., Liu, X., Jin, W., Zhao, X., Tang, J., Li, Q.: Graph trend filtering networks for recommendation. In: Proceedings Of The 45th International ACM SIGIR Conference On Research And Development In Information Retrieval, pp. 112–121 (2022)
7. Wu, J., et al.: Self-supervised graph learning for recommendation. In: Proceedings Of The 44th International ACM SIGIR Conference On Research And Development In Information Retrieval, pp. 726–735 (2021)
8. Yu, J., Yin, H., Xia, X., Chen, T., Cui, L., Nguyen, Q.: Are graph augmentations necessary? simple graph contrastive learning for recommendation. In: Proceedings Of The 45th International ACM SIGIR Conference On Research And Development In Information Retrieval, pp. 1294–1303 (2022)
9. Bayer, I., He, X., Kanagal, B., Rendle, S.: A generic coordinate descent framework for learning from implicit feedback. In: Proceedings Of The 26th International Conference On World Wide Web, pp. 1341–1350 (2017)
10. Liu, X., et al.: Self-supervised learning: generative or contrastive. IEEE Trans. Knowl. Data Eng. **35**, 857–876 (2021)
11. Yao, T., et al.: Self-supervised learning for large-scale item recommendations. In: Proceedings Of The 30th ACM International Conference On Information & Knowledge Management, pp. 4321–4330 (2021)
12. He, X., Deng, K., Wang, X., Li, Y., Zhang, Y. & Wang, M.: Lightgcn: simplifying and powering graph convolution network for recommendation. In: Proceedings of the 43rd International ACM SIGIR Conference on Research and Development in Information Retrieval, pp. 639–648 (2020)
13. Rendle, S., Freudenthaler, C., Gantner, Z., Schmidt-Thieme, L.: BPR: Bayesian personalized ranking from implicit feedback. ArXiv Preprint ArXiv:1205.2618 (2012)
14. Wang, X., Liu, N., Han, H., Shi, C.: Self-supervised heterogeneous graph neural network with co-contrastive learning. In: Proceedings Of The 27th ACM SIGKDD Conference On Knowledge Discovery & Data Mining. pp. 1726–1736 (2021)
15. Yi, Z., Wang, X., Ounis, I., Macdonald, C.: Multi-modal graph contrastive learning for micro-video recommendation. In: Proceedings of The 45th International ACM SIGIR Conference on Research and Development in Information Retrieval, pp. 1807–1811 (2022)
16. Wang, H., Zhang, F., Wang, J., Zhao, M., Li, W., Xie, X., Guo, M.: RippleNet: propagating user preferences on the knowledge graph for recommender systems. In: Proceedings of the 27th ACM International Conference on Information and Knowledge Management, pp. 417–426 (2018)
17. Zhang, F., Yuan, N., Lian, D., Xie, X., Ma, W.: Collaborative knowledge base embedding for recommender systems. In: Proceedings of the 22nd ACM SIGKDD International Conference on Knowledge Discovery and Data Mining, pp. 353–362 (2016)
18. Yu, X., et al.: Personalized entity recommendation: a heterogeneous information network approach. In: Proceedings of the 7th ACM International Conference on Web Search and Data Mining, pp. 283–292 (2014)

19. Wang, H., Zhao, M., Xie, X., Li, W., Guo, M.: Knowledge graph convolutional networks for recommender systems. In: The World Wide Web Conference, pp. 3307–3313 (2019)
20. Wang, H., et al.: Knowledge-aware graph neural networks with label smoothness regularization for recommender systems. In: Proceedings Of The 25th ACM SIGKDD International Conference On Knowledge Discovery & Data Mining, pp. 968–977 (2019)
21. Wang, X., He, X., Cao, Y., Liu, M., Chua, T.: KGAT: knowledge graph attention network for recommendation. In: Proceedings Of The 25th ACM SIGKDD International Conference On Knowledge Discovery & Data Mining, pp. 950–958 (2019)
22. Wang, X., et al.: Learning intents behind interactions with knowledge graph for recommendation. Proc. Web Conf. **2021**, 878–887 (2021)
23. Chen, Y., Yang, Y., Wang, Y., Bai, J., Song, X., King, I.: Attentive knowledge-aware graph convolutional networks with collaborative guidance for personalized recommendation. In: 2022 IEEE 38th International Conference On Data Engineering (ICDE), pp. 299–311 (2022)
24. Wang, H., Zhou, B., Zhang, L., Ma, H.: Recommendation method for contrastive enhancement of neighborhood information. Comput. Mater. Continua. **78**, 453–472 (2024)
25. Yan, F., Xu, X., Zhao, R., Sun, S., Ju, S.: Knowledge-aware recommender system with cross-views contrastive learning. Adv. Eng. Sci. **56**, 44–53 (2024)

Popularity-Aware Graph Neural Network with Global Context for Session-Based Recommendation

Xiangwei Zeng[1], Chao Chang[2,3(✉)], Feiyi Tang[2,3], Zhengyang Wu[1,2], and Yong Tang[1,2]

[1] School of Computer, South China Normal University, Guangzhou 510631, China
[2] Pazhou Lab, Guangzhou 510335, China
`changchao@m.scnu.edu.cn`
[3] School of Information Engineering, Guangzhou Panyu Polytechnic, Guangzhou 511483, China

Abstract. Session-based recommendation aims to predict the next interaction in an anonymous user's sequence and has gained significant attention. Most existing systems model user preferences from the current session using graph neural networks but overlook the varying importance of items with different popularity. To address this, we propose the Popularity-aware Graph Neural Network with Global Context (PGNN-GC), which models popularity features to better capture users' diverse preferences. By explicitly modeling popularity-aware embeddings and using attention mechanisms, PGNN-GC differentiates user preferences for items of varying popularity. Additionally, we enhance representations using a contrastive learning paradigm. Experiments on three open datasets show that PGNN-GC achieves state-of-the-art performance.

Keywords: Session-based recommendation · Graph neural network · Popularity-aware · Contrastive learning

1 Introduction

Session-based recommendation systems (SBRSs) have gained attention for their ability to capture user preferences within a specific session and provide real-time recommendations without relying on a user's entire history. While SBRSs effectively address data sparsity and cold start issues by modeling dynamic interest transitions, they often overlook the importance of item popularity. Graph neural networks (GNNs) have been extensively applied in SBRSs, demonstrating strong performance in modeling unstructured data and complex item relationships [3]. However, current GNN-based approaches fail to capture user preferences for items with varying popularity levels, limiting prediction accuracy [13]. Explicitly encoding item popularity as a vector is crucial for enhancing prediction models by differentiating items based on their popularity.

In this work, we propose a novel model names Popularity-aware Graph Neural Network with Global Context for Session-based Recommendation (PGNN-GC),

C. Jin et al. (Eds.): WISA 2024, LNCS 14883, pp. 163–171, 2024.
https://doi.org/10.1007/978-981-97-7707-5_14

which model and leverage the popularity of items by embedding their number of occurrences. In addition, to further improve item representations, we designed a contrastive learning loss to align similar items and separate dissimilar items.

2   Related Work

SBRSs aim to recommend items by a sequence of items a user interacts with. Recently, GNN-based methods have become the dominant approach for session-based recommendation systems [10]. Stem from the ability of graph neural networks to model non-structured data, GNN-based methods are able to capture hidden transition patterns in session sequences and high-order item transitions in the global graph. GCE-GNN [9] not only captures session-level item representations, but also proposed a novel global graph to learn the global-level item representations. DHCN [12] proposed a dual channel hypergraph convolutional network to model the complex high-order relationships between items and enhance the modeling by a contrastive learning schema. More recently, HyperS2Rec [1] combines hypergraph convolutional network and GRU to extract item embeddings from both consistency and sequential dependence between items.

3   Method

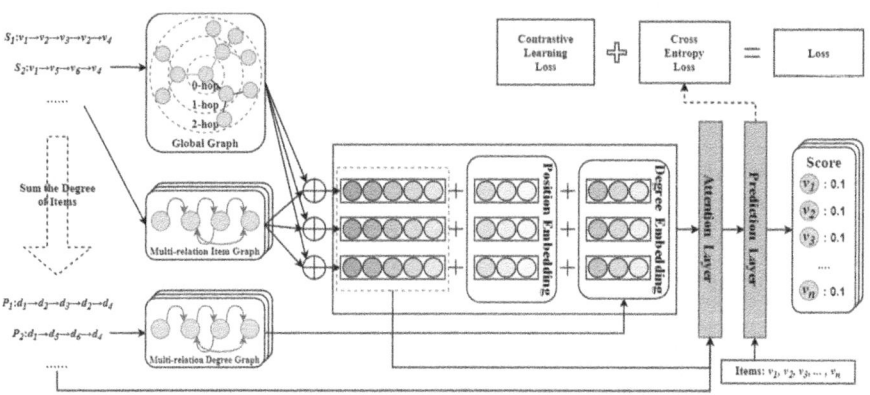

Fig. 1. An overview of our proposed PGNN-GC framework

In this section, the detail of the proposed model Popularity-aware Graph Neural Network with Global Context for Session-based Recommendation (PGNN-GC) will be presented. Figure 1 presents the architecture of PGNN-GC.

3.1   Global-Level Item Representation Learning Layer

For items, different neighbors should be considered of different importance during the message propagation. Adhering to this concept, we perform a weighted linear combination of neighbor items for an item to obtain its first-order neighbor features:

$$h_{\mathcal{N}_{v_i}^g} = \sum_{v_j \in \mathcal{N}_{v_i}^g} \pi(v_i, v_j) h_{v_j} \tag{1}$$

where $\pi(v_i, v_j)$ represents the importance weights of different neighbors of item v_i and h_j is the representation of v_j. Items that are closer to the current session should be considered more important. Therefore, $\pi(v_i, v_j)$ are estimated as follows:

$$\pi(v_i, v_j) = q_1^T \text{LeakyRelu}(W_1[(s \odot h_{v_j}) \parallel w_{ij}]) \tag{2}$$

where q_1 and W_1 are trainable parameters, and w_{ij} is the weight of edge (v_i, v_j) in global graph. s donates the features of current session, and we obtain it by averaging the items within the session:

$$s = \frac{1}{|S|} \sum_{v_i \in S} h_{v_i} \tag{3}$$

To enable comparability between the coefficients of different neighbors of v_i, we normalized the coefficients of all its neighbors using the softmax function:

$$\pi(v_i, v_j) = \frac{\exp(\pi(v_i, v_j))}{\sum_{v_k \in \mathcal{N}_{v_i}^g} \exp(\pi(v_i, v_k))} \tag{4}$$

Ultimately, the item representation h_v and the neighbor representation $h_{\mathcal{N}_v^g}$ are aggregated to obtain the global-level representation of items:

$$h_v^g = \text{relu}(W_2[h_v \parallel h_{\mathcal{N}_v^g}]) \tag{5}$$

where W_2 is trainable parameter and relu is chosen activation function.

3.2   Local-Level Item Representation Learning Layer

The session graph captures item transition patterns within the session, we differentiate the importance of these patterns by assessing the similarity between the current item and its various neighbors:

$$e_{ij} = \text{LeakyRelu}(a_{r_{ij}}^T(h_{v_i} \odot h_{v_j})) \tag{6}$$

where e_{ij} represents the importance weight of item v_j to item v_i, and $a_{r_{ij}}$ are trainable parameters for different item relationships.

Then we normalize the importance weight across all items within the current session applying the softmax function:

$$\alpha_{ij} = \frac{\exp(\text{LeakyRelu}(a_{r_{ij}}^T(h_{v_i} \odot h_{v_j})))}{\sum_{v_k \in \mathcal{N}_{v_i}^s} \exp(\text{LeakyRelu}(a_{r_{ik}}^T(h_{v_i} \odot h_{v_k})))} \tag{7}$$

By performing a linear combination of the items within the current session, we obtain the local-level representation of item v_i.

$$h_{v_i}^s = \sum_{v_j \in \mathcal{N}_{v_i}^s} \alpha_{ij} h_{v_j} \tag{8}$$

3.3   Session-Level Degree Representation Learning Layer

Mirroring local item transitions, we extract local-level degree representation within the current session as follows:

$$e_{ij} = \text{LeakyRelu}(a_{r_{ij}}^{T}(\text{d}_{v_i} \odot \text{d}_{v_j})) \tag{9}$$

$$\alpha_{ij} = \frac{\exp(\text{LeakyRelu}(a_{r_{ij}}^{T}(\text{d}_{v_i} \odot \text{d}_{v_j})))}{\sum_{v_k \in \mathcal{N}_{v_i}^s} \exp(\text{LeakyRelu}(a_{r_{ik}}^{T}(\text{d}_{v_i} \odot \text{d}_{v_k})))} \tag{10}$$

$$\text{d}_{v_i}^s = \sum_{v_j \in \mathcal{N}_{v_i}^s} \alpha_{ij} \text{d}_{v_j} \tag{11}$$

where $d_{v_i}^s$ is the embedding of the number of occurrences of item v_i across all sessions.

3.4   Session Representation Learning Layer

After extracting item representations within contexts, we design a soft attention mechanism to fuse these three representations as the final item representation. Specifically, we first combine global-level and local-level item representations to obtain item representation with two-level information by sum pooling:

$$\text{h}_v^{g,(k)} = \text{dropout}(\text{h}_v^{g,(k)}) \tag{12}$$

$$\text{h}_v' = \text{h}_v^{g,(k)} + \text{h}_v^s \tag{13}$$

Then we introduce a soft attention mechanism to aggregate the item representation with two-level information, its reversed position embedding and its session-level degree representation,

$$z_i = \tanh(\text{W}_3[h_{v_i^s}' \parallel \text{p}_{l-i+1} \parallel d_{v_i^s}^s] + b_3) \tag{14}$$

where W_3, b_3 is trainable parameters, and $\tanh(\cdot)$ is chosen activation function.

The session information of current session are calculated by two-level representations and degree representation of items within current session:

$$s' = \frac{1}{l} \sum_{i=1}^{l} (h_{v_i^s}' + d_{v_i^s}) \tag{15}$$

The attention weights of items within current session are calculated as follows:

$$\beta_i = \text{q}_2^{T}\sigma(\text{W}_4 z_i + \text{W}_5 s' + \text{b}_4) \tag{16}$$

where q_2, W_4, W_5, b_4 are trainable parameters, and $\sigma(\cdot)$ is chosen activation function.

Finally, we obtain the final session representation of current session:

$$\text{S} = \sum_{i=1}^{l} \beta_i h_{v_i^s}' \tag{17}$$

3.5   Contrastive Learning Layer

Current work typically applies contrastive learning to distinguish between local and global-level session representations, but often overlooks the contrast between items and their neighbors. To address this, we introduce a neighbor contrastive learning loss, which is formulated to enhance the precision of item representation by considering neighbor relationships, drawing items closer to their neighbors and pushing non-neighbors further apart in the embedding space. Specifically, the loss are formulated as:

$$\mathcal{L}_{con} = -\log\frac{\sum_S \sum_{v_i,v_j \in S, v_k \in \mathcal{N}^g_{v_j}} \exp(\mathrm{sim}(h_{v_i}, h_{v_k}))}{\sum_{v_i,v_a \in V} \exp(\mathrm{sim}(h_{v_i}, h_{v_a}))} \tag{18}$$

where $\mathrm{sim}(\cdot)$ denotes dot product, $\mathcal{N}^g_{v_j}$ are the neighbors of v_j sampling from the global graph, S are whole sessions in a batch.

3.6   Prediction Layer

After we obtain the final session representation S, the final recommendation probability $\hat{y}_i$ for each item v_i are calculated as the dot product of its embedding h_{v_i} and session representation S:

$$\hat{y}_i = \mathrm{Softmax}(S^T h_{v_i}) \tag{19}$$

The cross-entropy loss function is defined as follows:

$$\mathcal{L}_c = -\sum_{i=1}^{m} y_i \log(\hat{y}_i) + (1 - y_i)\log(1 - \hat{y}_i) \tag{20}$$

where y is the one-hot encoding vector of the ground truth item.

Ultimately, we combine the recommendation task and the contrastive task by calculating total loss as follows:

$$\mathcal{L} = \mathcal{L}_c + \lambda\mathcal{L}_{con} \tag{21}$$

where $\mathcal{L}_{con}$ is contrastive loss defined in Eqs. 18, $\mathcal{L}_{con}$ controls the magnitude of the contrastive learning task.

4   Experiments

In this section, to evaluate the performance of our proposed model, we conduct extensive experiments in real world datasets.

Table 1. Statistics of datasets used in the experiments

Dataset	Tmall	Nowplaying	Last.FM
# click	818,479	1,367,963	3,835,706
# train	351,268	825,304	2,837,330
# test	25,898	89,824	672,833
# items	40,728	60,417	38,615
avg. len.	6.69	7.42	11.78

4.1   Datesets and Preprocessing

We evaluate our proposed model on three widely used real-world datasets, including *Tmall*, *Nowplaying* and *Last.FM*. For each dataset, the first 80% of the data is treated as the training set. We filtered sessions containing only one interaction and items appearing less than 5 times. After then, we split a session into sub-sessions. For instance, giving a session sequence $s = [v_1, v_2, v_3, \ldots, v_n]$, sub-sessions and corresponding labels $([v_1], v_2), ([v_1, v_2], v_3), \ldots, ([v_1, \ldots, v_{n-1}], v_n)$ will be generated. Statistics for the three datasets after preprocessing are summarized in Table 1.

4.2   Baseline Algorithms

We compare our proposed model with nine representative SBRSs categorized into three types. Traditional SBRSs, represented by item-KNN [8] and FPMC [7], identify similar items based on features, interactions, or by integrating matrix factorization with Markov chains. Early neural network based SBRSs, including GRU4Rec [2], NARM [4], and STAMP [5], capture user preferences and sequential patterns using techniques like recurrent neural networks and attention mechanisms. GNN-based SBRSs, such as SR-GNN [11], FGNN [6], GCE-GNN [9], and HyperS2Rec [1], leverage graph neural networks to model item relationships, capturing both short-term and long-term user interests within sessions through various graph structures and attention mechanisms.

4.3   Overall Comparison

To assess the effectiveness of PGNN-GC, we compare its performance with several established baselines. As evident in Table 2, PGNN-GC outperforms all the baseline methods, except for the P@20 metric on Nowplaying.

Among all the traditional session-based recommendation approaches, FPMC and Item-KNN are two of the most representing works. However, convolutional methods have difficulty capturing the sequential dependencies within sessions. Early neural network based methods (i.e. GRU4Rec, NARM and STAMP) leverage the GRU architecture to model user preferences based on behavior sequences and achieve better performance. However, these methods rely solely on explicit

Table 2. The performance of PGNN-GC with other baseline methods over three datasets

Dataset	Nowplaying		Tmall		Last.FM	
	P@20	MRR@20	P@20	MRR@20	P@20	MRR@20
Item-KNN	15.94	4.91	9.15	3.31	9.89	3.24
FPMC	7.36	2.82	16.06	7.32	14.89	4.16
GRU4Rec	7.92	4.48	10.93	5.89	18.89	6.20
NARM	18.59	6.93	23.30	10.70	22.00	7.46
STAMP	17.66	6.88	26.47	13.36	21.53	7.77
SR-GNN	17.9618	7.6007	27.1565	12.9175	23.2973	8.3015
FGNN	18.78	7.15	25.24	10.39	22.61	7.67
GCE-GNN	22.425	<u>8.5261</u>	<u>32.4504</u>	15.1602	<u>24.2537</u>	<u>8.6206</u>
HyperS2Rec	**22.9933**	7.9985	29.6409	<u>16.1197</u>	22.4368	7.0647
PGNN-GC	<u>22.7667</u>	**8.6937**	**37.3697**	**16.9348**	**24.6903**	**8.846**
Improv. (%)	−0.9855	1.9657	15.1594	5.0565	1.8001	2.6147

user interactions and overlook implicit item transitions within sessions. While GNN-based methods (i.e. SR-GNN, FGNN, GCE-GNN and HyperS2Rec) are capable of modeling unstructured data and capturing implicit item transition patterns in session-based data, addressing limitations observed in RNN-based approaches, outperforming other approaches in comparison. However, the above works neglect to utilize the valuable information contained in the popularity of items, which limits the ability of the model to capture users' preferences towards items of different popularity, ultimately affecting the accuracy of predictions.

Compared with the above models, our proposed PGNN-GC constructs degree embedding to capture the popularity feature within the sessions and items, and improve the item representations by applying contrastive learning loss. As a result, our proposed model outperforms them on all datasets, especially on the *Tmall* dataset, where it achieves a 15.16% improvement in *P@20*, and a 5.06% improvement in *MRR@20*.

5   Conclusion

In this study, we propose a Popularity-aware Graph Neural Network with Global Context for Session-based Recommendation (PGNN-GC) for session-based recommendation, which captures the popularity semantic within items and applies the contrastive learning paradigm to improve the item representations. Extensive experiments on PGNN-GC demonstrate its decent recommendation performance on three benchmark datasets.

Training this model is computationally expensive due to the large item graphs and extensive data volume, especially during the contrastive learning. In the future, we aim to optimize network design and learning methods for faster training.

Acknowledgement. This work is supported in part by the National Natural Science Foundation of China under Grant 62377015, and the Collaborative Innovation Center for Intelligent Educational Technology of Guangzhou under grant 2023B04J0002.

References

1. Ding, C., Zhao, Z., Li, C., Yu, Y., Zeng, Q.: Session-based recommendation with hypergraph convolutional networks and sequential information embeddings. Expert Syst. Appl. **223**, 119875 (2023)
2. Hidasi, B., Karatzoglou, A., Baltrunas, L., Tikk, D.: Session-based recommendations with recurrent neural networks. arXiv preprint arXiv:1511.06939 (2015)
3. Huang, X., Kou, Y., Shen, D., Nie, T., Li, D.: Exploiting item relationships with dual-channel attention networks for session-based recommendation. In: Yuan, L., Yang, S., Li, R., Kanoulas, E., Zhao, X. (eds.) Web Information Systems and Applications, WISA 2023, LNCS, vol. 14094, pp. 198–205. Springer, Singapore (2023). https://doi.org/10.1007/978-981-99-6222-8_17
4. Li, J., Ren, P., Chen, Z., Ren, Z., Lian, T., Ma, J.: Neural attentive session-based recommendation. In: Proceedings of the 2017 ACM on Conference on Information and Knowledge Management, pp. 1419–1428 (2017)
5. Liu, Q., Zeng, Y., Mokhosi, R., Zhang, H.: Stamp: short-term attention/memory priority model for session-based recommendation. In: Proceedings of the 24th ACM SIGKDD International Conference on Knowledge Discovery & Data Mining, pp. 1831–1839 (2018)
6. Qiu, R., Li, J., Huang, Z., Yin, H.: Rethinking the item order in session-based recommendation with graph neural networks. In: Proceedings of the 28th ACM International Conference on Information and Knowledge Management, pp. 579–588 (2019)
7. Rendle, S., Freudenthaler, C., Schmidt-Thieme, L.: Factorizing personalized markov chains for next-basket recommendation. In: Proceedings of the 19th International Conference on World Wide Web, pp. 811–820 (2010)
8. Sarwar, B., Karypis, G., Konstan, J., Riedl, J.: Item-based collaborative filtering recommendation algorithms. In: Proceedings of the 10th international conference on World Wide Web, pp. 285–295 (2001)
9. Wang, Z., Wei, W., Cong, G., Li, X.L., Mao, X.L., Qiu, M.: Global context enhanced graph neural networks for session-based recommendation. In: Proceedings of the 43rd International ACM SIGIR Conference on Research and Development in Information Retrieval, pp. 169–178 (2020)
10. Wang, Z., et al.: Exploring global information for session-based recommendation. Pattern Recogn. **145**, 109911 (2024)
11. Wu, S., Tang, Y., Zhu, Y., Wang, L., Xie, X., Tan, T.: Session-based recommendation with graph neural networks. In: Proceedings of the AAAI Conference on Artificial Intelligence, vol. 33, pp. 346–353 (2019)

12. Xia, X., Yin, H., Yu, J., Wang, Q., Cui, L., Zhang, X.: Self-supervised hypergraph convolutional networks for session-based recommendation. In: Proceedings of the AAAI Conference on Artificial Intelligence, vol. 35, pp. 4503–4511 (2021)
13. Zeyu, H., Yan, L., Wendi, F., Wei, Z., Alenezi, F., Tiwari, P.: Causal embedding of user interest and conformity for long-tail session-based recommendations. Inf. Sci. **644**, 119167 (2023)

Enhancing Sentiment Analysis for Chinese Texts Using a BERT-Based Model with a Custom Attention Mechanism

Linlin Ding, Yiming Han, Mo Li$^{(\boxtimes)}$, and Dong Li

School of Information, Liaoning University, Shenyang 110036, China
{dinglinlin,limo,lidong}@lnu.edu.cn, hanyiming135@gmail.com

Abstract. The rise of social media has made automatic emotion recognition from extensive texts crucial in NLP(Natural language processing). Traditional sentiment analysis focuses on basic positive or negative sentiments, neglecting the broader spectrum of emotional complexity. Our innovative model addresses this by incorporating a pretrained BERT (Bidirectional Encoder Representations from Transformers) language model with an additional custom attention mechanism. This mechanism dynamically adjusts encoder layer weights to distinguish between semantically similar but emotionally distinct expressions, such as "anger" versus "sadness" or "happiness" versus "surprise." This approach enhances the model's sensitivity to emotional boundaries, allowing for more accurate identification and classification of complex emotions. Experimental results on two six-emotion datasets demonstrate superior performance in precision, recall, and F1-score compared to traditional models.

Keywords: Emotion Classification · Social Media Analysis · BERT

1 Introdction

With the rise of internet technology, platforms like Weibo have become crucial for daily expression, generating over 100 million messages daily. Understanding the emotional content in these messages is essential for gauging individual and social states [11]. Unlike simple positive or negative sentiments, emotions span a wide range, from shock to sadness, offering valuable insights for public policy and crisis management [6]. Multi-emotion analysis provides a deeper understanding of user emotions compared to traditional polarity analysis. However, analyzing emotions in Chinese texts is challenging due to their nuanced and context-dependent nature, complicating accurate classification [2]. Recent deep learning techniques like RNN [1], LSTM [13], Bi-LSTM [3], and CNN [12] have enhanced sentiment analysis by learning hierarchical features and improving classification accuracy. However, these models require substantial data, are prone to overfitting, and lack transparency. BERT [5] and its variants [8,10]

C. Jin et al. (Eds.): WISA 2024, LNCS 14883, pp. 172–179, 2024.
https://doi.org/10.1007/978-981-97-7707-5_15

have revolutionized sentiment analysis with robust context-capturing capabilities, achieving state-of-the-art results but requiring significant computational resources and expertise. Fine-grained multi-class sentiment analysis in Chinese faces challenges, especially in categorizing specific emotions like anger, joy, sadness, and surprise. Chinese sentiments are nuanced, leading models to overfit on common sentiments and neglect rare expressions. Chinese expressions are context-dependent, requiring precise context capture. The rapid evolution of Chinese internet language poses challenges for sentiment analysis models to adapt and recognize new expressions. This paper explores a multi-category Chinese sentiment analysis approach using the BERT model, enhanced with a custom attention mechanism for better classification of complex emotions. The main contributions of this work are as follows:

- We have enhanced our sentiment analysis model by integrating a pre-trained BERT model with a custom attention mechanism. This innovative approach improves the model's ability to identify and process words crucial to sentiment judgment, enhancing performance on complex textual data.
- Our model is tailored for Chinese texts, leveraging BERT's strong language understanding to adapt to Chinese linguistic features and semantic complexity. This adaptability extends its effectiveness beyond standard sentiment analysis tasks to handling emotional nuances in informal texts like social media.
- Experimental comparisons show that our model surpasses several baseline models on standard datasets (including SMP2020 and NLPCC) in accuracy, precision, recall, and F1-score.

2   The Proposed Model

The proposed model (Fig. 1) outlines a structured approach from preprocessing input text to predicting emotion categories, transforming raw text into a format suitable for BERT-based sentiment analysis. This process involves several key stages: preprocessing, feature extraction and capture, custom attention mechanisms, and loss analysis. Each stage enhances the model's ability to handle the complexities of Chinese text and perform accurate multi-category emotion classification.

2.1   Preprocessing and Feature Capture

Data preprocessing is crucial for effective sentiment analysis, as it directly affects the model's performance. In this study, we convert Chinese text into a format suitable for BERT, preserving semantic integrity. Using the pretrained BERT-Tokenizer, we tokenize text and encode it into input ids, attention masks, and token type ids via the encode plus method.

Feature extraction and capture are achieved through the BERT model, which is responsible for transforming text data into deep semantic features. The input

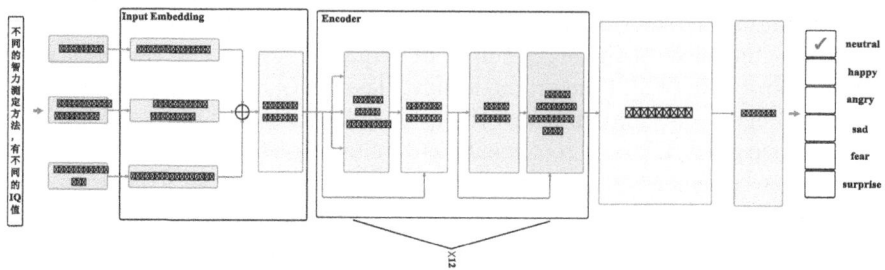

Fig. 1. Overview of the Model

embedding consists of three parts: Token Embedding, Positional Embedding, and Segment Embedding. The overall input embedding can be represented as:

$$E = E_{token} + E_{pos} + E_{segment} \tag{1}$$

where E_{token} represents the token embedding, E_{pos} represents the positional embedding, and $E_{segment}$ represents the segment embedding. The vector sequence E, processed by embeddings, is then fed into the encoder. The encoder consists of several layers, each including two main parts: Multi-Head Self-Attention and Feed Forward Network, where Multi-Head Self-Attention is defined as:

$$MultiHead(Q, K, V) = Concat(head_1, \ldots, head_h)W^O \tag{2}$$

where each head is computed as:

$$thead_i = Attention(QW_i^Q, KW_i^K, VW_i^V) \tag{3}$$

The basic attention function is defined as:

$$Attention(Q, K, V) = softmax\left(\frac{QK^T}{\sqrt{d_k}}\right)V \tag{4}$$

where W_i^Q, W_i^K, and W_i^V are weight matrices, and d_k is the dimension of the key vectors. The output of the multi-head attention is then passed through the feed forward network:

$$FFN(x) = \max(0, xW_1 + b_1)W_2 + b_2 \tag{5}$$

Custom Attention Mechanism. The custom attention layer in the model refines encoder-extracted features by reducing their dimensionality through a linear transformation, enhancing processing efficiency. This layer acts as an information bottleneck, emphasizing the most significant emotional features. The hyperbolic tangent activation function (tanh) introduces non-linearity, capturing complex emotional feature combinations. Regularization is added through a Dropout layer, enhancing model robustness by randomly dropping features. Another linear transformation serves as an emotion-aware filter, emphasizing crucial classification information. The final emotion representation is computed by multiplying Softmax weights with feature vectors and summing them up, expressed as:

$$O = \sum \left(softmax(\mathbf{W}_2(Dropout(\tanh(\mathbf{W}_1 X)))) \odot X\right) \tag{6}$$

Through this mechanism, the model not only captures obvious emotional signals in the text, such as vividly expressed emotions, but also recognizes more subtle and implicit emotional expressions, which is crucial for accurately classifying complex emotions.

2.2 Loss Analysis

During training, we utilize CrossEntropyLoss to minimize loss and employ regularization techniques to prevent overfitting. Unlike traditional approaches that add L2 regularization to the loss function using the Adam optimizer, the AdamW optimizer integrates weight decay directly into its mechanism, decoupling it from gradient updates. This strategy is particularly effective for complex deep learning models, as it enables precise decay of specific weights, especially those in layer normalization layers. Our implementation uses AdamW with L2 norm for regularization, streamlining our code and optimizing performance. The final formula is defined as follows:

$$\mathcal{L} = -\sum_{i=1}^{C} y_i \log(p_i) \tag{7}$$

where C represents the total number of classes, y_i is the one-hot encoding of the true labels, and p_i is the probability distribution predicted by the model.

Table 1. Number of Six Types of Emotional Texts in the SMP2020 Dataset

Table 2. Number of Six Types of Emotional Texts in the NLPCC Dataset

Label	Number
neutral	5749
happy	5379
angry	4990
sad	1220
fear	8344
surprise	2086

Label	Number
null	13993
like	6697
sad	5348
disgust	5978
anger	3169
happiness	4950

3 Experiments

We evaluate our Model against 5 competitors on 2 social media datasets. All experiments are conducted on a Windows server with an Intel(R) Xeon(R) Silver 4210R CPU @ 2.40 GHz, 2.39 GHz, 128 GB RAM, and an RTX 3090 GPU.

3.1   Details of Experimental Setting

In our study, we used two social media datasets for experiments. The SMP2020 dataset includes 27,768 Weibo posts from the SMP2020 Weibo Emotion Classification Technical Evaluation. The NLPCC dataset, from Tsinghua University, contains 40,133 posts. The specific statistica data are listed in Table 1 and Table 2.

We implemented our model using the PyTorch framework with a dropout rate of 0.1, batch size of 64, and learning rate of 0.0002 for all datasets. The experimental data were randomly split into training, validation, and testing sets in a 7:2:1 ratio. For the SMP2020 dataset, the text length distribution and proportion of each emotion category are illustrated in Fig. 2. To cover 90% of the data, we chose a maximum length of 87 based on a histogram. For the NLPCC dataset, the text length distribution and proportion of each emotion category are depicted in Fig. 3. To cover 90% of the data, we chose a maximum length of 64 based on a histogram.

For multi-class sentiment analysis, we tested various models including SVM, Bi-LSTM [3], TextCNN (Text Convolutional Neural Network) [7], CNN-LSTM (Convolutional Long Short-Term Memory networks) [12], DPCNN [4], BERT-BiLSTM [9], and RoBERTa-Base [10]. These models range from traditional machine learning to advanced deep learning, providing a comprehensive technical comparison to assess the effectiveness of our proposed model in complex sentiment analysis tasks.

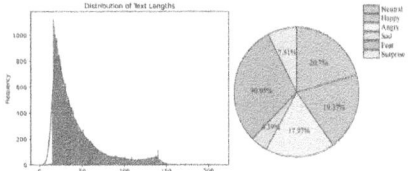 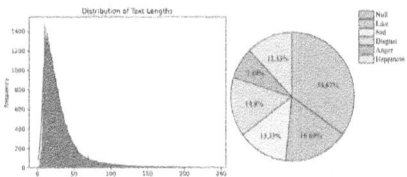

Fig. 2. Text Length Distribution(left) and Proportion of Each Emotion(right) in the SMP2020 Dataset

Fig. 3. Text Length Distribution and Proportion of Each Emotion in the NLPCC Dataset

3.2   Comparison and Ablation Study

Table 3 presents sentiment analysis results on the SMP2020 and NLPCC datasets. Our model outperforms other comparative models across all metrics. Specifically, it achieves 77.4% accuracy (Acc), 74.9% precision (P), 74.0% recall, and 74.3% F1-score, representing a significant improvement over baseline SVM and advanced models like BiLSTM, TextCNN, CNN-LSTM, RoBERTa-Base, DPCNN and BERT-BiLSTM. Compared to RoBERTa-Base, our model shows a 10% point increase in accuracy, highlighting the custom attention mechanism's importance in recognizing sentiments in Chinese texts. On the NLPCC dataset, our model achieves 57.3% accuracy, 55.2% precision, 54.3% recall, and 54.6% F1-score, surpassing all other models. The dataset's inherent characteristics or sample distribution may explain the lower performance compared to the SMP2020 dataset, indicating the dataset's complexity.

Table 3. Performance Results of Different Models on the SMP2020 and NLPCC Datasets

Model	SMP2020 Dataset				NLPCC Dataset			
	Acc	P	Recall	F1	Acc	P	Recall	F1
SVM	35.1%	0.488	0.351	0.249	48.7%	0.492	0.487	0.488
Bi-LSTM	70.0%	0.660	0.665	0.662	50.6%	0.474	0.478	0.475
TextCNN	69.6%	0.679	0.648	0.661	51.0%	0.482	0.454	0.464
CNN-LSTM	65.8%	0.625	0.642	0.630	49.6%	0.460	0.449	0.453
RoBERTa-Base	67.8%	0.638	0.639	0.637	52.4%	0.489	0.462	0.468
DPCNN	61.31%	0.605	0.598	0.606	46.20%	0.460	0.459	0.459
BERT-BiLSTM	71.72%	0.665	0.703	0.683	51.2%	0.491	0.511	0.501
Our Model	**77.4%**	**0.749**	**0.740**	**0.743**	**57.3%**	**0.552**	**0.543**	**0.546**

The custom attention mechanism significantly improves the BERT model's performance in Chinese sentiment analysis.Compared to traditional models like BERT-BiLSTM and TextCNN, our model shows superior adaptability and robustness, particularly with large datasets containing complex semantics and structures.

We conducted ablation experiments on the attention layer, dropout rate, and optimizer choice, as detailed in Tables 4 and 5. In the SMP2020 dataset, our model excelled in the "sad" (1220) category, achieving high precision and F1-score due to its comprehensive structure that effectively learns sparse features. For the largest "fear" (8344) category, simpler models without attention layers or with minimal dropout performed adequately, indicating that basic BERT suffices for larger datasets. In the NLPCC dataset, the model with a lower dropout rate performed best in the smallest "anger" (3169) category, maintaining high precision and F1-score, suggesting that less dropout helps preserve information in larger datasets. However, models performed poorly in the largest "null" (13993) category, especially those using the SGD optimizer, highlighting SGD's limitations with complex models and large sample sizes.

Table 4. Results of Ablation Experiments on the SMP2020 Dataset.

Model	neutral		happy		angry		sad		fear		surprise	
	P	F1	P	F1	P	F1	P	F1	P	F1	P	F1
no Attention	0.832	0.830	**0.771**	**0.786**	0.618	0.628	0.621	0.629	0.832	0.815	**0.638**	**0.628**
Dropout(0.1)	0.814	0.823	0.713	0.767	0.666	0.633	0.667	0.585	**0.857**	0.808	0.543	0.610
Dropout(0.5)	0.829	0.831	0.753	0.771	0.661	0.643	0.627	0.599	0.820	0.817	0.590	0.612
Optimizer_sgd	**0.872**	0.834	0.626	0.545	0.556	0.293	0.000	0.000	0.458	0.615	0.000	0.000
Ours Model	0.867	**0.873**	0.749	0.768	**0.721**	**0.668**	**0.750**	**0.738**	0.841	**0.842**	0.568	0.615

Table 5. Results of Ablation Experiments on the NLPCC Dataset.

Model	null		like		sad		disgust		anger		happiness	
	P	F1	P	F1	P	F1	P	F1	P	F1	P	F1
no Attention	0.635	**0.656**	0.535	0.578	0.548	0.530	**0.485**	0.423	**0.473**	**0.447**	0.619	**0.612**
Dropout(0.1)	**0.683**	0.615	0.495	0.577	0.510	**0.545**	0.482	0.447	0.444	0.433	0.574	0.581
Dropout(0.5)	0.640	0.651	**0.551**	0.567	**0.570**	0.524	0.455	**0.451**	0.453	0.434	0.590	0.605
Optimizer_sgd	0.372	0.540	0.464	0.217	0.516	0.029	0.177	0.010	0.000	0.000	**0.833**	0.112
Ours Model	0.641	0.655	0.531	**0.586**	0.531	0.503	0.444	0.382	0.421	0.433	0.615	0.594

4   Conclusions

This study developed a Chinese sentiment classification model based on BERT with a custom attention mechanism, effectively capturing complex emotional features. Tested on the SMP2020 and NLPCC datasets, the model accurately classified emotions like anger, sadness, fear, and surprise. The custom attention layer emphasized emotion-related features, enhancing classification. Preprocessing and encoding ensured consistent feature extraction. Future work will refine the model, explore deeper emotional categories, and consider broader contextual influences for increased precision. Applying this model to real-world social media platforms could aid businesses and organizations in understanding user-generated content.

Acknowledgement. This study was funded by the National Natural Science Foundation of China (No.62072220); Natural Science Foundation of Liaoning Province (2022-KF-13-06). Natural Science Foundation of Liaoning Province (Nos.2022-KF-13-06, 2022-BS-111); Liaoning Provincial Department of Education Youth Project (No. JYTQN2023189).

References

1. Bai, Q., Zhou, J., He, L.: PG-RNN: using position-gated recurrent neural networks for aspect-based sentiment classification. J. Supercomput. **78**(3), 4073–4094 (2022)
2. Deng, H., Li, Y., Ju, S., Liu, M.: Combines contrastive learning and primary capsule encoder for target sentiment classification. In: Yuan, L., Yang, S., Li, R., Kanoulas, E., Zhao, X. (eds) Web Information Systems and Applications, WISA 2023, LNCS,

vol. 14094, pp. 284–296. Springer, Singapore (2023). https://doi.org/10.1007/978-981-99-6222-8_24

3. Gan, C., Feng, Q., Zhang, Z.: Scalable multi-channel dilated CNN-BiLSTM model with attention mechanism for Chinese textual sentiment analysis. Futur. Gener. Comput. Syst. **118**, 297–309 (2021)

4. Johnson, R., Zhang, T.: Deep pyramid convolutional neural networks for text categorization. In: Proceedings of the 55th Annual Meeting of the Association for Computational Linguistics (Volume 1: Long Papers), pp. 562–570 (2017)

5. Kenton, J.D.M.W.C., Toutanova, L.K.: Bert: Pre-training of deep bidirectional transformers for language understanding. In: Proceedings of naacL-HLT, vol. 1, p. 2 (2019)

6. Khan, A.: Improved multi-lingual sentiment analysis and recognition using deep learning. J. Inf. Sci. (2023). https://doi.org/10.1177/01655515221137270

7. Kim, Y.: Convolutional neural networks for sentence classification. In: Moschitti, A., Pang, B., Daelemans, W. (eds.) Proceedings of the 2014 Conference on Empirical Methods in Natural Language Processing (EMNLP), pp. 1746–1751. Association for Computational Linguistics, Doha, Qatar, October 2014. https://doi.org/10.3115/v1/D14-1181, https://aclanthology.org/D14-1181

8. Lan, Z., Chen, M., Goodman, S., Gimpel, K., Sharma, P., Soricut, R.: Albert: a lite bert for self-supervised learning of language representations. arXiv preprint arXiv:1909.11942 (2019)

9. Li, X., Lei, Y., Ji, S.: Bert-and bilstm-based sentiment analysis of online chinese buzzwords. Future Internet **14**(11), 332 (2022)

10. Liu, Y., et al.: Roberta: a robustly optimized bert pretraining approach. arXiv preprint arXiv:1907.11692 (2019)

11. Van Duyn, E., Muddiman, A.: Emotion work on social media: differences in public and private emotions about politics and covid-19 on facebook. Soc. Media+ Soc. **9**(4), 20563051231207853 (2023)

12. Wang, J., Yu, L.C., Lai, K.R., Zhang, X.: Dimensional sentiment analysis using a regional CNN-LSTM model. In: Proceedings of the 54th Annual Meeting of the Association for Computational Linguistics (volume 2: Short papers), pp. 225–230 (2016)

13. Wang, J., Yu, L.C., Lai, K.R., Zhang, X.: Tree-structured regional CNN-LSTM model for dimensional sentiment analysis. IEEE/ACM Trans. Audio Speech Lang. Process. **28**, 581–591 (2019)

Contrastive Learning-Based Cross-Domain Data Augmentation for Aspect-Based Sentiment Analysis

Xiaoling Xue, Bin Xu$^{(\boxtimes)}$, Xiaodi Dong, Qihang Cai, and Kening Gao

Northeastern University, Shenyang 110169, China
xubin@mail.neu.edu.cn

Abstract. Cross-domain Aspect-Based Sentiment Analysis (ABSA) leverages unsupervised domain adaptation techniques to transfer knowledge from a source domain, which is rich in labeled data, to the target domain which lacks labeled data. Many recent studies have attempted to address this issue by generating a large amount of labeled target domain data, and the domain adaptive model DA^{2}LM has achieved state-of-the-art results. However, training this model requires the use of target domain data, which should be annotated with pseudo labels. Therefore, it is important to generate high-quality pseudo labels effectively. Furthermore, when a substantial amount of labeled data for the target domain is obtained using the generative model, it becomes essential to train an effective model to predict the label of test data from the target domain. In this work, we propose a novel Cross-Domain Data Augmentation approach based on Contrastive Learning, named CLCDDA. This approach combines the Multiple Kernel Maximum Mean Discrepancy and enhanced contrastive learning to obtain high-quality target domain data with pseudo labels, which is used to improve the performance of the generative model. In addition, contrastive learning is introduced to improve the performance of the sequence labeling model. The experimental results on four benchmarks show that CLCDDA significantly outperforms previous approaches in both cross-domain End2End ABSA and AE tasks.

Keywords: Cross-Domain · Aspect-Based Sentiment Analysis · Contrastive Learning

1 Introduction

Aspect-based Sentiment Analysis (ABSA) [1] [2] comprises Aspect Extraction (AE) and Aspect Sentiment Classification (ASC) [3]. A recent mainstream paradigm is to jointly extract aspects and determine their corresponding sentiment polarity in an end-to-end way [4]. Although supervised methods perform well, they are difficult to generalize due to the lack of labeled data in new domains. To address this issue, many studies use unsupervised domain adaptation to improve sentiment analysis on unlabeled domains. The latest research is DA^{2}LM [5], which demonstrates the effectiveness of generative model in cross-domain data augmentation. However, to effectively train the model, pseudo-labeled data of the target domain is involved. To address this issue, we

© The Author(s), under exclusive license to Springer Nature Singapore Pte Ltd. 2024
C. Jin et al. (Eds.): WISA 2024, LNCS 14883, pp. 180–188, 2024.
https://doi.org/10.1007/978-981-97-7707-5_16

propose a Cross-Domain Data Augmentation approach based on Contrastive Learning (named CLCDDA), which improves the performance of cross-domain ABSA through three stages. In summary, the main contributions of this paper are as follows:

- We propose a new Cross-Domain Data Augmentation approach, CLCDDA, based on Contrastive Learning to address the cross-domain ABSA task.
- We combine MK-MMD with enhanced contrastive learning to generate high-quality pseudo labels for unlabeled data in the target domain.
- We integrate contrastive learning to improve the performance of the domain data sequence labeling model.
- The experimental results on four benchmarks demonstrate the significant effectiveness of CLCDDA. Compared with the state-of-the-art methods, CLCDDA achieves an average improvement of 2.32% for cross-domain End2End ABSA task and 2.82% for cross-domain AE task on the Micro-F1 score.

2   Related Work

Aspect-based Sentiment Analysis relies on supervised models for effective results. However, applying these models directly to unlabeled data domains often yields poor performance. Therefore, many studies have proposed using domain adaptation techniques to solve this problem, mainly based on the following two paradigms [6]:

Feature-based Adaptation: This paradigm enables source domain features to capture shared information in two domains for the ABSA task. However, they ignore the target domain-specific features, which are important factors for target domain sequence labeling [5] [7].

Data Augmentation-based Adaptation: In order to solve the problems existing in the above paradigm, CDRG [7] proposes to directly obtain target domain data with fine-grained annotations. DA^2LM [5] is a cross-domain data augmentation approach based on a generative model to alleviate the sentence structure limitations and sentence coherence issues of previous methods.

3   Methodology

Problem Statement: Given a labeled source domain $\mathcal{D}^S = \{(x_i^s, y_i^s)\}_{i=1}^{N^S}$, and an unlabeled target domain $\mathcal{D}^T = \{x_i^t\}_{i=1}^{N^T}$, the objective is to predict the label of test data from the target domain. Following [8], we denote a sentence containing L words as $x = \{w_1, w_2, ..., w_L\}$, and the task is to predict the labels $y = \{y_1, y_2, ..., y_L\}$ of sentence x. For AE, $y_i \in \{B, I, O\}$, for End2End ABSA, $y_i \in \{B\text{-POS}, I\text{-POS}, B\text{-NEG}, I\text{-NEG}, B\text{-NEU}, I\text{-NEU}, O\}$.

Overview: CLCDDA consists of three stage, as shown in Fig. 1. In Stage 1, $\mathcal{D}^S$ and $\mathcal{D}^T$ are used as inputs to train a model that combines MK-MMD and enhanced contrastive learning to obtain target domain data $\mathcal{D}^{PT} = \{x_i^{pt}, y_i^{pt}\}_{i=1}^{N^T}$ with pseudo labels. In Stage 2, $\mathcal{D}^S$ and $\mathcal{D}^{PT}$ are used as inputs to train a generative model to generate a target domain dataset $\mathcal{D}^G = \{(x_i^g, y_i^g)\}_{i=1}^{N^G}$ with fine-grained annotations. In Stage 3, $\mathcal{D}^G$ is used as input to train a contrastive learning-based sequence labeling model, assigning labels to the test data in the target domain.

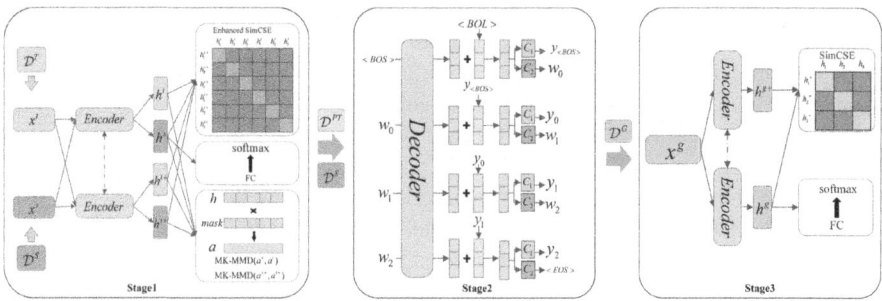

Fig. 1. Overview of Contrastive Learning-Based Cross-Domain Data Augmentation for Aspect-Based Sentiment Analysis (CLCDDA).

3.1   Stage 1: Generating Pseudo Labels for the Target Domain

We propose a novel cross-domain enhanced contrastive learning method that preserves specific discriminative information while incorporating multiple-kernel MMD [9] to mitigate the distribution discrepancy of aspect terms between the two domains. In this stage, we use the fine-tuning BERT [10] as the encoder to map x to a set of token embeddings $\{e_1, e_2, ..., e_L\} \in \mathcal{R}^{L \times d_e}$. The hidden representation of the sentence $h^d = g(e_i, ..., e_L)$ and the aspect term representation $a^d = g(e_i, ..., e_j)$, where g($\cdot$) denotes the meanpooling operation, and $d \in \{s, t\}$. For the source data $\mathcal{D}^S$, the aspect terms have been provided in the gold labels. For the unlabeled target data $\mathcal{D}^T$, we use BERT_E [11] as the base model to extract the aspect terms in $\mathcal{D}^T$, which can be viewed as the BERT version of Syntactic Bridge [12].

We apply unbiased MK-MMD [9] on aspect [5]. The objective is as follows:

$$L_{mmd} = \frac{2}{n} \sum_{i=1}^{n/2} [k(a_{2i-1}^s, a_{2i}^s) + k(a_{2i-1}^t, a_{2i}^t) - k(a_{2i-1}^s, a_{2i}^t) - k(a_{2i-1}^t, a_{2i}^s)] \tag{1}$$

where k can be expressed as the linear combination of m Gaussian kernels k_u, and k_u is defined as $\{k_u(x_i, x_j) = e^{-\|x_i - x_j\|^2 / \gamma_u}\}$, γ represents bandwidth.

Unsupervised Enhanced SimCSE. Inspired by DAEM [13], we propose a new cross-domain negative sampling strategy based on the basic contrastive learning framework SimCSE [14] to solve the cross-domain ABSA task. Each sentence from both the source and target domains is processed by the encoder twice to obtain two embeddings. For sentences $X^d = \{x_i^d\}_{i=1}^n$, two embeddings are $H^d = \{h_i^d\}_{i=1}^n$ and $H^{d+} = \{h_i^{d+}\}_{i=1}^n$. The objective is as follows:

$$l_i^s = -\log \frac{e^{sim(h_i^s, h_i^{s+})/\tau}}{\sum_{j=1}^n \left(e^{sim(h_i^s, h_j^{s+})/\tau} + e^{sim(h_i^s, h_j^{t+})/\tau}\right)} \tag{2}$$

$$l_i^t = -\log \frac{e^{sim(h_i^t, h_i^{t+})/\tau}}{\sum_{j=1}^n \left(e^{sim(h_i^t, h_j^{t+})/\tau} + e^{sim(h_i^t, h_j^{s+})/\tau}\right)} \tag{3}$$

$$L_{c_1} = \frac{1}{2n} \sum_{i=1}^n \left(l_i^s + l_i^t\right) \tag{4}$$

where n denotes a mini-batch, and $sim(h_i, h_i^+) = \frac{h_i^\top h_i^+}{\|h_i\| \cdot \|h_i^+\|}$.

Sequence Labeling. In training phase, only samples from $\mathcal{D}^S$ are used for the sequence labeling task. We use cross-entropy loss for optimization as follows:

$$L_{cls} = \sum_n \sum_{i=1}^L l(\text{softmax}(We_i + b), y_i) \tag{5}$$

where l is the cross-entropy loss function, and y_i is the gold label of the i-th token.

We calculate the distance between a^{s+} and a^{t+} in a similar way, denoted as L_{mmd}^+. Our ultimate optimization objective is to minimize the following loss function:

$$L = L_{cls} + \frac{\alpha}{2}(L_{mmd} + L_{mmd}^+) + \beta L_{c_1} \tag{6}$$

where α and β are hyperparameters.

3.2 Stage 2: Generating Target Domain Sentences

Following [5], for each sample $(x, y) \in \mathcal{D}^S \cup \mathcal{D}^{PT}$, the GPT-2 [15] model serves as a generative model to perform both token generation and sequence labeling simultaneously. For data initialization, $x = \{\langle BOS \rangle, w_0, w_1, w_2, \ldots, w_n, \langle EOS \rangle\}$, $y = \{\langle BOL \rangle, y_{\langle BOS \rangle}, y_0, y_1, y_2, \ldots, y_n, y_{\langle EOS \rangle}\}$, where $\langle BOS \rangle$ and $\langle EOS \rangle$ are the start and end marks of the sentence respectively, and $\langle BOL \rangle$ is the start mark of the label. $w_0 \in \{[\text{source}], [\text{target}]\}$. We obtain the representations of x and y as follows:

$$E^x = \text{TokenEmb}([\langle BOS \rangle, w_0, w_1, w_2, \ldots, w_n, \langle EOS \rangle]) \tag{7}$$

$$E^y = \text{LabelEmb}([\langle BOL \rangle, y_{\langle BOS \rangle}, y_0, y_1, y_2, \ldots, y_n, y_{\langle EOS \rangle}]) \tag{8}$$

$e_t = e_t^x + e_{t-1}^y$ is the feature representation at time step t, which is used to predict the next token w_{t+1} and the label of the current token y_t:

$$P(w_{t+1}|w_{\leq t}, y_{\leq t-1}) = \text{softmax}(W_w e_t + b_w) \tag{9}$$

$$P(y_t|w_{\leq t}, y_{\leq t-1}) = \text{softmax}(W_y e_t + b_y) \tag{10}$$

where W_w and W_y are trainable weight matrices, b_w and b_y are hidden unit biases.

3.3 Stage 3: Predicting Sequence Label for the Target Domain

After filtering the generated data according to the rules of [5], we obtain the final labeled dataset $\mathcal{D}^G = \{(x_i^g, y_i^g)\}_{i=1}^{N^G}$ for the target domain. We use SimCSE [14] with BERT$_E$ [11] as the encoder. For the sentence x_i^g, we pass it through the encoder twice, resulting in embeddings h_i^g and h_i^{g+} are a positive pair, h_i^g and other $\{h_j^{g+}\}_{j=1, j \neq i}^n$ in the mini-batch are negative pairs. The training objective is as follows:

$$L_c = \frac{1}{n} \sum_{i=1}^{n} (-\log \frac{e^{sim(h_i^g, h_i^{g+})/\tau}}{\sum_{j=1}^{n} e^{sim(h_i^g, h_j^{g+})/\tau}}) \tag{11}$$

The training objective for this stage is as follows:

$$L = L_{cls} + \gamma L_c \tag{12}$$

where γ is a hyperparameter.

4 Experiments

4.1 Experimental Settings

Table 1 shows information about the benchmark datasets [5]. For contrastive learning with BERT, we set attention dropout and hidden dropout to 0.3. All stages use the Adam optimizer, with a learning rate of 3e-4 for Stage 2 and 3e-5 for the other stages. Hyperparameters in Eq. 6 are $\alpha = 0.005$ and $\beta = 0.005$ for pairs R $\rightarrow$ D, R $\rightarrow$ L, S $\rightarrow$ D, and $\alpha = 0.03, \beta = 0.005$ for others. γ in Eq. 12 is set to 0.01.

Table 1. Statistics of the datasets.

Dataset	Domain	Sentences	Training	Testing
L	Laptop	3845	3045	800
R	Restaurant	6035	3877	2158
D	Device	3836	2557	1279
S	Service	2239	1492	747

4.2 Baselines and Main Results

• **Feature-based methods**
- BERT$_E$ [11]: a fine-tuned version of the pre-trained BERT model.
- BERT$_E$-DANN [10]: a method of combining BERT and adversarial learning to accomplish cross-domain ABSA.
- BERT$_E$-UDA [10]: a method that incorporates part-of-speech, dependency relations, and token-level instance weighting into the fine-tuning process of BERT.

- FMIM [16]: a method using maximum mutual information for cross-domain ABSA.
• **Data Augmentation-based methods**
- CDRG [7]: a method that includes converting the source domain data into domain-independent data and utilizing BERT to transform them into target domain data.
- DA^{2}LM [5]: a cross-domain data augmentation method based on generative models to capture the shared distribution of words and labels.

The results of cross-domain End2End ABSA and AE tasks are shown in Table 2. We speculate that the main reason is the high-quality pseudo-labeled target domain data obtained in the first stage is used to train the generative model. To verify this, we show the impact of the target domain pseudo-labels generated in DA^{2}LM and CLCDDA on the results, as shown in Fig. 2. It is worth noting that, in order to verify the quality of pseudo-labels for target domain training data, we use the gold labels when calculating the Micro-F1 score of the pseudo-labels. However, the gold labels are invisible during the entire model training process. We can observe that CLCDDA produces higher quality pseudo-labels, and the trend of the Micro-F1 score change of the sequence labling is basically consistent with the Micro-F1 score change of the pseudo-label.

Table 2. Comparison results for cross-domain End2End ABSA and AE tasks based on Micro-F1. The results marked by ★ are from [10], and the results marked by † are from [5].

End2End ABSA	Source → Target Pairs										
Model	S → R	L → R	D → R	R → S	L → S	D → S	R → L	S → L	R → D	S → D	AVG
BERT$_E$*	51.34	45.40	42.62	24.44	23.28	28.18	39.72	35.04	33.22	33.22	35.65
BERT$_E$-DANN*	50.31	47.39	42.20	28.35	26.69	28.77	38.83	34.29	33.42	37.14	36.74
BERT$_E$-UDA*	53.97	49.52	51.84	30.67	27.78	34.41	43.95	35.76	40.35	38.05	40.63
FMIM†	49.46	57.02	55.68	40.59	41.61	40.76	39.26	31.83	33.11	32.46	42.18
CDRG†	52.93	57.77	53.18	**43.07**	41.51	40.30	**44.70**	33.33	30.82	36.14	43.38
DA^{2}LM	57.90	**59.99**	58.61	36.86	36.49	38.42	44.48	34.47	40.78	38.50	44.65
Ours	**58.27**	59.67	**58.90**	36.64	**46.02**	**50.85**	39.24	**39.37**	41.42	**39.36**	**46.97**
AE	Source → Target Pairs										
Model	S → R	L → R	D → R	R → S	L → S	D → S	R → L	S → L	R → D	S → D	AVG
BERT$_E$*	57.56	50.42	45.71	26.50	25.96	30.40	44.18	41.78	35.98	35.13	39.36
BERT$_E$-DANN*	58.55	52.40	45.21	31.29	30.16	30.86	46.90	40.43	36.32	39.17	41.13
BERT$_E$-UDA*	59.07	55.24	56.40	34.21	30.68	38.25	54.00	44.25	42.40	40.83	45.53
FMIM†	57.43	68.67	61.64	47.60	51.68	49.53	50.57	39.14	36.11	35.26	49.76
CDRG†	60.20	68.63	57.51	**49.97**	51.07	43.19	55.50	39.49	34.89	38.59	49.90
DA^{2}LM	65.13	**70.15**	63.92	40.29	39.94	41.36	**57.54**	42.31	43.32	41.20	50.52
Ours	**65.34**	68.35	63.61	40.45	**51.77**	**54.99**	54.38	**49.27**	43.82	**41.41**	**53.34**

4.3 Ablation Study

To validate the effectiveness of CLCDDA, we conduct a series of ablation studies. The results are shown in Table 3. Firstly, we replace the first stage with the component DAPL of the aspect-level MMD in DA²LM, which leads to a decrease in the average F1 score. This indicates that introducing enhanced contrastive learning to increase discriminative information can effectively improve the performance of the model and assign high-quality pseudo labels to unlabeled target domain data. Secondly, we replace the third stage with BERT-CRF, and the average F1 score also showes a decreasing trend, indicating the effectiveness of contrastive learning in sequence labeling task. Finally, we replace both the first and third stages, which correspond to the original DA²LM approach. We observe a decrease of 2.32% and 2.82% in the average F1 scores for the End2End ABSA and AE tasks, respectively.

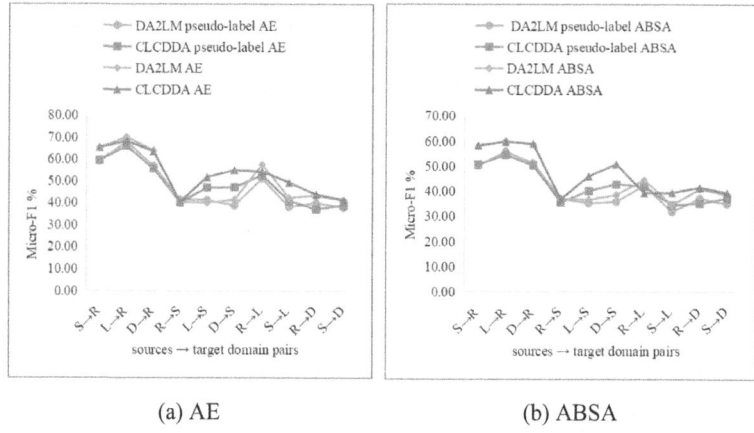

(a) AE   (b) ABSA

Fig. 2. The impact of pseudo-labels of target domain training data on the results.

Table 3. Ablation study of our CLCDDA.

Methods	End2End ABSA	AE
CLCDDA	**46.97**	**53.34**
- replace Stage 1 with DAPL	45.61	51.65
- replace Stage 3 with BERT-CRF	45.60	52.05
- simultaneously replace Stage 1 & Stage 3	44.65	50.52

5   Conclusion

In this paper, we propose a new Cross-Domain Data Augmentation approach based on Contrastive Learning, named CLCDDA. To verify the effectiveness of CLCDDA, we conduct experiments on four benchmarks. Experimental results show that our approach significantly outperforms previous methods in both cross-domain End2End ABSA and AE tasks. However, CLCDDA generates target domain labeled data through GPT2. With the development of Large Language Models, integrating the latest LLM is a direction for our future work.

Acknowledgement. This work was supported by the Liaoning Natural Science Foundation (2022-MS-119).

References

1. Zhang, W., Li, X., Deng, Y., Bing, L., Lam, W.: A survey on aspect-based sentiment analysis: tasks, methods, and challenges. IEEE Trans. Knowl. Data Eng. **35**(11), 11019–11038 (2023)
2. Liu, B.: Sentiment analysis and opinion mining. Synth. Lect. Hum. Lang. Technol. **5**(1), 1–167 (2012)
3. Deng, H., Li, Y., Ju, S., Liu, M.: Combines contrastive learning and primary capsule encoder for target sentiment classification. In: Yuan, L., Yang, S., Li, R., Kanoulas, E., Zhao, X. (eds.) Web Information Systems and Applications, WISA 2023, LNCS, vol. 14094, pp. 284–296. Springer, Singapore (2023). https://doi.org/10.1007/978-981-99-6222-8_24
4. Li, X., Bing, L., Zhang, W., Lam, W.: Exploiting bert for end-to-end aspect-based sentiment analysis. In: Proceedings of the 5th Workshop on Noisy User-generated Text (W-NUT 2019), pp. 34–41 (2019)
5. Yu, J., Zhao, Q., Xia, R.: Cross-domain data augmentation with domain-adaptive language modeling for aspect-based sentiment analysis. In: Proceedings of the 61st Annual Meeting of the Association for Computational Linguistics. pp. 1456-1470 (2023)
6. Li, J., Yu, J., Xia, R.: Generative cross-domain data augmentation for aspect and opinion co-extraction. In: Proceedings of the 2022 Conference of the North American Chapter of the Association for Computational Linguistics: Human Language Technologies, pp. 4219–4229 (2022)
7. Yu, J., Gong, C., Xia, R.: Cross-domain review generation for aspect-based sentiment analysis. In: Findings of the Association for Computational Linguistics: ACL-IJCNLP 2021, pp. 4767–4777 (2021)
8. Li, Z., Li, X., Wei, Y., Bing, L., Zhang, Y., Yang, Q.: Transferable end-to-end aspect-based sentiment analysis with selective adversarial learning. In: Proceedings of the 2019 Conference on Empirical Methods in Natural Language Processing and the 9th International Joint Conference on Natural Language Processing (EMNLP-IJCNLP), pp. 4590–4600 (2019)
9. Gretton, A., et al.: Optimal kernel choice for large-scale two-sample tests. In: Advances in Neural Information Processing Systems, vol. 25 (2012)
10. Gong, C., Yu, J., Xia, R.: Unified feature and instance based domain adaptation for aspect-based sentiment analysis. In: Proceedings of the 2020 Conference on Empirical Methods in Natural Language Processing (EMNLP), pp. 7035-7045 (2020)
11. Xu, H., Liu, B., Shu, L., Philip, S.Y.: Bert post-training for review reading comprehension and aspect-based sentiment analysis. In: Proceedings of the 2019 Conference of the North American Chapter of the Association for Computational Linguistics: Human Language Technologies, Volume 1 (Long and Short Papers), pp. 2324–2335 (2019)
12. Chen, Z., Qian, T.: Bridge-based active domain adaptation for aspect term extraction. In: Proceedings of the 59th Annual Meeting of the Association for Computational Linguistics and the 11th International Joint Conference on Natural Language Processing (Volume 1: Long Papers), pp. 317-327 (2021)
13. Bai, H., Shen, D., Dou, W., Nie, T., Kou, Y.: Domain-generic pre-training for low-cost entity matching via domain alignment and domain antagonism. In: 2023 International Joint Conference on Neural Networks (IJCNN), pp. 1–7 (2023)

14. Gao, T., Yao, X., Chen, D.: Simcse: simple contrastive learning of sentence embeddings. In: Proceedings of the 2021 Conference on Empirical Methods in Natural Language Processing, pp. 6894–6910 (2021)
15. Radford, A., Wu, J., Child, R., Luan, D., Amodei, D., Sutskever, I., et al.: Language models are unsupervised multitask learners. OpenAI Blog 1(8), 9 (2019)
16. Chen, X., Wan, X.: A simple information-based approach to unsupervised domain-adaptive aspect-based sentiment analysis. arXiv preprint arXiv:2201.12549 (2022)

Intelligent Computing

A Dynamic Convergence Criterion for Fast K-means Computations

Hui Yu[1], Yujie Du[2], Xiaoqi Zhang[1], Zhigang Wang[1(✉)], Ning Wang[1],
Juncheng Yi[3], Xiaodong Wang[1], Jie Nie[1], and Zhiqiang Wei[1]

[1] Ocean University of China, Qingdao, China
`yuhui1472@stu.ouc.edu.cn`,
`{wangzhigang,wangning8687,wangxiaodong}@ouc.edu.cn`
[2] Yantai Engineering and Technology College, Yantai, China
[3] Big Data Center, Qingdao, China
`yijuncheng@qd.shandong.cn`

Abstract. The K-Means algorithm has effectively promoted the development of intelligent systems and data-driven decision-making through data clustering and analysis. A reasonable convergence judgment directly determines when the model training can be terminated, which heavily affects the model quality. There are many researches for training acceleration and quality improvement, but few focus on the judgment. Currently, the convergence criteria still adopt a centralized judgment strategy for a single loss value. The criterion is simply copied between different optimized K-Means variants, typically like the fast Mini-Batch version and the traditional Full-Batch version. Our analysis reveals that such a design cannot guarantee that different variants converge to the same point, that is, it can result in abnormal situations such as false-positive and over-training. To perform a fair comparison and guarantee the model accuracy, we proposed a new dynamic convergence criterion VF (Vote for Freezing) and optimized version VF+. VF adopts a distributed judgment strategy where each sample can decide whether to participate in training based on the criterion (i.e., freezing itself) or not. Meanwhile, combined with the priority of samples, VF adaptively adjusts the sample freezing threshold which achieves asymptotic withdrawal of samples and accelerates model convergence. VF+ further introduced parameter freezing thresholds and freezing periods to eliminate redundant distance calculations, hence it improves the training efficiency. Experiments on multiple datasets validate the effectiveness of our convergence criterion in terms of training quality and efficiency.

Keywords: K-Means · Convergence Criterion · Priority

1 Introduction

As the most widely used clustering algorithm [1,4], the K-Means algorithm has been extensively explored. Several methods have been proposed to enhance

C. Jin et al. (Eds.): WISA 2024, LNCS 14883, pp. 191–202, 2024.
https://doi.org/10.1007/978-981-97-7707-5_17

clustering quality [9]. Besides, given the K-Means algorithm's high dependency on centroid initialization, optimizing the initial centroid is a good strategy to improve the model convergence speed [18]. The Canopy algorithm proposed by McCallum et al. [8] improved distance calculation and Pérez et al. [10] enhanced classification. Parallel computing and distributed algorithms [7,14] improved the efficiency of processing large-scale data.

The above optimization methods are all applied to two basic training policies: Full-Batch and Mini-Batch. Compared with Full-Batch, where centroids are updated after all samples are scanned, Mini-batch can perform frequent updates when partial samples are processed. Mini-Batch speeds up convergence because fresh information learned from previously processed samples can be quickly used to process subsequent samples. Nowadays, Mini-Batch has become a popular choice for model training.

The training process iteratively adjusts parameters until convergence. There are multiple convergence criteria including reaching a pre-set number of iterations, minimizing loss below a threshold, or finding the loss difference between two consecutive iterations below a given threshold. The thresholds of the first two criteria are difficult to set, and it is difficult to ensure model quality. The most commonly used convergence criterion is the last one.

Currently, when running the Mini-Batch training, researchers usually directly use the loss difference threshold in Full-Batch. However, such a design can lead to unfair comparisons. Under Mini-Batch, the accumulated change of the loss function per iteration is smaller than that in Full-Batch because the former only processes partial samples instead of all. Then, the difference threshold used in Full-Batch can be easily satisfied under Mini-Batch. This false-positive judgment will terminate training abnormally. A possible solution is normally updating the centroid after each Batch but only judging convergence after completing an Epoch (resembling Full-Batch). However, such a delayed judgment detection may result in redundant training, because the frequent update will largely change the centroids during an epoch. The loss difference between two epochs can be overly large, making it difficult to satisfy the threshold (but actually the absolute loss value has been significantly smaller than that achieved in Full-Batch).

We use two real-world datasets HIGGS and HepMASS to validate the false-positive and over-training phenomena on K-Means. Figure 1 shows the phenomenon of false-positive convergence. Table 1 shows the phenomenon where over-training generates 13.2% and 17.9% runtime degradation.

In order to solve the above problems, we introduce a new threshold-setting policy based on the "Vote-to-Halt" [6] mechanism and study a new convergence algorithm VF. The reason for the two phenomena is that the convergence judgment always resorts to the accumulated loss value. Such a centralized policy is very sensitive to the number of samples involved in the computation. Our solution is designing a decentralized policy where each sample can individually decide whether to participate in training (i.e., it can vote to halt if it's unnecessary to be continuously processed) or not.

Recently, researchers also noted that some non-critical samples do not need to participate in training models. The SlimML [5] framework eliminates non-critical data during iterations. It selects a small number of representative aggregated data samples based on a priority threshold for training. However, its prioritization method can't be directly applied to K-Means due to the absence of gradient computation. Besides, The threshold for excluding non-critical data is set manually, and each sample shares the same constant threshold. We change the prioritization method for K-Means and give a dynamic threshold.

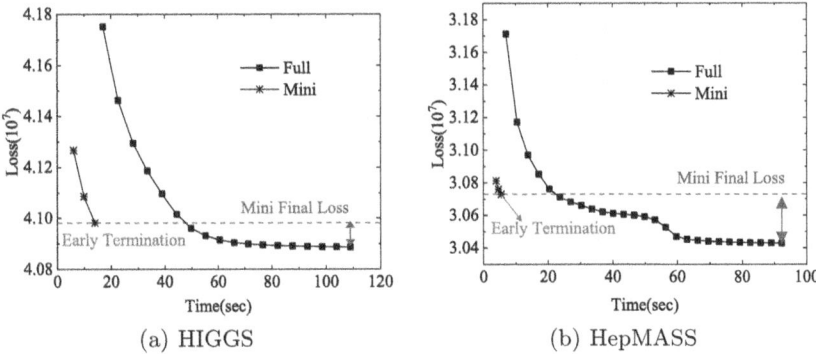

(a) HIGGS   (b) HepMASS

Fig. 1. Evaluation of false-positive. The blue dashed line represents the final loss value of the Mini-Batch when the Full-Batch threshold is directly used. (Color figure online)

Table 1. Evaluation of over-training. Truncated Time is the time when the Mini-Batch loss descends to match the Full-Batch for the first time. Complete Time is the final training time of the Mini-Batch.

Datasets	Full-Batch Loss(10^7)	Mini-Batch Loss(10^7)	Truncated Time(sec)	Complete Time(sec)
HepMASS	3.0429643	3.0426370	39.272	45.236
HIGGS	4.0886614	4.0883565	43	52.364

In addition to sample-level freezing, we are aware that some work also focuses on parameters, i.e., Adaptive Parameter Freezing (APF) [3], which avoids high communication costs in distributed environments by freezing stable parameters. The freezing period is adjusted using the TCP approach based on the stability of unfrozen parameters. We also borrow its idea to accelerate K-Means but the frozen threshold and loss effective variation requires careful design.

Inspired by "Vote-to-Halt" and APF, we propose new convergence criteria VF and VF+. Our main contributions can be summarized as follows:

– We propose a novel convergence criterion (VF) based on a distributed loss
  metric. This criterion replaces the commonly used centralized loss metric to
  eliminate avoid the unfairness caused by directly transferring loss thresholds
  from Full-batch to Mini-Batch.
– We propose a priority-based adaptive sample freezing strategy. We develop
  a simple but efficient prioritization strategy based on the contribution of
  each data sample to convergence. To boost model convergence efficiency, we
  gradually withdraw samples by dynamically adjusting the freezing threshold.
– An adaptive parameter freezing strategy for K-Means (VF+) is proposed.
  This strategy dynamically adjusts parameter freezing thresholds and freezing periods based on the characteristics of each parameter. By minimizing
  redundant sample-centroid distance calculations, we reduce model training
  time.

2 Method

In this section, we will provide a detailed explanation of our proposed new convergence criteria, VF and VF+.

A detailed introduction to VF will be given in Sect. 2.1. VF is a decentralized convergence criterion which samples with low priority are withdrawn from
training by voting. Section 2.2 will provide a detailed introduction to VF+. VF+
builds upon VF, adding a parameter freezing module.

2.1 VF: "Vote-to-Halt" for Samples

VF is a sample voting freeze algorithm that combines the "Vote-to-Halt" with
the freeze threshold for samples.

A. Vote-to-Halt. There are several graph processing systems [11–13,15] and
Vote-to-Halt is a part of the Pregel. In Pregel, each vertex has a binary identifier
vote which identifies its two states: active and inactive (i.e., participating in
computation or not). After all vertices vote their status as inactive, the graph
algorithm terminates (converges). We treat samples in K-Means as vertices in
the graph. In each iteration, these samples independently decide whether to join
the current training or not. This approach differs from the traditional method
where samples either fully participate or abstain from the computation.

VF sets a state, $vote_n$, to sample X_n. X_n also has two states: active and
inactive, changing dynamically during model training. Initially, all of the samples
are active and participate in iteration calculations. Every sample is assigned a
freezing threshold based on the Full-Batch threshold. If the point's loss-difference
between two consecutive iterations falls below this threshold, the point will make
little contribution to the model and the state of the sample will switch to inactive.

VF localizes the "Vote-to-Halt" technique specifically for K-Means. On the
one hand, VF does not adopt the mechanism of active messages. These messages

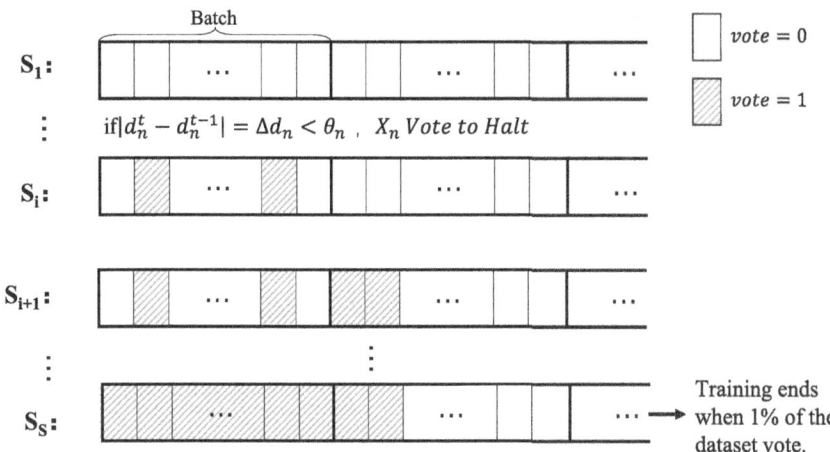

Fig. 2. Overall framework of "Vote-to-Halt" applied to K-Means

are only sent when the distance of the sample to the centroid changes significantly. Monitoring this requires checking the distance of each sample to the centroid at each iteration, regardless of convergence. Even though this approach may enhance convergence quality, it brings substantial computational costs. Experimental evaluation reveals that using messages only results in a quality improvement of 0.04% to 0.09%, with a time loss ranging from 11% to 64%.

On the other hand, we think the model converged when 1% of the samples have voted rather than all. Initially, with significant centroid variations, only a few samples vote, but the loss function fluctuates greatly. As the training progresses, centroids stabilize, leading to decreased distances between samples and centroids, thereby lowering the likelihood of cluster changes. After the loss function is stable, continuing to require all samples to converge will not greatly improve the accuracy of the model, so redundant training should be stopped in time. The voting framework of VF is illustrated in Fig. 2.

B. Freezing Threshold of Sample. The freezing threshold of sample is a crucial parameter. Next, we will provide a detailed explanation of how to determine the threshold. Simply, we can evenly distribute the Full-Batch threshold θ to all samples. However, using this threshold might not fully leverage the individual characteristics of each sample. Some samples might not exit at the right time, leading to inefficient resource usage and low training accuracy. Therefore, we introduce sample priorities to adjust the thresholds.

The distance from the sample to its centroid C_1 is d_1. The next closest centroid is C_2, and the distance is d_2. In most of time, sample change its centroid from C_1 to C_2. If the difference between d_1 and d_2 is not much large, it means that the sample is very easy to change cluster. These samples are labeled as high-priority due to their potential to impact model parameters. Others are labeled as low-priority. As shown in Fig. 3(a), x_1 is a high-priority sample while x_2 is a low-priority sample.

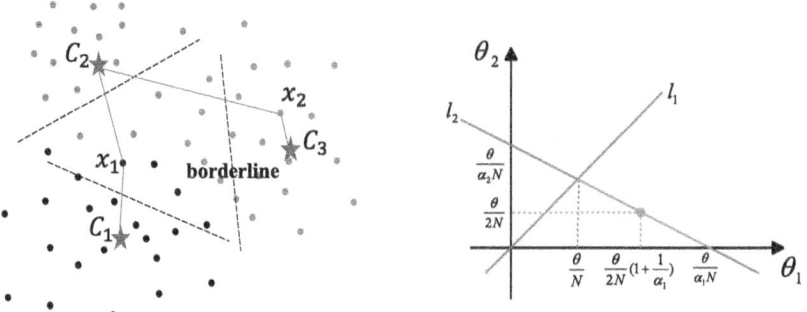

(a) Illustration of sample priority

(b) Illustration of threshold for different priority samples

Fig. 3. Illustration of priority.

To classify the samples, the key is to observe R, the ratio of d_2 to d_1. If R is less than the threshold α, it means that the gap between d_1 and d_2 is large and it is stable. Whether the α is reasonable or not directly determines the effectiveness of the method. The changes in the centroid gradually become stable, so α also should be reduced. The threshold of the jth iteration we designed is the mean of all sample ratios R in the previous epoch. N is the number of samples in the dataset:

$$\alpha_j = \frac{1}{N} \sum_{n=1}^{N} R_n^{j-1} \tag{1}$$

The freezing threshold assignment strategy should satisfy two conditions. First, samples with higher priority have lower freezing thresholds, which makes high-priority samples participate in more training. Second, the sum of sample thresholds must equal the Full-Batch threshold. The freezing thresholds should not be increased or decreased blindly. Otherwise, it will lead to underfitting and overfitting. We define the freezing threshold for low-priority samples as θ_1 and for high-priority samples as θ_2. To satisfy the above conditions, the constraints given by the thresholds are as follows:

$$s.t. \begin{cases} \theta_1 \varphi_1 N + \theta_2 \varphi_2 N = \theta \\ \theta_1 > \theta_2 \\ \theta_1, \theta_2 > 0 \\ \varphi_1 + \varphi_2 = 1 \end{cases} \tag{2}$$

Here, φ_1 represents the proportion of low-priority samples, and φ_2 represents the high-priority. θ is the Full-Batch threshold. According to Fig. 3(b), l_1 is the second expression in Eq. 2. l_2 is the first expression in Eq. 2. It is easy to see that the orange line satisfies the condition, thus we have:

$$
\begin{cases}
\theta_1 = \dfrac{\theta}{N} \left(1 - k + \dfrac{k}{\alpha_1}\right) \\[2mm]
\theta_2 = (1 - k)\dfrac{\theta}{N}
\end{cases}
\tag{3}
$$

k is used to control the difference between θ_1 and θ_2. Generally, we set $k=0.5$.

2.2   VF+: Freezing for Parameters

Inspired by APF, a parameter freezing module is added to VF to further improve the training efficiency, forming the VF+ algorithm which is shown in Fig. 4.

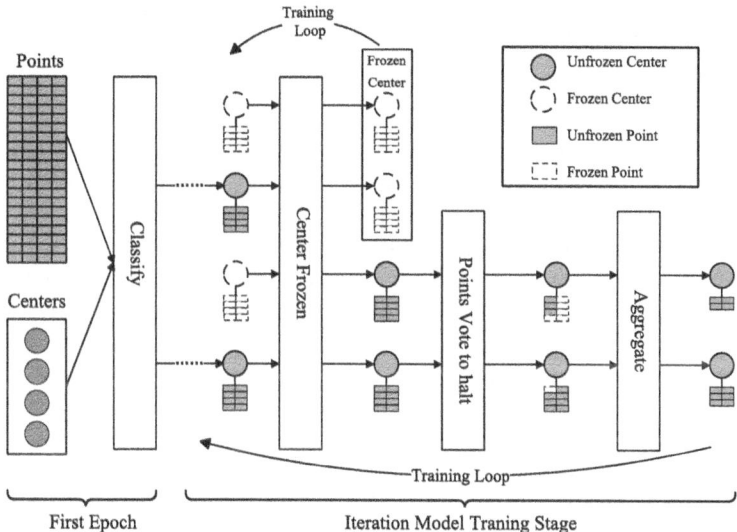

Fig. 4. Overview of VF+. Using four centroids as an example: After centroids are frozen, they can unfreeze during training, whereas samples, once frozen, will no longer be involved in calculations

In K-Means, the parameters of the model are the centroids. When the centroid undergoes slight changes, distances between samples and their centroids will not change greatly. We greedily think that few other centroids will undergo significant changes, which means the probability of centroid change is extremely low. Hence, it is of little significance to calculate sample points in clusters with smaller centroid changes. Especially, when large-scale datasets are applied to K-Means, there could be tens or even millions of samples in a cluster, and each iteration needs to calculate the distance of each sample point to all centroids. Therefore, freezing certain centroids and their samples at appropriate times can significantly reduce redundant distance calculations. However, frozen parameters must be unfrozen at the right time to maintain quality clustering. Although this may cause a loss in convergence accuracy, experiments show that the time saved outweighs this loss.

A. Frozen Threshold & Period. This section introduces how we apply some methods in APF to K-Means. Similar to samples, centroid C_i gradually stabilizes with iterations, requiring the freezing threshold ε_i^r to be dynamically adjusted. Freezing all parameters makes the mechanism useless, yet leaving all unfrozen would lack centroids for training. At first, we give an initial parameter freezing threshold. If most parameters are frozen, we halve the threshold to reduce the likelihood of all parameters being frozen. The freezing period also needs careful consideration. Too short, it will risk redundant detection; too long, it will slow convergence. Hence, we use TCP mode to adaptively adjust the freezing period. The parameters are tracked after they are unfrozen. If a parameter is frozen again, its threshold will be increased by 1. If they remain unfrozen, we greedily halve the freezing period.

B. Loss Function's Effective Variation. In APF, a parameter is a single value, making it easier to calculate positive and negative directions for effective perturbation. However, in K-Means, a parameter is a multi-dimensional vector, making it challenging to compute positive and negative directions. If we are to calculate the positive and negative directions of the centroids for each dimension independently, some centroid dimensions will be frozen, leading to biased clustering. Hence, in K-Means, we focus on the Euclidean distance between centroid positions to determine parameter variation. If the distance P_n^r is less than a threshold ε_i^r, the centroid is frozen.

$$P_n^r = \|C_n^r - C_n^{r-1}\|_2, (X_n \in C_i)$$

3   Experiments

In this section, we will present the experimental results of VF and VF+ on three datasets. In Sect. 3.1, we will introduce the datasets and our baseline. In Sect. 3.2, we will present a comparison between our methods and baselines to demonstrate the effectiveness of our approaches. In Sect. 3.3, we will conduct experiments to analyze the necessity of certain details in our method.

3.1   Experimental Setup

A. Datasets. The details of the three datasets we used are shown in Table 2.

Table 2. Dataset Summary

Datasets	Samples Size	Dimensions
Covtype [2]	581012	54
HIGGS [16]	11000000	28
HepMASS [17]	7000000	27

B. Baseline. We compare the models in terms of both accuracy and training time. In terms of model accuracy, our intention is to use SlimML as the baseline. However, the freezing threshold for samples in SlimML is selected through many experiments, and exact values for K-Means are not given. Therefore, we use two extreme values $\theta/(k*N)$, with $k = 0.1$ and $k = 0.9$ as the baseline. But users typically need the Full-Batch training loss, so we add a Full-Batch baseline, just in case. In terms of training time, we also use the two extreme values of SlimML and an adding one as the baseline. However, directly applying the Full-Batch threshold to Mini-Batch training will lead to unfairness. Therefore, it is not wise to directly use the training time of these two ways as baselines. In Mini-Batch training mode, we use the truncated time as the time baseline.

3.2 Overall Performance

Figure 5(a) shows the results of comparing the loss and runtime among VF, VF+, and baselines. For these three datasets, the maximum difference between VF and the three loss baseline is 0.97%, 0.05%, and 0.59%. The maximum difference between VF+ and the three loss functions baseline is 1.26%, 0.06%, and 0.73%. It can be seen that the loss value of our method and the three baselines can be bounded. Figure 5(b) shows that our method achieves a significant time gain compared to all three baseline methods on the HIGGS and HepMASS datasets. Due to the different orders of magnitude, the difference in Covtype is not fully reflected in Fig. 5(b). For VF, Covtype reduces the time by 18.61% over Full-Batch and by 31.65% and 58.26% over SlimML. For VF+, the time of Covtype is 30.55% shorter than Full-Batch, 41.68% and 64.38% shorter than SlimML.

3.3 Analysis of VF & VF+

A. Dynamic Sample Freezing Threshold. As iterations progress, centroids stabilize, leading to stabilized distances from samples. Therefore, it is evidently unreasonable to use a fixed sample freezing threshold and we have confirmed this suspect through experiments. We set the fixed sample freezing threshold as $\theta/(k*N)$, where k is given as 0.1, 0.3, 0.5, 0.7, and 0.9. Model accuracy and training time trend in opposite directions as k grows, and there is no satisfactory threshold for model accuracy and training time.

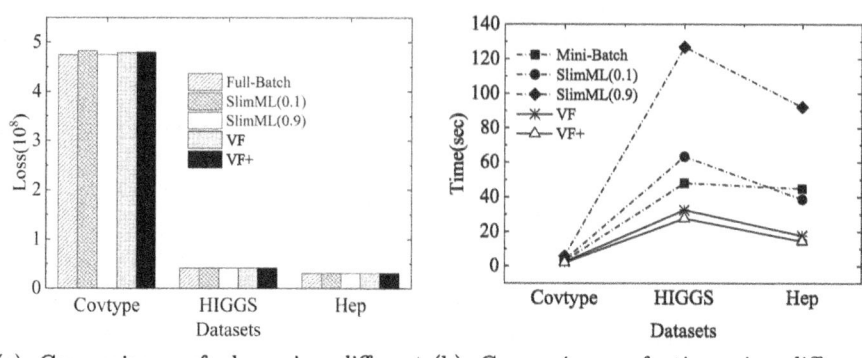

(a) Comparison of loss in different datasets

(b) Comparison of time in different datasets

Fig. 5. Overall performance.

B. Partial Samples Converge. In Sect. 2.1, we have given a detailed explanation for adopting partial samples convergence. Here we give specific experimental validation. Figure 6 illustrates the running results on VF but lets all samples converge across three datasets. The bar is the loss and the line is the frozen sample count. We can see that in the early stage of training, only a few samples are frozen, while the loss makes a huge change. In the later stages, although the number of frozen samples increases, the change of loss is almost negligible. so we truncate at 1% on the three datasets, and we get a loss function that differs from the fully converged loss function by only 1.06%, 0.03%, and 0.60%. But we get 60.72%, 71.94%, and 81.42% gains in time.

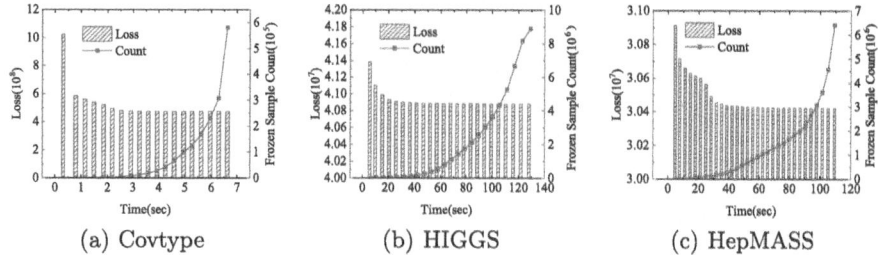

(a) Covtype   (b) HIGGS   (c) HepMASS

Fig. 6. Changing process of loss and frozen sample count with the entire dataset converged

C. Parameter Frozen Threshold. Since K-Means has fewer parameters, we've found through experiments that if the threshold is too large, all parameters may be frozen and our halving strategy won't have time to take effect. Consequently, samples will idle until a parameter is unfrozen. This will significantly weak the training efficiency. To find a better threshold, we gradually reduce the freezing threshold of the parameters while ensuring that no idling occurs.

We conduct experiments with 7 thresholds for each dataset, and the results are shown in Fig. 7. The loss value of model training changes relatively little, so we select the freezing thresholds with the shortest running time, which are 0.7, 0.0007, and 0.0014.

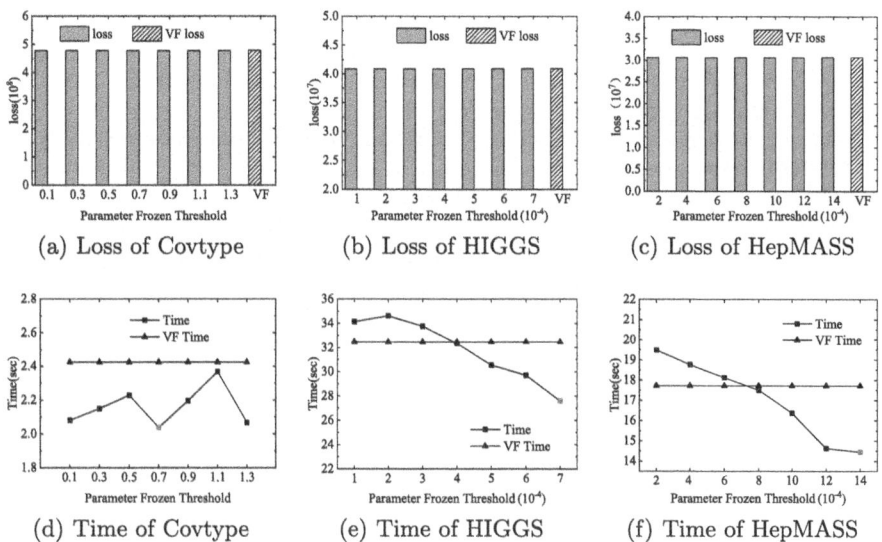

Fig. 7. Loss and training time with different freezing thresholds for parameters.

4   Conclusion

VF uses decentralized convergence and sample priority based on distance ratios to address unfairness when applying the loss threshold of Full-Batch to Mini-Batch. VF+ enhances efficiency with dynamically freezing parameter freezing. Experiments show that VF and VF+ improved efficiency and ensured quality.

Acknowledgement. This work was supported by the Key R&D Program of Shandong Province, China (No. 2023CXPT020), and the National Natural Science Foundation of China (No. U23A20320 and No. U22A2068).

References

1. Ahmed, M., Seraj, R., Islam, S.M.S.: The k-means algorithm: a comprehensive survey and performance evaluation. Electronics **9**(8), 1295 (2020)
2. Blackard, J.: Covertype. UCI Machine Learning Repository (1998). https://doi.org/10.24432/C50K5N
3. Chen, C., et al.: Communication-efficient federated learning with adaptive parameter freezing. In: 2021 IEEE 41st International Conference on Distributed Computing Systems (ICDCS), pp. 1–11. IEEE (2021)

4. Ghazal, T.M.: Performances of k-means clustering algorithm with different distance metrics. Intell. Autom. Soft Comput. **30**(2), 735–742 (2021)
5. Han, R., et al.: SlimML: removing non-critical input data in large-scale iterative machine learning. IEEE Trans. Knowl. Data Eng. **33**(5), 2223–2236 (2019)
6. Malewicz, G., et al.: Pregel: a system for large-scale graph processing. In: Proceedings of the 2010 ACM SIGMOD International Conference on Management of Data, pp. 135–146 (2010)
7. Mao, Y., Gan, D., Mwakapesa, D.S., Nanehkaran, Y.A., Tao, T., Huang, X.: A mapreduce-based k-means clustering algorithm. J. Supercomput. 1–22 (2022)
8. McCallum, A., Nigam, K., Ungar, L.H.: Efficient clustering of high-dimensional data sets with application to reference matching. In: Proceedings of the Sixth ACM SIGKDD International Conference on Knowledge Discovery and Data Mining, pp. 169–178 (2000)
9. Olukanmi, P., Nelwamondo, F., Marwala, T., Twala, B.: Automatic detection of outliers and the number of clusters in k-means clustering via Chebyshev-type inequalities. Neural Comput. Appl. **34**(8), 5939–5958 (2022)
10. Pérez, J., Martínez, A., Almanza, N., Mexicano, A., Pazos, R.: Improvement to the k-means algorithm by using its geometric and cluster neighborhood properties. In: Proceedings of ICITSEM, pp. 21–26 (2014)
11. Song, Z., et al.: ADGNN: towards scalable GNN training with aggregation-difference aware sampling. Proc. ACM Manag. Data **1**(4), 1–26 (2023)
12. Song, Z., Gu, Y., Qi, J., Wang, Z., Yu, G.: EC-graph: a distributed graph neural network system with error-compensated compression. In: 2022 IEEE 38th International Conference on Data Engineering (ICDE), pp. 648–660. IEEE (2022)
13. Wang, Z., Gu, Y., Bao, Y., Yu, G., Yu, J.X., Wei, Z.: HGraph: I/O-efficient distributed and iterative graph computing by hybrid pushing/pulling. IEEE Trans. Knowl. Data Eng. **33**(5), 1973–1987 (2019)
14. Wang, Z., et al.: FSP: towards flexible synchronous parallel frameworks for distributed machine learning. IEEE Trans. Parallel Distrib. Syst. **34**(2), 687–703 (2022)
15. Wang, Z., et al.: Lightweight streaming graph partitioning by fully utilizing knowledge from local view. In: 2023 IEEE 43rd International Conference on Distributed Computing Systems (ICDCS), pp. 614–625. IEEE (2023)
16. Whiteson, D.: HIGGS. UCI Machine Learning Repository (2014). https://doi.org/10.24432/C5V312
17. Whiteson, D.: HEPMASS. UCI Machine Learning Repository (2016). https://doi.org/10.24432/C5PP5W
18. Yu, C., Fei, L., Chen, F., Chen, L., Wang, J.: Heterogeneous graphs embedding learning with metapath instance contexts. In: Yuan, L., Yang, S., Li, R., Kanoulas, E., Zhao, X. (eds.) WISA 2023. LNCS, vol. 14094, pp. 149–161. Springer, Singapore (2023). https://doi.org/10.1007/978-981-99-6222-8_13

Efficient p-Biclique Query on Large Bipartite Networks

Zhizhi Gao[1], Deming Chu[2(✉)], Fan Zhang[1], Kai Wang[3], and Long Yuan[4]

[1] Guangzhou University, Guangzhou 510006, China
zhizhigao@e.gzhu.edu.cn, zhangf@gzhu.edu.cn
[2] University of New South Wale, Sydney 2052, Australia
deming.chu@unsw.edu.au
[3] Shanghai Jiao Tong University, Shanghai 200240, China
[4] Nanjing University of Science and Technology, Nanjing 210094, China
longyuan@njust.edu.cn

Abstract. The biclique is an important model in finding bipartite cohesive subgraphs. This paper proposes the model of p-biclique and aims to enumerate every p-biclique in a bipartite network. For each vertex u in the p-biclique, at least p fraction of u's neighbors are inside the biclique, and at most $1 - p$ fraction are outside. Compared with the traditional biclique model, our p-biclique model can guarantee the engagement level of vertices, thus providing bicliques with higher quality. We propose an efficient algorithm for enumerating maximal p-bicliques in a bipartite network. The algorithm incorporates three novel rules for graph reduction that can significantly improve empirical efficiency. The experiments on real-world networks demonstrate that our algorithm is efficient and our model is effective.

Keywords: Graph query · Bipartite graph · Cohesive subgraph

1 Introduction

Bipartite graphs are widely used to model the relationships between two classes of entities in various information systems, e.g., the customer-product bipartite graph in an e-commerce platform, the user-content bipartite graph in a social platform, and the gene-role bipartite graph in a biological process. People now are increasingly interested in bipartite graphs and are generating large-scale bipartite graphs in various real-world networks [18].

The biclique is a well-known model in bipartite cohesive subgraph mining, and it is used in various applications such as anomaly detection [2,16] and gene expression analysis [14,31]. Let $G = (L, R, E)$ be a bipartite graph, in which $L(G)$ and $R(G)$ represent the two disjoint sets of vertices, and $E(G)$ represent the edge set between the two vertex sets. Given G, a *biclique* is a subgraph C of G such that every pair of vertices $u \in L(C)$ and $v \in R(C)$ is adjacent. The *maximal biclique enumeration* [4,20,24,31,32] is a problem that aims to enumerate every biclique C that is not contained in any other biclique.

C. Jin et al. (Eds.): WISA 2024, LNCS 14883, pp. 203–214, 2024.
https://doi.org/10.1007/978-981-97-7707-5_18

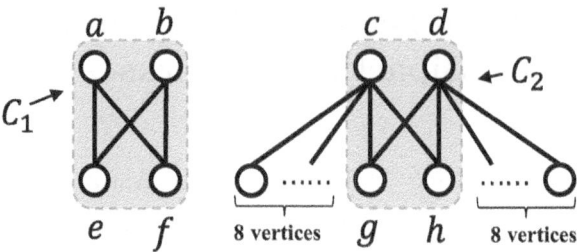

Fig. 1. An example of the p-biclique.

However, the existing works do not consider the node engagement in the bicliques. Intuitively, for any vertex u in the biclique, u is highly engaged when a large faction of its neighbors is inside the biclique while a small fraction is outside. In this paper, we propose and study the model of p-*biclique*, i.e., a biclique C where the degree of every u in C is at least p fraction of its degree in the whole graph. Consider the maximal bicliques C_1 and C_2 in Fig. 1. Although C_1 and C_2 have the same structures, the nodes in C_1 are better engaged in C_1 compared with the nodes in C_2, as C_2 is connected to 16 less-relevant nodes. Using the p-biclique model, we observe that C_1 is a 1-biclique while C_2 is a 0.2-biclique. That is, the p-biclique model can identify C_1 as a biclique with higher engagement.

This paper studies the problem of *maximal p-biclique enumeration*. Given a graph G and a threshold p, we aim to enumerate every maximal p-biclique in the network. In addition, we guarantee that any resulting biclique C satisfies the size constraints τ_L, τ_R, i.e., $|L(C)| \geq \tau_L$ and $|R(C)| \geq \tau_R$. We devise MPBE, an efficient algorithm for enumerating maximal p-bicliques. The algorithm incorporates three rules that can reduce the size of the bipartite graph while preserving the p-biclique that contains a particular node. For every vertex q in the graph, our MPBE enumerates the maximal p-cliques that contain q on a graph reduced by the rules. The experiments show that our MPBE algorithm can scale to large bipartite networks, and the query of p-bicliques can return meaningful results in different information systems.

The principal contributions of this paper are as follows:

1. To our best knowledge, we are the first to formulate and study the problem of the maximal p-biclique enumeration.
2. We propose MPBE, an algorithm for enumerating maximal p-bicliques. It includes three novel rules for reducing the graph.
3. The experiments on real-world datasets show that our algorithm can efficiently handle large-scale bipartite graphs, and that it can find insightful results in the real-world network.

Organization. Sect. 2 details the related works. Section 3 defines the problem of the maximal p-biclique enumeration. Section 4 presents our MPBE algorithm

Table 1. Summary of Notations.

Notation	Definition
G	a bipartite graph
τ_L, τ_R	the size thresholds of a biclique
$N(u)$	the set of neighbors of a node u
$N_2(u)$	the set of 2-hop neighbors of a node u
$d(u, C); d(u)$	the number of neighbors of u in C/in G
$f(u, C)$	the engagement ratio of u, i.e., $deg(u, C)/deg(u)$

for enumerating maximal p-biclique. Section 5 evaluates the performance of our algorithm. Finally, we conclude the paper in Sect. 6.

2 Related Works

Biclique Enumeration and Search. The problem of enumerating all maximal bicliques has been widely studied in the literature of information systems. Zhang et al. [31] design an efficient branch-and-bound algorithm for enumerating bicliques. Chen et al. [4] design the state-of-the-art algorithm for maximal biclique enumeration based on unilateral order and batch-pivot. Zhao et al. [32] and Wang et al. [24] investigate the maximal biclique enumeration on uncertain bipartite graphs. Das et al. [11] propose a dynamic maintenance algorithm for the maximal bicliques. On the other hand, the biclique search aims to find a single biclique with the desired properties. For example, Lyu et al. [16] develop an efficient algorithm for the maximum edge biclique problem, i.e., finding the biclique with the largest number of edges. Given a query node q, Wang et al. [23] aim to find the maximum edge biclique that contains q. Note that the algorithms above cannot solve our studied problem, because they do not consider the engagement ratio of vertices.

Engagement-Based Community Search. Zhang et al. [30] propose the (k, p)-core model. The (k, p)-core is the maximal subgraph where each vertex has at least k neighbors and at least p fraction of its neighbors in the subgraph. Morris et al. [17] study the p-cohesive model that aims to find a group of vertices whose members have at least p fraction of neighbors inside the group. The techniques in these works cannot be applied to solve our problem, as their models are based on unipartite graphs rather than bipartite graphs.

Other Models of Cohesive Subgraph Mining. Cohesive subgraph mining is an essential model in network analysis [7,9,25,26]. There is a plethora of literature on mining bipartite cohesive subgraphs, such as (α, β)-core [13,19], bi-plex [15,28], bi-truss [22,34]. In addition, the cohesive subgraph models on unipartite graphs include k-core [1,6,8,12,33], k-truss [10,21], and clique [5,29].

3 Preliminaries

In this section, we formally define the problem of maximal p-biclique enumeration. Table 1 lists the notations frequently used in the paper.

3.1 Problem Definition

We consider a bipartite graph $G = (L, R, E)$, where $L(G)$ and $R(G)$ denote two disjoint sets of vertices (i.e., the left-side and right-side vertices), while $E(G) \subseteq L(G) \times R(G)$ denotes the edge set in the graph. Given a subgraph C of G, we use $L(C)$ (resp. $R(C)$) to denote the left-side (resp. right-side) vertices from C.

Definition 1 (Biclique). *Given a bipartite graph G, a biclique is defined as a subgraph C of G such that every vertex pair between the two sides of C are connected, that is, $\forall (u \in L(C))(v \in R(C)) \ (u, v) \in E$.*

For each vertex u in G, we let $N(u)$ be the neighbor set of u, that is, $N(u) = \{v \mid (u, v) \in E(G)\}$. The degree of u is defined as $d(u) = |N(u)|$. Given a subgraph C of G, we define $d(u, C)$ as the degree of u in C, i.e., $d(u, C) = |\{v \mid (u, v) \in E(C)\}|$. Intuitively, given a subgraph C and a vertex u in the subgraph, u is highly engaged in the subgraph when most of its neighbors are contained in the subgraph. This concept can be formulated as the engagement ratio:

Definition 2 (Engagement Ratio). *Let C be a subgraph of G. Given a vertex u in C, the engagement ratio of u is defined as the ratio between the degree of u in C and in G, i.e., $f(u, C) = d(u, C)/d(u)$.*

The idea of the engagement ratio is widely adopted in existing works [17, 30]. Based on the definitions of biclique and engagement ratio, we can formally define the p-biclique:

Definition 3 (p-biclique). *Let G be a bipartite graph. A p-biclique is defined as a biclique C in G such that the minimum engagement ratio of C's left-side vertices is equal to p, i.e., $p = f(C) = min_{u \in L(C)} f(u, C)$.*

A p-biclique is *maximal* if it is not a subset of any other p-biclique. The engagement ratio constraint is assumed to be on the left-side, otherwise, we can swap the two sides before we run the algorithm. This paper aims to enumerate every p-biclique in a bipartite graph:

Problem Statement (Maximal p-Biclique Enumeration). Given a bipartite graph G, size thresholds τ_L and τ_R, and an engagement ratio threshold $p \in [0, 1]$, we study the problem of enumerating every biclique C that satisfies

1. **Engagement Constraint:** $f(C) \geq p$;
2. **Size Constraint:** $|L(C)| \geq \tau_L \wedge |R(C)| \geq \tau_R$;
3. **Maximality:** C is maximal.

To improve the quality of the enumeration, we guarantee that any resulting biclique C is maximal and the size of C satisfies the size constraints τ_L and τ_R, i.e., $|L(C)| \geq \tau_L$ and $|R(C)| \geq \tau_R$. Note that the size constraints are widely used in existing works [16, 23, 27], as they can avoid too small or too skewed bicliques.

Algorithm 1: $MPBE$ (G, τ_L, τ_r, p)

---

1 Initialize the set of resulting p-bicliques $S = \emptyset$;
2 **for** $q \in L(G)$ **do**
3     $S_q \leftarrow MPBE\text{-}Local$ $(G, \tau_L, \tau_R, p, q)$;
4   Insert into S the bicliques in S_q;

5 **return** S;

---

4 Proposed Solution

This section presents $MPBE$, an algorithm that borrows ideas from the Bron-Kerbosch algorithm [3]. Algorithm 1 lists the pseudo-code of our $MPBE$ algorithm. The algorithm first initializes the set of resulting p-bicliques S (Line 1). Then, for each left-side vertex $q \in L(G)$, it enumerates every p-biclique containing q using a function $MPBE\text{-}Local$, and the results S_q are put into S (Lines 3–4). After that, the algorithm returns S as the final result.

The main bottleneck of the algorithm is the local p-biclique enumeration (Line 3), i.e., enumerating the p-bicliques that contain q. To improve the efficiency of the local p-biclique enumeration, we propose three rules for local graph reduction, namely, the hop-based, degree-based, and engagement-based rules. These rules aim to reduce the size of the bipartite graph while preserving all p-biclique, and can significantly improve the efficiency of the algorithm.

In what follows, we first introduce our local p-biclique enumeration algorithm, and then the rules for reducing the graph in the local enumeration.

4.1 Local p-Biclique Enumeration Algorithm

Algorithm 2 presents the local p-biclique enumeration of our $MPBE$. The algorithm incorporates three rules for local graph reduction (i.e., HOP, ENG, DEG), and they will be discussed in the next section. Note that we omit the parameters of these rules, as the parameters are the same as the definitions in the next section (see Algorithm 3), e.g., HOP is equivalent to $HOP(G_q, q)$.

Lines 1–4 is the algorithm. Let a S_q be the set of p-bicliques that contain q. The algorithm starts with $S_q = \varnothing$, and initializes a reduced graph G_p with rules (Lines 1–2). Then, it enumerates the p-biclique that contains q on the reduced graph G_p (Line 3). The function $Enum$ will insert all qualified bicliques into S_q. After that, S_q is returned as the result.

Lines 5–4 describe the enumeration procedure. In a nutshell, the algorithm maintains a partial biclique $C = (L, R, L \times R)$, and recursively moves candidate vertices from P to L. The function $Enum$ requires five parameters (Line 5). In particular, L, R are the left- and right-side vertices of the partial biclique, P is the set of candidate vertices, Q is a set of vertices used to verify if C is maximal or not, and $d\_max = \max_{v \in L}\{d(v)\}$ is the maximum vertex degree in L. Note that the engagement ratio of C is equal to $f(C) = |R|/d\_max$.

Algorithm 2: *MPBE-Local* $(G, q, \tau_L, \tau_R, p)$

---

1 Initialize $S_q \leftarrow \varnothing$, the set of p-bicliques that contain q;
2 Copy G into G_q; Reduce G_q with *HOP*, *ENG* and *DEG*;
3 Run *Enum* $(\{q\}, N(q, G_q), N_2(q, G_q), \varnothing, d(q))$;
4 **return** S_q;

5 **def** *Enum* (L, R, P, Q, d_max) :
6   **if** $\nexists u \in P \cup Q$ s.t. $R \subseteq N(u) \wedge |R| \geq p \times d(u)$ **then**
7   **if** $|L| \geq \tau_L$ **and** $|R| \geq \tau_R$ **then**   Add $(L, R, L \times R)$ into S_q;
8   **for** $u \in P$ **do**
9          $d\_max' = max\{d\_max, d(q)\}$;
10         $L' = L \cup \{u\}$; $R' = R \cap N(u, G_q)$;
11         $P' \leftarrow P \cap N_2(u, G_q)$ $Q' \leftarrow Q \cap N_2(u, G_q)$;
12   **if** $|R'|/d\_max' \geq p$ **and** $|L'| + |P'| \geq \tau_L$ **and** $|R'| \geq \tau_R$ **then**
13   *Enum* (L', R', P', Q', d_max');
14         $P = P \setminus \{u\}$; $Q = Q \cup \{u\}$;

Next, we detail the procedure of *Enum*. When the algorithm finds a maximal p-biclique C with size constraints satisfied, it adds C into the result set S_q (Line 7). Note that Line 6 checks if C is maximal while Line 7 verifies the size constraints. After that, more vertices are added to L to enumerate bicliques further (Lines 9–14). For every vertex $u \in P$, we either move u into L (Line 13), or move u into Q (Line 14). When we insert u into L', the expansion is valid only when the engagement ratio of C is large enough (i.e., $|R'|/d\_max' \geq p$) and the sizes constraints are satisfied (i.e., $|L'| \cup |P'| \geq \tau_L$ and $|R'| \geq \tau_R$).

4.2 Rules for Local Graph Reduction

Assume that q is our target node for the local p-biclique enumeration. Then, we formalize the p-biclique that contains q:

Definition 4 (Local p-Biclique). *Given G, τ_L, τ_R, $p \in [0, 1]$, and a left-side node q, the local p-biclique is a biclique that contains node q and satisfies the conditions (i)-(iii) in the Problem Statement.*

For convenience, we simply add the size constraints and maximality constraint to the definition above. Given G and a subgraph G_q of G, we can reduce G to G_q if and only if G_q preserves all local p-similar bicliques that contain q. This idea can be formally defined as:

Definition 5 (Local p-Biclique Preserved Subgraph). *Let G be a graph, τ_L and τ_R be the size threshold, p be the engagement ratio threshold, and $q \in L(G)$ be a node. A subgraph G_q of G is the local p-biclique preserved subgraph, denoted by $G_q \overset{\tau, p, q}{\equiv} G$, if all local p-bicliques that contain q are the same on G and G_q.*

Then, it is immediate that the local p-biclique preserved subgraph is transitive, that is,

$$G_1 \stackrel{\tau,p,q}{\equiv} G_2 \wedge G_2 \stackrel{\tau,p,q}{\equiv} G_3 \implies G_1 \stackrel{\tau,p,q}{\equiv} G_3.$$

Based on the transitive property above, we can reduce a graph to its p-biclique preserved subgraph multiple times. We use $G \ominus u$ to denote the deletion of u from G, i.e., we eliminate u and all its incident edges.

Hop-based Rule (HOP). Let q be our target node for graph reduction. The hop-based rule removes any node u whose distance to q is more than 2, and it can be formalized as follows:

Lemma 1 (Rule HOP). *Let q be a node, and $N_2(q)$ be the two-hop neighborhood of q, then we have*

$$\forall u : \ u \notin N_2(q) \implies G \ominus u \stackrel{\tau,p,q}{\equiv} G.$$

Proof. Let u be a node outside the 2-hop neighborhood of q (i.e., $u \notin N_2(q)$) and C_q be the maximal p-biclique that contains q in G. Note that proving $G \ominus u \stackrel{\tau,p,q}{\equiv} G$ is equivalent to proving u is not contained in C_q. We prove this by contradiction. Suppose u is contained in C_q. By the fact that a biclique is a complete bipartite graph, the distance between u and q must be no more than 2. This contradicts the our assumption that $u \notin N_2(q)$.

Degree-Based Rule (DEG). Given the size thresholds τ_L and τ_R, the degree-based rule recursively removes any left-side (resp. right-side) node whose degree is less than τ_L (resp. τ_R) until no more nodes can be eliminated.

Lemma 2 (Rule DEG). *Let $d(u, G)$ be degree of u in a graph G, then we have*

$$\forall u \in L(G) : \ d(u, G) < \tau_L \implies G \ominus u \stackrel{\tau,p,q}{\equiv} G.$$
$$\forall u \in R(G) : \ d(u, G) < \tau_R \implies G \ominus u \stackrel{\tau,p,q}{\equiv} G.$$

Proof. Let C_q be the maximal p-biclique that contains q in G. Proving $G \ominus u \stackrel{\tau,p,q}{\equiv} G$ is equivalent to proving u is not contained in C_q. We prove this by contradiction. Assume that u is contained in C_q. If $u \in L(G)$ is a left-side vertex, then by the fact that $d(u, G) < \tau_L$ and C_q is a subgraph of G, we have $d(u, G_q) < \tau_L$. This contradicts the size constraint of the local p-biclique. When u is a right-side vertex, we can prove by contradiction using similar steps.

Engagement-Based Rule (ENG). Given an engagement ratio threshold p and a target node q, the third rule removes any left-side node u with $|N(q) \cap N(u)| < p \times \max\{d(u), d(q)\}$, as any biclique containing $\{u, q\}$ is not a p-biclique.

Lemma 3 (Rule ENG). *Let q be a node, and u be a left-side node, then*

$$\forall u \in L(G) : \ |N(q) \cap N(u)| < p \times \max\{d(u), d(q)\} \implies G \ominus u \stackrel{\tau,p,q}{\equiv} G.$$

Algorithm 3: graph reduction rules

---

1 **def** $HOP(G_q, q)$:
2 $\quad$ **for** $\forall u : u \notin N_2(u)$ **do** $\quad G_q \leftarrow G_q \ominus u$;

3 **def** $DEG(G_q, \tau)$:
4 $\quad$ **while** $\exists u \in L(G_q) : d(u, G_q) < \tau_L$ **or** $\exists u \in R(G_q) : d(u, G_q) < \tau_R$ **do**
5 $\quad\quad$ $G_q \leftarrow G_q \ominus u$;

6 **def** $ENG(G_q, q)$:
7 $\quad$ **for** $u \in L(G_q)$ **and** $|N(q) \cap N(u)| < p \times \max\{d(u), d(q)\}$ **do**
8 $\quad\quad$ $G_q \leftarrow G_q \ominus u$;

Proof. Let C_q be the maximal p-biclique that contains q in G. Proving $G \ominus u \overset{\tau,p,q}{\equiv} G$ is equivalent to proving u is not contained in C_q. We prove this by contradiction. Assume that u is contained in C_q. Then, we have $|R(C_q)| \leq |N(u) \cap N(q)|$. the engagement ratio of C_q is at most

$$f(C_q) \;\leq\; \frac{|R(C_q)|}{\max\{d(u), d(q)\}} \;\leq\; \frac{|N(u) \cap N(q)|}{\max\{d(u), d(q)\}} \;<\; p.$$

This contradicts the fact that C_q is a p-biclique.

Implementation. Algorithm 3 presents the pseudo-code of our rules for local graph reduction. Let G_q be a reduced graph for a node q. The algorithm starts with $G_q = G$, and then reduces G_q whenever necessary. Given size constraints τ_L, τ_R, we implement DEG with a BFS-like method. That is, we remove any vertex $u \in L(G_q)$ with $d(u, G_q) < \tau_L$ (resp. any $u \in R(G_q)$ with $d(u, G_q) < \tau_R$) and recursively remove any neighbor w of u if the degree of w drops below τ_R (resp. τ_L). The running time of all these rules is $O(|E(G_q)|)$ time.

5   Experiments

In this section, we first evaluate the efficiency of our algorithm on real-world networks. Then, we present a case study that compares our p-biclique model with existing models.

5.1   Experimental Setup

Datasets. Table 2 presents the real-world datasets that our experiments use. These datasets are widely used in the existing works related to biclique [4, 16, 23, 31], and they are publicly available on KONECT[1]. For each dataset, we remove all edge directions, duplicated edges, and self-loops.

---

[1] http://konect.cc/networks/.

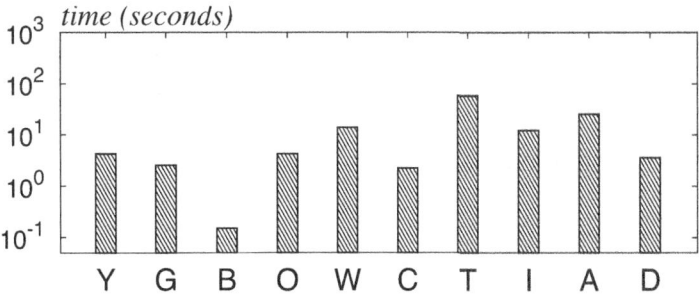

Fig. 2. The running time of our *MPBE* algorithm.

Parameters. By default, the size thresholds are set to $\tau_L = \tau_R = 5$ while the engagement ratio threshold is set to $p = 0.5$. We repeat each experiment three times and report the average result. The program is stopped when it runs for more than 24 h.

Environment. We perform the experiments on a server with an Intel Xeon Silver 4210R 2.1 GHz CPU and 256 GB memory. We implement all algorithms in C++ and compile the code with g++7.5.0 using O3 optimization.

5.2 Efficiency Analysis

Figure 2 presents the running time of our *MPBE* algorithm. The running time of our *MPBE* is no more than 58 s across different datasets. On the largest dataset DBLP (a bipartite graph with ten million edges), the algorithm only takes 3.58 s to enumerate all *p*-bicliques, while loading the network into the memory requires 8.34 s. Note that the state-of-the-art maximal biclique enumerate algorithm [4] would take 2,872 s on the BookCross dataset, while *MPBE* only takes 4.27 s on

Table 2. Statistics of Datasets.

Dataset	$|L(G)|$	$|R(G)|$	$|E(G)|$
YouTube (Y)	94,238	30,087	293,360
GitHub (G)	56,519	120,867	440,237
Bibsonomy (B)	767,447	5,794	801,784
BookCross (O)	105,278	340,523	1,149,739
WebUni (W)	6,202	200,148	1,948,004
CiteULike (C)	731,769	153,277	2,338,554
TVTropes (T)	64,415	87,678	3,232,134
IMDB (I)	303,617	896,302	3,782,463
Amazon (A)	2,146,057	1,230,915	5,743,258
DBLP (D)	1,953,085	5,624,219	12,282,059

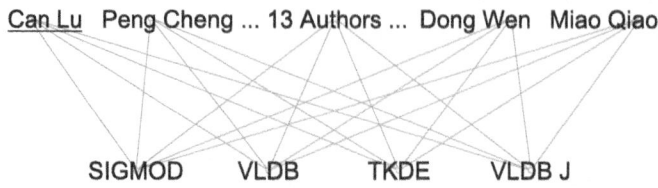

Fig. 3. A Local 0.4-Biclique (q = "Can Lu", $\tau_L = \tau_R = 4$).

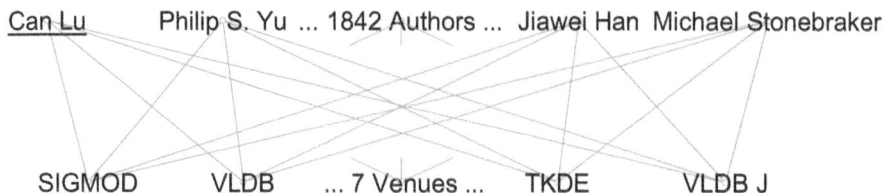

Fig. 4. The (α, β)-core that contains "Can Lu" ($\alpha = \beta = 4$).

the same dataset. Therefore, our *MPBE* algorithm is significantly faster than the existing works on maximal biclique enumeration [4, 31], as our p-biclique model can prune many unpromising bicliques and speed up the computation.

5.3 Case Study on Different Bipartite Cohesive Subgraph Models (DBLP)

This case study evaluates the effectiveness of our *MPBE* algorithm on a real-world DBLP network. In particular, we use the latest release of DBLP[2] and build an author-venue bipartite network, where an author is connected to a venue if the author has published a paper in the venue. This author-venue graph contains 3,509,155 authors, 16,751 venues, and 13,303,652 author-venue relationships. In what follows, we report the result of the p-biclique model on this DBLP network, and compare the results with other bipartite cohesive subgraph models.

Results of the Local p-Biclique. Figure 3 presents the local p-biclique that contains "Can Lu" ($\tau_L = \tau_R = 4, p = 0.4$), a fresh faculty member at Guangzhou University. The biclique contains 17 authors and 4 venues. More interestingly, these 17 authors are all young scholars in the field of data management, and they share an interest in prestigious venues in data management and mining. It takes 0.48 s to query the p-biclique mentioned above. This shows that our *MPBE* algorithm can efficiently query meaningful communities and provide insights into the network.

Results of (α, β)-Core [13]. Given size thresholds α, β, the (α, β)-core model seeks a bipartite subgraph C such that $|L(C)| \geq \alpha$ and $|R(C)| \geq \beta$. Liu et al. [13] propose the (α, β)-core model and an index-based algorithm for the problem.

---

[2] https://dblp.org/xml/release/dblp-2024-03-01.xml.gz.

This experiment reports the (α, β)-core connected component that contains a given query node q. Figure 4 demonstrates the $(4, 4)$-core that contains "Can Lu", and it consists of 1,846 authors and 11 venues. As the resulting (α, β)-core may be too large, it is hard to analyze the network with the (α, β)-core model. Besides, the (α, β)-core computation takes 1.8 s.

6   Conclusion

This paper proposes and studies the model of p-biclique, a novel variant of biclique that considers the engagement of vertices. We devise a novel and efficient algorithm *MPBE* for enumerating maximal p-bicliques in a bipartite network. The experiments on 10 real graphs validate the efficiency of *MPBE* and the effectiveness of the p-biclique model in real applications such as the cooperation analysis on co-author networks.

References

1. Batagelj, V., Zaversnik, M.: An o(m) algorithm for cores decomposition of networks. arXiv preprint cs/0310049 (2003)
2. Beutel, A., Xu, W., Guruswami, V., Palow, C., Faloutsos, C.: Copycatch: stopping group attacks by spotting lockstep behavior in social networks. In: WWW, pp. 119–130 (2013)
3. Bron, C., Kerbosch, J.: Algorithm 457: finding all cliques of an undirected graph. CACM **16**(9), 575–577 (1973)
4. Chen, L., Liu, C., Zhou, R., Xu, J., Li, J.: Efficient maximal biclique enumeration for large sparse bipartite graphs. PVLDB **15**(8), 1559–1571 (2022)
5. Cheng, J., Ke, Y., Fu, A.W.C., Yu, J.X., Zhu, L.: Finding maximal cliques in massive networks. TODS **36**(4), 1–34 (2011)
6. Chu, D., et al.: Finding the best k in core decomposition: a time and space optimal solution. In: ICDE, pp. 685–696. IEEE (2020)
7. Chu, D., et al.: Discovering and maintaining the best k in core decomposition. TKDE (2024)
8. Chu, D., Zhang, F., Zhang, W., Lin, X., Zhang, Y.: Hierarchical core decomposition in parallel: from construction to subgraph search. In: ICDE, pp. 1138–1151. IEEE (2022)
9. Chu, D., Zhang, F., Zhang, W., Zhang, Y., Lin, X.: Graph summarization: compactness meets efficiency. Proc. ACM Manag. Data **2**(3), 1–26 (2024)
10. Cohen, J.: Trusses: cohesive subgraphs for social network analysis. Natl. Secur. Agency Tech. Rep. **16**(3.1), 1–29 (2008)
11. Das, A., Tirthapura, S.: Incremental maintenance of maximal bicliques in a dynamic bipartite graph. IEEE Trans. Multi-Scale Comput. Syst. **4**(3), 231–242 (2018)
12. Khaouid, W., Barsky, M., Srinivasan, V., Thomo, A.: K-core decomposition of large networks on a single pc. PVLDB **9**(1), 13–23 (2015)
13. Liu, B., Yuan, L., Lin, X., Qin, L., Zhang, W., Zhou, J.: Efficient (α, β)-core computation: an index-based approach. In: WWW, pp. 1130–1141 (2019)
14. Liu, J., Wang, W.: Op-cluster: clustering by tendency in high dimensional space. In: ICDM, pp. 187–194. IEEE (2003)

15. Luo, W., Li, K., Zhou, X., Gao, Y., Li, K.: Maximum biplex search over bipartite graphs. In: ICDE, pp. 898–910. IEEE (2022)
16. Lyu, B., Qin, L., Lin, X., Zhang, Y., Qian, Z., Zhou, J.: Maximum biclique search at billion scale. PVLDB (2020)
17. Morris, S.: Contagion. Rev. Econ. Stud. **67**(1), 57–78 (2000)
18. Sahu, S., Mhedhbi, A., Salihoglu, S., Lin, J., Özsu, M.T.: The ubiquity of large graphs and surprising challenges of graph processing. PVLDB **11**(4), 420–431 (2017)
19. Shu, K., Liang, Q., Guo, H., Zhang, F., Wang, K., Yuan, L.: Finding introverted cores in bipartite graphs. In: Yuan, L., Yang, S., Li, R., Kanoulas, E., Zhao, X. (eds.) WISA 2023. LNCS, vol. 14094, pp. 162–170. Springer, Heidelberg (2023). https://doi.org/10.1007/978-981-99-6222-8_14
20. Sun, R., Wu, Y., Chen, C., Wang, X., Zhang, W., Lin, X.: Maximal balanced signed biclique enumeration in signed bipartite graphs. In: ICDE, pp. 1887–1899. IEEE (2022)
21. Wang, J., Cheng, J.: Truss decomposition in massive networks. PVLDB **5**(9) (2012)
22. Wang, K., Lin, X., Qin, L., Zhang, W., Zhang, Y.: Efficient bitruss decomposition for large-scale bipartite graphs. In: ICDE, pp. 661–672. IEEE (2020)
23. Wang, K., Zhang, W., Lin, X., Qin, L., Zhou, A.: Efficient personalized maximum biclique search. In: ICDE, pp. 498–511. IEEE (2022)
24. Wang, K., Zhao, G., Zhang, W., Lin, X., Zhang, Y., He, Y., Li, C.: Cohesive subgraph discovery over uncertain bipartite graphs. TKDE **35**, 11165–11179 (2023)
25. Wang, Y., Yuan, L., Chen, Z., Zhang, W., Lin, X., Liu, Q.: Towards efficient shortest path counting on billion-scale graphs. In: ICDE, pp. 2579–2592. IEEE (2023)
26. Xie, J., Chen, Z., Chu, D., Zhang, F., Lin, X., Tian, Z.: Influence maximization via vertex countering. PVLDB **17**(6), 1297–1309 (2024)
27. Yao, K., Chang, L., Yu, J.X.: Identifying similar-bicliques in bipartite graphs. PVLDB **15**(11), 3085–3097 (2022)
28. Yu, K., Long, C., Deepak, P., Chakraborty, T.: On efficient large maximal biplex discovery. TKDE **35**(1), 824–829 (2021)
29. Yuan, L., Qin, L., Lin, X., Chang, L., Zhang, W.: Diversified top-k clique search. VLDB J. **25**(2), 171–196 (2016)
30. Zhang, C., et al.: Exploring finer granularity within the cores: efficient (k, p)-core computation. In: ICDE, pp. 181–192. IEEE (2020)
31. Zhang, Y., Phillips, C.A., Rogers, G.L., Baker, E.J., Chesler, E.J., Langston, M.A.: On finding bicliques in bipartite graphs: a novel algorithm and its application to the integration of diverse biological data types. Bioinformatics **15**, 1–18 (2014)
32. Zhao, G., Wang, K., Zhang, W., Lin, X., Zhang, Y., He, Y.: Efficient computation of cohesive subgraphs in uncertain bipartite graphs. In: ICDE, pp. 2333–2345. IEEE (2022)
33. Zhou, Z., Zhang, W., Zhang, F., Chu, D., Li, B.: Vek: a vertex-oriented approach for edge k-core problem. World Wide Web **25**(2), 723–740 (2022)
34. Zou, Z.: Bitruss decomposition of bipartite graphs. In: Navathe, S.B., Wu, W., Shekhar, S., Du, X., Wang, X.S., Xiong, H. (eds.) DASFAA 2016. LNCS, vol. 9643, pp. 218–233. Springer, Cham (2016). https://doi.org/10.1007/978-3-319-32049-6_14

High-Dimensional Nearest Neighbor Search-Based Blocking in Entity Resolution

Kaiyu Zhang[1], Chenchen Sun[1,2(✉)], Derong Shen[3], Tiezheng Nie[3], and Yue Kou[3]

[1] Tianjin University of Technology, Tianjin 300384, China
suncc_db@163.com
[2] Chang'an University, Xi'an 710021, China
[3] Northeastern University, Shenyang 110169, China
{shendr,nietiezheng}@mail.neu.edu.cn, kouyue@cse.neu.edu.cn

Abstract. Entity resolution is a key task in data integration and fusion, aiming to find all records describing the same real-world entity from multiple data sources. Blocking is an important step in entity resolution tasks to address the secondary time complexity challenge. Existing blocking methods based on token-based keys result in many redundant comparisons, while learning-based blocking methods incur significant time overhead during blocking generation. Therefore, we propose a Blocking with Vector Similarity Search (B-VSS) framework, which is based on high-dimensional nearest neighbor search, aiming to balance the effectiveness and efficiency of blocking. B-VSS mainly consists of two key stages. First, in the record embedding stage, we utilize deep learning models to generate vector representations for records. Secondly, in the blocking generation stage, after building an index for the dataset, the generated index is used to quickly retrieve records similar to the query and cluster them into blocks, thus significantly reducing the computational complexity. Through experimental analysis, we compare various methods of index construction under the blocking framework on 9 datasets. The results show that our methods can guarantee the quality of generated blocks and improve the speed.

Keywords: Entity resolution · Unsupervised blocking · High-dimensional vector search · Data integration

1 Introduction

Entity resolution (ER) identifies records describing the same real-world entity across multiple data sources [1]. Typically, ER involves pairwise comparisons of all records, resulting in quadratic time complexity. Blocking is a crucial step that improves ER efficiency by grouping similar records into blocks and limiting comparisons within these blocks. As illustrated in Fig. 1, without blocking (left), ER requires 9 exhaustive comparisons. With blocking (right), only 4 intra-block comparisons are necessary, significantly reducing unnecessary comparisons.

© The Author(s), under exclusive license to Springer Nature Singapore Pte Ltd. 2024
C. Jin et al. (Eds.): WISA 2024, LNCS 14883, pp. 215–226, 2024.
https://doi.org/10.1007/978-981-97-7707-5_19

Traditional blocking methods use token blocking keys but still result in redundant comparisons. Machine learning has improved blocking quality but requires manual feature extraction, while deep learning-based methods often use brute-force search, impractical for large datasets. To address these issues, we propose a Blocking with Vector Similarity Search (B-VSS) framework. This framework leverages high-dimensional vector similarity search, combining vectors generated by deep learning models with nearest neighbor search technology. By building indexes [3], it improves blocking efficiency while maintaining quality. Extensive experiments were conducted to examine the impact of various indexing methods on block generation quality and efficiency.

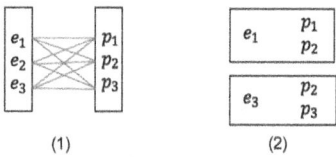

Fig. 1. The number of comparisons required for ER before and after blocking.

Our main contributions are as follows.

- We propose a Blocking framework with Vector Similarity Search, which improves the speed of block generation while ensuring the quality of generated blocks.
- Our framework utilizes 13 graph-based, 4 quantization-based, and 1 locally sensitive hashing-based methods for index construction and leverages these indexes to generate blocks via queries.
- Our framework validated on 9 datasets, achieves a balance between effectiveness and efficiency, nearing brute force retrieval performance. We also compare different components of B-VSS and their impact on block generation quality and efficiency, proposing recommended methods.

The structure of this paper is as follows: Sect. 2 introduces the definition of the problem and the overall framework. Section 3 introduces the components of B-VSS. In Sect. 4, a large number of experiments are conducted to compare different methods. Section 5 introduces the related work of blocking and high-dimensional vector nearest neighbor search. The paper is summarized in Sect. 6.

2 Preliminary

2.1 Problem Definition

The goal of ER is to find all record pairs describing the same real-world entity from multiple data sources. In more detail, it takes two record sets $E_1 = \{p_{11}, \ldots, p_{1n}\}$ and $E_2 = \{p_{21}, \ldots, p_{2m}\}$ as inputs and compares whether each pair of records in the record pairs $\{(p_{1i}, p_{2j}) | p_{1i} \in E_1, p_{2j} \in E_2, 1 \leq i \leq n \wedge 1 \leq j \leq m\}$ is the same real-world entity.

To extend ER to large entity collections, the blocking method computationally limits comparisons between similar records. In more detail, for deep blocking, deep learning techniques are first used to generate embedded V_1 and V_2 for each record in the two record sets E_1 and E_2. Then from V_2 for each record in V_1, the most similar K records are found and aggregated into blocks.

2.2 Deep Blocking Framework

We propose a Blocking framework with Vector Similarity Search, which is divided into two stages: record embedding stage and block generation stage. Record embedding stage: A deep learning model can generate a valid feature representation $V = M(E)$ for each record in a dataset, where M is the model used to record the embedding and E is the dataset. In Fig. 2, the record embedding phase takes index dataset E_I and query dataset E_Q as inputs, and uses a deep learning model M to generate embedded V_I and V_Q for each record in both datasets. Block generation phase: it puts similar records into the same block, which can be divided into two steps: index building and neighbor search. In Fig. 2, the framework adopts the high-dimensional vector indexing, taking the set of index vectors V_I as input to generate the corresponding index G. After the index is built, the nearest neighbor search methods can find the K-nearest neighbor record set $\{v_{d_{i1}}, \ldots v_{d_{iK}}\}$. Finally, the query vector K record collection and is assigned to the same block of $Block_i = \{\{v_{q_i}\}, \{v_{d_{i1}}, \ldots, v_{d_{iK}}\}\}$. We focus on the block generation phase.

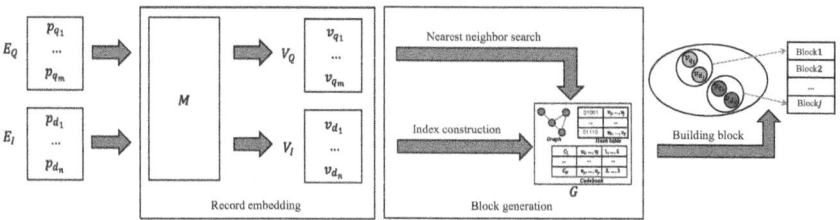

Fig. 2. Unsupervised blocking framework with Vector Similarity Search.

3 Blocking Generation Methods

We propose three block generation methods: quantization-based, graph-based and hash-based. Quantization-based block generation uses brute force retrieval, which compares every point in the query data set to every point in the index data set. It finds the K nearest neighbor points that are closest to the query point to generate a block. The graph-based block generation methods will build a graph with the original vector in the index construction stage. The block generation strategy is to use the nearest neighbor search algorithm to retrieve the K points nearest to the query point on the constructed index graph and gather them into blocks. The block generation method based on the local sensitive hash is based on the bucket and hash function generated in the index stage. It will first find several buckets that are close to the query point, and then use the brute force query strategy to find the K nearest neighbor points to the query vector to form a block.

3.1 Quantization-Based Block Generation Methods

The quantization-based methods map the original vector to a set of discrete vectors and represent the original vector with a set of compressed codes. The reason why quantization can speed up the query is that the original high-dimensional vector is represented by compressed code, which reduces the computation. We mainly introduce 4 quantization methods: PQ, OPQ, LOPQ, IMI.

Vector Quantization (VQ). VQ [7] is a function f that maps a given vector $x \in \mathbb{R}^D$ to a vector $q(x) \in C = \{c(i)\}$. The mapper is called a quantizer and is denoted as $x \rightarrow c(i(x))$. Each vector $c \in C$ is called the center of mass and C is called the codebook. Given a finite set X of dataset points in $\mathbb{R}^D$, the distortion E of quantizer q is $\sum_{x \in X} \|x - q(x)\|^2$.

No matter what codebook or quantizer is chosen, a quantizer that minimizes distortion should map the vector x to its nearest center of mass, and the k-means clustering algorithm can satisfy this condition.

Product Quantization (PQ). PQ [7] restricts codebook C to the Cartesian product of each subspace's sub-codebook, whose optimal quantizer q should also be shown as E of PQ. Finally, the nearest centroid of each vector x is $c = q(x) = \left(q^1(x^1), q^2(x^2), \ldots, q^m(x^m)\right)$, and eventually every optimal sub-quantizer q^i can be found using the k-means clustering algorithm.

Optimized Product Quantization (OPQ). OPQ [9] represents a spatial decomposition using an orthogonal matrix R that first performs an R transformation on a D-dimensional vector space and then performs a product quantization in the transformed space.

Locally Optimized Product Quantization (LOPQ). LOPQ [10] is a further extension of the locally optimized product quantization method. It consists of a coarse quantizer and several OPQs. The coarse quantizer is first used to map the data to the nearest centroid, and then an optimized product quantization is performed on the residuals on the cluster formed by each centroid.

Inverted Multi-Index (IMI). IMI [11] is a non-exhaustive search method based on quantization. It breaks down the data space into several subspaces and independently divides the units on each subspace to generalize IVF.

3.2 Graph-Based Block Generation Methods

Graph structures maintain original and similar information of data points, reducing computation by enabling approximate nearest neighbor search. Graph-based high-dimensional vector nearest neighbor search relies on four basic graphs: Delaunay Graph (DG), Relative Neighborhood Graph (RNG), K-Nearest Neighbor Graph (KNNG), and Minimum Spanning Tree (MST).

Block Generation Methods Based on DG and RNG. There are four block generation methods based on DG and RNG: NSW, HNSW, FANNG and NGT. (1) Navigable Small World graph (NSW) [12] Builds an undirected graph by continuously inserting data points into the graph. It initializes an empty graph and then traverses every point in the dataset, using the graph search method to find the m points in the current graph closest to the insertion point. (2) Hierarchical Navigable Small World graph (HNSW) [13] generates a hierarchical graph and fixes an upper limit on the number of neighbors for each vertex, considering not only its nearest neighbors but also the distribution of its neighbors. (3) Fast Approximate Nearest Neighbor Graph (FANNG) [14] mainly proposes an occlusion rule to remove redundant neighbors and reduce the complexity of the search. (4) Neighborhood Graph and Tree (NGT) [15, 16] has two construction methods, one is to convert KNNG into bi-KNNG, and the other is to construct incremental construction similar to NSW. NGT also proposes to use Vantage Point Tree to quickly locate near the query point in the search phase to improve search efficiency.

Block Generation Methods Based on KNNG. There are four block generation methods based on KNNG: SPTAG, KGraph and EFANNA, IEH. (1) The index construction of Space Partition Tree and Graph (SPTAG) [17] is divided into Tree and Graph. Tree construction is divided into K-Dimensional Tree and Balance k-Means Tree. In the Graph part, the TP-Tree will be used to build a tree, and then the KNN subgraph will be built by brute force against the points in the child nodes. (2) The idea of KGraph [18] is that a neighbor's neighbor is more likely to be its neighbor. It initially randomly assigns K neighbors to each point, then explores the neighbors of its neighbors, updates the K-neighbor list with the closest neighbors, and iterates until the K-neighbor graph no longer improves or the maximum number of iterations is reached. (3) Extremely Fast Approximate Nearest Neighbor Search Algorithm (EFANNA) [19] divides the dataset into many subsets by randomly truncating the KD-Tree and then creating the KNN graph in a violent way from the bottom up. (4) Iterative Expanding Hashing (IEH) [20] is a method that combines LSH and KNNG. It uses the existing hash method to generate a hash function, then constructs a hash index for each vector, and finally violently constructs an exact KNN graph. In the query phase, it returns the candidates of the bucket within a very small Hamming radius and uses the KNN table for expansion and lookup.

Block Generation Methods Based on KNNG and RNG. There are four block generation methods based on KNNG and RNG: DPG, NSG, NSSG and Vamana. (1) Diversified Proximity Graph (DPG) [21] is an innovation of KGraph. Its idea is to consider the similarity and distribution of neighbors so that neighbors can be distributed in all directions. (2) The design of the Navigating Spreading-out Graph (NSG) [22] is essentially the same as that of DPG. Based on KGraph, NSG considers the uniformity of neighbor distribution through the edge selection strategy of Monotonic RNG. (3) Navigating Satellite System Graph (NSSG) [23] is doing the same thing as navigating NSG and DPG. Based on KGraph, NSSG considers the diversity of neighbor distribution through the SSG edge selection strategy. (4) Vamana [24] is a combination of KGraph, HNSW and NSG. In the edge selection strategy, Vamana adds a regulating parameter α to HNSW.

Block Generation Method Based on MST. Hierarchical Clustering-based Nearest Neighbor Graph (HCNNG) [25] is the only method that uses the MST graph. It first performs hierarchical clustering until each cluster size is $< n$, and then uses a minimum spanning tree to concatenate edges. It is executed multiple times, merging all the edges.

3.3 Local Sensitive Hashing-Based Block Generation Method

The method based on local sensitive hashing (LSH) is to map similar points in the original vector into the same bucket. The block generation strategy is to search in similar buckets. The basic idea of Locality Sensitive Hashing [26] is that after the same transformation of two adjacent data points in the original data space, there is a high probability that the two data points are still adjacent in the new data space.

4 Comparative Experimental Evaluation and Analysis

4.1 Datasets

Table 1 shows the statistics of the datasets evaluated in our paper. There are seven single-source datasets and two dual-source datasets.

Table 1. Datasets statistics.

Dataset	#records	#attribute	Type	Domain
Cora	1295	12	Structured+Dirty	Bibliography
Notebook	1661	1	Textual	Product
Altosight	1993	5	Structured+Dirty	Product
Abt-Buy	1081+1092	3	Textual	Product
Amazon-Google2	1363+3326	2	Textual	Software
WDC_Cameras	1904	1	Textual	Product
WDC_Computers	2790	1	Textual	Product
WDC_Shoes	1876	1	Textual	Product
WDC_Watches	2316	1	Textual	Product

4.2 Baselines

We take the brute force query method as the baseline, that is, each record in the query dataset is similar to each record in the index dataset (Cartesian product), sorted by decreasing similarity, and the Top-K most similar records are found to generate blocks.

4.3 Implementation Details

All experiments were conducted on the Inter Core I5-12400 CPU @2.50GHz with 32GB of RAM. The record embed uses SBERT (sentence-bert), version all-mpnet-base-v2, and is not fine-tuned. The high dimensional vector nearest neighbor search based on the graph is implemented in C++. PQ, OPQ, IMIPQ, and LSH are implemented using the Faiss library provided by Facebook, and LOPQ is implemented using Python. The number of nearest neighbors K ranges from [10, 100] and the step size is 10. Euclidean distance is used in the similarity calculation.

4.4 Performance Measures

We use pair completeness (PC), reduction ratio (RR), Fα, search time (ST) and build time (BT) as its evaluation indicators. PC measures how many real matching candidate pairs are retained in the blocking results: $PC = |G \cap C|/|G|$, where G is the true matching pair, C represents the candidate pair. RR measures how many unnecessary comparisons are reduced: for dual-source datasets, $RR = 1 - |C|/(|D_A| \cdot |D_B|)$ and for single source datasets, $RR = 1 - 2|C|/(|D| \cdot (|D| - 1))$, where D represents all records in the single source dataset, and D_x represents all records in the dual-source dataset x. The Fα measures the overall performance of the blocking: $F\alpha = 2 \cdot PC \cdot RR/(PC + RR)$. The block generation time refers to the time spent from the input search vector to the output block.

4.5 Experimental Analysis

Overall Result Analysis

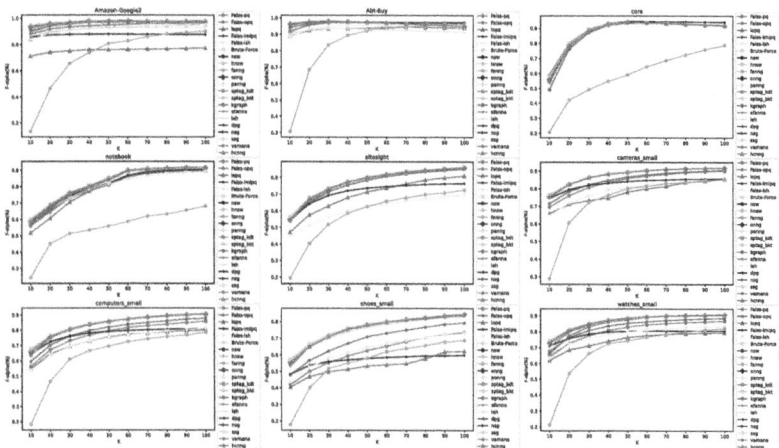

Fig. 3. Different blocking methods for Fα curves on different datasets.

In general, in the results shown in Fig. 3, almost all methods show an upward trend on each dataset as the number of nearest neighbors K increases. This is because the increase in the number of correct matches due to increasing K is more significant than the increase in the number of incorrect matches. It is worth noting that graph-based methods perform better in most cases compared to quantization and LSH-based blocking methods. Moreover, when $K \geq 60$, the Fα value tends to be stable and exceeds 90%, which highlights the importance of K size for balancing effectiveness and efficiency.

For graph-based blocking methods, all except FANNG achieve results close to brute-force queries, as they can identify the true K neighbors. Among quantization-based methods, OPQ outperforms others by better aligning the center of mass with the data distribution via R transformation. LSH-based methods perform poorly because LSH is probability-based, often failing to place matching records in the same or nearby buckets. Graph-based methods excel over quantization and LSH due to their preservation of original data and effective establishment of nearest neighbor relationships through the graph, while quantization loses data integrity and LSH's probabilistic nature can miss relevant buckets.

Blocking Efficiency Analysis

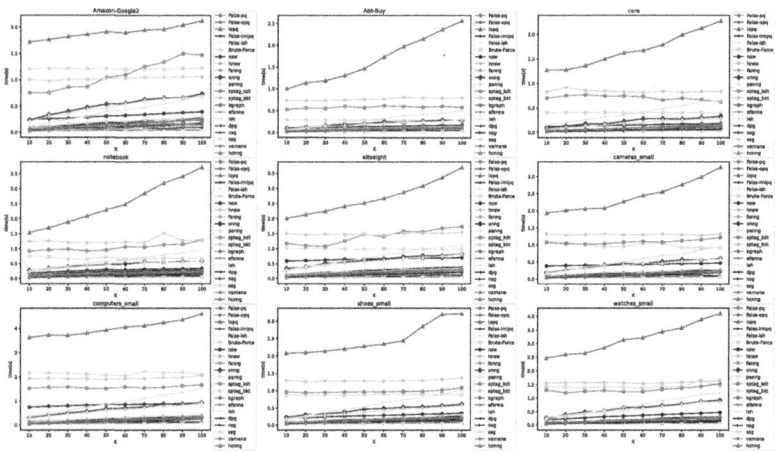

Fig. 4. Different blocking methods for block generation time curves on different datasets.

The comparison of block generation time results is shown in Fig. 4. The block generation time of most methods increases linearly with the continuous increase of K, and the time of many methods is less than the time spent on brute force search.

Most graph-based methods have less time than brute force search. However, the performance of SPTAG's BKT and KDT methods on many datasets is inferior to brute force search. The reason is that the dataset size is smaller but the dataset dimension is larger, and the initial time spent on BKT and KDT is more, so the efficiency is lower. And the Onng and Panng of NGT also have similar behavior to SPTAG. Except for LOPQ, quantization-based methods have low block generation time due to their smaller computational load and the fact that quantization-based methods are fast to perform full queries for each vector on small datasets. The LSH-based method is the fastest, but it comes at the cost of a certain amount of PC.

Figure 5 shows the time taken by different algorithms to build the index. As the dataset size increases, the construction time for most algorithms also increases. Although most algorithms have reasonable construction times, NSSG, NSG, SPTAG, and OPQ are notably slower. SPTAG's slower speed is due to the necessity of building a tree for high-dimensional vectors, impacting KDT and BKT's performance as well. NSG and NSSG are slow due to their complex edging processes. OPQ has high time complexity because it learns an orthogonal matrix R from the original dataset to transform the data. In contrast, other algorithms are relatively fast and scalable, making them suitable for large-scale datasets.

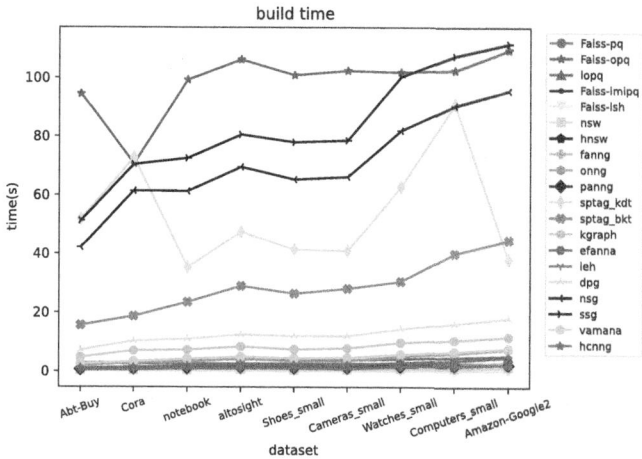

Fig. 5. Build times for different algorithms on different datasets.

Block Generation Methods Comprehensive Ranking. Table 2 shows the comprehensive ranking of all methods in each evaluation index. From a PC perspective, brute force retrieval means that true K-neighbors can be found for every query vector, so it ranks first on PC and SPATAG's method is second. It is oriented to the nearby points of the query vector through the Tree structure, and it uses space division to violently construct the KNN graph in each subspace, so its performance is also better. From the perspective of RR, FANNG and IMI have the largest reduction rates because they cannot always find the nearest K-neighbors of the query vector, which is at the cost of a poor PC. In terms of time efficiency, the LSH-based approach ranks first in both search time and build time, so it has the highest efficiency. In the comprehensive ranking (AR), IMI ranks first due to its favorable construction time and higher reduction rate, though it is not the top-performing algorithm overall. HNSW ranks second, offering fast index construction and high block quality, thus achieving the best balance between block quality and efficiency. NSW and PQ share the third position.

5   Related Work

The goal of ER is to find all records describing the same real-world entity from multiple data sources, and blocking is a key task of ER. The traditional block is mostly based on a token block key. Recently, a lot of attention has been paid to deep learning-based blocking methods, which obtain record embeddings through a deep learning model and then generate blocks based on the embeddings. DeepER [4] uses RNN or LSTM to acquire embeddings and block them using LSH. NLSHBlock [5] proposes a new method based on the LSH loss function to fine-tune the pre-trained language model to improve the quality of the block. DeepBlocker [2] uses fastText to take word embeddings and then aggregates them into record embeddings, generating candidate pairs from the cosine similarity between vectors.

Table 2. Blocking methods ranking.

Methods			PC	RR	ST	BT	AR
BF			1	18	19	–	12.67
LSH-based		LSH	20	22	1	1	11
Quantization-based		PQ	18	4	2	9	8.25
		OPQ	17	6	3	20	11.5
		IMI	19	2	4	6	**7.75**
		LOPQ	21	3	22	14	15
Graph-based	DG and RNG-based	NSW	5	10	15	3	8.25
		HNSW	10	8	12	2	8
		FANNG	22	1	7	7	9.25
		ONNG	8	7	17	12	11
		PANNG	9	5	16	8	9.5
	KNNG-based	S-KDT	4	21	20	18	15.75
		S-BKT	2	19	21	17	14.75
		KGraph	14	17	11	15	14.25
		EFANNA	15	16	8	5	11
		IEH	16	20	18	10	16
	KNNG and RNG-based	DPG	6	12	13	16	11.75
		NSG	7	9	6	19	10.25
		SSG	3	11	14	21	12.25
		Vamana	13	14	9	13	12.25
	MST-based	HCNNG	11	13	10	11	11.25

Combining deep learning models with high-dimensional vector search technology can improve the quality and efficiency of block generation. We focus on high-dimensional vector nearest-neighbor search technology. There are three main methods to high-dimensional vector nearest neighbor search: Locality Sensitive Hashing (LSH) based method, Product Quantization (PQ) based method, and Proximity graphs (PG) based method. In 2016, Andoni [6] et al. proposed a locally sensitive hashing method E2LSH based on Euclidean distance, which is a pioneering algorithm that applies Euclidean space to high-dimensional vector nearest neighbor search. In 2011, Jegou [7] et al. proposed a strategy based on product quantization to compress high-dimensional vectors. Wang [8] et al. analyzed and demonstrated the advanced graph-based methods in detail and proved the tradeoff between the effectiveness and efficiency of graph-based methods.

6 Conclusion

We introduce B-VSS, a blocking framework based on high-dimensional vector search, consisting of record embedding and block generation stages. Deep learning generates vector representations in the record embedding stage, while the block generation stage employs a high-dimensional vector nearest neighbor search. Our methods achieve performance comparable to brute force retrieval with significantly improved query efficiency and scalability across 9 datasets. Graph-based methods strike a balance between effectiveness and efficiency, while quantization and LSH-based methods offer the fastest search speeds. HNSW and NSW emerge as the top choice for balancing blocking quality, speed, and index construction. Future work may explore additional vector indexing techniques to enhance blocking quality and speed.

Acknowledgements. This work is supported by the National Natural Science Foundation of China (Grant Nos. 62002262, 62172082, 62072086, 62072084).

References

1. Du, J., Nie, T., Dou, W., Shen, D., Kou, Y.: SAREM: semi-supervised active heterogeneous entity matching framework. In: Zhao, X., Yang, S., Wang, X., Li, J. (eds.) WISA 2022. LNCS, pp. 77–88. Springer, Cham (2022). https://doi.org/10.1007/978-3-031-20309-1_7
2. Thirumuruganathan, S., et al.: Deep learning for blocking in entity matching: a design space exploration. Proc. VLDB Endow. **14**(11), 2459–2472 (2021)
3. Wang, J., Shen, D., Nie, T., Kou, Y.: A blockchain query optimization method based on hybrid indexes. In: Yuan, L., Yang, S., Li, R., Kanoulas, E., Zhao, X. (eds.) WISA 2023. LNCS, vol. 14094, pp. 455–466. Springer, Singapore (2023). https://doi.org/10.1007/978-981-99-6222-8_38
4. Li, B., Miao, Y., Wang, Y., Sun, Y., Wang, W.: Improving the efficiency and effectiveness for BERT-based entity resolution. In: AAAI 2021, pp. 13226–13233 (2021)
5. Wang, R., et al.: Neural locality sensitive hashing for entity blocking. CoRR abs/2401.18064 (2024)
6. Andoni, A., Indyk, P.: LSH algorithm and implementation (E2LSH) (2016). https://www.mit.edu/~andoni/LSH
7. Jégou, H., Douze, M., Schmid, C.: Product quantization for nearest neighbor search. IEEE Trans. Pattern Anal. Mach. Intell. **33**(1), 117–128 (2011)
8. Wang, M., Xu, X., Yue, Q., Wang, Y.: A comprehensive survey and experimental comparison of graph-based approximate nearest neighbor search. Proc. VLDB Endow. **14**(11), 1964–1978 (2021)
9. Ge, T., He, K., Ke, Q., Sun, J.: Optimized product quantization for approximate nearest neighbor search. In: CVPR 2013, pp. 2946–2953 (2013)
10. Kalantidis, Y., Avrithis, Y.: Locally optimized product quantization for approximate nearest neighbor search. In: CVPR 2014, pp. 2329–2336 (2014)
11. Babenko, A., Lempitsky, V.S.: The inverted multi-index. IEEE Trans. Pattern Anal. Mach. Intell. **37**(6), 1247–1260 (2015)
12. Malkov, Y., Ponomarenko, A., Logvinov, A., Krylov, V.: Approximate nearest neighbor algorithm based on navigable small world graphs. Inf. Syst. **45**, 61–68 (2014)

13. Malkov, Y.A., Yashunin, D.A.: Efficient and robust approximate nearest neighbor search using hierarchical navigable small world graphs. IEEE Trans. Pattern Anal. Mach. Intell. **42**(4), 824–836 (2020)

14. Harwood, B., Drummond, T.: FANNG: fast approximate nearest neighbour graphs. In: CVPR 2016, pp. 5713–5722 (2016)

15. Iwasaki, M., Miyazaki, D.: Optimization of indexing based on k-nearest neighbor graph for proximity search in high-dimensional data. CoRR abs/1810.07355 (2018)

16. Iwasaki, M.: Pruned bi-directed K-nearest neighbor graph for proximity search. In: Amsaleg, L., Houle, M., Schubert, E. (eds.) SISAP 2016. LNCS, vol. 9939, pp. 20–33. Springer, Cham (2016). https://doi.org/10.1007/978-3-319-46759-7_2

17. Chen, Q., et al.: SPTAG: a library for fast approximate nearest neighbor search (2018). https://github.com/Microsoft/SPTAG

18. Dong, W.: KGraph: a library for approximate nearest neighbor search (2011). https://github.com/aaalgo/kgraph

19. Fu, C., Cai, D.: EFANNA: an extremely fast approximate nearest neighbor search algorithm based on kNN graph. CoRR abs/1609.07228 (2016)

20. Jin, Z., Zhang, D., Yao, H., Lin, S., Cai, D., He, X.: Fast and accurate hashing via iterative nearest neighbors expansion. IEEE Trans. Cybern. **44**(11), 2167–2177 (2014)

21. Li, W., et al.: Approximate nearest neighbor search on high dimensional data - experiments, analyses, and improvement. IEEE Trans. Knowl. Data Eng. **32**(8), 1475–1488 (2020)

22. Cong, F., Xiang, C., Wang, C., Cai, D.: Fast approximate nearest neighbor search with the navigating spreading-out graph. Proc. VLDB Endow. **12**(5), 461–474 (2019)

23. Cong, F., Wang, C., Cai, D.: High dimensional similarity search with satellite system graph: efficiency, scalability, and unindexed query compatibility. IEEE Trans. Pattern Anal. Mach. Intell. **44**(8), 4139–4150 (2022)

24. Jayaram Subramanya, S., Devvrit, F., Simhadri, H.V., Krishnawamy, R., Kadekodi, R.: DiskANN: fast accurate billion-point nearest neighbor search on a single node (2019)

25. Muñoz, J.A.V., Gonçalves, M.A., Dias, Z., da Silva Torres, R.: Hierarchical clustering-based graphs for large scale approximate nearest neighbor search. Pattern Recogn. **96** (2019)

26. Datar, M., Immorlica, N., Indyk, P., Mirrokni, V.S.: Locality-sensitive hashing scheme based on p-stable distributions. In: SCG 2004, pp. 253–262 (2004)

Top-k Collective Spatial Keyword Approximate Query

Xiangfu Meng[1]([✉]), Zilun Zhang[1], Shuolin Cui[2], and Hongjin Huo[1]

[1] School of Electronic and Information Engineering, Liaoning Technical University, Huludao, China
marxi@126.com
[2] Glasgow International College, University of Glasgow, Glasgow, UK
3025865c@student.gla.ac.uk

Abstract. The rapid expansion of spatial textual data, covering location and textual information, has spurred extensive research and application of spatial keyword query technology. Traditional methods focus on identifying groups of spatial objects that satisfy spatial keyword queries but often overlook the relationships between these objects, such as social correlations. To address this problem, this paper proposes a top-k collective spatial keyword approximate query approach. Firstly, an association rule-based social relationship evaluation method for spatial objects is proposed. Then, we design a scoring function that combines the location distances and social relationships of spatial objects within a group. Secondly, a Vantage Point Tree (VP-Tree) based pruning strategy is proposed for quickly searching the local neighborhood of spatial objects. Finally, the top-k spatial object groups are selected as the query result by leveraging the scoring function to calculate the score of candidate object groups. The experimental results demonstrate that the proposed social relationship evaluation method can achieve high accuracy, the proposed pruning strategy has high execution efficiency, and the obtained top-k groups of spatial objects can further meet users' needs and preferences well.

Keywords: Spatial keyword query · Semantic similarity · Road network · VP-Tree

1 Introduction

With the development of GPS, mobile networks, and smart devices, the quantity of spatial textual objects has increased, enhancing Location-Based Services (LBS) [1–3]. Collective Spatial Keyword Query (CSKQ) [4–9] considers groups of spatial objects to identify the most cost-effective groups satisfying keyword queries, covering all query keywords, proximity to the query location, and spatial proximity within the group. However, CSKQ methods often yield only a single object, overlook social relationships among objects, and require substantial computational resources, impacting system responsiveness. To address these issues, this paper introduces Top-k Collective Spatial Keyword Approximate Query (Top-k CSKAQ), optimizing spatial and textual data handling to enhance query efficiency and practicality.

Fig. 1. Spatial objects and their textual descriptions

Table 1. Location information of spatial objects

Spatial Objects	Location Information
O_1	$(2, 8)$
O_2	$(4, 6)$
O_3	$(6, 1)$
O_4	$(9, 5)$
O_5	$(7, 7)$
O_6	$(1, 5)$
O_7	$(7, 2)$
O_8	$(2, 8)$

Table 2. User check-in records for the spatial objects in Fig. 1

Users	Check-in records
U_1	O_2, O_3, O_4
U_2	O_1, O_2
U_3	O_3, O_4, O_1
U_4	O_4, O_5, O_7
U_5	O_3, O_8
U_6	O_6, O_7
U_7	O_6, O_2, O_8
U_8	O_2, O_5

The basic idea of the approach is illustrated with a specific example. Figure 1, Table 1, and Table 2 show 8 spatial objects $\{O_1, O_2 \ldots O_8\}$ and their check-in records, respectively. Suppose a visitor stays at location $q.l = <5, 4>$, the query keywords $q.w =$ Shopping, Restaurant, Hotel. The top-1 result returned by the traditional CSKQ method is $\{O_3, O_8, O_7\}$ and the top-2 result denotes

$\{\{O_3, O_8, O_7\}, \{O_6, O_2, O_8\}\}$. However, if we use the Top-k CSKAQ method proposed in this paper. The top-1 result would be $\{O_6, O_2, O_8\}$ while the top-2 result would be $\{\{O_6, O_2, O_8\}, \{O_3, O_8, O_7\}\}$. According to the user check-in records shown in Table 2, it can be observed that O_6, O_2 and O_8 are more likely to be associated visited.

The contributions of this paper are summarized as follows:

- An associated rule based algorithm is presented to evaluate the social relationships between spatial objects.
- A VP-Tree based pruning strategy is proposed to prune spatial objects, reducing the amount of computation and speeding up the query execution of spatial object neighborhood matching.
- A new scoring function is designed to evaluate candidate spatial object groups. The candidate spatial objects groups are ranked, and the approximate results of top-k groups are selected.
- A large number of experiments are conducted on the Yelp dataset to evaluate the efficiency and effectiveness of the proposed Top-k CSKAQ approach.

2   Related Work

Traditional Spatial Keyword Query (SKQ) methods use the query location (q.l) and query keywords (q.w) to return matching spatial objects but often overlook user preferences and the semantic relationships. To address these limitations, Qian et al. [10–12] initially enhanced spatial proximity and textual similarity. They later developed semantic-aware SKQ methods that also consider semantic relevance [10].

Building on individual object retrieval, the field has shifted towards methods addressing groups of objects within road networks, where most practical applications occur. Cao et al. [13,14] introduced the Collective Spatial Keyword Query (COSKQ), which retrieves groups of objects whose keywords collectively cover the query keywords. Su et al. [15] further refined this approach by developing a method that operates without an index structure, simplifying the retrieval process.

The concept of personalization in spatial queries was advanced by Yu et al. [16], who proposed the Personalized Spatial Keyword Group Query (CKPQ), and Park et al. [17], who defined a Reverse Collective Spatial Keyword Query. These methods aim to better meet personalized user requirements by considering user preferences in the query process.

Recent studies have also explored the application of these methods to road networks, which are critical for practical implementations. Su et al. [18], along with Gao et al. [19–21], have specifically focused on the CSKQ problem within road networks. Their research enhances the accuracy of user query results, improves system efficiency, and increases user satisfaction by integrating social relationships into the querying process.

In summary, while traditional SKQ methods focus on identifying individual spatial objects, recent advancements in the field have emphasized the importance of spatial proximity, user preferences, and social interactions. This evolution reflects a broader trend towards developing more sophisticated, contextually aware querying systems that can cater to the complex needs of users in dynamic environments.

3   Definitions

Definition 1. *Top-k CSKAQ. Given a spatial object set D, user check-in records R, spatial keyword query q includes query position q.l, a group of query keywords q.w and k, Top-k CSKAQ returns the top-k group spatial objects with the smallest synthetic distance, the synthetic distance is determined by the scoring function.*

Definition 2. *Social relationships between spatial objects. Given a spatial objects set D and user check-in records R, the more times a group of objects in D is jointly accessed by users, the higher the social relationships between the objects in the group.*

Definition 3. *Scoring function. Given a set of spatial objects, the synthetic distance scoring function includes two parts: distance relationships and social relationships. The calculation method is shown in Eq. (1):*

$$\text{score}(h) = \alpha \left(\frac{\beta \max \text{dis}(q,c) + (1-\beta) \max \text{dis}(p_1, p_2)}{\max \text{dis}(p_1, p_2)} \right) + (1-\alpha)(1-GL) \quad (1)$$

where, h is the candidate spatial object group; α is the parameter which is used to balance the effects of distance and social relationships in the scoring function. maxdis (p_1, p_2) is the maximum Euclidean distance between spatial objects in the dataset; GL is the social relationships between the spatial objects in h; maxdis (q, c) represents the maximum Euclidean distance between q and spatial objects in h; maxdis (o_1, o_2) represents the distance part of the scoring function, which is denoted as the distance score, β is the parameter which is 0.5.

Definition 4. *Local neighborhood. Given a spatial objects set $D, Z(Z \subseteq D)$ and the neighborhood threshold σ, the σ-local neighborhood of Z is defined as,*

$$L(Z, D, \sigma) = \left\{ o \mid o \in D, \min_{o' \in Z} \{\text{dis}(o, o')\} \le \sigma \right\} \quad (2)$$

4   Accurate Selection Algorithm of Top-k CSKAQ

4.1   Evaluation of Social Relationships Between Spatial Objects

Apriori algorithm can find the potential association rule between spatial objects from users' check-in records R, so as to evaluate the social relationships. Apriori algorithm first scans the dataset R to generate an itemset C_1, and then scans C_1 to filter items that do not meet the minimum support and obtain frequent itemset L_1. According to the Apriori principle, that is, all non-empty subsets of frequent itemsets must be frequent. Therefore, in the second round of iteration, it is only necessary to make a new combination of the frequent itemsets generated in the previous round of iteration. Then, we need to filter the new combination that does not meet the minimum support requirements until no new combination can be generated. The algorithm generates a non duplicate itemset C_k through the frequent itemset L_{k-1}. It returns the itemset no less than the minimum support as the frequent k itemset L_k, and adds L_k to L. Finally, we select each frequent itemset in L circularly, and calculate the confidence conf for it, If conf $\geq$ minimum confidence, we add the itemset as a social relationships analysis result to the association rule set RL. The minimum confidence and the minimum support can be set according to the real application requirements.

According to the content shown in Fig. 1, Table 1, and Table 2, we analyze the social relationship between O_2 and O_6. The support of O_6 is 1/4, O_2 is 1/2, and $O_2 \cap O_6$ is 1/8; The social relationships between O_2 and O_6 is $GL(O_6 \mid O_2) = 1/8/1/2 = 0.25$. This indicates that 25% of users who have visited O_2 will visit O_6.

4.2 Accurate Selection Algorithm

The Exact Collective Spatial Keyword Approximate Query (CSKAQ-E) algorithm first deletes the objects that do not contain any query keywords in spatial object set D to obtain relevant spatial objects set RO. Spatial objects in RO are then sorted in ascending order according to the distance to query q. The spatial objects in RO are selected one by one in a sequential cycle, which are recorded as c. And then, the spatial objects are added whose distance from query q is less than dis(q, c) (the distance between c and q) to set S_1. dis (q, c) is the maximum distance between the objects in the group and q. Based on this, we make difference set between $q.w$ (query keywords set) and $c.w$ (keyword set in spatial object c) to get $s.w$. The objects in S_1 are grouped according to the keywords in $s.w$ to get the sets $sw_1, sw_2, \ldots, sw_i$ (i is the number of keywords in $s.w$), and these sets are sorted in ascending order according to the distance. Next, we perform mathematical combination operations on spatial objects in $sw_1, sw_2, \ldots, sw_i$ to obtain the spatial object group $x = \{o_1, o_2, \ldots, o_i\}$. The candidate spatial object group h is obtained by combining c and x, denoted as $h = \{c, o_1, o_2, \ldots, o_i\}$ and the score(h) is calculated by Eq. (1).

Example 1: According to the content shown in Fig. 1, Table 1, and Table 2. The $q.l = <5, 4>$ and the $q.w$ is "Shopping, Restaurant, Hotel". According to the above query conditions, we firstly delete the irrelevant object O_1 in the spatial object set shown in Table 1, obtain the related object set RO, and sort it in ascending order. The sorting result is $RO = \{O_2, O_8, O_7, O_3, O_5, O_4, O_6\}$. Then, we obtain top-3 results of the CSKAQ-E algorithm.

5   Approximate Selection Algorithm of Top-k CSKAQ

5.1   VP-Tree

For a given relevant spatial object set RO, VP-Tree is built in a top-down mode. It is necessary to select a vantage point, specify a distance function and a threshold. Firstly, according to the idea of literature [16], we randomly select a group of objects in RO as candidate vantage points, then use the remaining objects to evaluate them. Next, we select the object that can build a highly balanced VP-Tree as vantage point. Finally, the objects in the dataset whose Euclidean distance from the vantage point is not higher than the given threshold are divided into the left subtree. Otherwise, they are divided into the right subtree. The left and right subtrees are further divided recursively until the node contains only one object (leaf node). The time complexity is $O(|RO| \log |RO|)$. VP-Tree can be used for the nearest neighbor query and range query.

5.2   Local Neighborhood Search

Given a threshold σ, for any $r \in RO$, we calculate the σ-local neighborhood $L(\{r\}, D, \sigma)$. However, it will take a long time to calculate the σ-local neighborhood for each spatial object. Therefore, in order to improve the efficiency of the σ-local neighborhood search, this paper uses VP-Tree to accelerate the process.

Given an object $r \in RO$, the process of using VP-Tree to search the σ-local neighborhood of r is as follows. The algorithm compares the Euclidean distance between each node in VP-Tree and the given object from top to bottom. If an intermediate node in VP-Tree is in $L(\{r\}, D, \sigma)$, all subsequent objects of the node are in $L(\{r\}, D, \sigma)$. Ideally, the time complexity of VP-Tree search is $O(\log |RO|)$. Thus, VP-Tree can quickly obtain other spatial objects close to the given spatial objects, which provides support for quickly obtaining top-k group spatial objects.

5.3   Approximate Selection Algorithm of Top-k CSKAQ

This section proposes CSKAQ-A algorithm based on pruning strategy. The algorithm uses VP-Tree to search the local neighborhood of spatial object to prune the spatial objects, which reduces the possibility of combination and the amount of calculation.

The Approximate Collective Spatial Keyword Approximate Query (CSKAQ-A) algorithm first initializes J, h, x, J_{maxc}, and builds the relevant spatial object set RO into VP-Tree. Next, the spatial object whose distance from the query q is less than dis (q, c) is added to the set S_1. Moreover, we use VP-Tree to search the local neighborhood of c to obtain the σ-local neighborhood set S_2 of c. Based on this, the algorithm calculate $S_1 \cap S_2 = M$. This strategy can avoid repeated calculation. Then, we make difference set between $q.w$ (query keywords set) and $c.w$ (keyword set in spatial object c) to get $s.w$. The objects in M are grouped according to the keywords in $s.w$ to get the sets $sw_1, sw_2, \ldots, sw_i$ (i is the number of keywords in $s.w$), and these sets are sorted in ascending order according to the distance. The subsequent steps are the same as CSKAQ-E algorithm.

By $S_1 \cap S_2 = M$, the combination possibilities and computational load are reduced, avoiding redundant calculations. The σ-local neighborhood of c, denoted as $L(\{c\}, RO, \sigma)$, is processed by the CSKAQ-A algorithm to retain only spatial objects near c, reducing combinable objects. Assuming c_1 is within $L(\{c\}, RO, \sigma)$, we derive $L(\{c_1\}, RO, \sigma)$ using a VP-tree. The algorithm randomly forms combinations (o_1, o_2) from $L(\{c\}, RO, \sigma)$, cycles through these combinations, and pairs them with c to create spatial object group x. Here, c and c_1 are combined once. If c_1 is cyclically selected, $L(\{c_1\}, RO, \sigma)$ includes c, potentially causing repeated combinations of c and c_1, preventable by calculating $S_1 \cap S_2 = M$. S_1 includes objects closer to q than $\mathrm{dis}(q, c)$. If $\mathrm{dis}(q, c_1) > \mathrm{dis}(q, c)$, c_1 won't appear in $c's$ M set, but c will be in $c_1's$ M set, necessitating a single combination and calculation for c and c_1.

Example 2: According to the content shown in Fig. 1, Table 1, and Table 2, we obtain top-3 results of the CSKAQ-A algorithm. The processing process of CSKAQ-A algorithm is as follows:

Algorithm first builds a VP-Tree for the relevant spatial object set RO. Compared with CSKAQ-E algorithm, CSKAQ-A algorithm not only obtains the spatial object set S_1 whose distance from q is less than the distance from c to q, but also uses VP-tree to accelerate the search of σ-local neighborhood set S_2. Then, the algorithm calculates $S_1 \cap S_2 = M$. Then, we make difference set between $q.w$ (query keywords set) and $c.w$ (keyword set in spatial object c) to get $s.w$. The objects in M are grouped according to the keywords in $s.w$ to get the sets $sw_1, sw_2 \ldots sw_i$ (i is the number of keywords in $s.w$).

According to the above discussion, no candidate object combination is generated when $c = O_2, O_8$ and O_7. When $c = O_3$, $S_1 = \{O_2, O_8, O_7\}$, the σ-local neighborhood of c is $S_2 = \{O_8, O_7\}$, $S_1 \cap S_2 = M = \{O_8, O_7\}$, $c.w = $ "Shopping", $s.w = q.w - c.w = $ "Restaurant, Hotel". Next, the objects in M are grouped according to the keywords in $s.w$ to obtain sets sw_1 and sw_2 respectively, where sw_1(Restaurant) $= \{O_8\}$, sw_2(Hotel) $= \{O_7\}$. At this time, $x = \{O_8, O_7\}$, $h = \{\{O_3, O_8, O_7\}\}$, score $(\{O_3, O_8, O_7\}) = 0.933$. Similarly, we can obtain candidate spatial object groups when $c = O_5, O_4$, and O_6. The rest of the process is the same as the CSKAQ-E algorithm. And the top-3 results obtained by CSKAQ-A algorithm is: $\{\{O_4, O_7, O_3\}, \{O_6, O_8, O_2\}, \{O_3, O_8, O_2\}\}$.

6   Experiment

6.1   Data and Experimental Settings

The experimental data uses the Yelp dataset, which contains 20,290 spatial objects and 74,773 pieces of user comments. Each spatial object contains attributes such as ID, latitude and longitude, category, etc., and each review dataset contains user ID, spatial object ID and comment content.

Table 3 shows the parameter setting in this paper. During the experiment, the influence of one parameter on the experimental results is discussed by varying the value of it and fixing the value of other parameters. All experiments are implemented in Java. The computer is configured with 32.0 GB RAM of CPU-3.7 GHz and Windows 10 operating system.

The accurate selection algorithm (CSKAQ-E) and the approximate selection algorithm (CSKAQ-A) proposed in this paper are compared with the RC algorithm in the TkCoSKQ method (TkCOSKQ-RC) [18] and the RKD-search algorithm in the CKPQ method proposed in the latest related research work (CKPQ-RKD) [16].

Table 3. Default values for parameters

Parameters	Default Values
$|D|$	20290
k	10
|Keywords|	2
Threshold σ	0.025
parameter α	0.7

6.2 Analysis of Experimental Results

This section examines the impact of the parameters in Table 3 on the query effectiveness and efficiency of all algorithms using the same dataset. Each algorithm retrieves the top-10 groups of spatial objects for a given spatial keyword query q.

Query Efficiency Experiment. This experiment measures the query response time of four algorithms as parameters change under the same query conditions. The proposed approximate selection algorithm demonstrates the fastest query speed. Although the accurate selection algorithm is slower than the comparison algorithm, the difference is not significant.

(1) The influence of k on response time: Fig. 2(a) shows the query response times for each algorithm with different k. CSKAQ-E, which obtains social relationships between spatial objects, runs slower than TkCoSKQ-RC. CKPQ-RKD uses a pruning strategy, making it faster than both. CSKAQ-A prunes farther objects, resulting in the shortest response time.
(2) The influence of the number of keywords ($|q.w|$) on response time is evident from Fig. 2(b), showing that as $|q.w|$ increases, the query response time for all four algorithms also increases, and the disparity between TkCoSKQ-RC, CSKAQ-E, and CSKAQ-A widens. This is due to the increased number of spatial objects that need processing. Both CKPQ-RKD and CSKAQ-A employ a pruning strategy, which reduces their response times. However, CKPQ-RKD is slower than CSKAQ-A as it accounts for both the distance and the user preferences and accessibility of objects within the group.

(3) The impact of $|D|$ on response time: Fig. 2(c) shows that as the dataset grows, the query response time of all four algorithms increases due to more relevant spatial objects. For small datasets, TkCoSKQ-RC has the shortest response time as it doesn't require index structure. However, as $|D|$ increases, CSKAQ-A becomes significantly faster than CSKAQ-E and TkCoSKQ-RC, with the gap widening. Additionally, CSKAQ-A is faster than CKPQ-RKD.

(4) The influence of threshold σ on response time: Fig. 2(d) shows that the query response time of CSKAQ-A increases with σ due to the growing number of objects in the local neighborhood set, which raises the possibility of combinations and the amount of calculation.

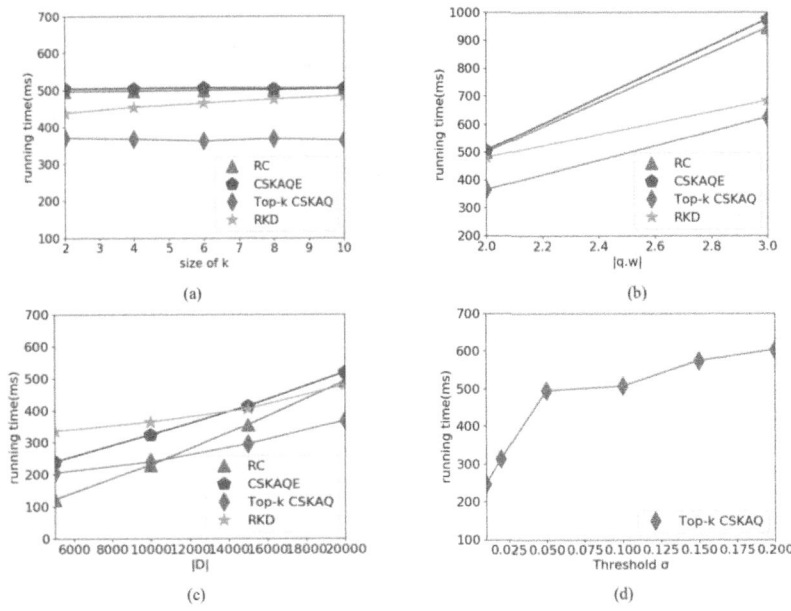

Fig. 2. The query response time $(k, |keywords|, |D|, \sigma)$

Query Performance Experiment. This experiment aims to use the Approximate Ratio (ARatio) and Accuracy to evaluate the query effect. Under the same query conditions, we can observe their Approximate Ratio and Accuracy with the change of parameters, and verify that the query results of the accurate selection algorithm proposed in this paper are the most accurate. The approximate selection algorithm also has high accuracy. Accuracy is used to evaluate query performance, and it is calculated as follows:

$$accuracy = \frac{|I(q)IR(q)|}{I(q)} \tag{3}$$

where, $I(q)$ is the accurate result set obtained by CSKAQ-E, and $R(q)$ is the result set returned by TkCOSKQ-RC, CSKAQ-A and CKPQ-RKD, respectively.

Approximate ratio is widely used to measure the performance of collective spatial keyword query [15, 22, 23]. The approximate ratio is calculated as follows:

$$\text{ARatio} = \frac{AVG(\text{distance})}{AVG(\text{Cost})} \tag{4}$$

where, AVG(distance) is the average of distance scores obtained by CSKAQ-E, CSKAQ-A and CKPQ-RKD, respectively. AVG(Cost) represents the average synthetic distance obtained by TkCOSKQ-RC. If ARatio is ≤ 1, the result is better.

(1) The influence of k on the query effect. Figure 3 shows the query results of each algorithm under different k. When $k < 6$, the approximate ratio of CSKAQ-E, CSKAQ-A and CKPQ-RKD is slightly greater than 1. However, when $k > 6$, the accuracy of TkCoSKQ-RC method decreases, but the approximate ratio of CSKAQ-E is less than 1, indicating that when the required query results increase, the performance of CSKAQ-E is better, and the query results are better than TkCoSKQ-RC in distance. The approximate ratio of CSKAQ-A is always slightly greater than 1, because CSKAQ-A method uses a pruning strategy to get approximate query results.

(2) The influence of the $|q.w|$ on the query effect. Figure 4 shows the approximate ratio and accuracy of the four algorithms with different number of query keywords. The accuracy of TkCoSKQ-RC is not particularly high, and the approximate ratio of CSKAQ-E is less than 1. This is because some better group results may be pruned away when TkCoSKQ-RC considers the upper and lower boundary conditions. It indicates that the results obtained by the CSKAQ-E are more accurate.

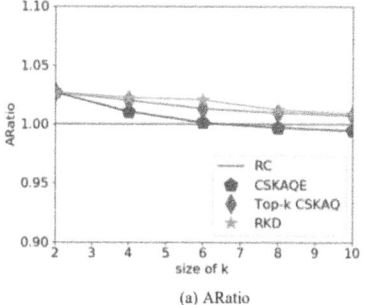
(a) ARatio

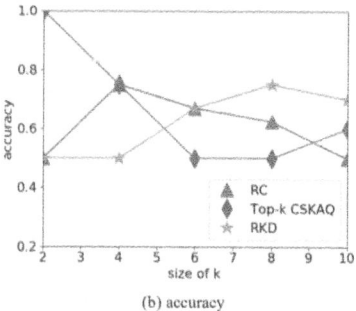
(b) accuracy

Fig. 3. The influence of k on the query effect

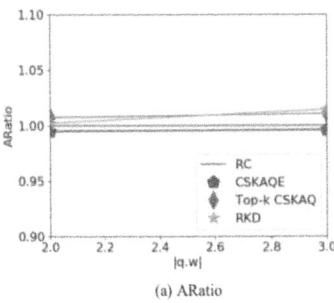
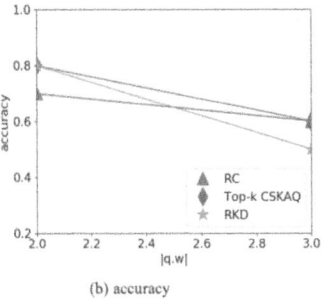

(a) ARatio   (b) accuracy

Fig. 4. The influence of the $|q.w|$ on the query effect

7   Conclusion

This paper proposes a new collective spatial keyword approximate query method, named Top-k CSKAQ. This method not only considers whether the spatial object group can cover all query keywords, but also synthetically considers the distance relationships (the distance between the objects and query q, and the distance between the spatial objects in the group) and social relationships between spatial objects in the group. We used the Apriori algorithm to evaluate social relationships between spatial objects. In addition, we proposed a pruning strategy based on the VP-Tree to optimize the solution and realize the approximate query. Finally, experimental results on the Yelp dataset demonstrated the performance of the algorithm proposed in this paper. In future work, we will avoid putting spatial objects with the same keyword in the same group, and consider the diversity of objects in the group.

References

1. Yu, X., Zhu, S., Ren, Y.: Continuous trajectory similarity search with result diversification. Future Gener. Comput. Syst. **143**, 392–400 (2023)
2. Luo, C., Wang, P., Li, Y., Zheng, B., Li, G.: Efficient time-interval augmented spatial keyword queries on road networks. Inf. Sci. **593**, 505–526 (2022)
3. Li, J., Xiong, X., Li, L., He, D., Zong, C., Zhou, X.: Finding top-k optimal routes with collective spatial keywords on road networks. In: 2023 IEEE 39th International Conference on Data Engineering (ICDE), pp. 368–380. IEEE (2023)
4. Liu, H., Sun, Y., Wang, G.: Continuous spatial keyword query processing over geo-textual data streams. World Wide Web **26**(3), 889–903 (2023)
5. Chen, L., Shang, S., Yang, C., Li, J.: Spatial keyword search: a survey. GeoInformatica **24**, 85–106 (2020)
6. Chen, Z., Zhao, T., Liu, W.: Time-aware collective spatial keyword query. Comput. Sci. Inf. Syst. **18**(3), 1077–1100 (2021)
7. Chen, Z., Chen, L., Cong, G., Jensen, C.S.: Location-and keyword-based querying of geo-textual data: a survey. VLDB J. **30**(4), 603–640 (2021)
8. Tong, Y., et al.: Hu-Fu: efficient and secure spatial queries over data federation. Proc. VLDB Endow. **15**(6), 1159 (2022)

9. Gong, Z., Li, J., Lin, Y., Wei, J., Lancine, C.: Efficient privacy-preserving geographic keyword Boolean range query over encrypted spatial data. IEEE Syst. J. **17**(1), 455–466 (2022)
10. Qian, Z., Xu, J., Zheng, K., Zhao, P., Zhou, X.: Semantic-aware top-k spatial keyword queries. World Wide Web **21**, 573–594 (2018)
11. Luo, S., Luo, Y., Zhou, S., Cong, G., Guan, J., Yong, Z.: Distributed spatial keyword querying on road networks. In: EDBT, pp. 235–246. Citeseer (2014)
12. Cho, H.-J., Kwon, S.J., Chung, T.-S.: ALPS: an efficient algorithm for top-k spatial preference search in road networks. Knowl. Inf. Syst. **42**, 599–631 (2015)
13. Zhang, D., Chan, C.-Y., Tan, K.-L.: Processing spatial keyword query as a top-k aggregation query. In: Proceedings of the 37th International ACM SIGIR Conference on Research & Development in Information Retrieval, pp. 355–364 (2014)
14. Haque, S., Eberhart, Z., Bansal, A., McMillan, C.: Semantic similarity metrics for evaluating source code summarization. In: Proceedings of the 30th IEEE/ACM International Conference on Program Comprehension, pp. 36–47 (2022)
15. Su, S., Zhao, S., Cheng, X., Bi, R., Cao, X., Wang, J.: Group-based collective keyword querying in road networks. Inf. Process. Lett. **118**, 83–90 (2017)
16. Park, S., Park, S.: Reverse collective spatial keyword query processing on road networks with g-tree index structure. Inf. Syst. **84**, 49–62 (2019)
17. Liu, H., Xu, J., Zheng, K., Liu, C., Du, L., Wu, X.: Semantic-aware query processing for activity trajectories. In: Proceedings of the Tenth ACM International Conference on Web Search and Data Mining, pp. 283–292 (2017)
18. Su, D., Zhou, X., Yang, Z., Zeng, Y., Gao, Y.: Top-k collective spatial keyword queries. IEEE Access **7**, 180779–180792 (2019)
19. Zheng, K., et al.: Interactive top-k spatial keyword queries. In: 2015 IEEE 31st International Conference on Data Engineering, pp. 423–434. IEEE (2015)
20. Gao, Y., Zhao, J., Zheng, B., Chen, G.: Efficient collective spatial keyword query processing on road networks. IEEE Trans. Intell. Transp. Syst. **17**(2), 469–480 (2015)
21. Duan, G., Ma, S., Wen, Y.: Exact query in multi-version key encrypted database via bloom filters. In: Yuan, L., Yang, S., Li, R., Kanoulas, E., Zhao, X. (eds.) WISA 2023. LNCS, vol. 14094, pp. 415–426. Springer, Singapore (2023). https://doi.org/10.1007/978-981-99-6222-8_35
22. Tang, J., Lu, X., Xiang, Y., Shi, C., Gu, J.: Blockchain search engine: its current research status and future prospect in Internet of Things network. Future Gener. Comput. Syst. **138**, 120–141 (2023)
23. Yao, J., Bao, X.: Interactively mining interesting spatial co-location patterns by using fuzzy ontologies. In: Yuan, L., Yang, S., Li, R., Kanoulas, E., Zhao, X. (eds.) WISA 2023. LNCS, vol. 14094, pp. 112–124. Springer, Singapore (2023). https://doi.org/10.1007/978-981-99-6222-8_10

SMSRD: A Streaming Graph Data Management System Based on Relational Database

Yi Luo[1], Peng Ren[2(✉)], Weifan Wang[3(✉)], Xianbo Liu[1(✉)], Yuhang Hu[1], Zeming Li[1], Xiangkuan Li[4], Wenyao Li[5], and Chunxiao Xing[2]

[1] School of Computer Science and Technology, Beijing Institute of Technology, Beijing 100081, China
{luoyi,liuxianbo,huyuhang_21,lizeming}@bit.edu.cn
[2] BNRist, DCST, RIIT, Tsinghua University, Beijing 100084, China
{renpeng,xingcx}@tsinghua.edu.cn
[3] School of Artificial Intelligence, Henan University, Zhengzhou 450046, China
henuwwf@henu.edu.cn
[4] Room 602, Unit 2, Building 19, Yuquanxincheng, Laoshan Street, Shijingshan District, Beijing 100049, China
[5] School of Software, Henan University, Kaifeng 475004, China

Abstract. Graphs are one of the most widely used data structures, and with the explosion of data volume, static graph analysis and storage models have become inadequate to meet application demands. This has led to the emergence of dynamic graphs, where graphs exist in the form of streams, leading to the requirements for managing streaming graph data. Existing systems face challenges in simultaneously providing high throughput, data accessibility, and comprehensive streaming data management capabilities, especially under the pressure of handling large data volumes. To address these challenges, we presents a streaming data management system designed upon relational databases. Based on the current application requirements of streaming graph computation and data management, we design the entire framework and unify the top interface for applications, making significant optimizations for high throughput from the aspect of asynchronous and parallel data writing, data storage and sharding strategy. Experimental results validate the system based on relational database outperforms system based on graph database, effectively addressing the demands of streaming graph data management.

Keywords: Streaming graph · Data management · Relational database

1 Introduction

In the modern information age, networks are integral to human society, with graphs serving as their computational representation at the computer level. In

Supported by the National Natural Science Foundation of China under Grant 62076027.

real-world applications, graphs undergo continuous updates, with edges arriving sequentially over time, shaping an evolving dynamic graph. Streaming graph data epitomizes this abstract concept, illustrating interactions among graph nodes at different timestamps. The temporal connections between these nodes form edges, thereby organizing streaming graph data chronologically.

Streaming graphs are often of considerable magnitude, prompting the development of novel frameworks for their management [10]. This management encompasses two fundamental facets: timely storage of rapidly incoming graph data and furnishing high-performance support for analytical operations. However, existing graph management systems frequently falter in delivering efficient performance across these domains. Certain dynamic graph systems solely accommodate batch data updates [7], whereas Neo4j, despite its prowess in static analysis, demonstrates sluggish insertion rates in practical tests, as delineated in Sect. 2.

In contrast, relational databases possess excellent read-write performance, intuitive storage models, and mature product features, making them ideally suited for storing and managing streaming graph data. Therefore, this paper aims to address the challenge of streaming graph data management by implementing a streaming graph data management system based on the relational database MySQL, using the increasingly popular Go programming language. Our system offers rapid data write speeds and is designed with different graph query interfaces tailored to the characteristics of streaming data, thus tackling the issue of streaming graph data management.

Our approach has two innovations:

1. Our management system implements timely caching of graph data as it arrives at high speeds.
2. Our management system provides high-performance support for analytical tasks, including asynchronous writing, concurrent querying, and significant optimizations of read-write operations leveraging unique features of the Go programming language.

We are the first to extend the work of relational databases into the context of streaming graph data management, with establishing an experimental framework for streaming data management, offering both relational database MySQL and graph database Neo4j as optional backend databases. We conduct experiments to verify the performance of these databases across various tasks. The results demonstrate the superior performance of the streaming graph management system based on relational databases in various tasks.

Roadmap: In the rest of the paper, we begin by discussing related work in Sect. 2. Section 3 gives the problem formulation, while Sect. 4 talks about system structure and outline optimization strategies for analytical tasks. The experimental results and analysis on streaming graph data will be presented in Sect. 5, followed by the final conclusions in Sect. 6.

2   Related Work

In the domain of streaming graph data management, two focal points arise: the structuring of data storage, which encompasses the foundational organization and storage format, and data read-write optimization, aimed at enhancing computational analytical tasks.

Within the domain of graph data management, Neo4j stands as a prominent choice owing to its maturity as a graph database, fortified by its robust Cypher query language and formidable static analysis capabilities. However, due to the strong consistency guarantee provided by Neo4j [2], while there are instances of Neo4j being combined with Kafka to create middleware for streaming data [3], the throughput capacity of system is a critical consideration for evaluating a streaming system. Consequently, Neo4j is not ideally suited as a backend database for managing streaming data.

Hence, consideration can be given to relational databases boasting excellent read-write performance. PostgreSQL released an extension called "AGE" in July 2020 [4], supporting graph storage. It translates Cypher queries into SQL queries and executes them at the lowest level of PostgreSQL, demonstrating the feasibility of storing graphs in relational databases.

Currently, the closest work to our objectives is presented by Kumar et al. [5]. They proposed the GraphOne system, which integrates edge and adjacency tables to implement an in-memory database system, achieving efficient read and write capabilities for graph data. The paper mainly presents two views: a streaming view and a static view, suitable for different analytical tasks. However, there are two unresolved issues in their work. First, it is based on edge and adjacency tables, where disk storage lacks rapid query capabilities. Second, the main data in the adjacency table, with varying numbers of neighboring nodes for each point, lacks the ability to persist to disk in an organized format, becoming a bottleneck when dealing with large data volumes. Another work [9] proposes an efficient strategy supporting historical data query, providing further motivation for us to design a comprehensive streaming data management system.

3   Problem Formulation

For a static graph G, comprised of sets of edges and vertices, we can represents it in tuple form (V, E), where $V = \{v_1, ..., v_n\}$ denotes the set of all vertices, and $E = \{e_1, ..., e_m\} \subseteq V \times V$ represents the set of edges. The sizes of the two sets are denoted by n and m respectively. N_v represents all points adjacent to vertex v. If G is a directed graph, N_v is the set of points formed by the endpoints of the outgoing edges from v; if G is an undirected graph, N_v is the set of endpoints of the edges starting or ending at v. Each edge is represented as (s, d), denoting the start and end points of the edge. If the graph G is weighted, the edges are represented as (s, d, ω), where ω is the weight of the edge. In practical applications, edge attributes may be represented by labels instead of weights. For the sake of brevity, this paper discusses the definition of graphs in the weighted form (s, d, ω).

Therefore, we can define a streaming graph as $S = (e_1, e_2, e_3, ..., e_n)$, where $e_i = (\overrightarrow{s, d}; t, \omega)$. A streaming graph, along with its data, represents a sequence of tuples, where $\overrightarrow{s, d}$ denotes an edge pointing from s to d, and ω represents the edge weight. In certain scenarios, edge weights may be substituted with labels. Each edge arrives at time t, denoted as the edge timestamp. A streaming graph can be viewed as a specific representation of a graph, where graphs can take various forms such as adjacency matrices, adjacency lists, or streaming formats. Dynamic graphs, as a conceptual framework, are distinguished from static graphs. Streaming graph data denotes a particular type of data structure, specifically a sequence of edge tuples.

This paper primarily addresses dynamic graphs in streaming formats, considering them synonymous. Computational tasks are uniformly treated as streaming computation engines, disregarding redundancy from maintaining distinct graph representations. In query tasks, the goal is to include all updates preceding the query timestamp. Despite separate threads for queries and writes, concurrent execution is possible. To optimize query performance, transactions are directly submitted within query tasks rather than awaiting completion of all write transactions.

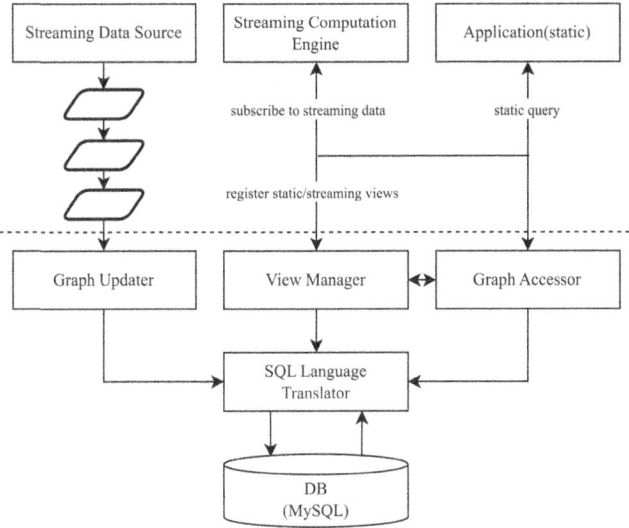

Fig. 1. Abstract structure of the streaming graph data management system based on relational database. Data source acquires streaming data from external applications. Graph accessor controls querying methods, distinguishing between static and streaming queries. View manager facilitates query execution by translating into SQL language and identifies relevant views for streaming queries. Streaming computation engine registers streaming view and retrieves current data stream via graph accessor. Unaccessed data streams can be continuously obtained from view manager through subscriptions.

4   System Structure and Optimization Strategies

This section delves into the structure of our management system and outlines three optimization strategies employed in the implementation of our streaming data management system: data storage, asynchronous and parallel data writing, and sharding strategy.

4.1   System Structure

The proposed management system integrates two edge tables: one in-memory for buffering incoming data and the other in a relational database for persistent storage and query support. Incoming data is appended to the buffer, which is asynchronously read and consumed by other threads. When the buffer reaches a threshold or a timer cycle completes, a database write operation is triggered. Leveraging database support for concurrent access, multiple asynchronous threads submit transactions simultaneously.

For querying graph data, a B+ tree index is established, and two querying methods are defined. Static queries treat the database data as a static graph, allowing node and edge queries. Streaming queries transform data from specific time periods into streams for access by a streaming computation engine. The engine continuously obtains data from the view manager, providing access to the latest data streams. The system structure, depicted in Fig. 1, is based on MySQL as the underlying database, with the frontend developed in Go to interface with it. This abstraction allows seamless adaptation to different databases through corresponding drivers.

The storage architecture of streaming graph data management system based on the relational database is primarily partitioned into two domains: memory and disk, as depicted in Fig. 2(a). A distinctive aspect of our method is the migration of a portion of the memory system to disk, facilitated by the underlying database's page-based file system. While the edge table may not offer an optimal storage solution and may entail some data redundancy compared to adjacency lists, the relatively economical nature of disk space compared to memory space allows for the temporary disregard of surplus capacity.

Newly arriving edges are appended to the end of table, preserving chronological order within pages. Leveraging pages as cache units enhances efficiency for streaming queries over specific time periods. However, static queries face challenges due to the uneven data distribution across pages. Hence, we adopt an adjacency list for memory caching, facilitating efficient queries for static queries (queries for nodes, edges, successor nodes, and predecessor nodes). Given limited memory space, caching the entire graph is infeasible. Instead, records for current nodes undergo replacement based on a cache replacement algorithm (e.g., LRU). If no match is found, a database query updates the cache. Additionally, the system monitors edge count, removing nodes when it exceeds a predefined threshold, typically set as the maximum edge count. To ensure timely updates during concurrent database queries and updates, checks are conducted to update relevant cache data.

4.2   Data Storage

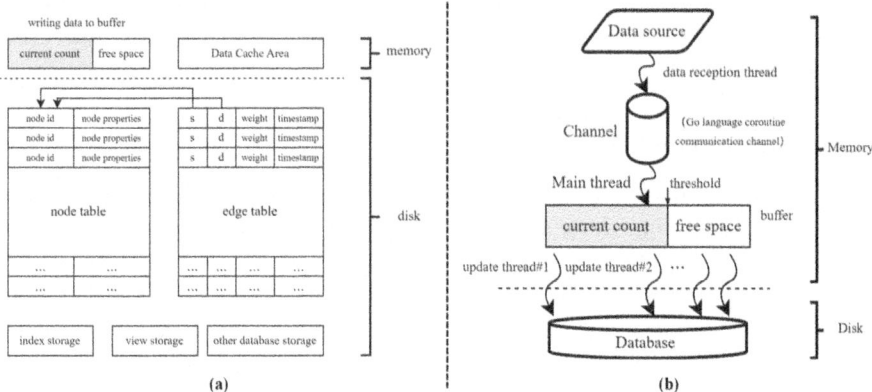

(a)   (b)

Fig. 2. (a) Data storage structure. Disk storage includes an edge table for primary data retrieval, alongside a node table for node attributes. The edge table enforces foreign key constraints between edge start and end points and node IDs. This setup ensures edge validity and facilitates associated edge deletion. Additionally, disk space is allocated to database indexes and view sections. (b) Data writing process. Data is received from the data source and sent to the data communication channel (Go Channel). The main thread serves as another endpoint of the communication channel, responsible for receiving data, buffering, and initiating write threads to store data in the disk database.

4.3   Asynchronous and Parallel Data Writing

In the streaming graph management system with relational databases, the data writing process exhibits asynchronous and concurrent characteristics. As shown in Fig. 2(b), data reception from the source and its transmission via the data communication channel (Go Channel) resemble a multi-producer, multi-consumer model inherent in concurrent programming. Buffer management employs a single main thread operation, utilizing a basic queue implementation. The buffer is configured with size, threshold, and timer parameters. Initial monitoring of the channel signal indicates new data presence. Upon surpassing the threshold, a data write operation is triggered, clearing the buffer; otherwise, data continues appending. Additionally, expiration of the timer, signifying a data write-free period, prompts a write operation to uphold database result timeliness.

4.4   Sharding Strategy

Presently, default page size of MySQL is set at 16KB [1], enabling a B+ tree with a height of 3 to accommodate millions of records. For the purpose of this discussion, let's assume that querying an index requires 3 page I/O operations. During a successor node query confined to a single page, the erratic distribution of nodes in the edge table may scatter the queried rows across multiple pages, thus requiring multiple disk page I/O operations.

To address this issue, we proposes a sharding strategy. As data volume increases, the height of the B+ tree may also increase. However, the primary objective is not to minimize the added number of pages due to increased access depth. Instead, the focus is on leveraging the parallel page reading capability of the current disk (SSD). While a database query operates in a single thread by default, sharding allows for the initiation of multiple query threads simultaneously accessing data tables. This multi-threaded approach aggregates all results, balancing space cost against time cost.

4.5   Interface Design

In this paper, we present three primary external interfaces: data updating interface (GraphUpdater), analysis interfaces (GraphAccessor), and view management interfaces (ViewManager), by using interface feature of Go language. In the implementation, descriptions of user-accessible interfaces are defined first, followed by the development of different driver codes for various underlying databases (e.g., MySQL, Neo4j), ensuring a unified approach to accessing external interfaces.

The GraphUpdater interface encompasses operations for updating database data, including node insertion (update), and edge insertion. All data is treated as a weighted model, allowing deletion of edges by inserting negative-weight edges.

The GraphAccessor interface provides access to the graph, offering both static and streaming access. Static access includes point queries, edge queries, predecessor, and successor node queries, each requiring a handle parameter for either static or streaming access.

View handles are managed within the ViewManager, facilitating registration and retrieval of relevant handles for use.

5   Experiments

This section outlines our method to evaluating the effectiveness and rationality of our system design and optimization strategies. Focused on implementing a streaming data management system using the MySQL relational database (RDB), our objective is efficient persistence and retrieval of large data volumes. To benchmark, we compare our system with Neo4j (GDB), affirming the efficacy and efficiency of relational database implementation in streaming graph data contexts.

5.1   Datasets

In our experiments, three open-source streaming datasets [6] were utilized: DBLP, MathOverflow, and YouTube datasets. These datasets were formatted as edge quadruples $(s, d, label, timestamp)$, representing individual edges within dynamic graphs. Table 1 outlines the general statistics of these datasets. In the context of streaming, leveraging these datasets, we evaluated the performance of two types of databases (RDB and GDB) across various tasks and data scales. The evaluation was categorized into two main sections: write and read (query). Write tasks encompassed database initialization and performance testing of write speed with varying buffer threshold sizes. Read tasks included six types of randomly generated basic queries, range queries, and queries after sharding.

Table 1. General statistics of DBLP, MathOverflow and YouTube datasets.

Properties	DBLP	MathOverflow	YouTube
Number of nodes	12590	24818	3223589
Number of edges	49759	506550	9375374
Maximum out-degree	617	5931	83292
average out-degree	16	25	8
maximum in-degree	227	5378	61556
average in-degree	4	23	3

5.2   System Write Performance Testing

This subsection tests the write speed of different databases. Here, we simulate streaming data arrival by reading data from files.

Initialization Process and Write Performance Testing. In the experimental setup, database initialization precedes the process of loading all edge data from files into memory, followed by a single-threaded insertion of all data directly into the underlying database. Table 2 presents a performance comparison of insertion tests across various datasets using two distinct databases (MySQL and Neo4j). Within the table, "-" signifies operations that exceeded a 5-min duration. Across all datasets, Neo4j consistently exhibits significantly slower speeds in both initialization and insertion times compared to MySQL. This discrepancy underscores inability of Neo4j to meet the demanding requirements for high-speed data writes in the context of extensive streaming graph data.

Asynchronous and Multithreaded Write Performance Testing. During the file reading process, data is concurrently loaded into the buffer. Once the number of data entries in the buffer reaches a predefined threshold, a write transaction is initiated, with each transaction corresponding to a database thread. Leveraging multithreading, the system aims to enhance the writing speed. Figure 3(a) illustrates the variations in total edge writing time across

Table 2. Comparison of Initialization and Insertion Time Across Different Databases.

Dataset-Database	Initialization Time	Insertion Time	Total Time
DBLP-RDB	0.02 s	4.04 s	4.06 s
DBLP-GDB	5 min 50 s	56 min 55 s	1 h 2 min
MathOverflow-RDB	0.14 s	40.92 s	41.06 s
MathOverflow-GDB	11 min 15 s	12 h 14 min 3 s	12 h 25 min
YouTube-RDB	1 s	11 min 38 s	11 min 39 s
YouTube-GDB	–	–	–

three datasets as the buffer threshold is adjusted. Notably, a smaller buffer threshold implies an increased number of concurrent write transactions within the database, consequently reducing insertion time.

Across all datasets, consistent results emerge, indicating minimal performance enhancements beyond 16 threads (write transactions). This bottleneck predominantly stems from disk speed limitations. As the number of read-write threads escalates, performance gains are curtailed by disk IO constraints rather than further improvements. Notably, a nearly fivefold increase in writing speed has been observed across datasets of varying scales.

Regarding the remaining optimization strategies, specifically data storage and sharding strategy testing, the cache is implemented using the internal data structure Map of the Go language, achieving a constant writing speed of $O(1)$. The cache eviction algorithm LRU incurs negligible overhead, given its swift execution for memory operations, thereby exerting minimal impact on database writing speed. Moreover, sharding strategy does not affect the writing speed. Table creation is a very fast process, and the time involved can be neglected.

5.3   System Query Performance Testing

In this subsection, we mainly evaluate the query efficiencies of different databases under static queries, streaming queries, and sharding strategies.

Comparison of Static Query Performance. In static queries, we tested the access performance of three types of tasks: edge queries, successor node queries, and predecessor node queries on three datasets using two types of databases. Each type of task has two query strategies. For edge queries, one strategy involves randomly selecting 1000 edges from existing edges to generate a query load, labeled as EEQ (ExistedEdgeQuery), while the other strategy involves randomly selecting start and end points to generate a query load (where most queries do not exist), labeled as ERQ (EdgeRandomQuery). For successor node queries, one strategy involves querying the lowest 1000 nodes by out-degree, labeled as SLQ (SuccessorLowQuery), while the other involves querying the highest 1000 nodes by out-degree, labeled as SHQ (SuccessorHighQuery). The strategies for

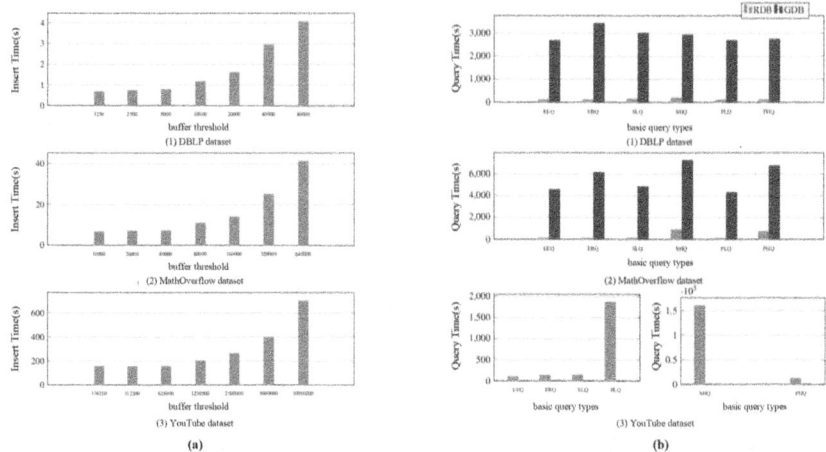

Fig. 3. (a) Illustrates the relationship between buffer threshold and insertion time in the multithreaded writing experiment. Here, we set the threshold to exceed the total amount of edge data, effectively simulating a single-threaded scenario. (b) Comparison of different query tasks across three datasets.

predecessor node queries are the same as above, labeled as PLQ (Predecessor-LowQuery) and PHQ (PredecessorHighQuery) respectively.

Figure 3(b) shows the stark performance contrast between Neo4j and MySQL across all tasks. In MySQL, the differences between EEQ and ERQ, as well as between SLQ and PLQ, are minimal. EEQ slightly outperforms ERQ due to fewer page IO operations. Conversely, SHQ and PHQ queries involve multiple pages, resulting in longer processing times. Due to inefficient writing of Neo4j, read query results for the YouTube dataset are missing.

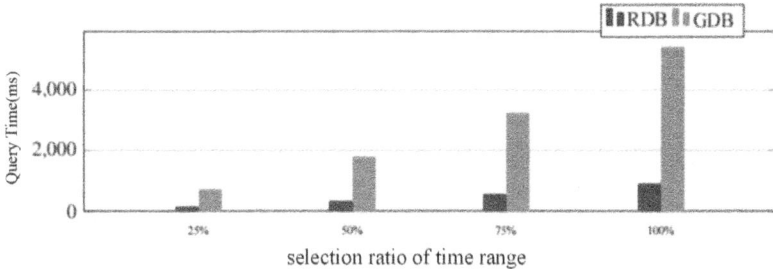

Fig. 4. The comparison between the selection ratio of time range and the query time consumed by RDB and GDB.

Comparison of Streaming Query Performance. Another set of tests regarding queries focuses on streaming queries. The difference between this query method and the aforementioned method lies in how the database stores data, which is organized into pages. The final results static queries are randomly distributed across pages in relational databases, whereas data stored by timestamp is continuous. We test the performance of different databases in streaming queries by varying the selection ratio of time range. From Fig. 4, we observe that the performance of RDB exceeds that of GDB.

Comparison of Sharded Query Performance. The sharding strategy enables concurrent processing of sequential page queries, facilitating parallel read-write operations on multiple tables. In the case of SLQ queries, which typically yield few results per node, multi-table queries may incur unnecessary disk IO congestion due to redundant page reads and writes, as illustrated in Fig. 5. Decreasing the number of shards correlates with reduced processing time, albeit with diminishing returns. Conversely, for queries with extensive results like SHQ, sharding significantly boosts performance. Stability in performance is observed when sharding into 5 or 10 tables, attributing to potential disk IO limitations as previously discussed.

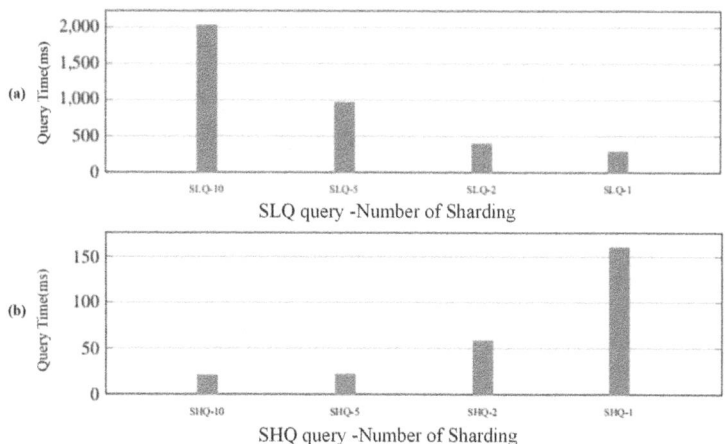

Fig. 5. (a) Successor node queries on low-degree nodes. (b) Successor node queries on high-degree nodes.

The experiments reveal a stark performance contrast between Neo4j and MySQL, with the latter outperforming in both write speed and query tasks. While Neo4j boasts superior traversal capabilities thanks to its traversal framework [8], MySQL based systems demonstrate significant advantages in streaming graph data applications due to their robust read and write performance. The conducted additional experiments on data storage and sharding strategy, further substantiating the effectiveness of optimizing the streaming graph data management system based on relational database.

6 Conclusion

In this work, we investigate the expansion of relational databases in managing streaming graph data. We extend the functionality of relational databases for streaming graph data management. Utilizing the popular Go programming language, we define interfaces tailored for streaming graph data. Additionally, we propose optimizations, such as asynchronous write strategies and data storage solutions, to enhance performance further. Experimental results demonstrate the superior performance of systems based on relational databases (e.g., MySQL) compared to those based on graph databases (e.g., Neo4j). However, our current work still has limitations. For instance, relational databases, despite their good performance, incur unnecessary page reads and writes, consuming significant system resources, making them suboptimal for graph storage.

Future research directions could focus on designing storage structures optimized for streaming graph databases, leveraging hierarchical storage, and exploring methods to fully exploit system I/O performance from a lower-level perspective. Integration with in-memory data structures could also be explored to achieve efficient read and write performance.

References

1. MySQL 8.0 reference manual 15.5.1 buffer pool (2016). https://dev.mysql.com/doc/refman/8.0/en/innodb-file-space.html
2. Neo4j (2016). https://neo4j.com/
3. Streaming graphs: Combining Kafka and Neo4j (2019). https://neo4j.com/blog/streaming-graphs-combining-kafka-neo4j/
4. Agensgraph: Powerful graph database (2020). https://github.com/bitnine-oss/agensgraph
5. Kumar, P., Huang, H.H.: Graphone: a data store for real-time analytics on evolving graphs. ACM Trans. Storage (TOS) **15**(4), 1–40 (2020)
6. Kunegis, J.: KONECT: the Koblenz network collection. In: Proceedings of the 22nd International Conference on World Wide Web, pp. 1343–1350 (2013)
7. Macko, P., Marathe, V.J., Margo, D.W., Seltzer, M.I.: Llama: efficient graph analytics using large multiversioned arrays. In: 2015 IEEE 31st International Conference on Data Engineering, pp. 363–374. IEEE (2015)
8. Robinson, I., Webber, J., Eifrem, E.: Graph Databases: New Opportunities for Connected Data. O'Reilly Media, Inc. (2015)
9. Xiangyu, L., Yingxiao, L., Xiaolin, G., Zhenhua, Y.: An efficient snapshot strategy for dynamic graph storage systems to support historical queries. IEEE Access **8**, 90838–90846 (2020)
10. Zhai, H., Cao, X., Sun, P., Shen, D., Nie, T., Kou, Y.: Rule-enhanced evolutional dual graph convolutional network for temporal knowledge graph link prediction. In: Yuan, L., Yang, S., Li, R., Kanoulas, E., Zhao, X. (eds.) WISA 2023. LNCS, vol. 14094, pp. 64–75. Springer, Singapore (2023). https://doi.org/10.1007/978-981-99-6222-8_6

PLIS: Persistent Learned Index for Strings

Yu Zhang[1], Shiyu Yang[1(✉)], Wenlei Zhong[1], Guojie Ma[2], Jianye Yang[1], and Weihong Zhou[1]

[1] School of Guangzhou University, Guangzhou, China
{2112233028,2112233026,2112106080}@e.gzhu.edu.cn,
{syyang,jyyang}@gzhu.edu.cn
[2] School of Shanghai University of International Business and Economics, Shanghai, China
mgj0616@suibe.edu.cn

Abstract. Persistent memory, emerging as a potential replacement for the next generation of main memory, is gradually gaining prominence. Presently, index structures based on persistent memory primarily focus on B+ trees, hash tables, and indexes, with significant performance improvements observed in recent research on learned indexes. However, most efforts concentrate on reducing the update costs of learned indexes, leaving inadequate support for string key types. Therefore, this paper aims to create a persistent memory string learned index structure capable of handling variable-length string keys, reducing write amplification, and ensuring crash consistency. We evaluate SLIP cost effectiveness ratio using real and synthetic datasets, results show outperforming state-of-the-art string learned indexes SIndex and SLIPP across various workloads.

Keywords: Persistent memory · Learned index · String key

1 Introduction

In the traditional storage hierarchy, dynamic random access memory (DRAM) typically functions as the primary storage medium, serving as a bridge between the processor and storage devices. Despite being volatile, its faster read/write speeds and latency advantages fill the gap between storage (HDD and SSD) and CPU caches built with SRAM [1], meeting the high-speed computational demands of processors. However, as data volumes surge and computational requirements continue to rise, DRAM faces challenges. Scalability is constrained, and with chip manufacturing technology nearing physical limits, increasing DRAM density and capacity becomes increasingly difficult [29]. Furthermore, DRAM prices do not linearly scale with capacity, the larger its capacity, the more expensive it is, compelling users to weigh cost and performance considerations when upgrading or configuring memory [2]. DRAM is a volatile storage, meaning data stored within it is lost once power is cut off. Hence, frequent backups to non-volatile storage media are required, increasing the complexity and cost of system recovery.

To address the aforementioned issues, various non-volatile memory (NVM) technologies such as phase change memory (PCM) [3], memristors [4], and resistive random access memory [5] (RRAM) have been successfully developed. In 2019, Intel introduced the Optane DC Persistent Memory product, leveraging 3D XPoint [6] technology, which bridges the performance and cost gap between traditional DRAM and storage devices like SSDs. Notably, the average price per gigabyte (GB) of Optane DC Persistent Memory is \$4, compared to \$7.6 for DRAM [11]. Its non-volatility during power loss significantly reduces data synchronization costs, and its byte-addressability allows for flexible and fine-grained data management. Additionally, its read/write performance rivals that of DRAM. By combining DRAM's byte-addressability with the high capacity and persistence of hard drives, it effectively bridges the performance gap between DRAM and SSDs. In recent years, numerous research endeavors have emerged focusing on persistent memory indexes, predominantly targeting tree structures [7,8], hash [9,10,21], and learned indexes [11–13]. Recent studies have shown that the latest learned indexes perform significantly better than tree structures, so we focus on learned index.

The research proposed by Krasa et al. [14] in 2018 on learned indexes has introduced a novel approach to constructing sorted data indexes. Given a dataset, learned indexes utilize machine learning models to learn the data distribution and predict the position of lookup keys within the dataset. This can be achieved by training with the cumulative distribution function (CDF) of the dataset. As the model's predictions of final positions may not be precise, learned indexes require employing binary search around the predicted position within a bounded range to search for keys. Despite the high spatiotemporal efficiency of learned indexes, they still face certain challenges: the high cost of index updates and constraints on the types of searchable keys.

Recent optimizations of learned indexes targeting persistent memory have primarily focused on numerical keys, exemplified by APEX [11], PLIN [12], and WIPE [13]. While research on learned indexes for string keys exists, exemplified by SIndex [15], SLIPP [27], it remains relatively scarce for persistent memory. Given the widespread need for string key indexing in databases, this paper aims to enhance and extend existing learned index techniques, focusing on building an efficient string learned index on persistent memory.

In order to verify the advantages of indexing structure on persistent memory, we propose a new cost model to measure the memory overhead cost. Previous cost model studies mainly focus on the memory access cost which is affected by bandwidth and latency [25]. However, we mainly concern the performance and price cost of the storage medium, since the performance of persistent memory is close to that of DRAM. We observe the difference between them in the same experimental environment by measuring the memory value cost overhead between the two.

The contributions of this paper are as follows:

- We construct a learned index adapted for strings, named PLIS, on persistent memory. We split the strings into 8-byte slices and align them with atomic storage of persistent memory to ensure atomicity of operations.

- We provide a detailed explanation of specific operations on PLIS, including point lookup, range lookup, insert, delete and update.
- We proposed a cost effectiveness ratio that intuitively demonstrates the cost effectiveness of PM. Experiments were conducted under different workloads using both synthetic and real datasets, and comparisons were made with SIndex and SLIPP. Experimental results show that PLIS is less costly for the same workloads

2   Related Work and Challenges

2.1   Persistent Memory

Non-Volatile Memory (NVM) [26], also referred to as persistent memory, can be connected to the CPU Integrated Memory Controller (IMC) via the memory bus. It allows direct access to the CPU through load and store instructions. However, when designing and developing index structures on persistent memory, two aspects need to be considered: Firstly, the asymmetry of read and write operations on persistent memory necessitates reducing the software overhead of index data structures [20]. In persistent memory, the overhead of persistence is comparable to that of read and write operations. Hence, when designing index structures, efforts should be made to minimize additional operational overhead while ensuring index accuracy as much as possible. Secondly, despite providing non-volatility, persistent memory still faces crash consistency [8,22] issues. Modern 64-bit CPUs can ensure atomic writes for 8-byte operations on persistent memory. However, writes larger than 8 bytes require the use of cache line flush instructions like CLWB (Cache Line Write Back) or memory barrier instructions like MFENCE (Memory Fence) [26] to ensure crash consistency.

Despite encountering challenges related to economic scale and market acceptance, Intel announced the cessation of Optane Persistent Memory production in 2022. However, its key technologies continue to exert a profound influence on future storage and memory solutions [16].

2.2   Persistent Memory Based Learned Index

Learned-based indexes leverage machine learning models to predict data positions, significantly reducing lookup path lengths and frequencies. This capability manifests in efficient query performance, particularly when handling large datasets. Consequently, optimization efforts for learned-based indexes have garnered widespread attention, with most optimizations targeting DRAM. Notable examples include ALEX [17], LIPP [18], and XIndex [19], which focus primarily on optimization strategies for insertion policies, data fitting models, and location searches. Today, persistent memory emerges as a promising alternative to DRAM, offering larger capacities, lower access latencies [23], and byte-addressable persistence. Despite this potential, research on constructing learned-based indexes on persistent memory remains relatively limited. Examples such as APEX [11], PLIN [12], and WIPE [13] primarily optimize for persistent memory, ensuring durability and rapid recovery while mitigating write amplification to some extent.

Through a comprehensive analysis of the current database indexing landscape, it is evident that while learned-based indexes exhibit outstanding spatiotemporal efficiency, they still suffer from notable limitations: high update costs and limited support for key types. Although a significant amount of subsequent research has been dedicated to improving the update mechanisms of learned-based indexes, these studies predominantly focus on numerical keys. Presently, only a few studies, such as SIndex [15], have explored the feasibility of constructing learned-based indexes on string keys. However, for many database applications, the ability to index large volumes of string keys is indispensable. Therefore, this paper aims to delve deeper into enhancing and extending the technology of building learned-based indexes on persistent memory, developing an efficient and precise string learned index structure constructed on persistent memory.

2.3 Challenges

Building a string learned index on persistent memory presents three challenges: (1) Efficiently handling variable-length string keys to adapt them to learning models and improve model accuracy, given the diversity and irregularity of string keys; (2) Addressing the asymmetry in read and write operations inherent to persistent memory, necessitating index structure optimizations to reduce write operations and prevent write amplification. Ensuring data consistency and integrity in the event of system crashes, despite the non-volatility of persistent memory; (3) Special consideration is required in design to guarantee atomic writes for data on persistent memory, leveraging its byte-addressable nature to ensure crash consistency.

3 PLIS

In this section, we introduce string preprocessing methods and learning models, then describe the index structure of PLIS and how do we build PLIS on persistent memory in detail.

3.1 String Data Pre-processing

In database indexing, the essence of an index lies in organizing and sorting data to enhance retrieval efficiency. The index proposed in this paper divides strings into segments of 8 bytes each. This segmentation is based on the fact that in Intel persistent memory hardware, atomic persistent storage is 8 bytes. Thus, during the transmission process aligned with persistent memory, if a system crash occurs, either the entire 8-byte content is new or it remains old when recovered. As shown in Fig. 1, "example1" is the stringslice, and the index converts this fragment into a 64-bit unsigned integer (represented by uint64_t), serving as input for a one-dimensional linear model, and creates persistent pointers accordingly.

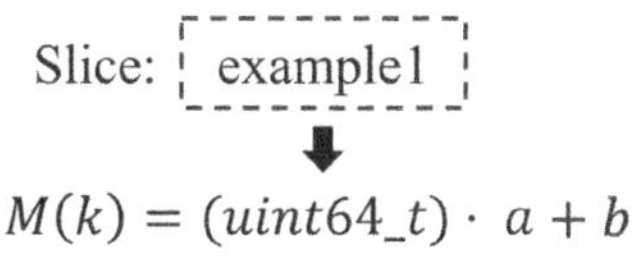

Slice: example1

$$M(k) = (uint64\_t) \cdot a + b$$

Fig. 1. String slice model

3.2 Learning Model

The linear model serves as a simple yet effective approach, suitable for handling one-dimensional data with monotonic increasing patterns, requiring storage of only two parameters: slope and intercept. While linear models exhibit limitations in fitting complex data distributions, particularly cumulative distribution functions (CDFs), which may lead to significant prediction errors on certain datasets, efficient index structure design can alleviate issues of inadequate prediction accuracy. Consequently, in learning-based indexing research, linear models find widespread application, with many advanced learning-based index structures favoring the simplicity of linear models over more complex models like neural networks. Assuming M represents a given monotonic increasing model, it should satisfy the following equations: Eq. 1, 2.

$$key_i \leq key_j \rightarrow M(key_i) \leq M(key_j) \tag{1}$$

$$M(k) = \begin{cases} 0, & \text{if } [a\dot{k} + b] < 0 \\ L - 1, & \text{if } [a\dot{k} + b] \geq L \\ [a\dot{k} + b], & \text{otherwise} \end{cases} \tag{2}$$

3.3 Index Structure

Drawing inspiration from LIPP, this paper introduces PLIS (Persistent Learned Index for Strings), a learned index designed for persistent memory (PM). By ingeniously leveraging PM to construct index structures, PLIS enhances read performance, ensures persistence, and minimizes unnecessary writes on persistent memory, thereby mitigating write amplification. The structure of PLIS is illustrated in the accompanying Fig. 2.

3.4 Index Construction

As shown in Fig. 2, in Intel architectures, there are typically three levels of CPU caches: L1, L2, and L3. Each level is closer to the CPU and operates at a higher speed, with L1 being the closest and fastest, albeit with the smallest capacity. The capacities of L2 and L3 caches increase progressively, but their speeds are relatively slower. Communication with DRAM and Persistent Memory (PM) is conducted through the Memory Controller (IMC). PM utilizes Asynchronous

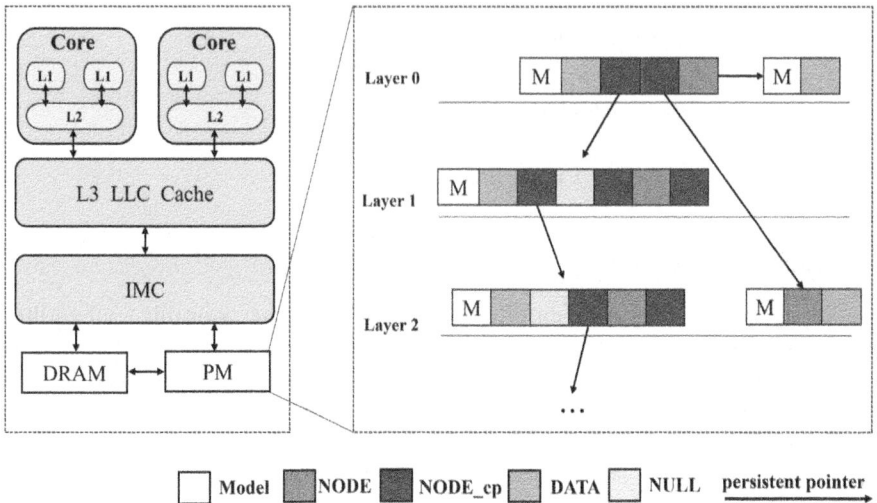

Fig. 2. Overall structure of the PLIS.

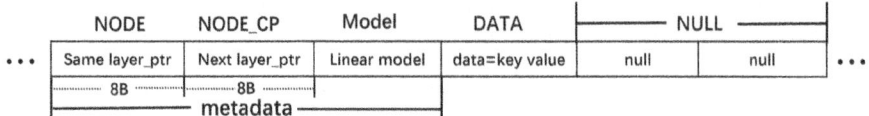

Fig. 3. Model node structure layout.

DRAM Refresh (ADR) to flush the write-protected data buffer, ensuring that data remains safe and consistent on the persistent memory by placing DRAM into a self-refresh state.

When building an index, it is necessary to include the header files of the persistent memory library and link the corresponding library files, using functions and data structures related to persistent memory. Next, initialize the persistent memory by mapping the persistent memory file to the virtual address space of the process using the "pmem_map_file" function, which returns a pointer to the persistent memory region. Then, use the "pmem_memcpy_persist" function to copy the value of "current_slice" to "entry->key_slice", ensuring that the data is persisted in persistent memory. Utilize the "TX_BEGIN" and "TX_END" macros to define the start and end points of a transaction, ensuring that all operations will be executed as an atomic operation upon transaction commit.

PLIS is a tree-like structure composed of several array nodes, where each node is responsible for indexing a corresponding slice of the string key based on its layer. Inside each node, there is an array and a linear model M. The linear model M receives inputs and calculates their positions within the array. There are four types of entries in the array, as shown in Fig. 3.

NULL: Represents an empty entry. When data is mapped to this entry, it is directly inserted into this position, and the type of this entry is converted to DATA.DATA: Represents an entry responsible for storing the string key and value.NODE: Represents an entry that stores a persistent pointer pointing to a child node in the same layer. In persistent memory, it is represented by "pmem::obj::persistent_ptr" and referred to as Same "layer_ptr". When a new data is mapped to a DATA-type entry during insertion, and the slice of the new data's key is different from that of the entry's key, indicating a position conflict, PLIS will create a new child node and allocate new entries to accommodate these conflicting data.NODE_cp: Represents an entry where the slot stores a persistent pointer Next "layer_ptr" pointing to a child node in a different layer, whose subtree contains keys with a common prefix. When some data cannot be distinguished at the current layer, PLIS creates a new child node for the corresponding layer to store this data, and sets the type of the parent node's corresponding entry to NODE_cp.

4   PLIS Operations

4.1   Search

The indexing search operations in PLIS encompass Point Lookup and Range Lookup. Point Lookup is the most fundamental search operation in the index, providing a mapping from a single key to its position and returning the corresponding data as needed. Range Lookup operations accept an arbitrary range and return all data whose keys fall within that range. This functionality supports advanced database retrieval capabilities, such as "finding all keys containing a certain prefix". First, we describe the Point Lookup operation.

Point Lookup. For a given key, the search begins at the root node of the index structure. Based on the current node's layer, a slice of the key is retrieved. The linear model of the current node is then used to calculate the position of the given key's data within the array. The next step involves determining the type of the array element at the current position and performing different operations according to the type of the element. If the element type is DATA, the key of the element is checked to see if it matches the key being searched for; if it matches, the item is returned, otherwise, it indicates that the item does not exist. For NODE, the search continues to the child node using the pointer and the process is repeated; for NODE_cp, the key being searched is compared with the stored common prefix, if it matches, the search continues to the child node, if not, the item does not exist. Regardless of whether the item is found or not, the Point Lookup operation returns the respective item.

Range Lookup. In the PLIS index, when performing a range scan operation within the interval $[u, v]$, the operation starts by pinpointing the start key u and the end key v through point lookup. As the keys in the index are sorted, this allows for a direct scan from u to v. During the scan, bitmaps within the nodes are utilized to swiftly skip all entries of type NULL, thereby enhancing the scan speed. To further increase efficiency, the position of v within the index is predetermined, rather than comparing it with every encountered key. When entries of type NODE or NODE_cp are encountered, the scan follows pointers to explore their child nodes to ensure a comprehensive traversal of the related subtrees. If the end of the entry array in a node is reached before finding v, the scan backtracks to the parent node to continue the scan. This backtracking is crucial for completely covering the specified range, especially when spanning multiple nodes.

4.2   Insert

The insertion process relies on Point Lookup operations to determine the position in the index array for the new data. The entry type at this position may be NULL, DATA, or NODE_cp. Depending on the entry type, the insertion process takes different steps. If the entry type is NULL, indicating a vacant slot for insertion, the new data is directly inserted into this position, and the entry type is updated to DATA.

If the entry is of type DATA, it indicates a conflict between the new data and the existing data at that position. This may be due to prediction errors or shared key prefixes. The insertion algorithm resolves conflicts by identifying unique key fragments and creating a new child node in the corresponding layer to store them. The original node's entry will point to this new child node, and the type of the original entry will be updated to either NODE or NODE_cp based on whether the layer of the new child node matches the layer of the original node.

If the entry is of type NODE_cp, it indicates a conflict between the new data's key and a common prefix of some existing dataset in the index. PLIS handles this similarly to the DATA type conflict by creating a new child node in the corresponding layer to store the conflicting data. However, in this case, it needs to store not only the new data but also the subtree saved in the original NODE_cp entry, including the pointers and common prefix.

4.3   Delete and Update

Deletion operations in the PLIS index also rely on the search operation described earlier. This process involves locating the relevant entry, removing the stored key and value, and marking the entry as NULL. Additionally, PLIS's range scanning feature supports rapid range deletion, such as removing all data containing a specific prefix.

In data updates, if only the value changes, concise operations can be achieved by directly searching for and updating the corresponding entry for the key. However, if the update involves a change in the key, it is necessary to delete the existing key-value pair and insert the new data to ensure the freshness of the data and the cleanliness and consistency of the index structure.

5   Cost Model

To comprehensively assess the cost-effectiveness of memory systems, we propose a novel cost-benefit metric CER (Cost Effectiveness Ratio) to define memory overhead costs. In computer system design, memory cost is a crucial factor, especially in large-scale data processing and storage domains. Previous work has also proposed a cost model for query processing [28]. Traditionally, memory value is usually measured in terms of the price per gigabyte (GB) of memory. However, this approach overlooks the actual usage and performance of memory. In contrast, our cost model integrates both operational throughput and memory pricing, thus providing a more comprehensive evaluation of memory costs. Such an approach allows us to gain a more accurate understanding of the actual costs associated with memory systems and provides more effective guidance for system design and selection. As illustrated in Eq. (3). The term "OPS/s" denotes the number of operations performed per second, while "Cost/GB" indicates the price per gigabyte of memory (for instance, the cost for PM is \$4/G, and for DRAM, it is \$7.6/G). A higher CER signifies lower memory cost overhead per unit of storage cost.

$$CER = \frac{\text{OPS/s}}{\text{Cost/GB}} \tag{3}$$

6   Evaluation

6.1   Experimental Setup

Environment. All experiments were conducted on a server equipped with actual persistent memory (Intel Optane DCPMM). The server comprises eight 256GB Optane DCPMM, eight 16GB DRAM, and two Intel Xeon Platinum 8336C 2.3GHz two sockets. Optane DCPMM is configured in APP Direct mode, and the PMDK development kit is utilized to allocate the persistent memory's dataset space.

Datasets. The experiments utilized two types of datasets. The first dataset consisted of 100 million randomly generated 64-byte strings. The second type comprised real-world data, including 74 million unique URLs collected from Memetracker [24]. These URLs ranged in length from 64 to 128 bytes, with an average length of 86 bytes.

Baselines. This study compares the PLIS index with the existing state-of-the-art SIndex and SLIPP indexes. SIndex is a concurrent learned index that supports string keys, reducing model inference and data access costs by grouping keys with common prefixes and training models using the unique parts of each key. SLIPP, on the other hand, is a string learned index that features precise predictive positioning and can be updated.

Workload. To comprehensively evaluate the performance of the indexes, this experiment was conducted under four different workloads: Read-Only,

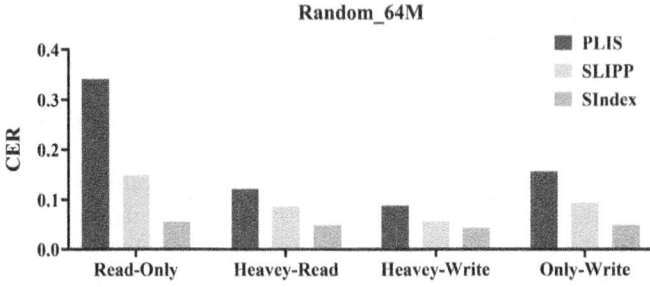

Fig. 4. Different workloads in Random64_100M dataset

Read-Heavy, Write-Heavy, and Write-Only. Each workload consists of 60 million operations, with the following proportions of write operations for the four workloads: 0%, 10%, 90%, and 100%, respectively.

6.2   Overall Performance

Random64_100M. In the Read-Only workload on the Random64_100M dataset, SLIP exhibits a cost effectiveness ratio that is 2.28 times and 6.12 times higher than SLIPP and SIndex, respectively. This superior cost effectiveness of SLIP is due to its structural characteristics and the low cost of persistent memory. Although the throughput of persistent memory can be affected by an increase in write operations, SLIP still demonstrates good cost effectiveness under Read-Heavy, Write-Heavy, and Write-Only workloads (Fig. 4).

URL_74M. On the URL_74M dataset, due to the more complex data distribution in the URL dataset, the uneven distribution of data with common prefixes leads to position conflicts in SLIP, affecting throughput. However, under the Read-Only workload, the cost effectiveness of SLIP is 2.26 times and 5.17 times higher than that of SLIPP and SIndex, respectively. Similarly, although the throughput of SLIP is affected by an increase in write operations, it still demonstrates a significant cost effectiveness advantage over SLIPP and SIndex (Fig. 5).

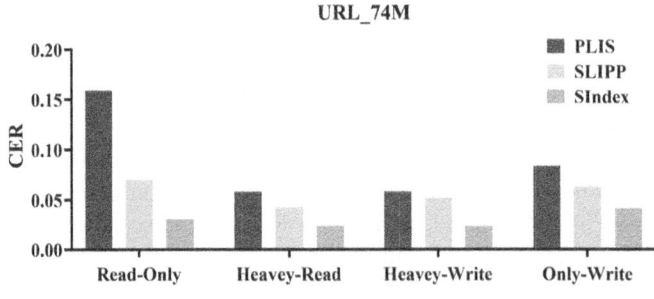

Fig. 5. Different workloads in URL_74M dataset

PM and DRAM CER Gap. We compared the CER (cost effectiveness ratio) performance of SLIP on PM (Persistent Memory) and DRAM across various datasets and workloads. PM shows a significantly higher performance benefit than SLIPP and SIndex under Read-Only and Heavy-Read workloads. Despite being affected by write operations, the cost effectiveness ratio of SLIP remains substantial (Fig. 6).

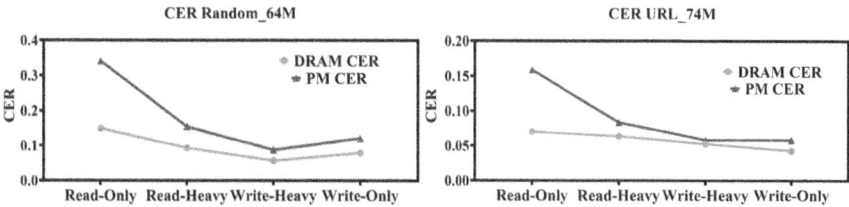

Fig. 6. Performance gap between PM and DRAM under different workloads.

Crash Recovery. When the system crashes and restarts due to a failure, the information in DRAM needs to be reconstructed, but PM employs special persistence operations, such as persistent pointers and persistent memory pools, allowing it to quickly recover to the pre-crash state upon a system crash. We tested its fault recovery capability by simulating a system crash: we first loaded the Random64_100M dataset into the system, then terminated the process, and simultaneously initiated another process for read-write operations to test if the system could function normally. We observed that the throughput of the test process was essentially consistent with the results before the process termination, indicating that the index structure built on PM could immediately recover to its pre-crash state after a system crash.

7   Conclusion

In this paper, we explore the challenges of building string learned indexes on persistent memory, and optimise the structure of PLIS to address these issues, using a slicing approach to handle string keys so that their sizes are aligned with the persistent memory atomic storage to ensure the atomicity of the operations, and optimising the index structure to improve the read/write performance and ensure crash consistency so that even if the system crashes, it can be recovered immediately. The throughput under different workloads on synthetic and real datasets is higher than that of SIndex and SLIPP, achieving good performance.

Acknowledgment. Shiyu Yang is supported by National Key R&D Program of China (2022YFB3103701) and Guangzhou Basic and Applied Basic Research Foundation (202201020131).

References

1. Huang, K., Wang, T.: Indexing on Non-volatile Memory, 1st edn. Springer, Cham (2023)
2. Huang, K., Imai, D., Wang, T., Xie, D.: SSDs striking back: the storage jungle and its implications to persistent indexes. In: CIDR, vol. 22, pp. 1–8 (2022)
3. Raoux, S., Burr, G.W., Breitwisch, M.J., et al.: Phase-change random access memory: a scalable technology. IBM J. Res. Dev. **52**(4.5), 465–479 (2008)
4. Strukov, D.B., Snider, G.S., Stewart, D.R., et al.: The missing memristor found. Nature **453**(7191), 80–83 (2008)
5. Hong, X.L., Loy, D.J.J., Dananjaya, P.A., et al.: Oxide-based RRAM materials for neuromorphic computing. J. Mater. Sci. **53**, 8720–8746 (2018)
6. Hady, F.T., Foong, A., Veal, B., et al.: Platform storage performance with 3D XPoint technology. Proc. IEEE **105**(9), 1822–1833 (2017)
7. Arulraj, J., Levandoski, J., Minhas, U.F., et al.: BzTree: a high-performance latch-free range index for non-volatile memory. Proc. VLDB Endow. **11**(5), 553–565 (2018)
8. Yang, J., Wei, Q., Chen, C., et al.: NV-tree: reducing consistency cost for NVM-based single level systems. In: 13th USENIX Conference on File and Storage Technologies (FAST 2015), pp. 167–181 (2015)
9. Wang, C., Hu, J., Yang, T.Y., et al.: SEPH: scalable, efficient, and predictable hashing on persistent memory. In: 17th USENIX Symposium on Operating Systems Design and Implementation (OSDI 2023), pp. 479–495 (2023)
10. Hu, D., Chen, Z., Che, W., et al.: Halo: a hybrid PMem-DRAM persistent hash index with fast recovery. In: Proceedings of the 2022 International Conference on Management of Data, pp. 1049–1063 (2022)
11. Lu, B., Ding, J., Lo, E., et al.: APEX: a high-performance learned index on persistent memory. Proc. VLDB Endow. **15**(3), 597–610 (2021)
12. Zhang, Z., Chu, Z., Jin, P., et al.: PLIN: a persistent learned index for non-volatile memory with high performance and instant recovery. Proc. VLDB Endow. **16**(2), 243–255 (2022)
13. Wang, Z., Ding, C., Song, F., et al.: WIPE: a write-optimized learned index for persistent memory. ACM Trans. Archit. Code Optim. **21**(2), 1–25 (2024)
14. Kraska, T., Beutel, A., Chi, E.H., et al.: The case for learned index structures. In: Proceedings of the 2018 International Conference on Management of Data, pp. 489–504 (2018)
15. Wang, Y., Tang, C., Wang, Z., et al.: SIndex: a scalable learned index for string keys. In: Proceedings of the 11th ACM SIGOPS Asia-Pacific Workshop on Systems, pp. 17–24 (2020)
16. Handy, J., Coughlin, T.: Persistent memories without optane, where would we be. In: Storage Developer Conference (SDC). SNIA (2022)
17. Ding, J., Minhas, U.F., et al.: ALEX: an updatable adaptive learned index. In: Proceedings of the 2020 ACM SIGMOD International Conference on Management of Data, pp. 969–984 (2020)
18. Wu, J., Zhang, Y., Chen, S., et al.: Updatable learned index with precise positions. arXiv preprint arXiv:2104.05520 (2021)
19. Tang, C., Wang, Y., Dong, Z., et al.: XIndex: a scalable learned index for multicore data storage. In: Proceedings of the 25th ACM SIGPLAN Symposium on Principles and Practice of Parallel Programming, pp. 308–320 (2020)

20. Gugnani, S., et al.: Understanding the idiosyncrasies of real persistent memory. In: Proceedings of the VLDB Endowment, pp. 626–639 (2020)

21. Chen, Z., Hua, Y., Ding, B., et al.: Lock-free concurrent level hashing for persistent memory. In: Proceedings of the 2020 USENIX Annual Technical Conference, pp.799–812 (2020)

22. Alguliev, R.M., et al.: CDDS: constraint-driven document summarization models. Expert Syst. Appl. **40**(2), 458–65 (2013)

23. Yang, J., Kim, J., Hoseinzadeh, M., et al.: An empirical guide to the behavior and use of scalable persistent memory. In: File and Storage Technologies (2020)

24. Leskovec, J., Backstrom, L., Kleinberg, J.: Meme-tracking and the dynamics of the news cycle. In: Proceedings of the 15th ACM SIGKDD International Conference on Knowledge Discovery and Data Mining, pp. 497–506 (2009)

25. Manegold, S., Boncz, P., Kersten, M.L.: Generic database cost models for hierarchical memory systems. In: Proceedings of the 28th International Conference on Very Large Databases VLDB, pp. 191–202 (2022)

26. Venkataraman, S., Tolia, N., Ranganathan, P., et al.: Consistent and durable data structures for non-volatile byte-addressable memory. In: Proceedings of the 9th USENIX Conference on File and Storage Technologies. FAST (2011)

27. Zhou, W., Yang, S.: SLIPP: a space-efficient learned index for string keys. In: Proceedings of the 2024 6th International Conference on Big-data Service and Intelligent Computation (2024)

28. Qi, X., Wang, M., Wen, Y., Zhang, H., Yuan, X.: Weighted cost model for optimized query processing. In: Zhao, X., Yang, S., Wang, X., Li, J. (eds.) WISA 2022. LNCS, vol. 13579, pp. 473–484. Springer, Cham (2022). https://doi.org/10.1007/978-3-031-20309-1_42

29. Mandelman, J.A., Dennard, R.H., Bronner, G.B., et al.: Challenges and future directions for the scaling of dynamic random-access memory (DRAM). IBM J. Res. Dev. **2**(3), 187–212 (2002)

Dataset Construction for Fine-Grained Emotion Analysis in Catering Review Data

Junling Liu[1,2], Tianyu Chang[1], Xinyun Shi[1], Huanliang Sun[1,2(✉)],
and Jingke Xu[1,2,3]

[1] School of Computer Science and Engineering, Shenyang Jianzhu University,
Shenyang 110168, China
`sunhl@sjzu.edu.cn`
[2] Liaoning Province Big Data Management and Analysis Laboratory of Urban
Construction, Shenyang 110168, China
[3] Shenyang Branch of National Special Computer Engineering Technology Research
Center, Shenyang 110168, China

Abstract. Emotion datasets serve as the foundation for training and evaluating emotion analysis models. By analyzing emotion datasets, we can gain users' emotional tendencies and feedback in specific field. In this paper, we construct a large-scale fine-grained emotion classification dataset in catering field, named CateringEmo17, to enhance service quality and improve user experience. We employed an emotion annotation method that combines semi-automatic annotation and manual annotation to label 44,158 reviews of catering field with 17 different emotions.In the experiments we first trained a BERT-based model on the CateringEmo17 dataset to verify the accuracy of fine-grained emotion classification. Them we mapped fine-grained emotions to coarse-grained emotion to test the accuracy of coarse-grained emotion classification through a transformation experiment. Finally, we compared the model trained on a general emotion dataset, GoEmotion, with the same dataset for classification results. In the dataset anaysis, we validated the fine-grained emotion model on two restaurant dimensions including type, location. Experimental results indicate that our proposed fine-grained emotion classification model can provide potential information in the catering field, while traditional coarse-grained emotion classification models cannot capture this information.

Keywords: catering field · emotion detection · fine-grained emotion classification · deep learning · natural language processing

1 Introduction

With the widespread popularity of online platforms in the catering field, a large amount of review data has been generated in the consumption, and using machine learning to analyze text emotions has become an important research topic. Existing datasets in the catering field typically contain 3 categorical attributes including *positive, negative,* and *neutral,* such as the Yelp dataset [9], while actual user reviews contain rich expressions of emotions. Demszky et al. selected 28 common emotions and constructed the GoEmotion dataset [2]. However, the dataset is on general field but not applicable to the catering field. We have identified biases in the distribution of emotions by Goemotion's 28 emotions in the catering field. For example, there are a large number of comments for *admiration,* while *admiration* can be divided into several fine-grained emotions including *approval, love* and *joy.* While the comments for several emotions such as *pride* are few, these emotions should be combined. We propose a fine-grained emotion dataset construction method in the catering field, and use it to analyze user emotions in the catering field.

Due to the large scale of catering review data, the determination of fine-grained emotions is a complex task. First, most existing emotion is not all apply to the catering field, so to determinate the source of emotions and filter the fine-grained emotions is a challenge. Secondly, in the labeling process, different annotators may label different emotions for the same review data, to deal with the subjective influence of annotators is another challenge. Third, due to the large amount of data and the personal preferences of users, there are often more positive reviews than negative ones, so to solve the imbalance of data and data bias is the third challenge.

To capture a comprehensive range of emotions and leverage the extensive user base and rich emotional content on social platforms, we used the emotions extract from short texts and videos. From these, we selected the 17 most frequent emotions as part of the standard dataset. We used the multi-person consensus strategy to determine the emotions, and used the syntax trees and clustering techniques to get preprocessing labels. Regarding to the data bias problem, we randomly sample from 100k pieces of data, then pre-annotate the sample to find the high frequency emotions. Finally, we use the 17 fine-grained emotions to label the reviews. The contributions of this paper are summarized as follows:

(1) We construct a fine-grained emotion dataset in the catering field, named CateringEmo17. The dataset is at https://github.com/Changchang12345/The-cateringEmo17-Dataset.
(2) We use the BERT-based model to fine-tune and train the dataset, and experimental results shows that the proposed dataset is suitable for multi-classification tasks in the catering field.
(3) We analyze the fine-grained emotion of the catering field from three aspect including beverage categories, location and time. The experimental results show that CateringEmo17 can provide potential information, while traditional coarse-grained emotion classification models cannot reflect.

2   Related Work

In this section, we introduce the related works including emotion dataset construction, emotion classification and emotion classification model.

2.1   Emotion Dataset Construction

Emotion datasets can contribute to a deeper understanding of social emotional phenomena and help us better address social emotional issues. Some datasets has been proposed including: Ghaz et al. constructed a sentence dataset based on the frame net [7]; Ohman et al. constructed a movie subtitle dataset [10]; a large-scale high-quality DiaASQ dataset in both chinese and english languages [14]; Wenmeng Yu et al. constructed a chinese single and multi-modal emotion analysis dataset [24]; Ranganathan et al. studied the bio-Reactions and faces for emotion-based personalization for AI systems (BIRAFFE2) dataset [17]; Yan Wang et al. built a large-scale multi-scene dataset for facial expression recognition in videos, coined as FERV39k [21]; Taylan K. Sen et al. built a dataset for discerning benefits of audio, textual, and facial expression features in competitive debate speeches [20]. Most of these emotion datasets are manually constructed and tend to be relatively small in scale. However, Bostan and Klinger pointed out that this dataset contains a lot of noise [13]. Demszky et al. created a large dataset called GoEmotion, which focuses on conversational data and includes 58,000 human-annotated Reddit comments labeled with 28 emotions [2].

In summary, existing datasets mostly consist of general sentiment data or cover only a few types of emotions and there is no fine-grained emotion datasets in the catering domain were available. While we have constructed a large scale fine-grained emotion dataset with 44158 reviews in the catering domain called CateringEmo17 using a semi-automatic human annotation method. During the construction process we improve the fine-grained classification accuracy of the emotions classification model and enhance the diversity of emotion analysis.

2.2   Emotion Classification

Firstly, in discrete emotion theory, H. Lillie et al. studied the role of discrete emotions in the engagement and dissemination of political news on facebook [15]. Ekman et al. proposed the six basic emotions including *anger, disgust, fear, happiness, sadness,* and *surprise* [5]. Ekman et al. explains that each emotion has specific characteristics expressed by varying degrees [6]. So we should first determine the main discrete categories of emotions, we define the discrete categories of emotions in the catering field.

Secondly, emotions can be grouped according to dimensions. The best-known two-dimensional models are the circumplex model, the vector model, and the positive-negative activation (PANA) model [18]. Xiaoyang Gong et al. examined the dimensions of achievement emotions and tested a model in which achievement emotions mediated the relationship between chemistry self-efficacy and classroom engagement [8]. In 1954, Harold Schlossberg proposed another method in which the three dimensions are pleasant-unpleasant, attention-rejection, and activation levels [19]. In 1897, Wilhelm Max Wundt proposed a method of describing emotions using three dimensions including *pleasant* and *unpleasant, excited* or *inhibited,* and *tense* or *relaxed* [11].

Based on the emotional dimensional models described above, we expanded the initially identified basic emotions into a more comprehensive emotions list, forming a finer-grained list of 17 emotions in the catering domain.

2.3   Emotion Classification Model

Generally, the models of emotion classification are divided into three categories including rule-based emotion classification models [23], neural network-based emotion classification models [12], and deep learning-based emotion classification models. Rule-based classification models mainly use manual constructed bags of words or feature rules for analyzing. Yiran Liu et al. proposed a rule-based classification model from big data to extract revisit intentions [16], so it requires a large amount of reliable data or a mature classification model to generate a classifier. The model based on neural network needs to be provided important features of the data for training, such as the word embedding model Word2Vec [1]. The training performance is great for pictures and sequences, for table format, machine learning classification models cannot accurately find commonalities in these attributes. Deep learning-based emotion classification models can automatically identify important features that are helpful for classification. Zhang Wenxuan et al. proposed RILA model for exploring the effectiveness of leveraging multimodal interactions between intermediate representations of deep pretrained converters to achieve end-to-end emotion recognition [22]. Diao Yuxuan et al. proposed a multi-label imbalanced data classification method ESP based on label partition integration make all binary classification models are integrated into a multi-label classification model [4].

We use a deep learning-based transformer model BERT [3], which has the characteristics of pre-training language model and it has achieved the best results in several recent NLP performance tests which include emotion classification. We use the CateringEmo17 dataset to train a model with multi-emotion classification, and the model has been fine-tuned based on the pre-training model of BERT. We performed multi-classification emotion distribution analysis according the performance testing results.

3   Construction of Fine-Grained Emotion Datasets in the Catering Field

We constructed a fine-grained emotional datasets in the catering field. The dataset is constructed using natural language processing technology combined with artificial annotation, the construction process is shown in Fig. 1, and the steps including: (1) Sampling and cleaning of original data. (2) Filtering the universal emotions extracted from social platforms. Then we analogous our emotions to GoEmotion's 28 general emotion and obtaining 17 pre-labeled emotions. The adapter is designed to perform a sample check on the emotion of the review data. (3) Since the aspects in the catering field are relatively distinct, we can cluster the data to obtain specific aspects, thereby narrowing down the large space of emotions. (4) Identifying the key words in the sentence through POS tagging and adding key words to the labeling sample table, so annotators can quickly identify subjective emotions. (5) Analyzing the relationship between words in the sentence through the grammar tree, it also helps us to determine the subjective emotions of the sentence. (6) Setting the Preprocessed five attributes for each data according to the above five steps(original data, pre-labeled emotion, aspect of the data, key subjective words, key modifying relationships) and forming a labeling sample table, Table 1 provides an example of the labeling sample table. (7) labeling review data according to the labeling sample table.

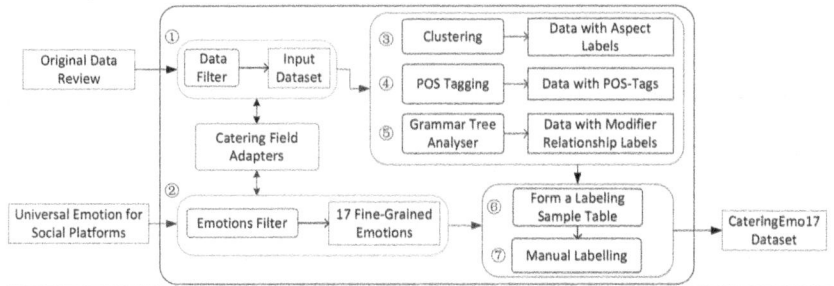

Fig. 1. The flow chart of dataset construction

3.1   Choice of Data and Emotions

We choose Yelp dataset as the data source, since the data does not require superfluous user information, to ensure the accuracy of the dataset, we focused on selecting data from the business table and user table. Specifically, we selected reviews with restaurant tags and a minimum of 1000 data per location. In total, we gathered a dataset consisting of 100k comments. We then selected 1000 data for pre-labeling to obtain an overall emotional distribution of the sample. Based on our previous research, we have identified 17 emotions that are relevant to our research topic. In addition, we analogous the 28 general emotions in Goemotion dataset to our proposed emotions, and for the features of catering field we merged some emotions that are not commonly observed in the catering field. At the same time, we have introduced a new emotion called *certainty*, which often appears in the catering field to represent the awakening of certain memories.

3.2   Dataset Clustering Methods

We have clustered the datasets based on different aspects. This is because emotions related to the same aspect tend to be similar. For example, when an emotion comes to dishes, consumers often describe it as delicious or unpalatable. while an emotion comes to service, consumers often describe it as attentive or enthusiastic. This helps us focus on labeling the emotions that are more relevant and commonly expressed in the same aspect. Common aspects in the catering field are location, service, environment, price, and dishes. We used the Fast-Text model to transform the sentences to be trained into word vectors, and then obtained aspect labels of each comment data. By using this classification methods, annotation does not spend too much time assigning comments emotion labels according the aspect label of the comment.

3.3 POS Tagging

We can also help annotators to identify a certain word and to find its relevant modification relationship quickly through POS tagging. For example, "I need a tasty apple", the sentence is neutral. Another example is "This apple is so tasty", the sentence also contains apples and delicious, but the emotions of this sentence are positive. The key words help annotators to observe the components of sentences and then find the structure that can reflect the subjective emotions, so we put the words into the labeling sample table. Through the above example, we put "tasty" in the labeling sample table, the annotator can know the strycture needs to be focused on in the sentence is around "tasty". It can be quickly distinguished the review is positive, and it is easy to find that the main modifier in the sentence is "apple". Through POS tagging, we can also find out whether the sentence has negative modification, so as we can find the corresponding negative emotion. We use the Spacy module to process existing data into a combination of sentences and parts of speech to help annotators label data.

3.4 Syntax Trees Construct for Word Modifier Relationships

A grammatical dependency tree represents the relationships between words in a sentence, uncovering various connections and dependencies between them. By analyzing these relationships, we can determine the modifying relationships between words and detect potential subjective emotions. In general, a subjective emotion is usually an adjective and its variant, so we should find a word and its reduced part of speech. Usually, the prototype of a word is an adjective and several adverbs. Taking "Other than that, I was extremely disappointed at the food quality and service." as an example. "At" and "quality" are Prepositional object relationships, "was" and "disappointed" are adjectival complement relationships. The grammar tree allows you to see the relationship between words and thus we can find out the subjective emotions of the sentence.

3.5 Manual Annotation Stage

We adopt the manual annotation method to label a part of data. The grouping strategy of data annotators is to divide 20 people into two groups, each group of 10 people. Each group of annotators takes 1000 pieces of data. The data to be labeled contains five attributes including data text, aspect, subjective word, modification relationship, coarse-grained emotions (4 classification), and fine-grained emotions(17 classification). The annotator obtains the data in the form of Table 1, and people annotate data through observing the labeling sample table. We set the label 0–16 for each emotion. For reducing the labeling error caused by the subjective influence, two people labeled the same sentence for comparison. If the two labels are not exactly consistent, we would use the overlapping emotion and discard the incorrect emotion. Then put this data back into the dataset and use existing data table to obscure it, and new labelers are arranged to relabel the data until the data labels are completely consistent, and if there is still a multi-person labeling inconsistency, the comment should be discarded.

Table 1. Example of a labeling sample table for labeling

Data Text	Aspect	Subjective Word	Modification Relation-ship	Coarse-grained Emotions	Fine-grained Emotions
I need a tasty apple	Food	tasty	Apple - tasty	positive	Approval
Unusual comments	–	–	–	–	–
I just walk to here	–	–	–	neutral	neutral
I'm so anger to this waiter Server	Service	anger	Anger - waiter	negative	Anger

4   Experiments

In order to verify the validity of the dataset, this section designs two experiments based on the dataset established in Sect. 3.

4.1   Data Preparation and Model Establishment

We obtained a dataset of reviews from the catering field annotated with fine-grained emotion labels. We fine-tuned BERT pre-trained model to obtain a fine-grained emotion classification model. Each review in the dataset has 1–3 emotion labels. The dataset is divided into three portions: 80% as the training set, 10% as the test set, and the remaining 10% as the validation set. *Neutral* emotions are the most common, as they are mostly used for narrative commentary. *Positive* emotions are the second most prevalent, with the most common being *approval*, which is usually expressed to recommend something. *Negative* and *confusion* emotions are less common, as users usually do not express their opinions on unsatisfactory restaurants. The distribution of the training set, test set, and validation set is the same since the sample is representative of the population. The trained model is evaluated using precision, recall, and F1-score. The F1-score of each individual model is calculated using the macro-average method, and then the overall MacroF1 score is computed. The goal of this research is to develop a multi-classification fine-grained emotion analysis model.

4.2 Conversion Experiments

We trained the 4 classification model using the same experiment settings as the 17 classification above. The classification results are shown in Table 2. The experimental results show that 90.53% F1 score can be obtained through the *positive* emotion after mapping, indicating that the amount of data has a greater impact on the classification result of the model, the classification accuracy increases with the increase of the amount of data, and the *neutral* emotion remains the same as that of 17 classification, because *neutral* emotion does not have a corresponding classification map. For *fuzzy* emotion and *negative* emotion, since the data volume comparison is not as much as *positive* emotion, the feature acquisition of the model is not as comprehensive as *positive* emotion, so it has a certain impact on the accuracy rate, but it is still above 69%, so we got a good result. Experimental results show that the 4 classification model can also obtain high scores. The recall rate is more than 68% and an F1 score is more than 69%, which fully shows that the model performs well on the dataset.

Table 2. The experimental results of 4 classification

Emotion	Precision/%	Recall/%	F1/%
Ambiguous	64.09	76.32	69.67
Negative	71.54	68.72	70.10
Neutral	85.26	85.93	85.59
Positive	88.85	92.28	90.53

4.3 Comparative Experiments

We use a comparison experiment between our model and the GoEmotion model based on a public dataset SemEval2014 in the catering domain. In this experiment, the aspects were hidden, only the polarity in the sentence is retained. The goal is to detect whether our dataset can find more hidden polarity than the generic dataset. This experiment only retains emotions and discards aspects. The performance data of GoEmotion model and our trained model on the SemEval2014 dataset are shown in Table 3, and the GoEmotion model is a multi-classification emotion model, it fine-tuned the BERT pre-trained model on the

GoEmotion dataset. The results show that the model trained on our dataset can obtain an improvement in F1 score of about 10% compared with the data of GoEmotion general model in the catering field. It shows that the model trained by the special domain data can be well adapted to other domain data, and can obtain good classification effect. The model achieves an F1 score of 87% in the classification of *positive* emotion, indicating that the data quality is reliable. Through different data training models to test the same catering field dataset SemEval2014, our model can complete some data classification in the same field well.

Table 3. Comparative results on SemEval2014 dataset

GoEmotion model				Our model			
Emotion	Precision/%	Recall/%	F1/%	Emotion	Precision/%	Recall/%	F1/%
Negative	71.30	44.57	54.85	Negative	83.47	54.89	66.23
Neutral	20.60	74.19	32.24	Neutral	33.04	80.65	46.88
Positive	84.74	72.51	78.15	Positive	90.70	83.63	87.02

5   Dataset Analysis

We design natural language processing technology combined with artificial labeling and a variety of auxiliary labeling methods. We establish a multi-emotion fine-grained catering field dataset, and use the latest NLP pre-training model BERT for training. Good experimental results were achieved through testing, and it also proved the validity of the data. On this basis, two extended analysis experiments of type and location were carried out.

5.1   Type Analysis of Reviews

Table 4. Catering field review data analysis on type (4 classification)

Emotion	American/%	Bar/%	Fast food/%	Italian/%
Ambiguous	7.01	7.26	6.15	6.67
Negative	18.17	14.79	29.04	18.28
Neutral	2.06	1.67	3.93	1.95
Positive	72.74	76.27	60.86	73.09

Table 5. Catering field review data analysis on type (17 classification)

Emotion	American/%	Bar/%	Fast food/%	Italian/%
Approval	39.51	39.17	31.47	38.98
Caring	12.82	13.24	11.26	13.35
Love	7.73	9.14	7.42	8.07
Disapproval	7.26	6.21	7.81	6.88
Excitement	5.16	5.64	4.83	5.22
Surprise	4.97	5.19	3.11	4.59
Joy	4.17	5.31	2.81	3.9
Annoyance	3.63	2.9	7.85	4.01
Anger	3.01	2.27	7.29	2.77
Disappointment	2.75	2.31	3.03	3.04
Desire	2.57	2.72	2.79	2.85
Neutral	2.06	1.67	3.93	1.95
Disgust	1.52	1.1	3.06	1.58
Curiosity	1.09	1.09	1.66	1.04
Confusion	0.95	0.98	1.38	1.04
Relaxed	0.59	0.83	0.12	0.47
Certainty	0.19	0.22	0.16	0.25

Table 6. Catering field review data analysis on location (4 classification)

Emotion	MD/%	MA/%	MS/%	AR/%
Ambiguous	6.41	6.14	5.44	6.35
Negative	16.14	16.24	26.62	27.29
Neutral	1.5	1.72	2.60	2.66
Positive	75.95	75.9	65.34	63.7

This section conducts fine-grained emotion analysis in the catering field on the type dimension, we explore whether the emotions between different types of catering are different, the test data of this part is untrained data of Yelp in catering field. For reduction fitting, we select 4 types of catering, namely American restaurant, Bar, Fast-food, Italian restaurant. The experimental settings are 10,000 pieces of each type, and the results of the experiment are shown in Tables 4 and 5. The data is dominated by positive emotions, supplemented by negative emotions. However, there are many negative emotions in the field of fast food, and we can discover more details in Table 9.

Table 7. Catering field review data analysis on location (17 classification)

Emotion	MD/%	MA/%	MS/%	AR/%
Approval	40.12	39.93	34.62	33.41
Caring	14.24	14.11	11.25	10.52
Love	9.05	9.31	8.97	8.82
Disapproval	6.21	5.96	7.12	7.77
Excitement	5.66	5.41	5.33	5.44
Surprise	4.64	4.54	4.51	5.67
Joy	3.3	3.41	3.21	3.06
Annoyance	3.21	3.28	4.78	3.95
Anger	2.44	2.5	4.52	4.31
Disappointment	3.24	3.54	7.65	8.63
Desire	2.93	2.65	1.44	1.9
Neutral	1.5	1.72	2.60	2.66
Disgust	1.04	0.96	2.55	2.63
Curiosity	1.26	1.25	0.72	0.5
Confusion	0.51	0.35	0.21	0.18
Relaxed	0.45	0.53	0.33	0.25
Certainty	0.20	0.55	0.19	0.3

5.2   Location Analysis of Reviews

This section provides fine-grained emotion analysis on the location dimension. We select the data on four states in the United States, namely Maryland(MD), Massachusetts(MA), Mississippi(MS), Arkansas(AR). Each state selects 10,000 data. The experimental results are set to 4 and 17 classification, as shown in Tables 6 and 7. Through the observation of Table 7, we can see the *approval* emotion of the two economically backward states is low by 5% compared with the two continents with a high degree of economic development, and *Caring* is relatively low. Indicating that *Caring* is not felt in this scene. But the *Excitement* emotion is very close in four states, it shows that *excitement* is not affected by the location. The main reason for this is that consumer expectations for this place are generally high, and in order to be able to have a better experience, it is usually done in a relatively positive state.

6   Conclusions

Our analysis of existing emotion classification models and datasets in the catering field reveals their inadequacy in performing effective fine-grained emotion analysis. To address this problem, we conduct data sampling and propose a set of 17 emotions that are specific to the catering field. By the experiments on the fine-grained emotion model we proposed, we demonstrate that our model is capable of identifying hidden emotions. This validates the model trained on the dataset we proposed can accurately classify emotions in the catering field. Through analysis of real data, we are able to provide actionable insights and recommendations, demonstrating the practical value of our fine-grained emotion analysis approach in the catering field.

Acknowledgements. This work is supported by the National Natural Science Foundation of China (62073227), the Project of the Educational Department of Liaoning Province (JYTMS20231596).

References

1. Church, K.W.: Word2vec. Nat. Lang. Eng. **23**(1), 155–162 (2017)
2. Demszky, D., Movshovitz-Attias, D., Ko, J., Cowen, A., Nemade, G., Ravi, S.: GoEmotions: a dataset of fine-grained emotions. In: Proceedings of the 58th Annual Meeting of the Association for Computational Linguistics, pp. 4040–4054. Association for Computational Linguistics (2020)
3. Devlin, J., Chang, M.W., Lee, K., Toutanova, K.: BERT: pre-training of deep bidirectional transformers for language understanding (2018)
4. Diao, Y., Sun, Z., Zhou, Y.: A multi-label imbalanced data classification method based on label partition integration. In: Yuan, L., Yang, S., Li, R., Kanoulas, E., Zhao, X. (eds.) WISA 2023. LNCS, vol. 14094, pp. 14–25. Springer, Singapore (2023). https://doi.org/10.1007/978-981-99-6222-8_2
5. Ekman, P.: Are there basic emotions? [comment] (1992)
6. Ekman, P.: Emotions Revealed. Emotions Revealed (2004)
7. Ghazi, D., Inkpen, D., Szpakowicz, S.: Detecting emotion stimuli in emotion-bearing sentences. In: Gelbukh, A. (ed.) CICLing 2015. LNCS, vol. 9042, pp. 152–165. Springer, Cham (2015). https://doi.org/10.1007/978-3-319-18117-2_12
8. Gong, X., Bergey, B.W.: The dimensions and functions of students' achievement emotions in Chinese chemistry classrooms. Int. J. Sci. Educ. **42**, 835–856 (2020)
9. Hicks, A., Comp, S., Horovitz, J., Hovarter, M., Miki, M., Bevan, J.L.: Why people use yelp.com: an exploration of uses and gratifications. Comput. Hum. Behav. **28**(6), 2274–2279 (2012)
10. Hman, E., Kajava, K., Tiedemann, J., Honkela, T.: Creating a dataset for multilingual fine-grained emotion-detection using gamification-based annotation. In: Proceedings of the 9th Workshop on Computational Approaches to Subjectivity, Sentiment and Social Media Analysis (2018)
11. Jastrow, J.: Outlines of psychology. Science **5**(127), 882–884 (1897)
12. Lejiang, Z., Yuhan, H., Xiaokun, L., Weihua, Z., Zhongying, Z., Hongxu, C.: Application of neural network model in sentiment classification under explicit and implicit features. Intell. Comput. Appl. (2020)
13. Leonova, V.: Review of non-English corpora annotated for emotion classification in text. In: Robal, T., Haav, H.-M., Penjam, J., Matulevičius, R. (eds.) DB&IS 2020. CCIS, vol. 1243, pp. 96–108. Springer, Cham (2020). https://doi.org/10.1007/978-3-030-57672-1_8
14. Li, B., et al.: DiaASQ: a benchmark of conversational aspect-based sentiment quadruple analysis. ArXiv abs/2211.05705 (2022)

15. Lillie, H.M., Chernichky-Karcher, S., Venetis, M.K.: Dyadic coping and discrete emotions during Covid-19: connecting the communication theory of resilience with relational uncertainty. J. Soc. Pers. Relat. **38**, 1844–1868 (2021)
16. Liu, Y., Beldona, S.: Extracting revisit intentions from social media big data: a rule-based classification model. Int. J. Contemp. Hospit. Manage. (2021)
17. Ranganathan, H., Chakraborty, S., Panchanathan, S.: Multimodal emotion recognition using deep learning architectures. In: 2016 IEEE Winter Conference on Applications of Computer Vision (WACV), pp. 1–9 (2016)
18. Rubin, D.C., Talarico, J.M.: A comparison of dimensional models of emotion: evidence from emotions, prototypical events, autobiographical memories, and words. Memory **17**(8), 802–808 (2009)
19. Schlosberg, H.: Three dimension of emotion. Psychol. Rev. **61** (1954)
20. Sen, T.K., et al.: DBATES: dataset for discerning benefits of audio, textual, and facial expression features in competitive debate speeches. IEEE Trans. Affect. Comput. **14**, 1028–1043 (2023)
21. Wang, Y., et al.: FERV39k: a large-scale multi-scene dataset for facial expression recognition in videos. In: 2022 IEEE/CVF Conference on Computer Vision and Pattern Recognition (CVPR), pp. 20890–20899 (2022)
22. Wu, Y., Zhang, Z., Peng, P., Zhao, Y., Qin, B.: Leveraging multi-modal interactions among the intermediate representations of deep transformers for emotion recognition. Association for Computing Machinery, New York (2022)
23. Yanping, W., Zhu, M., Jiaming, T., Jiazhen, G., Fang, Z.: A sentiment classification method that combines sentiment rules and machine learning. J. Sci. Sci. **40**(6), 5 (2020)
24. Yu, W., et al.: CH-SIMS: a Chinese multimodal sentiment analysis dataset with fine-grained annotation of modality. In: Annual Meeting of the Association for Computational Linguistics (2020)

Database Parameters Tuning via Bayesian Optimization with Domain Knowledge

Zhongwei Yue and Peng Cai[✉]

East China Normal University, Shanghai, China
52195100010@stu.ecnu.edu.cn, pcai@dase.ecnu.edu.cn

Abstract. Bayesian optimization has gained widespread adoption in database knob tuning due to its theoretical advantages in balancing exploration and exploitation. Yet, a significant drawback of existing Bayesian optimization-based approaches is typically their failure to incorporate domain knowledge related to databases when searching for the optimal configuration. This limitation often leads to the recommendation of low-utility configurations that violate domain knowledge, thereby affecting its tuning efficiency. To address this issue, we propose DKTune, which seamlessly integrates Bayesian optimization with domain-specific database knowledge. DKTune leverages the inherent dominant relationships between database knobs to enhance the surrogate model used in Bayesian optimization. Additionally, it considers constraint relationships between knobs, competitive interactions among knobs, and the dynamic characteristic of knobs to assist the acquisition function in evaluating the utility of each configuration. We evaluated DKTune on two popular open-source database systems, and the experimental results demonstrate that DKTune significantly improves the efficiency of database knob tuning and the final tuning results.

Keywords: Bayesian optimization · Domain knowledge · Tuning efficiency

1 Introduction

Modern Database Management Systems (DBMSs) feature numerous knobs (or parameters) that affect many aspects of the database, including query performance, resource allocation, and concurrency control. Traditionally, Database Administrators (DBAs) rely on their domain knowledge to tune these knobs, aiming to boost system performance. Yet, navigating the multitude of knobs and accommodating the recurring shifts in workload often present formidable challenges during the tuning process. Consequently, various methods have been developed to aid DBAs in the sophisticated task of tuning database knobs.

Present approaches to database knob tuning entail continuous sampling. Each sampling serves as an iteration, primarily assessing the database's performance under specific knob configurations. Each sampling session typically requires a considerable amount of time, ranging from several minutes to tens of minutes. In the context of database knob tuning, there is a shared expectation of achieving satisfactory results within a few hours [15]. This requirement calls for an efficient tuning method capable of identifying high-quality configurations with a small number of samples. Bayesian optimization [12], renowned for its theoretical advantages in balancing exploration and exploitation, can deliver promising results within a limited number of iterations. In light of this, it has found widespread application in various database tuning methods, such as iTuned [15], OtterTune [17], and LlamaTune [8]. Furthermore, the Bayesian optimization method, SMAC [6], has been verified as a state-of-the-art approach for database knob tuning [20].

Table 1. The number of samples violating domain knowledge rules among 100 samplings for different tuning methods.

Methods	TPC-C	Wikipedia
iTuned	9	10
OtterTune	10	9
LlamaTune	8	9

While the current Bayesian optimization-based methods have demonstrated effectiveness in tuning database knobs with limited samples, their inability to integrate domain-specific knowledge related to databases may lead to choosing configurations that contradict principles within this domain. For instance, when conducting 100 samplings (or iterations) on the MySQL database using three different tuning methods iTuned, OtterTune, and LlamaTune respectively, the experimental results revealed that each of these methods produced a significant number of samples that violated domain knowledge rules (as shown in the Table 1). The domain knowledge rules mentioned above includes dominant relationships, constraint relationships, competitive relationships, and dynamic characteristics. Samples that violate established rules typically lead to poor database performance or result in redundant sampling. This scenario prompts tuning methods to undertake attempts that have relatively low utility. As a result, such a process diminishes the efficiency of the entire tuning process.

Given this, we propose DKTune, a solution that combines domain-specific database knowledge with Bayesian optimization techniques. This integration not only enhances the precision of the surrogate model but also enables the acquisition function to effectively avoid sampling less promising configurations. Therefore, it can improves the efficiency of database knob tuning. The following is the list of our main contributions.

– DKTune enhances the accuracy of Bayesian optimization's surrogate models significantly by leveraging dominant relationships between knobs to directly ascertain database performance under some unsampled configurations.

- DKTune employs competitive relationships among knobs, constraint relationships between knobs, and the dynamic characteristics of knobs to assist Bayesian optimization's acquisition functions in evaluating unsampled configurations, thereby facilitating the recommendation of more promising knob settings.
- We implemented the DKTune in two widely-used DBMSs, MySQL and PostgreSQL, and conducted tests across two distinct workloads: TPC-C and Wikipedia. Our experimental results show that DKTune greatly improves the efficiency of database knob tuning.

2 Preliminary

Database Knob Tuning. Consider a scenario with n database knobs to be tuned, denoted as $knob_1, knob_2, \ldots, knob_n$. Each knob has its own value domain, $\Theta_1, \Theta_2, \ldots, \Theta_n$, which can be continuous, discrete or categorical. The configuration space is thus defined as $\Theta = \Theta_1 \times \Theta_2 \times \ldots \times \Theta_n$. We assume a performance metric y determined by the objective function $f : \Theta \to \mathbb{R}$, assigning performance values (e.g., throughput or average latency) to each configuration $x \in \Theta$. The aim of database knob tuning is to find the optimal configuration x_*, that maximizes (or minimizes) this metric, formalized as $x_* = \arg\max_{x \in \Theta} f(x)$.

Bayesian Optimization. Bayesian optimization emerges as a potent technique for global optimization, particularly suited for high-cost, non-analytical blackbox functions. It comprises two key components: surrogate models and acquisition functions.

Surrogate Models. Surrogate models serve to approximate the objective function f. These models rely on historical samples to predict function values in unsampled configurations. For each configuration x, they not only estimate the expected value (i.e., mean $\mu(x)$) but also evaluate the uncertainty associated with this value (i.e., variance $\sigma^2(x)$). Gaussian Processes (GPs) [14] and Random Forests (RFs) [1] are commonly employed as surrogate models in Bayesian optimization, with the latter's Bayesian optimization known as SMAC [6].

GPs. GPs are a machine learning method rooted in Bayesian theory. They make use of the assumption that data points adhere to a Gaussian distribution, leveraging this to predict values for new data points. In the context of a historical dataset denoted as (X, Y), where X represents all sampled knob configurations and Y includes corresponding performance metrics. The mean $\mu(x)$ for a new configuration x is computed using the equation:

$$\mu(x) = k(X, x)^T [k(X, X) + \epsilon^2 I]^{-1} Y \tag{1}$$

In this equation, k signifies the kernel function, which measures the similarity between configurations. Common kernel functions are the Radial Basis Function [2] and the Linear Kernel. The ϵ^2 represents noise. The variance $\sigma^2(x)$ is given by:

$$\sigma^2(x) = k(x, x) - k(X, x)^T [k(X, X) + \epsilon^2 I]^{-1} k(X, x) \tag{2}$$

SMAC. SMAC utilizes a random forest as its surrogate model to establish the relationship between configurations and performance metrics. This model consists of multiple decision trees, with each tree trained on unique random subsets of historical data. For a given configuration x, every tree generates an output, and the mean $\mu(x)$ is computed as the average of these outputs. The variance $\sigma^2(x)$ is determined from the variances of outputs across all trees.

Acquisition Functions. In Bayesian optimization, the acquisition function plays a pivotal role in the optimization process. It assesses the utility of each configuration and guides the selection of the next optimal configuration. By theoretically balancing exploration and exploitation, this function enables an efficient discovery of the global optimum. Common acquisition functions encompass Expected Improvement (EI), Probability of Improvement (PI), and Upper Confidence Bound (UCB).

3   Integration of Domain Knowledge with Bayesian Optimization

Domain Knowledge. Several methods exist for acquiring domain knowledge in the field of databases, including learning from DBAs, consulting official database manuals, and utilizing artificial intelligence chatbots like ChatGPT[1]. DKTune primarily addresses the issue of effectively utilizing existing domain knowledge within Bayesian Optimization. Consequently, it assumes that the necessary database knowledge has been acquired. Regarding database domain knowledge, we focus on four categories: competitive relationships, dominant relationships, constraint relationships, and dynamic characteristics. This is because these four types of domain knowledge are prevalent in numerous databases.

3.1   Dominant Relationships

Dominant relationships between knobs are common in database domain knowledge. A dominant relationship refers to a situation where the adjustment of one knob can significantly affect the effectiveness of other knobs. For instance, setting the knob innodb_adaptive_hash_index to 0 will cause the knob innodb_adaptive_hash_index_parts to become ineffective. Table 2 provides examples of dominant relationships in MySQL and PostgreSQL. The lack of understanding regarding dominant relationships may introduce significant biases into the surrogate model of Bayesian optimization, thereby adversely affecting the evaluation of the acquisition function.

Therefore, we employ Algorithm 1 to seamlessly integrate Bayesian optimization with dominant relationships between knobs. Upon receiving a new sample denoted as (x, y), comprising the latest knob configuration x and its corresponding performance metric y, the algorithm first checks whether the knob values in configuration x render other knobs ineffective (line 3). When

---

[1] https://openai.com/chatgpt.

Table 2. Examples of dominant relationships in MySQL and PostgreSQL.

Database	Dominant Relationships between Knobs
MySQL	innodb_deadlock_detect **dominants** innodb_lock_wait_timeout
	innodb_thread_concurrency **dominants** innodb_concurrency_tickets
PosgrSQL	track_counts **dominants** autovacuum_vacuum_threshold
	bgwriter_lru_maxpages **dominants** bgwriter_delay

such a situation arises, the algorithm examines the value domain of the ineffective knobs. For knobs with discrete or categorical value domains (line 4), the algorithm iterates through every possible value within the knob's value domain (line 5). It then combines each possible value with other knob values in x to generate new knob configurations x' (line 6). Since the ineffective knob does not impact performance (i.e., the database performance for configuration x' remains y), the algorithm utilizes (x', y) to update the Bayesian optimization's surrogate model (line 7). In the case of an ineffective knob with a continuous value domain, the algorithm samples a predefined number of equidistant values from the knob's domain (line 11). Each sampled value is paired with other knob values in x to create new parameter configurations x' (line 12). Then, the surrogate model is updated with (x', y) (line 13). Through this algorithm, the surrogate model effectively incorporates dominant relationships during tuning, resulting in a positive impact on the tuning process.

Algorithm 1: Integrating Bayesian optimization with dominant relationships between knobs

---

Input: x: the knob configuration of the latest sample; y: the performance metrics corresponding to the configuration x; m: the number of samples taken for a continuous value domain

1   **for** $x_i \in x$ **do**
2   **for** $x_j \in x$ **do**
3   **if** x_i *causes* x_j *to become invalid* **then**
4   **if** Θ_j *is discrete or categorical domain* **then**
5   **for** $v \in \Theta_j$ **do**
6             $x' \leftarrow x \setminus \{x_j\} \cup \{v\}$
7             $Update\_Surrogate\_Model(x', y)$

8   **else**
9           $increment = (max(\Theta_j) - min(\Theta_j))/m$
10   **while** $m > 0$ **do**
11             $v \leftarrow (min(\Theta_j) + increment)$
12             $x' \leftarrow x \setminus \{x_j\} \cup \{v\}$
13             $Update\_Surrogate\_Model(x', y)$
14             $m = m - 1$

Compared to the long runtime of workloads (typically several hundred seconds), the update time of DKTune's surrogate model is short, usually taking only a few seconds per update (as detailed in Sect. 4). Therefore, the model updates caused by dominant relationships will not affect the tuning efficiency of DKTune.

3.2   Competitive Relationships

In database systems, it's common to encounter multiple knobs competing for the same resources, such as memory and disk. Table 3 provides a list of knobs in MySQL and PostgreSQL that consume memory resources. Given the typically limited server resources, there's a competition for these resources. Consider three knobs, $knob_1, knob_2, knob_3$, each consuming memory resources in a server with a total available memory R. Suppose the domain of $knob_1$ is the closed interval $[min_1, max_1]$, of $knob_2$ is $[min_2, max_2]$, and of $knob_3$ is $[min_3, max_3]$. To prevent database startup issues or errors, the domain of these knobs should satisfy the condition $max_1 + max_2 + max_3 \leq R$.

In such cases, correct allocation of domains for each knob becomes crucial. An incorrect division of knob domains can significantly impact database performance. For instance, workloads with heavy write operations often require a larger value for the knob innodb_log_buffer_size of MySQL. If the maximum value of the knob's domain is set too low, it prevents the knob value from being appropriately configured, thereby impeding potential database performance improvements. Currently, the assignment of knob value ranges heavily relies on the experience of DBAs. However, accurately defining knob domains poses challenges for DBAs as it necessitates a comprehensive understanding of the database's workload and data distribution. Furthermore, setting domains for each knob while ensuring that the total of their maximum values remains within the server's resource limits can result in suboptimal memory utilization. Specifically, if the value of the knob is significantly lower than the maximum value of its domain, then the unused resources of this knob cannot be reallocated to other knobs that consume the same type of resource. This inefficiency arises because the value range of each knob has been predefined and cannot be dynamically adjusted during the tuning.

Table 3. Examples of memory-consuming knobs in MySQL and PostgreSQL.

PostgreSQL Knobs	MySQL Knobs
shared_buffers	innodb_buffer_pool_size
work_mem	sort_buffer_size
maintenance_work_mem	read_buffer_size

To help DBAs address these problems, the following strategies are implemented: the value domains of knobs $knob_1, knob_2, knob_3$ are set to $[min_1, R]$,

$[min_2, R]$, and $[min_3, R]$ respectively. When using the acquisition function to evaluate the utility $U(x)$ of each configuration, the evaluation criterion defined in Eq. 3 is employed.

$$U(x) = \begin{cases} 0, & a + b + c > R \\ U_c(x), & \text{otherwise} \end{cases} \quad (3)$$

In this equation, a, b, and c represent the values of $knob_1$, $knob_2$, and $knob_3$, respectively. U_c denotes the utility evaluated using constraint relationships (as detailed in Subsect. 3.3).

3.3 Constraint Relationships

Table 4. Examples of constraint relationships in MySQL and PostgreSQL.

Database	Constraint Relationships between Knobs
MySQL	innodb_io_capacity_max *constrains* innodb_io_capacity
	thread_pool_size *constrains* thread_cache_size
PosgrSQL	max_worker_processes *constrains* max_parallel_workers
	seq_page_cost *constrains* random_page_cost

In database configurations, constraint relationships, where one knob's value must not be less than another, are common. Table 4 provides examples of constraint relationships in MySQL and PostgreSQL. For instance, in PostgreSQL, the value of max_worker_processes should surpass that of max_parallel_workers to make full use of system resources for parallel queries. Unfortunately, current Bayesian optimization methods often overlook these constraint relationships. This oversight can lead to the acquisition function occasionally recommending configurations that violate these constraints, ultimately resulting in a sampling of little significance.

To ensure that the recommended configurations comply with constraint relationships, DKTune incorporates constraint relationships into the utility calculation of the acquisition function. Specifically, configurations that violate the constraint relationships are assigned a utility value of zero in their acquisition function. For example, consider a constraint relationship between knobs $knob_1$ and $knob_2$, which requires the value of $knob_1$ to be not less than that of $knob_2$. In this scenario, the acquisition function can be formulated as:

$$U_c(x) = \begin{cases} 0, & \text{if } a < b \\ U_d(x), & \text{otherwise} \end{cases} \quad (4)$$

In this equation, a and b correspond to the values of $knob_1$ and $knob_2$ respectively. U_d represents the utility evaluated based on the dynamic characteristic of knobs, as elaborated in Subsect. 3.4.

3.4   Dynamic Characteristics

Table 5. Examples of dynamic knobs and static knobs in PostgreSQL.

Static Knobs	Dynamic Knobs
shared_buffers	bgwriter_delay
wal_level	bgwriter_lru_maxpages
fsync	wal_buffers

Based on whether modification of knob values requires a database restart, database knobs can be categorized into two groups: dynamic knobs and static knobs. Dynamic knobs are those whose values take effect immediately after modification without requiring a database restart. In contrast, static knobs refer to those that necessitate a database restart to become effective after their values are changed. Table 5 presents examples of dynamic knobs and static knobs in PostgreSQL. However, current Bayesian optimization-based methods do not differentiate between dynamic knobs and static knobs during knob tuning. This lack of distinction leads to unnecessary database restarts in cases where only dynamic knob values are modified, thereby reducing the efficiency of database tuning.

Therefore, when evaluating the utility of knob configurations using the acquisition function, it's essential to consider the restart time of the database. Assume the runtime of the workload is t_1, and the restart time of the database is t_2. In this context, the calculation formula for the acquisition function can be represented as follows:

$$U_d(x) = \begin{cases} \frac{\alpha(x)}{t1}, & \text{when not involving the modification of static knobs.} \\ \frac{\alpha(x)}{(t1+t2)} & \text{when involving the modification of static knobs.} \end{cases} \quad (5)$$

Here, α represents the acquisition function. Thus, Eq. 5 ensures a balance between the sampling time and potential utility for each knob configuration, enabling the enhancement of tuning efficiency.

4   Experiment

Setting. We have deployed DKTune across two distinct database environments: MySQL 8.0 and Postgresql 12.12. Each of these databases is hosted on a dedicated server, configured with the CentOS-7.5 operating system. These servers each boast a 16-core CPU, 8 GB of RAM, and a 200G HDD for storage. Since Bayesian optimization require initial samples for model construction, we initially utilized the Latin Hypercube Sampling (LHS) method [11] to gather 5 samples. The sampling number m is set to 30 for those knobs that have a continuous value range and are dominated by other knobs.

Comparative Methods. In this section, we conduct a detailed evaluation of DKTune's performance and engage in an in-depth comparison with existing methods: iTuned, OtterTune, and LlamaTune. Given that SMAC has been verified as a state-of-the-art approach for database knob tuning, we utilize LlamaTune technology to optimize SMAC, denoted as LlamaTune in this paper. In the experiments, we first integrated LlamadaTune, iTuned, and Ottertune with domain knowledge. This integration resulted in the methods being named DKTune(LlamadaTune), DKTune(iTuned), and DKTune(Ottertune), respectively.

Workload. In our experiments, we selected two unique workloads: TPC-C and Wikipedia. Regarding the database scales, we chose configurations comprising 100 warehouses for TPC-C, and 4,0000 articles for Wikipedia. For database performance metrics, each workload was required to run for 150 s and collect its throughput.

4.1   Experiments on MySQL

Knobs. A total of 70 knobs are selected for tuning. Among these, 6 knobs compete for memory resources. 6 knobs have dominant relationships, 6 knobs have constraint relationships and 18 knobs are dynamic knobs.

Execution Time Breakdown. The tuning process of DKTune comprises five main stages: database restart, workload execution, model updating, and database configuration adjustment. Restarting the MySQL database takes 20 s, while workload execution is fixed at 200 s. Updating the Gaussian Process model requires 5.2 s, and updating the SMAC model takes 0.67 s. Recommendations for configuration parameters take less than 0.2 s, and applying these parameters to the database takes no more than 0.01 s.

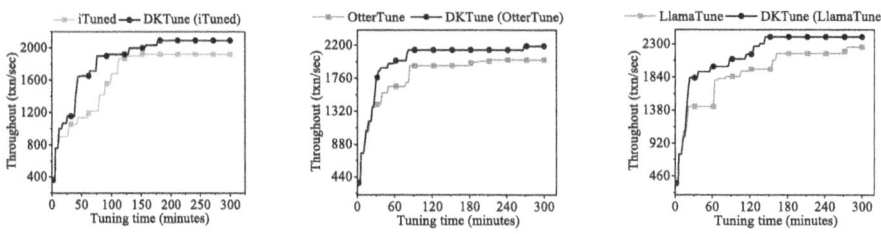

Fig. 1. Comparison of tuning results on TPC-C (MySQL).

Firstly, we compared the tuning results on TPC-C workloads (as shown in Fig. 1). The experimental results indicate that integrating domain knowledge with tuning methods can accelerate the tuning efficiency. Moreover, compared to iTuned, OtterTune, and LlamaTune, the final tuning outcomes of DKTune (iTuned), DKTune (OtterTune), and DKTune (LlamaTune) improved by approximately 12%, 13%, and 10%, respectively. The final recommended value of the knob innodb_buffer_pool_size by DKTune (LlamaTune) is higher than the maximum value in this knob domain of LlamaTune. This occurred because LlamaTune' memory allocation mechanism failed to utilize the remaining memory resources of other knobs.

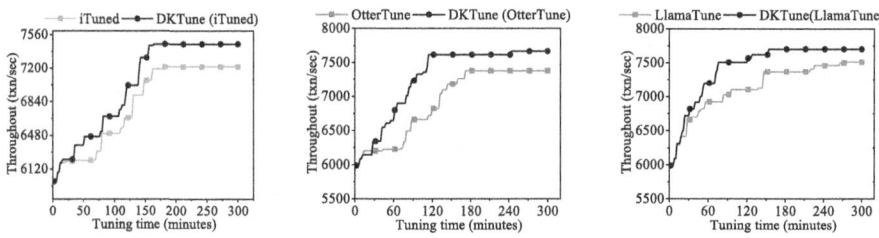

Fig. 2. Comparison of tuning results on Wikipedia (MySQL)

Subsequently, we conducted a comparison of the tuning results on Wikipedia (as shown in Fig. 2). The tuning results also demonstrated that integrating domain knowledge with tuning methods outperformed the original methods in terms of tuning efficiency and final optimization outcomes. During the tuning process, we observed that DKTune(iTuned) fully leveraged the dynamic nature of knobs, eliminating the need for database restarts in 6 sampling instances.

4.2 Experiments on PostgreSQL

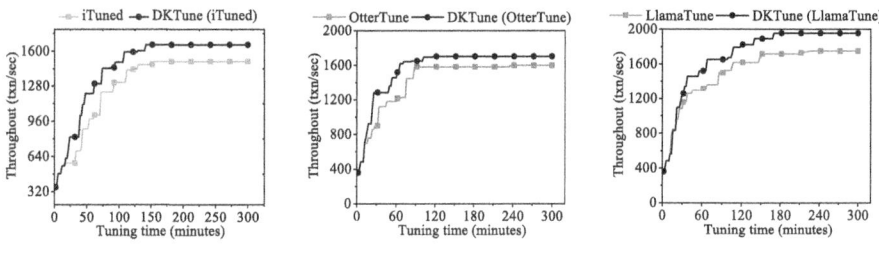

Fig. 3. Comparison of tuning results on TPC-C (PostgreSQL)

To validate the generality of DKTune across different databases, we applied DKTune to PostgreSQL.

Knobs. A total of 75 knobs are selected for tuning. Among these, 5 knobs contend for memory resources, 6 knobs exhibit dominant relationships, 6 knobs have constraint relationships, and 22 knobs are dynamic knobs.

Execution Time Breakdown. In the process of tuning PostgreSQL parameters, restarting the database takes 22 s. Workload execution is fixed at 200 s.

Updating the Gaussian Process model requires 5.98 s, while updating the SMAC model necessitates 0.69 s. Recommendations for configuration parameters typically take less than 0.2 s, and applying these parameters to the database takes no more than 0.01 s.

We conducted a comparison of the tuning results on TPC-C for PostgreSQL database (as shown in Fig. 3). The experimental results indicate that, DKTune outperformed other tuning methods in terms of tuning efficiency and final optimization outcomes. We observed that iTuned, OtterTune, and LlamaTune sampled 9, 8, and 11 low-utility configurations, respectively, that violated the rules. Furthermore, in the DKTune(OtterTune) tuning process shown in Fig. 3, when the tuning time had reached approximately 31 min or so, the Gaussian Process model began to be updated based on the dominance relationship between seq_page_cost and random_page_cost.

5 Releted Work

Currently, machine learning methods for tuning database knobs are of great interest, with Bayesian optimization and reinforcement learning being the two most widely applied approaches. In this context, we will primarily focus on introducing reinforcement learning-based methods and Bayesian optimization-based methods for database knob tuning.

Gaussian Process-based Tuning. iTuned [15] pioneered Gaussian process-based modeling the relationship between knobs and database performance. By leveraging the sampling data gathered through the LHS method, it constructs a preliminary tuning model and employs the expected improvement method to balance exploration and exploitation [13]. Conversely, OtterTune [17] builds an initial Gaussian model using historical tuning data, and applies Lasso to identify key knobs. It then utilizes an incremental approach [5] to dynamically increase the number of tuning knobs throughout the process. ResTune [21] also employs a Gaussian process, but aims to minimize system resource utilization while maintaining DBMS performance. Taking into account the influence of various factors such as operating system and Java virtual machine on database performance, CGPTuner [4] employs a Contextual Gaussian Process Bandit Optimization to tune knob of the entire IT stack of the database for performance maximization. In contrast to these offline tuning strategies, ONLINETUNE [22] offers an online approach, using contextual Bayesian optimization for adaptive database tuning in ever-changing cloud environments. The widely utilization of Gaussian processes [14] in database knob tuning is attributed to their theoretical capability to balance exploration and exploitation.

Reinforcement Learning-based Tuning. Unlike OtterTune, which segments the tuning process into various phases and relays the optimal solution from one stage to the next, CDBTune [19] introduces a more unified, end-to-end solution. Utilizing reinforcement learning for database knob tuning, CDBTune employs DDPG [10] as an agent, with the knob value as the action, the database as the environment, the internal state of the database as the state, and changes in database performance as the reward. On the other hand, QTune [9] also incorporates reinforcement learning but with a unique perspective, considering query statements during model training. Addressing this cold start problem, HUNTER [3] proposes a solution that integrates genetic algorithms with reinforcement learning.

At present, there are some other studies related to database knob tuning. BestConfig [23] is a heuristic tuning method designed to tune database knobs. LlamaTune [8] focuses on enhancing the sampling efficiency of existing optimizers. Studies such as [16] and [18] propose methods for extracting tuning rules from text data. [7] concludes, based on relevant experiments, that database knob tuning often requires adjusting only a few knobs.

6 Conclusion

This paper introduces a new tuning system, DKTune, which integrates domain knowledge of databases with Bayesian optimization. Specifically, DKTune enhances the accuracy of the surrogate model by utilizing the dominant relationships and aids the acquisition function in evaluating knob configurations through competitive relationships, constraint relationships, and dynamic characteristics. We evaluated DKTune on two popular open-source database systems, and the experimental results demonstrate that DKTune significantly improves the efficiency of database knob tuning and the final tuning results.

Acknowledgements. We thank the anonymous reviewers for their insightful comments and feedback. This work is supported by grants from the National Natural Science Foundation of China (U22B2020). Peng Cai is the corresponding author.

References

1. Breiman, L.: Random forests. Mach. Learn. **45**, 5–32 (2001)
2. Buhmann, M.D.: Radial basis functions. Acta Numer. **9**, 1–38 (2000)
3. Cai, B., et al.: HUNTER: an online cloud database hybrid tuning system for personalized requirements. In: SIGMOD, pp. 646–659 (2022)
4. Cereda, S., Valladares, S., Cremonesi, P., Doni, S.: CGPTuner: a contextual gaussian process bandit approach for the automatic tuning of it configurations under varying workload conditions. Proc. VLDB Endow. **14**(8), 1401–1413 (2021)
5. Danna, E., Perron, L.: Structured vs. unstructured large neighborhood search: a case study on job-shop scheduling problems with earliness and tardiness costs. In: Rossi, F. (ed.) CP 2003. LNCS, vol. 2833, pp. 817–821. Springer, Heidelberg (2003). https://doi.org/10.1007/978-3-540-45193-8_59
6. Hutter, F., Hoos, H.H., Leyton-Brown, K.: Sequential model-based optimization for general algorithm configuration. In: Coello, C.A.C. (ed.) LION 2011. LNCS, vol. 6683, pp. 507–523. Springer, Heidelberg (2011). https://doi.org/10.1007/978-3-642-25566-3_40

7. Kanellis, K., Alagappan, R., Venkataraman, S.: Too many knobs to tune? Towards faster database tuning by pre-selecting important knobs. In: 12th USENIX Workshop on Hot Topics in Storage and File Systems (2020)
8. Kanellis, K., Ding, C., Kroth, B., Müller, A., Curino, C., Venkataraman, S.: LlamaTune: sample-efficient DBMS configuration tuning. Proc. VLDB Endow. **15**(11), 2953–2965 (2022)
9. Li, G., Zhou, X., Li, S., Gao, B.: QTune: a query-aware database tuning system with deep reinforcement learning. Proc. VLDB Endow. **12**(12), 2118–2130 (2019)
10. Lillicrap, T.P., et al.: Continuous control with deep reinforcement learning. arXiv preprint arXiv:1509.02971 (2015)
11. McKay, M.D., Beckman, R.J., Conover, W.J.: A comparison of three methods for selecting values of input variables in the analysis of output from a computer code. Technometrics **42**(1), 55–61 (2000)
12. Mockus, J.: The Bayesian approach to global optimization. In: Drenick, R.F., Kozin, F. (eds.) System Modeling and Optimization. LNCIS, vol. 38, pp. 473–481. Springer, Heidelberg (2005). https://doi.org/10.1007/BFb0006170
13. Molnar, C.: Interpretable Machine Learning: A Guide For Making Black Box Models Explainable, chap. 8. Independently published (2022)
14. Ru, B., Alvi, A., Nguyen, V., Osborne, M.A., Roberts, S.: Bayesian optimisation over multiple continuous and categorical inputs. In: ICML, pp. 8276–8285 (2020)
15. Thummala, V., Babu, S.: iTuned: a tool for configuring and visualizing database parameters. In: SIGMOD, pp. 1231–1234 (2010)
16. Trummer, I.: The case for NLP-enhanced database tuning: towards tuning tools that "read the manual". Proc. VLDB Endow. **14**(7), 1159–1165 (2021)
17. Van Aken, D., Pavlo, A., Gordon, G.J., Zhang, B.: Automatic database management system tuning through large-scale machine learning. In: SIGMOD, pp. 1009–1024 (2017)
18. Zeng, J., Chen, T.: Interactive model and application of joint knowledge base question answering and semantic matching. In: Yuan, L., Yang, S., Li, R., Kanoulas, E., Zhao, X. (eds.) WISA 2023. LNCS, vol. 14094, pp. 206–217. Springer, Singapore (2023). https://doi.org/10.1007/978-981-99-6222-8_18
19. Zhang, J., et al.: An end-to-end automatic cloud database tuning system using deep reinforcement learning. In: SIGMOD, pp. 415–432 (2019)
20. Zhang, X., et al.: Facilitating database tuning with hyper-parameter optimization: a comprehensive experimental evaluation. Proc. VLDB Endow. **15**(9), 1808–1821 (2022)
21. Zhang, X., et al.: ResTune: resource oriented tuning boosted by meta-learning for cloud databases. In: SIGMOD, pp. 2102–2114 (2021)
22. Zhang, X., Wu, H., Li, Y., Tan, J., Li, F., Cui, B.: Towards dynamic and safe configuration tuning for cloud databases. In: SIGMOD, pp. 631–645 (2022)
23. Zhu, Y., et al.: BestConfig: tapping the performance potential of systems via automatic configuration tuning. In: Socc, pp. 338–350 (2017)

Attribute Multiplex Network Graph Clustering: Joint Contrastive And High-Order Proximity

Fei Ye[1], Hao Zhong[1], Xijia Lin[1], Lu Yu[1], and Chengjie Mao[1,2(✉)]

[1] South China Normal University, Guangzhou 510631, Guangdong, China
[2] Pazhou Lab, Guangzhou 510330, Guangdong, China
maochj@qq.com

Abstract. In this paper, we propose an improved subspace clustering algorithm, with the aim to improve accuracy and smoothness. To enhance the discriminative power of the coefficient matrix and broaden its applicability to various types of data, the algorithm incorporates contrastive loss and higher-order proximity, aiming to explore distinguishable subspace structure between data points. We first use graph convolution to design a low-pass filter to generate a smooth representation of the attribute view. Next, to focus on the multi-view consistency and the cross-view diversity, we introduce the contrastive loss function and high-order neighbor relations as regularizers. Once the optimal representation coefficient matrix Z has been determined, Z is used to construct the affine matrix for spectral clustering. After experimental verification on multiplex graphs datasets, our method shows higher accuracy and smoothness than current algorithms.

Keywords: Graph filtering · High-order structure · Contrastive learning · Clustering

1 Introduction

One of the fundamental operations of data mining is graph clustering, which employs topological relationships and node attributes to identify potential clustering structures [1]. Due to the limitations of a single view when describing objects, several researchers are turning to exploit diverse information for improving clustering result. Over the past decades, numerous scholars have proposed numerous graph clustering technologies. However, when employing a single-view clustering method on a shared latent representation obtained across different views, such a naive learning strategy tends to pay no attention to correlations between views [3].

To enhance the development of multi-view information, mainstream clustering algorithms can be categorised into two types: Graph Neural Networks (GNN) and Subspace Clustering (SC). The advantage of GNN-based methods

C. Jin et al. (Eds.): WISA 2024, LNCS 14883, pp. 290–297, 2024.
https://doi.org/10.1007/978-981-97-7707-5_25

lies in their ability to learn graph structure representation from topological relationship, allowing the model to better capture the complex structure of the data and improve the accuracy of the clustering. However, these methods require the adjustment of a large number of hyper-parameters. The main idea of SC is to discover potential subspace structures of high-dimensional data based on the self-expression properties of the data, but the results of the clustering depend so much on the learned structure of the shared subspace. Early SC methods based on prior knowledge or assumptions accurately captured the global structure by enhancing the robustness to noise and outliers. However, these models were predominantly designed for the single views.

To address the challenges outlined above, we design a general and effective attribute multiplex graph clustering framework, called Joint Contrastive with High-order Proximity (JCHP). Similar to most shallow methods, JCHP adpot the architecture of SC as the backbones. To solve issues related to noise and outliers whthin the original feature space, we employ the Laplacian filter to eliminate the high-frequency signals from the node attributes. The resulting smooth representation enhances the model's ability to identify nodes with similar characteristics. For making SC algorithm more comprehensive and well adapted to the manifold graph data, we also incorporate the contrastive learning constraint and high-order proximity. Our contributions are summarised below:

- An improved SC algorithm is proposed to solve the multiview clustering problem, in which the graph filter efficiently deals with the noise and outliers of the original feature attributes.
- We propose two kinds of graph regularization to constrain the learning of the common coefficient matrix: noise contrastive estimation and high-order proximity, which facilitates the recovery of the underlying subspace structure in multiplex networks.
- The proposed model is able to perform accomplish the subtasks of learning smooth representation, constructing similarity graphs, and synthesising the coefficient matrix within a unified framework.

2 Related Work

Over the past years, several multi-view SC methods have been proposed. Kang et al. [4] employed the strategy of K-means based anchor graph construction to extend SC to large-scale networks, thereby reducing the time complexity to linear time. Sun et al. [9] proposed a unified framework for solving the fact that the graph learning and clustering separation generate weakly discriminative graph structure. In addition, Liu et al. [8] proposed a novel end-to-end learning paradigm that uses an anchor strategy to accelerate the learning of the coefficient matrix by learning multiple common subspace representations and fusing them together. Yu et al. [11] employed a threefold aggregation strategy and utilized metapaths to enhance the quality of node embeddings learned from heterogeneous graphs. However, these methods often fail to learn robust discriminative node representations due to their high susceptibility to noise and outliers (Fig. 1).

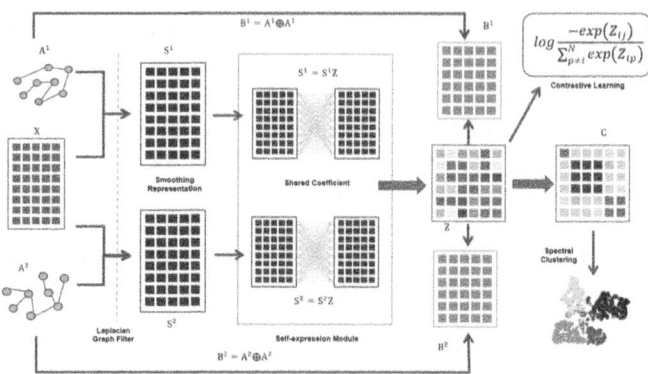

Fig. 1. The architecture of training the JCHP framework. X is the attribute feature, A^v is the adjacency matrix, S^v is the smooth feature matrix, B^v is the high-order proximity matrix, Z is the common coefficient matrix, and C is the similarity matrix.

3   Methodology

3.1   Graph Filtering

In a graph structure, the more similar the features of neighboring nodes are, the smoother the change in graph signal will be as it passes through adjacent nodes [12]. The Laplacian-Beltrami operator proves that the smoothness of the graph signal is positively correlated with the eigenvalues $\Lambda = diag(\lambda_1, ..., \lambda_n)$ derived from the eigendecomposition of the graph Laplacian matrix $L_s = I - D^{-\frac{1}{2}}AD^{-\frac{1}{2}} = U\Lambda U^{-1}$.

Given a attributed multiplex network $G^v = \{A^v, X\}$, each column x_i of attribute matrix $X = [x_1, .., x_N]$ can be regarded as a graph signal. the connection between node within view v be represented by adjacency matrix A^v. To cope with the impact of noise and anomalous data values present in the original feature set, we design a low-pass filter based on the Laplacian matrix. According to the adjacency matrix A^v, we can obtain the normalization adjacency matrix $A^v = D^{-\frac{1}{2}}(A^v + I)D^{-\frac{1}{2}}$ and graph Laplacian matrix $L^v = I - A^v$ for the view v. Finally, we calculate the smoothed representation for each view as follows:

$$S^v = F^v X = (I - \frac{1}{2}L^v)^m X \tag{1}$$

3.2   Graph Learning

After obtaining the smoothed representation through Eq. 1, we employ the self-expression-based SC algorithm to decompose the high-dimensional smoothed representation into multiple low-dimensional subspace structures, which can be formulated as the following optimization problem:

$$L_1 = \min_Z \sum_{v=1}^{V} \alpha^v \|E\|_F^2 + \lambda_1 \|Z\|_F^2 \qquad \text{s.t. } S^v = S^v Z + E, Z > 0 \tag{2}$$

where the term $\|E\|_F^2$ is the reconstruction loss. α^v is used to distinguish the varying degrees of information implied in different views and λ_1 is a trade-off parameter. $\|Z\|_F^2$ is regularization term, which is designed to prevent elements represented by themselves from causing the emergence of trivial solutions.

In the real world, nodes are often connected by multiple hops. However, the initial topology only reflects first-order proximity, and low-rank or sparse data may hinder the model's ability to find the inherent correlation of each data individually. We investigate high-order proximity matrices using the Hadamard product and enforce that the learned common coefficient matrix remains consistent with them. The loss function for the high-order regularizer is defined as follows:

$$L_2 = \min_Z \sum_{v=1}^{V} \|Z - B^v\|_F^2 \qquad \text{s.t. } B^v = A^1 \oplus A^2 ... \oplus A^p \qquad (3)$$

Inspired by the excellent performance of graph contrastive learning in unsupervised tasks, we aim to leverage the underlying principles to construct positive and negative samples for shrinking coefficients of correlated data and promoting group effect. Unlike other contrastive learning methods, this paper applies the approach to the common coefficient matrix Z instead of node embeddings F. In terms of sample selection strategy, we do not employ data augmentation to generate positive and negative samples. Instead, each node and its k nearest node (KNN) are treated as a positive pair, and other nodes are considered negative samples. The loss function for the contrastive regularizer is defined as follows:

$$L_3 = \min_Z \sum_{v=1}^{V} \sum_{n=1}^{N} \sum_{j \in N_i^v} log \frac{-exp(Z_{ij})}{\sum_{p \neq i}^{N} exp(Z_{ip})} \qquad (4)$$

where N_i^v represents the KNN of the node i in the v-th view. We combine Eq. 2–4 and set λ_2, λ_3 to trade-off parameter, the final loss function is defined as follows:

$$\min_{Z, \alpha^v} L_1 + \lambda_2 L_2 + \lambda_3 L_3 + \sum_{v=1}^{V} (\alpha^v) \qquad (5)$$

3.3   Optimization

We adopt an alternating direction method of multipliers for solving such a convex optimization problem.

S-Problem. We optimize S with Gradient Descent algorithm, the objective function can be expressed as:

$$\min_Z \sum_{v=1}^{V} \alpha^v \|E\|_F^2 + \lambda_1 \|Z\|_F^2 + \lambda_2 \|Z - B^v\|_F^2 + \lambda_3 \sum_{n=1}^{N} \sum_{j \in N_i^v} log \frac{-exp(Z_{ij})}{\sum_{p \neq i}^{N} exp(Z_{ip})} \qquad (6)$$

Considering the derivatives are too long and complicated, the results were presented in two parts:

$$\nabla_1^{(t)} + \nabla_2^{(t)} \tag{7}$$

$$\nabla_1^{(t)} = 2 \sum_{v=1}^{V} \alpha^v \left(- S^v S^{vT} + S^v S^{vT} Z^{(t-1)} \right) + \lambda_1 \left(Z^{(t-1)} - B^v \right) + \lambda_2 Z^{(t-1)} \tag{8}$$

$$\nabla_2^{(t)} = \begin{cases} \sum_{v=1}^{V} \alpha^v \left(1 - \dfrac{n \times exp(Z_{ij}^{t-1})}{\sum_{p \neq i}^{N} exp(Z_{ip}^{t-1})} \right) & if \quad j \in N_i^v \\ \sum_{v=1}^{V} \alpha^v \left(\dfrac{n \times exp(Z_{ij}^{t-1})}{\sum_{p \neq i}^{N} exp(Z_{ip}^{t-1})} \right) & otherwise \end{cases} \tag{9}$$

where $\nabla_2^{(t)}$ is the derivative consequence of L_3, n represents the total number of neighbors. The cost to compute Z is $O(V(n^3 + 2n^2d + 2n^2))$.

α^vProblem. The solution of α^v can be easily obtained by setting its first-order derivative to zero, which yields

$$\min_{\alpha^v} \sum_{v=1}^{V} \alpha^v M^v + \sum_{v=1}^{V} \alpha^{v^\gamma} \tag{10}$$

$$\alpha^v = \left(\frac{-M^v}{\gamma} \right)^{\frac{1}{\gamma-1}} \tag{11}$$

where $M^v = \left\| S^{vT} - S^{vT} Z \right\|_F^2$. As each sub-problem is strictly convex, the objective value will consistently decrease until either reaching the minimum or satisfying the convergence condition. The updating of α^v needs $O(n^2d)$.

4 Experiment

Table 1. Datasets statistics.

Datasets	Nodes	Features	Graph and Edges	Clusters
ACM	3025	1830	Co-Subject (29,281)	3
			Co-Author (2,210,761)	
IMDB	4780	1232	Co-Actor (98,010)	3
			Co-Director (21,018)	
SCHOLAT	2302	477	Co-Team (139,004)	11
			Co-Class (70,226)	
			Co-Friends (11,393)	

4.1 Experiment Setting

We evaluate JCHP using four public multiplex datasets: ACM, IMDB and SCHOLAT, as indicated in Table 1. For a comprehensive evaluation, we compare JCHP with several representative clustering algorithms. They are: GCN [6], VGAE [5], HAN [10], MvAGC [7], SMC [8], O2MA and O2MAC [2]. Among them, GCN and VGAE are selected as representatives of single-view methods, while the rest belong to multi-view attribute methods.

In the experiments, the Adam optimizer is employed to optimize variables. Through parameter analysis, we established $\gamma = -4, k = 4, p = 2$, and $n = 10$, while fine-tuning the values of λ_1 and λ_2. This configuration has been identified as yielding optimal results across all datasets. We certify the performance of our proposed method using the following 4 evaluation measures: ACC, NMI, ARI and F1. For all metrics, a higher value denotes better utility.

Table 2. Clustering performance on ACM, IMDB and SCHOLAT datasets. The best results are represented by bold value.

Datasets \ Methods		GCN	VGAE	HAN	O2MA	O2MAC	MvAGC	SMC	JCHP
ACM	ACC	62.05	54.61	88.23	88.80	90.42	89.75	88.49	**91.73**
	NMI	43.45	41.09	58.81	65.15	69.23	67.35	63.97	**71.59**
	ARI	41.02	34.27	**88.23**	69.87	73.94	72.12	69.29	76.89
	F1	53.18	56.36	88.44	88.94	90.53	89.86	88.69	**91.85**
IMDB	ACC	42.27	42.97	55.47	45.02	46.97	56.33	56.86	**59.95**
	NMI	00.52	02.75	**09.86**	04.21	05.24	03.71	03.56	08.80
	ARI	00.32	02.00	08.56	05.64	07.53	09.40	09.98	**15.70**
	F1	35.02	39.72	41.52	41.59	42.29	37.83	40.48	**42.75**
SCHOLAT	ACC	24.11	31.62	59.08	39.35	40.22	58.25	63.24	**80.32**
	NMI	39.68	41.17	68.11	30.83	32.37	34.66	36.12	**68.49**
	ARI	23.75	31.62	**60.93**	24.03	25.37	38.99	45.54	56.68
	F1	21.16	30.25	58.47	13.89	15.66	15.15	20.69	**71.38**

4.2 Compare Experiments

Based on the performance comparisons presented in Table 2, we have made the following observations: Compared to single-view methods, the GNN-based multi-view clustering approaches exhibit superior performance. This confirms the effectiveness of using multilayer nonlinear feature extractors when dealing with multiplex relational data. However, the stability and accuracy of these models appear to be lower. We believe this may be due to their failure to account for noise or deficiencies in the raw data. The results of subspace-based anchor learning algorithms, such as MvAGC and SMC, are comparable to GNN-based methods for most indicators. However, they utilize anchor graph technology to reduce computational costs, this may limit their ability to fully explore topological data. In conclusion, the JCHP proposed in this paper is demonstrated to exhibit promising performance compared to other methods, thereby demonstrating model superior discriminative capability.

4.3 Ablation Study And Visual Analytics

To further explore the effectiveness of each module proposed in our method, we conduct an ablation study on each module. Specifically, we test this by alternately removing graph filters, L_2 and L_3 while preserving other components, namely $JCHP_F$, $JCHP_C$ and $JCHP_H$, respectively. The results are shown in Fig. 2. Compared to the $JCHP_C$, JCHP shows significant improvement across two datasets, demonstrating the effectiveness of the contrastive module in exploiting the rich complementary information from multi-relational graph data. Additionally, we observed noticeable improvements resulting from the graph filter, emphasizing the superiority of graph filter for facilitating coefficient matrix learning. The absolute improvements on ACM are not as pronounced as those on SCHOLAT. This disparity arises from the denser nature of the ACM dataset compared to SCHOLAT, where the abundance of first-order proximity relationships dilutes the impact of higher-order proximity relationships.

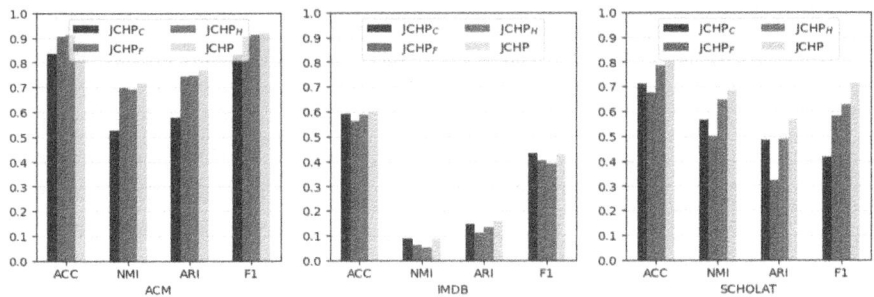

Fig. 2. Ablation study on ACM, IMDB and SCHOLAT datasets.

5   Conclusion

In this paper, we mainly exploit two regularization term, constrative loss and high order proximity, to explore the intrinsic geometric structure of the data points in the coefficient matrix. The experimental results show that the proposed method provides significant improvements over the baseline methods in terms of accuracy. In the future, additional research will be conducted to identify alternative methods for reducing computational complexity.

Acknowledgments. This work is supported in part by the National Key Research and Development Program of China (Research and Demonstration Application of Key Technologies for Personalized Learning Driven by Educational Big Data) under Grant 2023YFC3341200.

References

1. Abdolali, M., Gillis, N.: Beyond linear subspace clustering: a comparative study of nonlinear manifold clustering algorithms. Comput. Sci. Rev. **42**, 100435 (2021)
2. Fan, S., Wang, X., Shi, C., Lu, E., Lin, K., Wang, B.: One2Multi graph autoencoder for multi-view graph clustering. In: Proceedings of the Web Conference 2020, pp. 3070–3076 (2020)
3. Fang, U., Li, M., Li, J., Gao, L., Jia, T., Zhang, Y.: A comprehensive survey on multi-view clustering. IEEE Trans. Knowl. Data Eng. (2023)
4. Kang, Z., Zhou, W., Zhao, Z., Shao, J., Han, M., Xu, Z.: Large-scale multi-view subspace clustering in linear time. In: Proceedings of the 34th AAAI Conference on Artificial Intelligence, vol. 34, pp. 4412–4419 (2020)
5. Kipf, T.N., Welling, M.: Variational graph auto-encoders. arXiv preprint arXiv:1611.07308 (2016)
6. Kipf, T.N., Welling, M.: Semi-supervised classification with graph convolutional networks. In: Proceedings of the 5th International Conference on Learning Representations (2017)
7. Lin, Z., Kang, Z.: Graph filter-based multi-view attributed graph clustering. In: Proceedings of the 30th International Joint Conference on Artificial Intelligence, pp. 2723–2729 (2021)
8. Liu, L., Chen, P., Luo, G., Kang, Z., Luo, Y., Han, S.: Scalable multi-view clustering with graph filtering. Neural Comput. Appl. **34**(19), 16213–16221 (2022)
9. Sun, M., et al.: Scalable multi-view subspace clustering with unified anchors. In: Proceedings of the 29th ACM International Conference on Multimedia, pp. 3528–3536 (2021)
10. Wang, X., Ji, H., Shi, C., Wang, B., Ye, Y., Cui, P., Yu, P.S.: Heterogeneous graph attention network. In: The World Wide Web Conference, pp. 2022–2032 (2019)
11. Yu, C., Fei, L., Chen, F., Chen, L., Wang, J.: Heterogeneous graphs embedding learning with metapath instance contexts. In: Web Information Systems and Applications, pp. 149–161 (2023)
12. Zhang, X., Liu, H., Li, Q., Wu, X.M.: Attributed graph clustering via adaptive graph convolution. In: Proceedings of the 28th International Joint Conference on Artificial Intelligence, pp. 4327–4333 (2019)

Reliable Community Search
over Dynamic Bipartite Graphs

Mo Li, Zhiran Xie, and Linlin Ding$^{(\boxtimes)}$

School of Information, Liaoning University, Shenyang, China
dinglinlin@lnu.edu.cn

Abstract. Bipartite graphs naturally represent relationships between two different entities, such as people-location networks, author-paper networks. In dynamic scenarios like product recommendation systems, where purchase behavior and interests evolve, dynamic bipartite graphs are emerged. Despite extensive research on community search in (dynamic) unipartite graphs, dynamic bipartite graphs remain unexplored. While the duration of a community reflects its temporal continuity, managing its size is also crucial. Over time, a community's prolonged existence tends to decrease in size, posing a practical challenge. To fill this research gap, we introduce a reliable community model over dynamic bipartite graphs that accounts for time span, size and degree constraints. Then, we propose an efficient RCSearch algorithm to solve the reliable community search, leveraging properties of reliable (α, β)-communities and dynamic programming strategies. Furthermore, effective optimization strategies are devised to accelerate this process. Finally, extensive experiments conducted over 7 real-world graphs demonstrate the effectiveness and efficiency of our proposed methods.

1 Introduction

In many real-world applications, the relationships between different entities are modeled into bipartite graphs, such as user-item networks, user-location networks, and collaboration networks. Community structures naturally exist in these practical networks, and community search have garnered widespread attention and achieved success in various applications, such as personalized recommendations [1], fraud detection [4] and team formation [5].

Generally, the goal of community search is to identify densely connected structures related to query vertices. Most existing community search works primarily consider static network structures. For example, Liu et al. [3] stored the (α, β)-core by establishing an efficient BiCore index, significantly enhancing retrieval speed. Wang et al. [9] primarily focused on searching the (α, β)-core that satisfies weighted requirements on a edge-weighted bipartite graph. Yao et al. [10] introduced the concept of similar-bicliques on a bipartite graph. On the other hand, some other research works consider the temporal information, which mainly focus on the unipartite graphs, overlooking the bipartite graph

C. Jin et al. (Eds.): WISA 2024, LNCS 14883, pp. 298–307, 2024.
https://doi.org/10.1007/978-981-97-7707-5_26

structure. For instance, Li *et al.* [2] introduced a novel model to explore a persistent community in temporal graphs, while [7] introduced the concept of periodic communities by introducing the notion of periodicity.

Previous studies have evaluated community quality by examining either the structural cohesiveness of communities at discrete timestamps or in dynamic unipartite scenarios. However, these approaches fall short in capturing the evolving structure of communities on bipartite graphs over time. For example, targeting different types of advertisements to user communities requires understanding their evolving preferences, which may vary in time span and product preferences. Existing methods also overlook community reliability. As communities expand temporally, they may shrink in size, potentially losing practical significance. To address this, we propose the **$\underline{R}$*eliable* (α, β)-$\underline{C}$*ommunity*** (RC), which considers both the time span and size of the community.

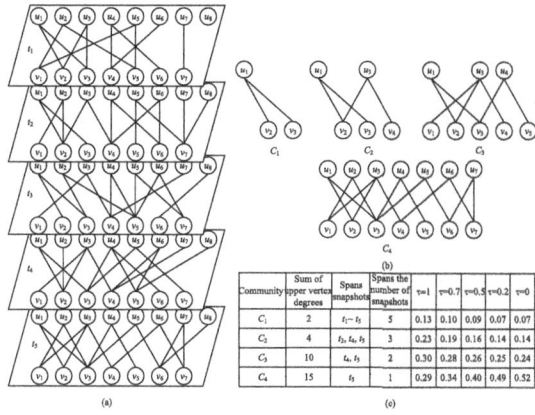

Fig. 1. A dynamic bipartite graph with five timestamps

In detail, our reliable (α, β)-community emphasizes three key constraints: i) **Cohesiveness:** it ensuring the cohesiveness of each snapshot within the community. Specifically, we define an (α, β)-community such that the degree of each vertex in the upper and lower layers is not less than α and β, respectively. ii) **Continuity:** the (α, β)-community extends over a period of time, ensuring temporal continuity. iii) **Size:** the result community size is relatively meaningful.

Additionally, we introduce a novel formula to compute a reliable score, which balances continuity and community size. A crucial parameter, τ, acts as a trade-off rate in this formula. The relationship between community size and the time span it covers is vital. A community spanning a longer time frame might decrease in size, potentially losing practical significance. Our formula accounts for both factors, providing a comprehensive score that enhances understanding and evaluation of the community's inherent characteristics and value.

Example 1. Figure 1 shows a dynamic online social network with five timestamps, where edges represent user-product interactions over time. Using $\alpha = 2$

and $\beta = 1$, τ balances community size and time span in queries. With $\tau = 1$, both factors weigh equally, leading to community C_3 with a degree sum of 10 over timestamps t_4 and t_5. Similarly, for $\alpha = 2$, $\beta = 1$, and $\tau = 0.5$, community C_4 is returned, highlighting greater sensitivity to degree sum and community size. Adjusting τ helps companies identify optimal ad placements effectively.

To address the challenge of finding the optimal reliable community, a naive approach involves listing all possible communities, scoring each, and selecting the highest-scoring one. However, this is time-intensive, potentially incurring exponential costs. To overcome this, we introduce the efficient RCSearch algorithm. It leverages the cohesiveness of reliable communities to prune surplus vertices and time snapshots, employing dynamic programming principles for pre-computational operations. Additionally, by pre-filtering candidates from post-pruning snapshots, we expedite query processing, enhancing computational speed. Experimentation on seven real datasets demonstrates the effectiveness and efficiency of our approach.

To sum up, our main contributions can be summarized as follows: i) We propose a Reliable (α, β)-Community search problem over dynamic bipartite graphs. ii) We devise an efficient algorithm RCSearch to solve the reliable community search over dynamic bipartite graphs. To further improve the efficiency, we also design an optimization strategy. iii) Extensive experiments on several real datasets have been conducted to evaluate the effectiveness and efficiency of our proposed algorithms.

2   Problem Definition

A bipartite graph $\mathcal{G}$ is defined as $\mathcal{G}(\mathcal{V} = (\mathcal{U}, \mathcal{L}), \mathcal{E})$, where $\mathcal{U}(\mathcal{G})$ and $\mathcal{L}(\mathcal{G})$ denote the upper and lower layer vertices, respectively. $\mathcal{U}(\mathcal{G}) \cap \mathcal{L}(\mathcal{G}) = \emptyset$, $\mathcal{V}(\mathcal{G}) = \mathcal{U}(\mathcal{G}) \cup \mathcal{L}(\mathcal{G})$ is the vertex set, and $\mathcal{E}(\mathcal{G}) \subseteq \mathcal{U}(\mathcal{G}) \times \mathcal{L}(\mathcal{G})$ is the edge set. An edge between vertices u and v in $\mathcal{G}$ is (u, v). If time information is present, it is (u, v, t) or (v, u, t), where $u, v \in \mathcal{V}$ and t is the interaction time. The neighborhood of a node u in $\mathcal{G}(\mathcal{V} = (\mathcal{U}, \mathcal{L}), \mathcal{E})$ is $N_u(\mathcal{G}) = \{(u, v) | (u, v) \in \mathcal{E}\}$. The degree of u in $\mathcal{G}$, denoted by $\deg(u, \mathcal{G})$ or $\deg(u)$, is the number of its neighbors. A time interval is $T = [t_s, t_e]$, where $|T|$ is the number of timestamps within it. A subgraph C of $\mathcal{G}$ is $C = (\mathcal{V}_C = (\mathcal{U}_C, \mathcal{L}_C), \mathcal{E}_C)$. A dynamic bipartite graph is defined as a sequence of time-variant bipartite graph snapshot, i.e., $G = \mathcal{G}_{t_1}, ..., \mathcal{G}_T$. And G_t represents a snapshot of G at t. When vertices and edges from two dynamic bipartite graphs intersect, the resulting graph is called a span graph.

Definition 1 ((α, β)-core and $\alpha, \beta)$-community).

1. (α, β)-***core:*** *A subgraph $R(\alpha, \beta)$ is an (α, β)-core if it is connected and each vertex $u \in U(R)$ satisfies $\deg(u, R) > \alpha$ and each vertex $v \in L(R)$ satisfies $\deg(v, R) > \beta$.*
2. (α, β)-***community:*** *An (α, β)-core containing the query vertex q.*

Definition 2 (Reliable (α, β)-community). *Given a dynamic bipartite graph $G = \mathcal{G}t_1, ..., \mathcal{G}T$, degree constraints α and β, the trade-off parameter τ, and a time interval $T_C = [t_s, t_e]$, a Reliable (α, β)-community is a subgraph $C = (V_C = (U_C, L_C), E_C)$ that spans the snapshots $Gt_s, ..., Gt_e$. For each timestamp $t_n \in T_C$, the subgraph induced by E_C from G_{t_n} is an (α, β)-core. In other words, $\forall t_n \in T_C$, $G_{t_n}[E_C]$ is an (α, β)-core of G_{t_n}.*

Definition 3 (Score of RC). *Given a dynamic bipartite graph $G = \{\mathcal{G}_{t_1}, ..., \mathcal{G}_T\}$, a query time interval $T_Q = [t_i, t_j]$, and an RC $C = (V_C = (U_C, L_C), E_C)$ for the timestamps $T_C = \{t_s, ..., t_e\}$, where $\forall t_n \in T_C, t_n \in T_Q$, the score $S(C)$ is the harmonic mean of the normalized number of time snapshots and the sum of the degrees of vertices on one side (upper or lower).*

$$S(C) = (1 + \tau^2) \cdot \frac{(R(D) \cdot R(T))}{((\tau^2 \cdot R(D)) + R(T))} \tag{1}$$

Where $R(D)$ is the ratio of the sum of vertex degrees in the upper or lower part of the subgraph to the total degrees in the search graph, and $R(T)$ is the ratio of the number of time snapshots in the subgraph to the total snapshots in the search time range. Adjusting τ indirectly controls the time span and degree size of vertices on one side in the output subgraph.

Problem Statement (Optimal Reliable Community Search). Given a dynamic bipartite graph $G = \{\mathcal{G}_{t_1}, ..., \mathcal{G}_T\}$, a query vertex q, parameters α, β, τ, and a query time interval $T_Q = [t_i, t_j]$, find the RC C and corresponding timestamps T_C, maximizing $S(C)$ subject to $q \in V_C$, $deg(u, C) \geq \alpha$ for all $u \in U_C$, $deg(v, C) \geq \beta$ for all $v \in L_C$, $e \in E_C \subseteq G$, and $t \in T_C \subseteq T_Q$.

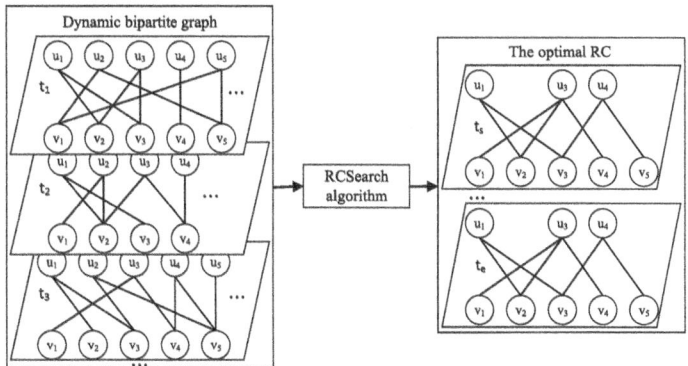

Fig. 2. Solution overview

3   RCSearch Algorithm

To solve the optimal community search problem efficiently, core decomposition is applied to each snapshot to identify the highest-scoring span graph. This approach faces exponential complexity, around 2^n combinations for n snapshots. To handle this, we propose the RCSearch algorithm.

Solution Overview. Figure 2 depicts the RCSearch algorithm, consisting of three main components. Firstly, the Pruning algorithm removes irrelevant vertices and snapshots. Next, pre-computation estimates the degree loss after synthesis into a span graph. Finally, it iteratively synthesizes the dynamic bipartite graph with minimal degree reduction.

3.1   Dynamic Bipartite Graph Reduction

This subsection simplifies dynamic bipartite graphs to improve computational efficiency by reducing irrelevant vertices and snapshots.

The Pruning algorithm checks each snapshot in the dynamic bipartite graph. If a snapshot does not intersect with the query vertex or is outside the query time period, it is removed. The algorithm iteratively removes vertices with degrees less than α in the upper vertex set and less than β in the lower vertex set until no more vertices need to be removed. This process continues until the graph is reduced according to the criteria.

3.2   Pre-computed Algorithm

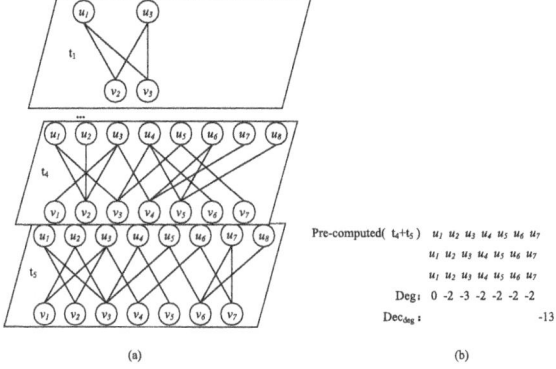

Fig. 3. About t_5 and t_4 snapshot synthesis precompute

This subsection introduces using dynamic programming principles for pre-computations, aiming to predict score changes post-simplification in advance.

Algorithm 1 Pre-computed algorithm

Input: Two dynamic bipartite graphs G_1 and G_2 prepared for synthesis, as well as a degree limited deg Deg_{Thresh}

Output: Dec_{deg}, N, Deg;

1: $Dec_{deg} = 0; N \leftarrow \emptyset; Deg \leftarrow \emptyset$;
2: **for** $u_1 \in U(G_1)$ ($U(G_1)$ is the set of upper vertices of G_1) **do**
3:    $G_{Deg} = |N_{u_1}(G_1)|$;
4:    **for** $u_2 \in U(G_2)$ ($U(G_2)$ is the set of upper vertices of G_2) **do**
5:      **if** $u_1 = u_2$ **then**
6:        $Q = N_{u_1}(G_1) \cap N_{u_2}(G_2); pre_{deg} = |Q|$;
7:      **if** $pre_{deg} < Deg_{Thresh}$ **then**
8:        $pre_{deg} = 0$;
9:      $N = N \cup u_1; Dec_{deg}+ = pre_{deg} - G_{deg}; Deg = Deg \cup (pre_{deg} - G_{Deg})$;
10: **return** Dec_{deg}, N, Deg;

Example 2. In Fig. 1, with $\alpha = 2$, $\beta = 1$, $\tau = 1$, and u_1 as the query vertex, we simplify the graph using the Pruning algorithm, yielding the simplified graph in Fig. 3. We then identify G_{t_5} with the highest score by sorting snapshot scores. Next, we apply the Pre-computed algorithm to G_{t_5} and G_{t_4}. From Fig. 3, we determine the degree loss by combining G_{t_5} with G_{t_4} into a span graph.

3.3 RCSearch Algorithm

After introducing the previous three sub-algorithms, we will now present our RCSearch algorithm.

$$R(D) = \frac{S(C) \cdot R(T)}{(1 + \tau^2)R(T) - \tau^2 S(C)} \qquad (2)$$

We start by sorting each dynamic bipartite graph by score and selecting the highest-scoring one as Max_{SG}. Using the Pre-computed algorithm, we merge Max_{SG} with other graphs to minimize degree loss Dec_{min}. Formula 2 helps us compute the minimum degree necessary to sustain the score post-merger.

4 Experiments

We implement the following algorithm: i) **RCSearch algorithm:** the algorithm presented in this article. ii) **Genetic algorithm** [6]: It adaptively adapts from a genetic algorithm. iii) **Naive algorithm:** This algorithm combines all time snapshots to search for the optimal RC. All algorithms are coded in C++ and run on a computer featuring an AMD R7-5800H 3.2 GHz CPU with 16 GB RAM, operating on Windows 10.

Datasets. We utilized seven real-world dynamic bipartite networks from KONECT[1], as outlined in Table 1. Where U represents editors and V repre-

---

[1] http://www.konect.cc/networks/.

Algorithm 2 RCSearch Algorithm

Input: A list of snapshots $List_{SG}$, α, β, τ, a query vertex q, total degree $total_{deg}$, and total snapshots $total_{temp}$

Output: Max_{SG}

1: **while** $|List_{SG}| \neq 0$ **do**
2:    $Dec_{min} \leftarrow \infty$; Sort $List_{SG}$ by some criterion; $Max_{SG} \leftarrow List_{SG}.pop()$;
3:    **for** $SG \in List_{SG}$ **do**
4:       $Dec_{deg}, N, Deg \leftarrow$ Pre-computed(Max_{SG}, SG, α);
5:       **if** $Dec_{deg} < Dec_{min}$ **then**
6:          $Dec_{min} \leftarrow Dec_{deg}$;
7:       $deg_2, temp_2 \leftarrow$ searchDegTemp(Max_{SG}); $temp_{pre} \leftarrow total_{temp} + temp_2$; $R'_D \leftarrow$ eq2$(\tau, total_{deg}, total_{temp}, temp_{pre}, Max_{SG}.score)$;
8:       **if** $deg + Dec_{min} > R'_D \cdot total_{deg}$ **then**
9:          $SG \leftarrow$ sysSpanGraph(SG, Max_{SG}); $List_{SG}.push(SG)$;
10:         **if** $SG.score > Max_{SG}.score$ **then**
11:            $Max_{SG} \leftarrow SG$;
12: **return** Max_{SG};

---

sents edited entries. Due to the scarcity of dynamic bipartite graph datasets, we treated interactions within a 10-second window as a snapshot for processing.

Table 1. Summary of Datasets.

Dataset	$|U|$	$|L|$	$|E|$	$|snapshots|$
npt	2,149	39,968	140,326	5,973
mvo	255	25,155	181,632	4,776
qar	1,544	128,295	185,970	4,709
vls	1,824	17,510	229,103	5,547
moc	426	44,745	243,608	5,310
my	2,747	71,541	290,218	5,318
ckb	3,691	110,477	398,182	4,851

Exp 1. In this experiment, we observed the runtime of the RCSearch algorithm with different parameter sizes using the vls and qar datasets (Fig. 4(a) and (b)). The runtime remains stable as the degree varies, consistent with the expected time complexity. Figure 4(c) shows that expanding the query time range significantly increases runtime due to the additional snapshots and vertices, which require more time for the Pruning algorithm.

Exp 2. This experiment aimed to verify the results of Exp 1 using the qar dataset. From Fig. 5, we observed that changes in degree had negligible impact on runtime. However, expanding the search time range noticeably increased the runtime of the Pruning algorithm, which supports the findings from Exp 1.

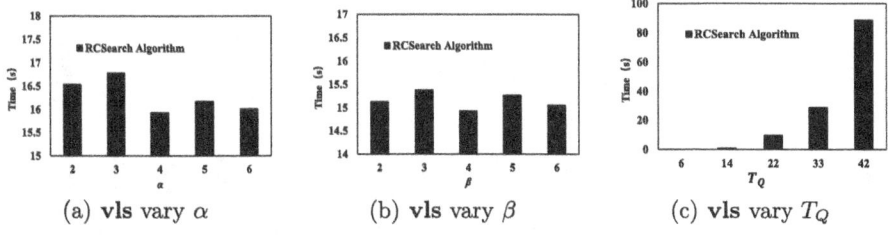

Fig. 4. The running time on the *vls* dataset with different parameters

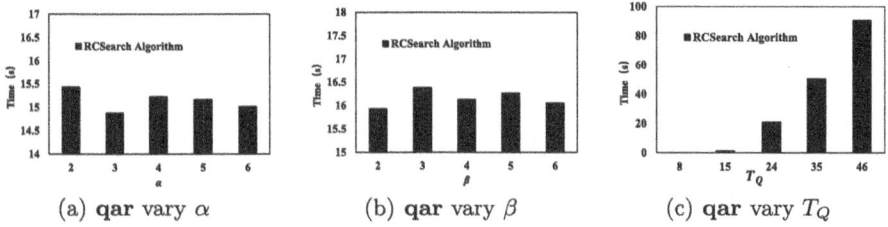

Fig. 5. The running time on the *qar* dataset with different parameters

Exp 3. In this experiment, we assessed the runtime of the Pruning algorithm on dynamic bipartite graphs of different sizes. From Fig. 6(a), it's clear that as the number of edges in the input graph increases, so does the algorithm's runtime. Moreover, there's a noticeable disparity in runtime between the Pruning algorithm and RCSearch algorithm. Particularly, the Pruning algorithm consumes a substantial portion of the overall runtime compared to RCSearch.

Exp 4. In this experiment (Fig. 6(b)), the Naive algorithm has a much longer runtime than the other two algorithms. As the dataset size increases, this gap widens, consistent with our discussion on the Naive algorithm's time complexity of $O(2^n)$.

5   Related Work

This paper explores community search in dynamic bipartite graphs with a focus on continuity and degree size. In related work on dynamic bipartite graphs, Liu et al. [3] developed efficient index maintenance techniques for bipartite graphs. Shu et al. [8]propose a novel cohesive subgraph model to find introverted communities in bipartite graphs, named (α, β, p)-core. Li et al. [2] proposed the PC model for temporal graphs, emphasizing inter-temporal cohesion, and Qin et al. [7] studied periodic community queries in evolving networks.

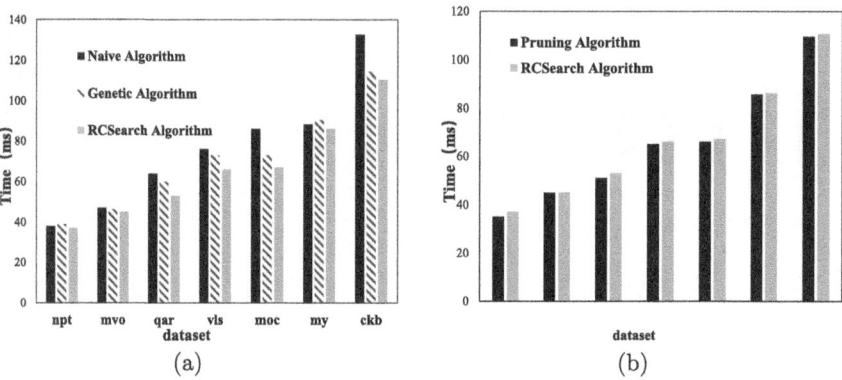

Fig. 6. The comparison of algorithm running time

6   Conclusion

In this paper, we explored the problem of finding the optimal RC in dynamic bipartite graphs. To solve this, we devised an algorithm called RCSearch. Initially, we simplified the dynamic bipartite graph. Using our Pre-computed algorithm, we assessed the feasibility of synthesizing each simplified graph before actual synthesis. Extensive experiments on real-world datasets confirmed the efficiency and effectiveness of our approach.

References

1. Haldar, N.A.H., et al.: Top-k socio-spatial co-engaged location selection for social users. TKDE **35**(5), 5325–5340 (2023)
2. Li, R.H., Su, J., Qin, L., Yu, J.X., Dai, Q.: Persistent community search in temporal networks. In: ICDE, pp. 797–808 (2018)
3. Liu, B., Yuan, L., Lin, X., Qin, L., Zhang, W., Zhou, J.: Efficient (α, β)-core computation: an index-based approach. In: The WWW Conference, pp. 1130–1141 (2019)
4. Lyu, B., Qin, L., Lin, X., Zhang, Y., Qian, Z., Zhou, J.: Maximum and top-k diversified biclique search at scale. VLDB J. **31**(6), 1365–1389 (2022)
5. Ma, Y., Yuan, Y., Zhu, F., Wang, G., Xiao, J., Wang, J.: Who should be invited to my party: a size-constrained k-core problem in social networks. J. Comput. Sci. Technol. **34**(1), 170–184 (2019)
6. Mirjalili, S., Mirjalili, S.: Genetic Algorithm. Evolutionary Algorithms and Neural Networks: Theory and Applications, pp. 43–55 (2019)
7. Qin, H., Li, R.H., Wang, G., Qin, L., Cheng, Y., Yuan, Y.: Mining periodic cliques in temporal networks. In: ICDE, pp. 1130–1141 (2019)
8. Shu, K., Liang, Q., Guo, H., Zhang, F., Wang, K., Yuan, L.: Finding introverted cores in bipartite graphs. In: Yuan, L., Yang, S., Li, R., Kanoulas, E., Zhao, X. (eds.) WISA 2023. LNCS, vol. 14094, pp. 162–170. Springer, Singapore. (2023). https://doi.org/10.1007/978-981-99-6222-8_14

9. Wang, K., Zhang, W., Lin, X., Zhang, Y., Qin, L., Zhang, Y.: Efficient and effective community search on large-scale bipartite graphs. In: ICDE, pp. 85–96 (2021)
10. Yao, K., Chang, L., Yu, J.X.: Identifying similar-bicliques in bipartite graphs. In: PVLDB, vol. 15, pp. 3085–3097 (2022)

A Hierarchical Structure Explanation Method for Complex Tables

Fangnuo Liu, Sainan Tong, Derong Shen[✉], Tiezheng Nie, and Yue Kou

Computer Science and Technology, Northeastern University, Shenyang 110819, Liaoning, China
2001810@stu.neu.edu.cn, {shenderong,nietiezheng,kouyue}@cse.neu.edu.cn

Abstract. Tables are used to organize and manage data. It is important to extract valuable information from tabular data. Existing methods for interpreting table structure focus on relational tables, but have limitations. In our paper, we propose a method for interpreting complex tables hierarchically. We use cell functional classification(CFCT) and a tree-based region detection model (RDT) to classify cells accurately. We also introduce a table layout prediction model(LPT) based on relational reasoning to explore the spatial structure of tables. Experiments on real datasets show the effectiveness of our approach.

Keywords: Table · Table Structure Interpretation · Region Detection · Layout Prediction

1 Introduction

The internet industry is rapidly growing. Understanding and utilizing table data is important. Table understanding involves extracting semantic information from tables through five steps: table detection, structure recognition, function analysis, structure analysis, and semantic interpretation. In recent years, significant progress has been made in table processing, including region detection, classification, structure recognition, and semantic understanding. However, existing methods mainly focus on relational tables, and their ability to interpret complex tables is limited. In addition, the problem of data imbalance also constrains the performance of deep learning models [1].

DeepDeSRT [2] and TableSense [3] have introduced deep learning-based methods for table region detection, and CFCT [4] is a cell function classification method related to the technology in this paper. Other works [5,6] have focused on structure recognition and semantic understanding. Our goal is to reveal the semantic content of tables, identify different layout structures, infer implicit information, and enable intelligent processing of table data.

The main contributions of this paper are: (1) Introducing a tree-based region detection model (RDT) that categorizes tables into regions based on functional roles to group cells with the same role together and correct classification errors.

(2) Presenting a table layout prediction model that relies on relationship inference to predict edges using block embeddings and regions identified by the region detection model as input. (3) Validating the effectiveness of the proposed approach through comparative and ablation experiments on real datasets.

2   Problem Definition

Definition 1 (Table). Typically, a table with N rows and M columns is denoted as: $T = \{C_{i,j}; 1 \leq i \leq N, 1 \leq j \leq M\}$, where $C_{i,j}$ represents the cell in the i-th row and j-th column of the table.

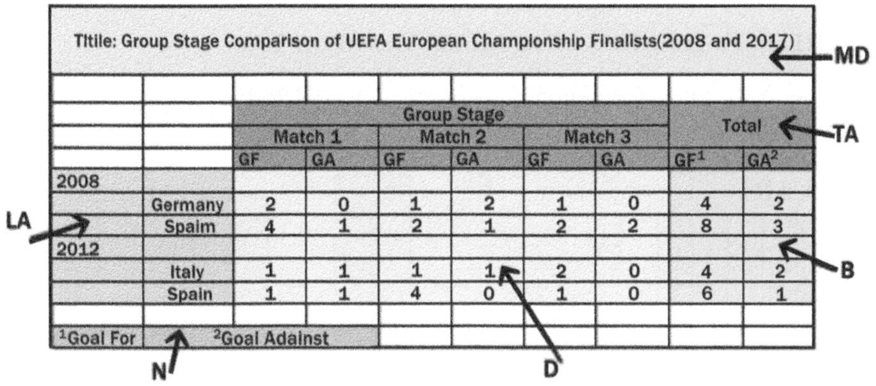

Title: Group Stage Comparison of UEFA European Championship Finalists(2008 and 2017)								←MD
		Group Stage						Total ←TA
		Match 1		Match 2		Match 3		
		GF	GA	GF	GA	GF	GA	GF¹	GA²
2008								
Germany	2	0	1	2	1	0	4	2
Spain	4	1	2	1	2	2	8	3
2012								←B
Italy	1	1	1	1	2	0	4	2
Spain	1	1	4	0	1	0	6	1
¹Goal For ²Goal Against								

LA →   N   D

Fig. 1. Table layout examples, different colors represent different cell types

Definition 2 (Cell Role Classification) involves assigning a specific label to each cell: metadata (MD), top attribute (TA), left attribute (LA), data (D), derived (B), and footnotes (N). This is illustrated in Fig. 1.

Definition 3 (Region Detection) is defined as the process of identifying a collection of non-overlapping rectangular blocks B= B_1, B_2, B_3, B_4, where each block B_i is represented as $\langle t_i, l_i, b_i, r_i \rangle$ with t_i, l_i, b_i, r_i denoting the top row, left column, bottom row, and right column, respectively. Additionally, each block B_i should also be assigned a label L_i.

Definition 4 (Layout Prediction) seeks to establish the relationships between every pair of blocks and assign a layout relation R_i to each pair. In this paper, we define five types of relations for a given block pair $\langle B_1, B_2 \rangle$: parent-child relation (R_1), above relation (R_2), left relation (R_3), global relation (R_4), and null relation (R_5).

The hierarchical structure explanation method proposed in this paper is mainly divided into three parts: cell function classification of tables (CFCT), region detection of tables (RDT), and layout prediction model of tables (LPT). Below is a detailed introduction to each module.

3   Proposed Method

This paper proposes a table structure interpretation method based on deep learn-ing, which consists of three main modules: Cell Function Classification (CFCT) [4], Region Detection (RDT), and Layout Prediction (LPT). The RDT module is based on the results of CFCT, uses a tree structure to segment the table into functional regions, and uses probability methods to assign functional labels to each region. The LPT module utilizes the Transformer architecture to predict the relationships between regions and reveal the layout structure of the table. These three modules collaborate with each other to achieve a deeper understanding and structural interpretation of complex tables.

3.1   RDT: Region Detection Model Based on Tree

The general workflow for region detection is depicted in Fig. 2.

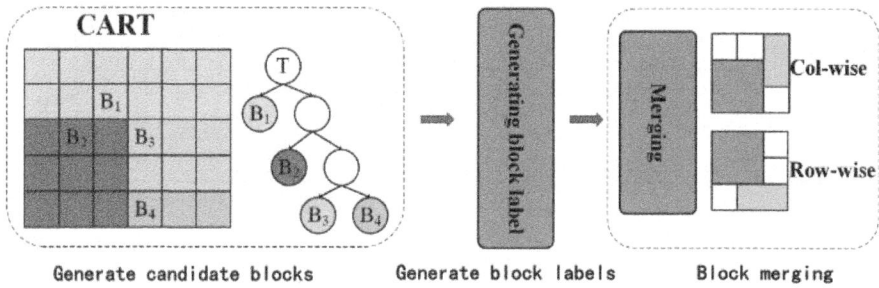

Fig. 2. The regional detection model

(1) Build a CART decision tree recursively

The overall process begins with the table T as input to the CART module, where it serves as the initial active node (each step i corresponds to an active node B_{ij}). The split probability P_{split} is used to determine if the active node can be further split. If the calculated value is below a specified threshold, the splitting process stops, and the node becomes a leaf. Otherwise, if the value exceeds the threshold, the node is split based on a rule-based selection of rows or columns. The resulting blocks then become new active nodes.

After multiple iterations, the system will generate n candidate trees. The weight function $\boldsymbol{W}_{\text{ent}}$ of these trees is used to determine their weights and select the final tree. The weight function $\boldsymbol{W}_{\text{ent}}$ of these trees uses entropy to evaluate the weight of candidate trees and selects the tree with the lowest entropy as the final result.

(2) Classification Method for the Functional Roles of Initial Blocks

Given the output of CART and the classification of units as known, we use a probabilistic approach to determine the functional role of blocks, ensuring that all units within a block have the same functional role type. The following section provides a detailed description of the probabilistic approach:

Initially, the unit classification model assigns functional role labels to all cells and delineates block boundaries using the CART method. Subsequently, it analyzes the data distribution within each block. In the case of block B1, the distribution is outlined as: $D = [d_1, d_2, \ldots, d_i]$, where $d_i \in \{\text{MD}, \text{TA}, \text{LA}, \text{D}, \text{B}, \text{N}\}$, with the block encompassing i cells. The occurrences of MD, TA, LA, D, B, and N are denoted as $x_1, x_2, x_3, x_4, x_5, x_6$, with $i = x_1 + x_2 + x_3 + x_4 + x_5 + x_6$. Following this, the method computes the probabilities for each functional role type within the block, that is, $P_1 = x_1/i, P_2 = x_2/i, P_3 = x_3/i, P_4 = x_4/i, P_5 = x_5/i, P_6 = x_6/i$. Lastly, it identifies the functional role type with the highest probability, selecting from $\{P_1, P_2, P_3, P_4, P_5, P_6\}$ as the definitive functional role for the block.

(3) Merging Candidate Blocks with the Same Function

The main merger strategies include:

Direct Merge: When two blocks are completely adjacent and have the same functional role, they can be merged directly into a larger block.

Vertical merge: Vertically merge blocks located in the same column, adjacent to the top or bottom.

Horizontal merge: Merge blocks located on the same row, adjacent to the left or right, horizontally.

Interleave Merge: Merge vertically or horizontally staggered blocks by adjusting the boundaries.

3.2 LPT: A Table Layout Prediction Model Based on Relational Inference

The overall architecture of the proposed method is depicted in Fig. 3.

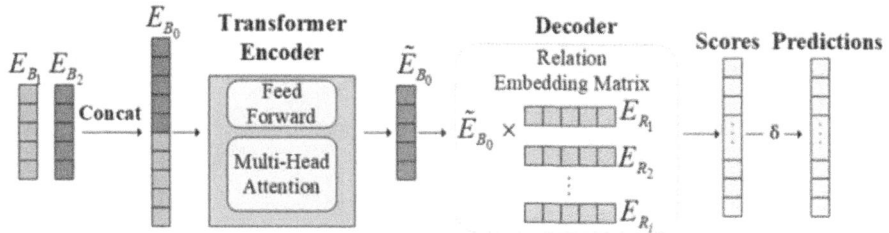

Fig. 3. The framework of the layout prediction

First, we generate the embedding representation E_B for each block based on block boundaries and unit embeddings, using the pointwise addition method and assign an embedding representation E_R for each relation. Next, we concatenate the embedding representations of any two blocks and input them into the Transformer encoder to produce the hidden vector H_o. Subsequently, H_o is passed to the decoder for decoding, calculate the score of R_i using inner product. Finally, convert the score of R_i into probability.

(1) Embedding Representations of Blocks

Due to the unequal number of cells in each block, the embedding dimension of each block is different. To ensure compatibility for tensor operations, it is crucial that every block and relation share the same embedding dimension, denoted as $d_{B_i} = d_{B_i} = d_{R_i} = d$. In this paper, we employ the element-wise addition approach. Specifically, for a block B_i with i cells and existing cell feature vectors $\{E_1, E_2, \ldots, E_i\}$, where $E_i \in R^d$. The fusion of these vectors is achieved through direct element-wise addition.

$$E_{B_i} = E_1 + E_2 + \ldots + E_i \tag{1}$$

$$E_{B_i} = \{x_n \mid x_n = E_1[n] + E_2[n] + \ldots + E_i[n], n = 1, \ldots, d\} \tag{2}$$

(2) Encoding Phase

Solution1: Input two blocks. For each triplet (B_i, B_j, R_i), where the embedding of B_i is represented as $E_{B_i} \in R^d$, the embedding of B_j is represented as $E_{B_j} \in R^d$, and the embedding matrix for the block pair (B_i, B_j) consists of query $Q \in R^{2 \times d}$, key $K \in R^{2 \times d}$ and value $V \in R^{2 \times d}$. Assuming this is a single-head attention mechanism, the formula for computing the attention matrix $A_{B_i, B_j} \in R^{2 \times 2}$ is:

$$A_{B_i B_j} = \mathrm{softmax}\left(\frac{QK^T}{\sqrt{d_k}}\right) \tag{3}$$

The output representation of the block-wise operation (B_i, B_j) on the matrix $H \in R^{2 \times d}$ is:

$$H = A_{B_i B_j} V \tag{4}$$

The attention matrix A_{B_i, B_j} serves as a guide for the model to balance its focus between intra-block and inter-block information.

Solution2: Input Blocks and Relations. The computation formula for each triplet (B_i, B_j, R_i), $A_{B_i R_i} \in R^{2 \times 2}$ is as follows:

$$A_{B_i R_i} = \mathrm{softmax}\left(\frac{QK^T}{\sqrt{d_k}}\right) \tag{5}$$

The output representation matrix $H \in R^{2 \times d}$ for the block and relation (B_i, R_i) is:

$$H = A_{B_i R_i} V \tag{6}$$

(3) Decoding Phase

The decoding phase aims to evaluate each triple (B_i, B_j, R_i) based on the relationship type and inter-block information to determine whether there is a relationship between blocks and what kind of relationship exists. The hidden output of the encoder is: $H_o = \tilde{E}_{B_n}, \tilde{E}_{B_n} \in R^d$.

We employ a method of score calculation that involves computing the bi-directional inner product.

$$\phi\left(B_i, B_j, R_i\right) = \tilde{E}_{B_0}^T E_{R_i} \tag{7}$$

Finally, in order to ascertain the ultimate type of relationship, we apply the sigmoid function to the scores obtained from the inner product calculation, computing $R_i \in \{R_1, R_2, R_3, R_4, R_5\}$ as the probability of the block pair (B_i, B_i) being in the layout relation. Decoding method for solution 2:

$$\phi\left(B_i, B_j, R_i\right) = \tilde{E}_{B_0}^T E_{B_j} \tag{8}$$

4 Experiments

We evaluates the effectiveness of our proposed methods on four different datasets in the fields of finance, business, and healthcare: DeEx (444 files) [7], SAUS (223 files) [8], CIUS (269 files), DG (431 files) [9].

4.1 Experimental Results and Analysis of Region Detection

(1) Evaluation Mertics

To assess the effectiveness of the proposed RDT, this section utilizes two evaluation metrics: (1) F1 score, which evaluates the predictive performance of block labels; and (2) Error-of-Boundary (EoB), which measures the precision of aligning detected region boundaries with actual boundaries. A smaller EoB value signifies less boundary error. The computation method for EoB is as follows:

$$\text{Eo}\,B\left(B^{g^t}, B^p\right) = \max\left(\left|B_t^{gt} - B_t^p\right|, \left|B_b^{gt} - B_b^p\right|, \left|B_l^{gt} - B_l^p\right|, \left|B_r^{gt} - B_r^p\right|\right)$$

In this paper, we use the average EoB over all blocks, as a table may contain multiple blocks. The top, bottom, left, and right boundaries of the ground-truth block B^{gt}, denoted by $B_t^{gt}, B_b^{gt}, B_l^{gt}$, and B_r^{gt} respectively, are annotated in the same way for the predicted block B^p as for B^{gt}.

$$EoB_{avg} = \sum_{1 \leq i \leq N, 1 \leq j \leq M} \frac{1}{\left| B^{gt^{ij}} \cap B^{p^{ij}}\right|} \left| EoB\left(B^{gt^{ij}}, B^{p^{ij}}\right) \right. \tag{9}$$

where N represents the number of rows, M stands for the number of columns, $B^{gt^{ij}}$ denotes the block to which the cell in the i-th row and j-th column belongs, and $B^{p^{ij}}$ carries the same implication as $B^{gt^{ij}}$.

(2) Comparative Method

In order to validate the effectiveness of the region detection model proposed in this study, we conducted a comparative analysis with previous methods. The specific methods are detailed below: PSL (RNN) [9], PSL (RF) [9], RBA [10], AC (RNN) [11], AC (RF) [10].

Table 1. Average EoB scores of all models on the DG dataset.

Method	EoB $_{avg}$
PSL (RNN)	2828
PSL (RF)	1995
RBA	63565
AC (RNN)	296
AC (RF)	423
RDT (F)	315
RDT (F+S)	**282**

The average EoB results for all models are presented in Table 1. Specifically, RDT (F) denotes the outcomes derived from constructing trees based on cell functional roles as rule distributions, while RDT (F+S) signifies the results obtained by incorporating both cell functional roles and style features as rule distributions. The experimental findings highlight the superior performance of the proposed RDT (F+S) model over the other models, indicating its ability to align blocks more effectively with the ground-truth blocks.

4.2   Experimental Results and Analysis of Layout Prediction

To validate the effectiveness of the layout prediction model, we compared the proposed LPT with RF, CRF, and PSL [11].

Table 2. Layout prediction results on the DG dataset (%).

Model	R1	R2	R3	R4	R5	Avg
RF	0	1.1	2.1	22.7	81.7	21.5
CRF	0	33.7	32.2	40.0	88.5	38.9
PSL	25.6	**70.3**	32.8	43.0	89.6	52.3
LPT (B^2)	**33.5**	68.4	**33.1**	**44.2**	**90.3**	**53.9**
LPT (BR)	32.7	67.2	32.5	43.6	89.6	53.1

We utilize the blocks generated by RDT(F+S) to execute various layout prediction models. For each block B^p, we match it with the ground-truth block

B^{gt} that shares the most cells with B^p. We incorporate all predicted relations of B^p into B^{gt} and juxtapose them against the ground-truth relations. The experimental findings are detailed in Table 2. LPT (B^2) denotes the input of two blocks, while LPT (BR) signifies the input of a block and a relation. The results consistently indicate that in most cases, the LPT (B^2) approach outperforms its counterparts. The main reason is that compared to other models, Multi-head Attention can better capture inter block information.

5   Conclusion

The interpretation of table structures is a crucial step in the process of understanding tables. In this paper, we propose a method that utilizes pre-trained cell embeddings and a Transformer-based approach for table structure interpretation. By applying deep learning techniques, this method accurately captures the textual style features of tables and performs cell-level classification, region detection and layout prediction tasks. Experimental evaluations on real benchmark datasets demonstrate the effectiveness of our method compared to other approaches. Future work will focus on developing unsupervised or semi-supervised methods to reduce the amount of manually labeled data without sacrificing performance.

Acknowledgments. This work was supported by the National Natural Science Foundation of China (62172082, 62072084,62072086), the Fundamental Research Funds for the central Universities (N2116008 No.).

References

1. Diao, Y., Sun, Z., Zhou, Y.: A multi-label imbalanced data classification method based on label partition integration. In: Yuan, L., Yang, S., Li, R., Kanoulas, E., Zhao, X. (eds.) WISA 2023. LNCS, vol. 14094, pp. 14–25. Springer, Singapore (2023). https://doi.org/10.1007/978-981-99-6222-8_2
2. Gatos, B., Danatsas, D., Pratikakis, I., Perantonis, S.J.: Automatic table detection in document images. In: Singh, S., Singh, M., Apte, C., Perner, P. (eds.) ICAPR 2005. LNCS, vol. 3686, pp. 609–618. Springer, Heidelberg (2005). https://doi.org/10.1007/11551188_67
3. Hassan, T., Baumgartner, R.: Table recognition and understanding from pdf files. In: Proceedings of ICDAR, pp. 1143–1147 (2007)
4. Tong, S., Shen, D., Kou, Y., et al.: CFCT: the cell function classification method for complex tables. In: HPCC/DSS/SmartCity/DependSys, pp. 2206–2213 (2022)
5. Harit, G., Bansal, A.: Table detection in document images using header and trailer patterns. In: Proceedings of Computer Vision,Graphics and Image Processing, pp. 1–8 (2012)
6. e Silva, A.C.: Learning rich hidden Markov models in document analysis: table location. In: Proceedings of Document Analysis and Recognition, pp. 843–847 (2009)
7. Eberius, J., Werner, C., Thiele, M., et al.: DeExcelerator: a framework for extracting relational data from partially structured documents. In: Proceedings of Information & Knowledge Management, pp. 2477–2480 (2013)

8. Chen, Z., Cafarella, M.: Integrating spreadsheet data via accurate and low-effort extraction. In: Proceedings of SIGKDD, pp. 1126–1135 (2014)
9. Sun, K., Rayudu, H., Pujara, J.: A hybrid probabilistic approach for table understanding. In: Proceedings of AAAI, vol. 35, no. 5, pp. 4366–4374 (2021)
10. Koci, E., Thiele, M., Romero, O., Lehner, W.: Cell classification for layout recognition in spreadsheets. In: Fred, A., Dietz, J., Aveiro, D., Liu, K., Bernardino, J., Filipe, J. (eds.) IC3K 2016. CCIS, vol. 914, pp. 78–100. Springer, Cham (2019). https://doi.org/10.1007/978-3-319-99701-8_4
11. Sun, K., Wang, F., Chen, M., et al.: Tabular functional block detection with embedding-based agglomerative cell clustering. In: Proceedings of Information & Knowledge Management, pp. 1744–1753 (2021)

Large Language Model

Low-Parameter Federated Learning with Large Language Models

Jingang Jiang, Haiqi Jiang, Yuhan Ma, Xiangyang Liu, and Chenyou Fan$^{(\boxtimes)}$

South China Normal University, Guangzhou, Guangdong, China
2022024923@m.scnu.edu.cn, fanchenyou@scnu.edu.cn

Abstract. We study few-shot Natural Language Understanding (NLU) tasks with Large Language Models (LLMs) in federated learning (FL) scenarios, which is challenging due to limited data and mobile device constraints. Recent studies show LLMs can handle tasks like sentiment analysis and arithmetic reasoning. However, their large sizes lead to high computation and communication costs, making traditional FL impractical. To address this, we propose Low-Parameter Federated Learning (LP-FL), which combines LLM prompt learning with efficient communication and federating techniques. LP-FL enables clients to assign soft labels to unlabeled data, expanding the labeled set during FL. It uses Low-Rank Adaptation (LoRA) for cost-efficient parameter construction, local model fine-tuning, and global model federation. LP-FL performs outstandingly in sentiment analysis of various FL scenarios and can be comparable to centralized training in a small number of scenarios. Notably, in a semi-supervised context, LP-FL demonstrates more robustness than FP-FL. This is attributed to the utilization of fewer parameters in LP-FL, which renders it less vulnerable to the adverse effects of overfitting caused by error noise in semi-supervised scenarios, resulting in superior performance compared to FP-FL.

Keywords: Federated Learning · Large Language Model · Prompt Learning · Low-Rank Adaptation · Sentiment Analysis

1 Introduction

The advent of Large Language Models (LLMs) such as GPT-3 [1] has a profound impact on the research landscape of not only Natural Language Processing (NLP) studies but also the entire AI and Big Data communities. The fine-tuning of Large Language Models (LLMs) has proven highly effective across a multitude of tasks [15]. However, the immense scale of parameters in LLMs entails significant computational costs for fine-tuning. Performing fine-tuning directly on these devices is impractical due to the formidable requirement of computation power. Hence, the pursuit of an effective fine-tuning method that achieves desirable outcomes with minimal parameter fine-tuning has become imperative.

Additionally, not all mobile devices can collect a sufficient amount of data. Typically, mobile device users only have processing permissions for the data

C. Jin et al. (Eds.): WISA 2024, LNCS 14883, pp. 319–330, 2024.
https://doi.org/10.1007/978-981-97-7707-5_28

generated during their own usage, and this data is often unlabeled. For example, mobile device users provide reviews for movies, hotels, or restaurants, and we can obtain a small amount of labeled data based on their positive or negative feedback. However, there is still a significant amount of comment data, such as those on TikTok or Twitter, where we cannot ascertain the sentiment. Fortunately, there is a substantial user base of mobile devices with a significant amount of data available, though dispersed among various devices. Hence, distributed learning with few-shot labeled data has become an emerging topic.

Federated learning (FL) enables distributed training of a global model on decentralized data [11,12]. Hence, we propose leveraging FL to collaboratively train these devices and achieve effective fine-tuning of a global model. Additionally, we explore optimal utilization of labeled and unlabeled samples under the FL scenarios. One approach, PET [18], rephrases input examples using diverse prompts to aid the LLMs' understanding of the task. Fine-tuning is performed on each prompt using a LLM. The resulting models assign soft labels to unlabeled data, expanding the labeled dataset for standard supervised training. However, PET's multi-task setup in a federated environment incurs high computational costs and communication overhead, posing challenges for resource-constrained clients like mobile devices and sensors.

Our contributions are summarized as follows:

1. We consider an under-studied task of fine-tuning LLMs with distributed devices with limited communications and local computational powers.
2. We fine-tune the LLMs by adding task descriptions to the input examples for text sentiment classification. We start with a small number of labeled samples and then use a semi-supervised method to augment the dataset and enhance the fine-tuning process.
3. We introduce a low-parameter methodology called Low Parameter Federated Learning, abbreviated as LP-FL, for efficiently fine-tuning a small subset of the local model parameters then federate averaging over all clients.
4. We demonstrate that our method achieves comparable or even better performance than Full-Parameter Federated Learning, while greatly reducing computational costs and communication requirements on individual devices.

2   Related Work

Federated Learning (FL). FL [7,14,22] is a distributed learning method that aims to train a global model on decentralized data while preserving data privacy. Our research focuses on federated few-shot learning, addressing the challenge of training an effective global model in a federated environment with limited client-side labeled data. Several approaches have been proposed to tackle this scenario. FedFSL [6] performs classification of unseen data classes using only a small number of labeled samples. FedSSL [5] utilizes semi-supervised learning to fully leverage labeled and unlabeled data sources for training. pFedFSL [23] identifies well-performing models on specific clients (without revealing local data to the server or other clients) and selects suitable clients for collaboration, enabling personalized and distinctive feature space learning for each client.

Large Language Models (LLMs).   LLMs like GPT-3 [1], PaLM [3], LaMDA [20] and LLaMa [21] exhibit remarkable few-shot capabilities by leveraging natural language prompts and task demonstrations as contextual input. However, these capabilities are built upon LLMs with parameter counts often exceeding 10 billion, making their application in real-world scenarios challenging. Consequently, researchers have begun investigating prompt performance on smaller-scale language models. Relevant approaches include PET [18], which reformulates input examples into cloze-style phrases to facilitate the language model's comprehension of the given task. AutoPrompt [19] selects a subset of discrete characters as triggers through gradient-based search and constructs templates to predict the probability of corresponding label words using models. LM-BFF [8] introduces a generative method for pattern construction, followed by a search-based technique to derive the associated verbalizers.

Parameter-Efficient Fine-Tuning (PEFT).   PEFT enables efficient adaptation of LLMs to various downstream tasks without the need to fine-tune all parameters of the LLMs. Notably, advanced PEFT techniques achieve performance comparable to full fine-tuning. Adapter Tuning [9,16,17] involves introducing new network layers or modules within the internal network layers of the LLMs to adapt to downstream tasks. Low-Rank Adaptation (LoRA) [10], while freezing the original model parameters, incorporates additional network layers and exclusively trains the parameters of these newly added layers to achieve results similar to full-model fine-tuning. QLoRA [4] reduces the memory footprint significantly by loading the model itself with a 4-bit representation. Numerical values are first dequantized to bf16 before being utilized for training. Additionally, the technique incorporates LoRA fine-tuning, focusing solely on training a limited set of LoRA parameters, thereby substantially reducing the GPU memory requirement.

3   Approach

We describe the training workflow of Low-Parameter Federated Learning (LP-FL) in Fig. 1. Assuming M as a Large language model (LLM) with a vocabulary size of V and the mask token [MASK] $\in V$. Now we consider a federated environment with K participating clients. As shown in Fig. 2, each client k possesses a labeled dataset T_k comprising n_k data instances, along with a much larger unlabeled dataset U_k with u_k instances, where $n_k \ll u_k$. For each client, given an input sequence example $x = (s_1, ..., s_n)$ where each word $s_j \in V$, we can employ the $P(x)$ to add a task description phrase with a mask token to the input sequence, allowing the LLM to predict the word for the mask position. Using the mapping $L \to V$, we associate the label $l \in L$ of x with a word $v \in V$.

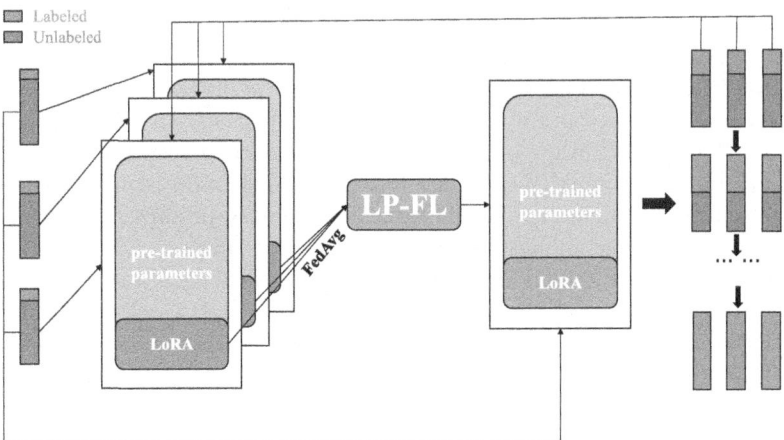

Fig. 1. The training workflow of LP-FL: (1) Each client conducts LoRA fine-tuning of the global model with local data. (2) Then the server performs FedAvg on the LoRA parameters. (3) Each client utilizes the updated global model to extend their local labeled dataset by annotating unlabeled data.

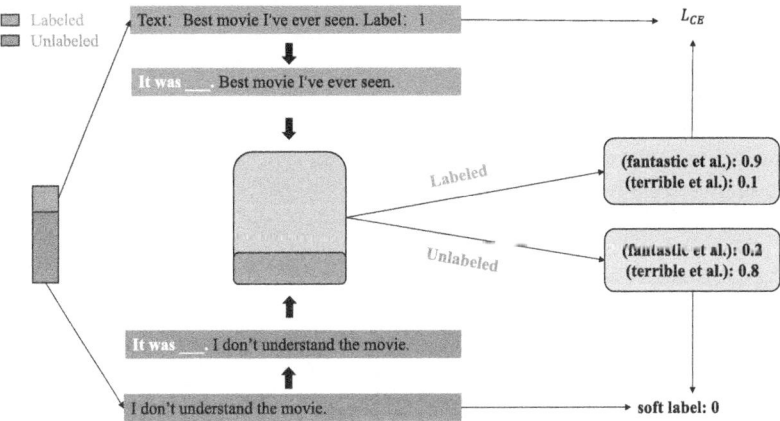

Fig. 2. LP-FL Fine-tuning and Data Annotation Process: For each data point in the local dataset, we process it using a task description resembling "It was [MASK]." with a mask token. Subsequently, we utilize the LLM to predict the logits of all words in the label mapping at the masked position and convert them into classification logits. In the case of labeled data, these logits can be employed to compute the cross-entropy loss for fine-tuning the LLM. Conversely, for unlabeled data, soft labels can be generated to expand the labeled dataset.

Our goal is to leverage the LLM M to predict the probability of each $v \in V$ at the position of the mask token in $P(x)$. For an input sequence example x, we compute the score of x with label $l \in L$ as follows:

$$s_P(l|x) = \sum_{v \in V} M(v|P(x)). \tag{1}$$

Upon acquiring the probability distribution over the labels through a softmax function, we can compare the predicted probability of the true word and measure the prediction with a standard cross-entropy loss, given below:

$$L_{CE} = \frac{1}{n} \sum_{i=1}^{n} (-\log \frac{e^{s_P(l|x)}}{\sum_{l' \in L} e^{s_P(l'|x)}}). \tag{2}$$

Give CE loss in Eq. (2), we can fine-tune the LLM M.

For each client, we utilize multiple task descriptions $P \in \mathcal{P}$ to fine-tune the local model with the aforementioned approach. Furthermore, to address the communication burden arising from the substantial parameter size of the LLM, we utilize the Parameter-Efficient Fine-Tuning (PEFT) technique known as Low-Rank Adaptation (LoRA) [10] during the fine-tuning phase.

LoRA involves preserving the parameters of the LLM itself without modification. Specifically, LoRA introduces a fine-tuning-only bottleneck module which forms a residual connection to its original parameters W_0 to the original LLM. As shown in Eq. (3):

$$\begin{aligned} W_0 + \Delta W = W_0 + BA, \\ s.t. \quad B \in \mathcal{R}^{d \times r}, A \in \mathcal{R}^{r \times k}, r \ll min(d, k). \end{aligned} \tag{3}$$

this bottleneck module is composed of two matrices A and B. Matrix A reduces the input dimension from d to r, then matrix B restores the output dimension from r to k, thereby simulating the concept of intrinsic rank. Throughout the training period, the pre-trained W_0 remains fixed and does not undergo gradient updates. The trainable parameters are the bottleneck parameter A and B. In our work, we fixed the r as 8, which the trainable parameters will be about 0.23% of the all parameters. Following the completion of a global training round, each client exclusively uploads ΔW to the server for parameter averaging. We adopt FedAvg in Eq. (4) as the method for parameter averaging:

$$\min_{\Delta W \in \mathbb{R}^{d \times k}} \sum_{k=1}^{K} \frac{n_k}{n} F_k(\Delta W)$$

$$s.t. \quad F_k(\Delta W) = \frac{1}{n_k} \sum_{i \in P_k} f_i(\Delta W). \tag{4}$$

in which $f_i(w)$ is the loss function for client i, and n is the total number of clients in the system. We employ an iterative approach wherein models are trained on continuously expanding datasets over multiple epochs. Each client assigns soft

---

Algorithm 1: LP-FL overview. M is a global LLM; T is the local labeled data, U is the local unlabeled data ($|T| \ll |U|$); l is step size, G is the number of global rounds; E is the number of epochs.

---

1 **Server executes:**
2   Initialize global LLM M with LoRA bottleneck modules, and only the parameters of LoRA bottleneck modules w can be trained;
3     $l \leftarrow 5 \times 10^{-5}$; $G \leftarrow 5$; $E \leftarrow 5$;
4   **while** $g \leq G$ **do**
5     $\quad$ **for** each client k *in K clients* **in parallel do**
6     $\quad\quad$ $w_{g+1}^k \leftarrow$ ClientUpdate (k, w_g)
7     $\quad$ $w_{g+1} \leftarrow FedAvg(w_{g+1}^{1:K})$
8     $\quad$ Server sends w_{t+1} back to clients
9     $\quad$ $g \leftarrow g + 1$
10   Return M
11
12 **ClientUpdate**(k, w): // *Run on client k*
13   T is the local labeled data, U is the local unlabeled data
14   **if** $g < 5$ **then**
15    $\quad$ $S \leftarrow$ label the 25% of the U
16    $\quad$ $T \leftarrow T + S$
17    $\quad$ $U \leftarrow U - S$
18   **else**
19    $\quad$ continue
20    $\mathcal{B} \leftarrow$ (split T and U into batches of size B)
21   **for** each local epoch i from 1 to E **do**
22    $\quad$ **for** batch b $\in \mathcal{B}$ **do**
23    $\quad\quad$ $w \leftarrow w - \eta \nabla \ell(w; b)$
24   Return w

---

labels to a subset of their local unlabeled data using the global model after FedAvg. When labeling the unlabeled data $x \in U$, the prediction accuracy a_P of each task description P on the validation set is utilized as a weight. The soft label for x is then obtained by performing a weighted average in Eq. (5):

$$s(l|x) = \frac{1}{Z} \sum_{P \in \mathcal{P}} a_P \cdot s_P(l|x) \quad s.t. \quad Z = \sum_{P \in \mathcal{P}} a_P. \tag{5}$$

The model is subsequently fine-tuned on both the original labeled data and the newly annotated data, iterating through all rounds of training. We provide the algorithm details in Algorithm 1 to illustrate the training procedures.

4   Experiments

We verify our approach on the widely used IMDB, Yelp and Sent140 datasets for evaluating sentiment analysis task. We performed an analysis of our method in both federated and centralized training settings, as well as an analysis of Low-Parameter Federated Learning (LP-FL) and Full-Parameter Federated Learning (FP-FL) in the federated environment. In LP-FL, we conducted experiments not only on Low-Rank Adaptation (LoRA) but also on QLoRA.

4.1 Datasets and Settings

IMDB dataset is a benchmark dataset for sentiment classification task. [13] It contains tens of thousands of textual reviews specifically related to movie evaluations. IMDB provides a well-balanced collection of movie reviews for sentiment classification tasks, which is widely compared in recent studies.

Yelp dataset is widely used for sentiment classification task [8,18] collected from the Yelp platform. The dataset contains business details such as names, locations, categories, and star ratings. It also consists of user profiles with IDs, names, and review histories. Our study focuses on the yelp-polarity dataset, which is a binary sentiment classification dataset.

Sent140 [2] is a public dataset for sentiment analysis. It comprises short text messages collected from Twitter, encompassing a diverse range of topics and contexts, ranging from personal emotions to commentary on global events.

Our method is validated on the IMDB, Yelp and Sent140 datasets. The complete sets of 25,000 training and testing samples are used for IMDB, while we randomly select 25,000 samples from the training and testing sets for Yelp and Sent140. The Large Language Model (LLM) employed in our experiments is BERT-Large model with 336M parameters. Within a federated setting, we retain our Low-Rank Adaptation (LoRA) parameter settings unchanged to guarantee model consistency between deployments on the server and clients. Through a series of experiments and the application of common configurations, we establish a fixed rank of 8 for LoRA.

Table 1. LP-FL results(LoRA/QLoRA)

Dataset	Labeled Data	Centralized	Our LP-FL (client num)		
			2	5	10
IMDB	1%	0.850/0.847	+0.008/+0.009	−0.010/+0.002	−0.028/−0.040
	5%	0.869/0.868	−0.002/−0.003	−0.008/−0.004	−0.016/−0.014
	10%	0.878/0.876	−0.005/−0.006	−0.014/−0.009	−0.019/−0.019
Yelp	1%	0.913/0.905	+0.009/−0.002	−0.010/−0.029	−0.013/−0.064
	5%	0.914/0.914	+0.013/−0.008	+0.005/−0.003	−0.003/−0.029
	10%	0.919/0.921	+0.010/−0.005	+0.006/−0.008	+0.000/−0.018
Sent140	1%	0.791/0.786	−0.022/−0.051	−0.024/−0.058	−0.057/−0.057
	5%	0.802/0.798	−0.004/−0.009	−0.012/−0.011	−0.014/−0.024
	10%	0.810/0.807	−0.005/−0.007	−0.010/−0.009	−0.013/−0.014

Considering previous research and practical considerations, we choose the following hyperparameter values: a batch size of 8, local epochs as 5, global training rounds as 5, a learning rate of 5×10^{-5}, and a maximum sequence length of 128. We conduct experiments in federated settings with 2, 5, and 10 participating parties. During the training process, we selectively employ a fraction of the training dataset, specifically 1%, 5%, and 10%, as labeled datasets. Consequently, each federated participant receives an equal distribution of total 250/1250/2500 labeled samples. Following each global update round, clients randomly pick 25% of the data from their local dataset for annotation.

4.2   Experimental Results of LP-FL

Result of LP-FL. Table 1 illustrates the outcomes of our method in the context of sentiment analysis task. We have the following observations.

- *A comparative analysis is conducted between our method under the federated environment and the centralized training approach.* Remarkably, our method exhibits minimal decrease in prediction accuracy when there are only two federated participants, demonstrating nearly identical performance to the centralized training approach.
- *Moreover, when five and ten client participants engage in the federated training, the prediction accuracy only decrease by approximately 1% and 2% respectively.*
- *In scenarios involving ten federated participants, each contributing a mere 0.1% of the data (equivalent to 24 data instances), the sample size adequately corresponds to the labeled samples commonly available on mobile devices in practical settings. Nevertheless, our approach consistently yields satisfactory results, with the final model's prediction accuracy only decrease mildly compared to centralized training.*

Table 2. Communication cost of LP-FL and FP-FL

Method	Tuned Params	Checkpoint	Training Memory
LP-FL(LoRA)	**0.78M**	**3.03 MB**	5.21 GB
LP-FL(QLoRA)	**0.78M**	**3.03 MB**	3.43 GB
FP-FL	336M	1.24 GB	8.86 GB

Efficiency Communication of LP-FL. Table 2 presents a comparison of communication costs between LP-FL and FP-FL. It clearly indicates that the total number of parameters to be transmitted in LP-FL accounts for only 0.23% (Tuned Params: 0.78M vs. 336M and Checkpoint: 3.03 MB vs. 1.24 GB) of FP-FL. As only the LoRA parameters need to be exchanged during LP-FL, the communication cost is notably lower. With more federated participants, this advantage becomes even more pronounced given the increased number of communication that the entire system undergoes during each FedAvg. More importantly, due to dequantized to bf16 before being utilized for training, LP-FL with QLoRA significantly reduces the required memory during training, which means more mobile devices can participate in federated learning.

Table 3. Comparison with FP-CT and FP-FL

Dataset	Labeled	LP-		FP-	
		CT	FL(2-clients)	CT	FL(2-clients)
IMDB	1%	0.850	**0.858**	0.855	0.850
	5%	0.869	0.867	**0.872**	0.864
Yelp	1%	0.913	**0.922**	0.914	0.916
	5%	0.914	**0.926**	0.922	0.924
Sent140	1%	**0.791**	0.769	0.772	0.766
	5%	**0.802**	0.798	0.789	0.776

Convergence and Stability of LP-FL. Throughout the experiments, we closely monitor the evolution of the global model's prediction accuracy during each round of global training.

Notably, we show in Fig. 3 an extreme case of using a minimal of 1% subset of the IMDB and Yelp datasets. Although the accuracy was low in the initial 1–2 rounds, it quickly rises to a reasonable level by just a few more rounds of training. This nice property of quick convergence can be partly attributed to our low-parameter methodology of training. That is, with limited training samples, we seek to fine-tune only about 0.23% of the total model parameters, preventing over-fitting to the tiny number of samples.

By using our iterative and semi-supervised methodology, we can enhance the utilization of unlabeled data in the presence of scarce labeled data. We establish a criterion that the prediction accuracy of the global model after FedAvg exceeds 70% before using it to annotate the unlabeled data. This approach allows us to capitalize on a greater quantity and higher quality of data for fine-tuning purposes. Furthermore, we ensure the completion of annotating all unlabeled data prior to initiating the final global update, thus ensuring consistency in the training data volume and fully exploiting all available local data.

As shown in Table 1, the final prediction accuracy of the global model of LP-FL, when compared to centralized training (third column), exhibits a performance loss of 2.8% (0.822 vs. 0.850) on IMDB, 1.3% (0.900 vs. 0.913) on Yelp and 5.7% (0.734 vs. 0.791) using 1% subset. In contrast, if we increase the labeled data to 10%, the performance loss diminishes to 1.9% (0.859 vs. 0.878) on IMDB, 1.3% (0.797 vs. 0.810) on Sent140 and becomes negligible on Yelp. These trends align with our expectations and falls within an acceptable range, thereby demonstrating the stability of LP-FL.

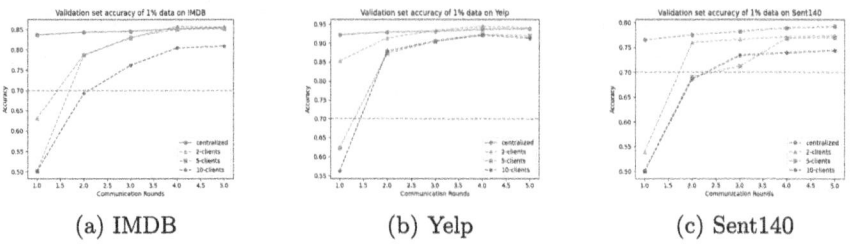

(a) IMDB   (b) Yelp   (c) Sent140

Fig. 3. Validation set accuracy of 1% data on LP-FL

4.3 Ablation Studies

We compare the performance of our method with Full-Parameter Centralized Training (FP-CT) and Full-Parameter Federated Learning (FP-FL) in Table 3. Based on the comparisons, we draw the following conclusions:

LP-CT is Equal or Better than FP-CT. As expected, the prediction accuracy of LP-CT is slightly lower than that of FP-CT on IMDB and Yelp, although the decrease is not substantial. The tinyest observed difference is only 0.1% (0.913 vs. 0.914), while the discrepancy between LP-CT and FP-CT is less than 0.8%, as shown in Table 3 column of LP-CT and FP-CT. The reason behind this is simple: FP-CT utilizes the complete set of model parameters for fine-tuning, enabling it to acquire more knowledge compared to LP-CT, which utilizes only approximately 0.23% of the parameters. Consequently, FP-CT exhibits superior performance. However, we also observe that the prediction accuracy of LP-CT is higher than that of FP-CT on Sent140 (0.791 vs. 0.772 and 0.802 vs. 0.789). We attribute this to FP-CT being influenced by a greater number of samples with labeling errors.

LP-FL is Equal or Better than LP-CT. The presence of noise induced by erroneous annotations, derived from the semi-supervised nature, is also observable in LP-CT and LP-FL. It has been noted in our investigations that the prediction accuracy of LP-CT is also lower than that LP-FL in certain settings (0.850 vs. 0.858 on IMDB, 0.913 vs. 0.922 and 0.914 vs. 0.926 on Yelp) as shown in Table 3 column of LP-CT and LP-FL. Although LP-CT reduces model complexity and mitigates the impact of erroneous annotations, the quantity of erroneous annotations still affects prediction accuracy. Centralized training models have access to a larger unlabeled dataset, leading to more erroneous annotations after labeling. In contrast, federated participants have tinyer local unlabeled datasets, resulting in fewer erroneous annotations. Given the same model complexity, a larger number of erroneous annotations increases the likelihood of over-fitting to incorrect labels, resulting in lower prediction accuracy.

FP-FL is Equal or Better than FP-CT. FP-FL demonstrates prediction accuracy that is comparable to, and sometimes even surpasses, that of FP-CT. This finding provides additional evidence that in a semi-supervised setting, a greater number of erroneous annotations result in increased noise, thereby impacting the model's prediction accuracy. Due to the relatively fewer erroneous annotations encountered by individual models in the federated environment compared to centralized training, FP-FL occasionally outperforms FP-CT.

We can also observe an unexpected conclusion in the table: the prediction accuracy of the LP-FL consistently outperforms the FP-FL in the federated environment (0.858 vs. 0.850 and 0.867 vs. 0.864 on IMDB, 0.922 vs. 0.916 and 0.926 vs. 0.924 on Yelp). This observation contradicts the majority of related work in centralized training. We attribute these findings to the semi-supervised nature of our approach. Although we annotate the unlabeled dataset only after achieving a high level of model prediction accuracy, there are still some erroneous samples present. Additionally, the large parameter size of the language model contributes to its complexity, making it more prone to memorizing noise. Under FP-FL, the erroneous annotations introduce more noise to the model, adversely affecting prediction accuracy. On the other hand, LP-FL only fine-tunes around 0.23% of the parameters, reducing the model's complexity and minimizing the impact of erroneous annotations.

Conclusion

We have validated that pre-trained large language models (LLMs) can attain reasonable classification accuracy in sentiment classification tasks with minimal labeled samples across different clients with federated learning paradigm. Our proposed Low-Parameter Federated Learning (LP-FL) permits federated clients to assign soft labels to unlabeled data, utilizing the evolving knowledge from the global model. Given the inherent constraints of mobile devices, such as limited computational resources and unreliable network environments, we achieve comparable, and sometimes superior performance to Full-Parameter Federated Learning (FP-FL) by fine-tuning and federating only a fraction of the model parameters at each local client. Our FP-FL approach establishes an effective learning framework for leveraging LLMs in the FL environment and shows promising applications over mobile devices.

References

1. Brown, T., et al.: Language models are few-shot learners. Adv. Neural. Inf. Process. Syst. **33**, 1877–1901 (2020)
2. Caldas, S., et al.: LEAF: a benchmark for federated settings. arXiv preprint arXiv:1812.01097 (2018)
3. Chowdhery, A., et al.: PALM: scaling language modeling with pathways. arXiv preprint arXiv:2204.02311 (2022)
4. Dettmers, T., Pagnoni, A., Holtzman, A., Zettlemoyer, L.: QLORA: efficient fine-tuning of quantized LLMs. Adv. Neural. Inf. Process. Syst. **36** (2024)
5. Fan, C., Hu, J., Huang, J.: Private semi-supervised federated learning. In: IJCAI, pp. 2009–2015 (2022)

6. Fan, C., Huang, J.: Federated few-shot learning with adversarial learning. In: 2021 19th International Symposium on Modeling and Optimization in Mobile, Ad hoc, and Wireless Networks (WiOpt), pp. 1–8. IEEE (2021)
7. Gao, M., Zuo, F., Wang, G.: Efficient differential privacy federated learning mechanism for intelligent selection of optimal privacy protection levels. In: Zhao, X., Yang, S., Wang, X., Li, J. (eds.) WISA 2022. LNCS, vol. 13579, pp. 603–614. Springer, Cham (2022). https://doi.org/10.1007/978-3-031-20309-1_53
8. Gao, T., Fisch, A., Chen, D.: Making pre-trained language models better few-shot learners. arXiv preprint arXiv:2012.15723 (2020)
9. Houlsby, N., et al.: Parameter-efficient transfer learning for NLP. In: International Conference on Machine Learning, pp. 2790–2799. PMLR (2019)
10. Hu, E.J., et al.: LoRA: low-rank adaptation of large language models. arXiv preprint arXiv:2106.09685 (2021)
11. Li, J., Wang, G., Hu, J., Wang, H., Zhang, H.: An IoT service development framework driven by business event description. In: Yuan, L., Yang, S., Li, R., Kanoulas, E., Zhao, X. (eds.) WISA 2023. LNCS, vol. 14094, pp. 527–538. Springer, Cham (2023). https://doi.org/10.1007/978-981-99-6222-8_44
12. Liu, X., Pang, T., Fan, C.: Federated prompting and chain-of-thought reasoning for improving LLMs answering. In: Jin, Z., Jiang, Y., Buchmann, R.A., Bi, Y., Ghiran, A.M., Ma, W. (eds.) KSEM 2023. LNCS, vol. 14120, pp. 3–11. Springer, Cham (2023). https://doi.org/10.1007/978-3-031-40292-0_1
13. Maas, A.L., Daly, R.E., Pham, P.T., Huang, D., Ng, A.Y., Potts, C.: Learning word vectors for sentiment analysis. In: ACL (2011)
14. McMahan, H.B., Moore, E., Ramage, D., Hampson, S., y Arcas, B.A.: Communication-efficient learning of deep networks from decentralized data. In: AISTATS (2017)
15. Pang, T., Tan, K., Yao, Y., Liu, X., Meng, F., Fan, C., Zhang, X.: REMED: retrieval-augmented medical document query responding with embedding fine-tuning. IJCNN (2024)
16. Pfeiffer, J., Kamath, A., Rücklé, A., Cho, K., Gurevych, I.: AdapterFusion: non-destructive task composition for transfer learning. arXiv preprint arXiv:2005.00247 (2020)
17. Rücklé, A., et al.: AdapterDrop: on the efficiency of adapters in transformers. arXiv preprint arXiv:2010.11918 (2020)
18. Schick, T., Schütze, H.: Exploiting cloze questions for few shot text classification and natural language inference. arXiv preprint arXiv:2001.07676 (2020)
19. Shin, T., Razeghi, Y., Logan IV, R.L., Wallace, E., Singh, S.: AutoPrompt: eliciting knowledge from language models with automatically generated prompts. arXiv preprint arXiv:2010.15980 (2020)
20. Thoppilan, R., et al.: LaMDA: language models for dialog applications. arXiv preprint arXiv:2201.08239 (2022)
21. Touvron, H., et al.: LLaMA: open and efficient foundation language models. arXiv preprint arXiv:2302.13971 (2023)
22. Zhao, Y., Li, M., Lai, L., Suda, N., Civin, D., Chandra, V.: Federated learning with non-IID data. arXiv preprint arXiv:1806.00582 (2018)
23. Zhao, Y., et al.: Personalized federated few-shot learning. IEEE Trans. Neural Netw. Learn. Syst. (2022)

The Journey of Language Models in Understanding Natural Language

Yuanrui Liu[1,2], Jingping Zhou[3], Guobiao Sang[2], Ruilong Huang[1], Xinzhe Zhao[1], Jintao Fang[2], Tiexin Wang[1], and Bohan Li[1(✉)]

[1] College of Artificial Intelligence and Computer Science and Technology, Nanjing University of Aeronautics and Astronautics, Nanjing 211106, China
bhli@nuaa.edu.cn
[2] Beijing Shenzhou Aerospace Software Technology Co., LTD., Beijing 100094, China
[3] Beijing Institute of Telemetry, Beijing 100083, China

Abstract. Since the Turing Test was proposed in the 1950s, humanity began exploring artificial intelligence, with an aim to bridge the interaction gap between machines and human language. This exploration enables machines to comprehend how humans acquire, produce, and understand language, as well as the relationship between linguistic expression and the world. The paper explores the basic principles of natural language representation, the formalization of natural language, and the modeling methods of language models. The paper analyzes, summarizes and compares the mainstream technologies and methods, including vector space-based, topic model-based, graph-based, and neural network-based approaches. And how to improve the development trend and direction of language model understanding ability is predicted and further discussed.

Keywords: Artificial intelligence · Natural language understanding · Vector space model · Topic model · Neural network · Deep learning

1 Introduction

In this interdisciplinary scientific era, many scientific achievements, especially the development of artificial intelligence (AI), have brought tremendous changes to human society. Particularly with the emergence and rapid growth of deep learning technology, language models have entered the stage of large-scale industrial application. Early research in language models primarily focused on learning algorithms. Traditional machine learning mainly relied on hand-designed features and statistical methods. Deep learning can automatically learn task-related features from data [1], but it also poses higher demands and challenges on computational capabilities. The technology for intelligent processing of natural language by computers aims to enable machines to understand and generate human language, facilitating smooth and equal communication with humans [2].

Human language is a medium for expressing thoughts and feelings, communicated through dialogue, writing, and other means. Formalizing natural language

© The Author(s), under exclusive license to Springer Nature Singapore Pte Ltd. 2024
C. Jin et al. (Eds.): WISA 2024, LNCS 14883, pp. 331–363, 2024.
https://doi.org/10.1007/978-981-97-7707-5_29

by using various data models, theories, and algorithms from different disciplines allows handling the complexity and variety in natural language. For instance, vectors, matrices, sets in mathematics; strings, linked lists, trees, graphs in computer science; neurons, neural networks in biology; propositions, predicates, productions, inference rules in logic; all can be used to formalize and analyze natural language [3]. These data models serve as symbolic systems representing the semantics and structure of natural language. Formalizing natural language also helps to reveal its essence and rules, improving human cognitive ability and communication efficiency.

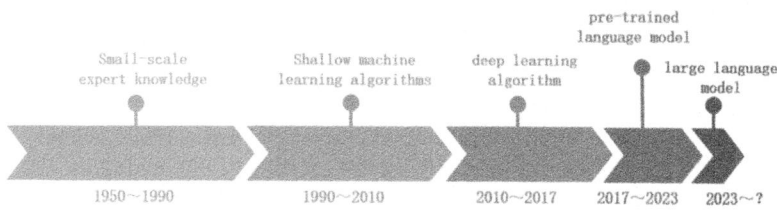

Fig. 1. Evolution of the natural language processing research paradigm

Since its inception, natural language processing has experienced five paradigm shifts (as shown in Fig. 1). Before the 1980s, rule-based language models were used, which were also known as grammar models [4–6]. Similar to programming language processing mechanisms, experts would induce and organize the syntactic rules of natural language, and computers would parse language texts based on these rules to judge their legality or generate legal texts according to the rules. From the late 1980s to the 2010s, statistical language models (SLM) became the mainstream method [7,8], viewing the appearance of linguistic units (such as words) in text sequences as random events, with the probability of a text sequence reflecting its adherence to natural language rules. In 2010, the earliest recurrent neural network (RNN) based language model, RNNLM, was proposed by Mikolov et al. [9]. An essential improvement of recurrent neural network language models was the introduction of attention mechanisms, allowing the model to focus on different historical information based on the current word, thereby enhancing the model's expressiveness and flexibility, leading to significant performance improvement. In 2017, Vaswani et al. introduced the Transformer model [10], a sequence-to-sequence model based entirely on self-attention and encoder-decoder structure, without using recurrent or convolutional neural networks. The Transformer model addressed issues such as inability to parallel computation, difficulty in capturing long-distance dependencies, vanishing or exploding gradients in traditional recurrent neural network models, breaking through the limitations of non-parallel computation in RNNML models, and improving training and inference efficiency. Compared to CNNML, the number of operations needed to compute the association between two positions does not grow with distance, better capturing long-distance dependencies in the sequence.

Self-attention generates more interpretable models, allowing inspection of attention distributions to analyze the model's focus on different positions within the input sequence.

In recent years, large models built on Transformers and other pre-training models have drawn inspiration from the abstraction and iteration ideas of speech and image, providing architectural and universal text representation and applications for natural language processing. Following this, pre-training techniques have made a series of breakthroughs in the NLP field, including models like BERT [11] and T5 [12], which have demonstrated significant performance improvement in many downstream tasks after training. Research on GPT-3 showed that scaling up the model can greatly enhance task-agnostic, few-shot performance, sometimes even comparable to previous state-of-the-art fine-tuning methods [13]. The substantial discoveries of the GPT-3 model have stimulated research on large-scale models and related technologies. The parameter scale of large models has quickly grown from billions to trillions and continues on a steep upward trend.

The remainder of this paper is organized as follows. In Sect. 2, we summarize the different forms of representation in language models, and in Sect. 3, we provide an overview of the types of language models. Section 4 introduces different model structures. We discuss the training process of LMs in Sect. 5. Section 6 introduces common evaluation methods, including intrinsic and extrinsic evaluations. Section 7 points out promising research directions for the future.

2 Representation of Language Models

Language models can be divided into different representations, specifically discrete representation and continuous representation. Discrete representation can only express shallow information of natural language, such as individual words or characters, or even radicals in Chinese, overlooking the contextual relationship of text objects [14, 15]. Continuous representation, also known as distributed representation, is fundamentally based on the distributional hypothesis theory proposed by Harris in 1954: "Words with similar contexts have similar semantics", and Firth's elaboration and demonstration of this theory in 1957: "The meaning of a word is determined by its context". Representing an object by other objects nearby in the text is a significant innovation in modern natural language processing. Human language expression focuses on semantic information; a word and its context together constitute precise semantic information. Extracting deep and complex features from the text usually involves the relationship between the text content objects and their context, and may even include auxiliary knowledge.

2.1 Discrete Representation in Language Models

The discrete representation of language models is a localized method, transforming text into a sequence of textual content objects. According to the target task, text objects are identified, and these objects are quantified as discrete features.

A notable example of discrete representation is the Bag-of-Words model (BOW), also referred to as the Vector Space Model (VSM), which represents documents or feature items as vectors. This model is intuitive and computationally simple, facilitating applications in natural language processing such as similarity measurement, text classification, and information retrieval. The simplest form of this is the "one-hot" encoding. One-hot representation of a word digitizes it into a vector, with the primary procedure including: first establishing a vocabulary, which can be generated from a text corpus or imported externally; then representing each word as a vector with dimensions equal to the length of the vocabulary, where the position corresponding to the word is set to 1, and all other positions are 0. Sentences or texts in one-hot representation also form a vector, with dimensions equal to the length of the dictionary, and based on the digitization of text features, they are mainly divided into two forms: boolean representation and count-based representation. The former consists of elements with values of 0 or 1, where in the digitization process, for the k-th element of the vector, its value is 1 if the k-th word in the vocabulary appears in the sentence or text, otherwise 0. The latter has non-negative integer values, where for the k-th element of the vector, its value corresponds to the occurrence count of the k-th word in the vocabulary within the sentence or text.

2.2   Continuous Representation in Language Models

Continuous representation language models are primarily divided into two mainstream methods: Matrix-based Distributed Representation and Neural Networkbased Distributed Representation.

Matrix-based Distributed Representation: The most common form of matrixbased distributed representation is the construction of a co-occurrence matrix M of size $V \times C$, where V is the vocabulary size, and C is the defined small context scope. For instance, in the representation of a word W, the context could be other words within a specific window, the sentence containing W, or the document containing it. If a sentence or chapter serves as the window scope, each row of the co-occurrence matrix M is the vector representation of the corresponding word W, and each column is the vector representation of the sentence or chapter. Examples include models such as Latent Semantic Analysis (LSA) and Latent Dirichlet Allocation (LDA), leading to different textual representations. Besides building a co-occurrence matrix [16] of text objects and their context, it is also possible to construct matrices representing grammatical or semantic relationships between text objects and their context.

Neural Network-based Distributed Representation: Unlike the algebra-based distributed representation that relies on matrix decomposition, neural networkbased distributed representation utilizes the nonlinear transformations of neural networks [17] to convert text into dense, low-dimensional, and continuous vectors. This approach can represent different text granularities, such as words, sentences, chapters, etc. The Neural Network Language Model (NNLM) is one of the earliest and most classic distributed representation models [18], which laid the foundation for Word2Vec [19], leading to derivatives like GloVe, fast-Text, ELMo, Transformer, GPT, and BERT. In recent years, advances in natural language processing, such as seq2seq models, attention mechanisms, particularly the combination of pre-training and fine-tuning techniques, have effectively integrated text representation learning with subsequent natural language processing tasks. This has been a milestone improvement both in text representation and specific natural language processing tasks [20], which will be elaborated in Sect. 5.

3   Types of Language Models

Language models in traditional language modeling (CLM) are primarily rule-based, employing vector spaces, thematic concepts, and graph-structured language models. Constructed on the statistical foundation of text features, these models are simple yet effective. Pre-trained language models (PLM), on the other hand, leverage neural networks to pre-train language, fine-tuning for different downstream tasks. PLM can utilize word vectors and deep learning techniques to capture complex linguistic characteristics. The following sections will analyze, summarize and summarize common language models.

3.1   Rule-Based Language Models

Rule-based language models refer to those that utilize manually defined grammatical rules and dictionaries to generate or parse natural language. These models typically rely on the knowledge and experience of linguistic experts and can make use of features such as punctuation, keywords, demonstratives, and more to construct rule templates. Language generation or understanding is then achieved through pattern matching. Advantages of rule-based language models include their ability to handle complex linguistic structures, ensure grammatical correctness to a certain extent, and avoid data sparsity issues that may be encountered in statistical models. However, there are also drawbacks to this approach. Rule-based models often require substantial human labor and maintenance costs, struggle to cover all linguistic phenomena, and have difficulties in dealing with ambiguous and uncertain situations.

3.2   Vector Space

The Vector Space Model (VSM) is a simple and effective language model, first proposed by Salton of Harvard University [21]. It's a discrete language model. First, the sentences in the corpus are tokenised, then the Bag of Words (BOW) is built, and the sentence sequence forms a Bag of Sentences (BOS) [22]. A document can be represented as a high-dimensional vector, where each dimension corresponds to a feature term, which can be a word or sentence in the dictionary. Term weights can be calculated using different formulas, such as Term Frequency (TF) or Inverse Document Frequency (IDF), to reflect the importance of the feature terms in the document. The dimension is the length of the feature term set, also known as the dictionary. In VSM, the smallest indivisible language units can be characters, words, phrases, etc., determined manually or automatically extracted or learned. From a set perspective, a document's content is a set of feature terms, represented as:

$$D = \{t_1, t_2, t_3, \ldots, t_n\} \tag{1}$$

where t_k is a feature term, and the term weight reflects its importance in the document. Thus, a document can be represented as a vector of feature terms and their corresponding weights, in the form: $Doc((t_1, w_1), (t_2, w_2), \ldots, (t_n, w_n))$, or simply $D = \{w_1, w_2, w_3, \ldots, w_n\}$, where w_k is the weight of feature term t_k.

In VSM, numerical weighting includes the boolean or one-hot model, where the feature weight can only take the binary values 0 or 1, and the document's vector dimension is the length of the dictionary. If the feature term's frequency in the document is considered, such as Term Frequency (TF), it is known as a document vector space representation based on TF. The vector weight can be absolute frequency or normalized frequency count, and the document's vector dimension is the dictionary's length. Similar representations include those based on TF-IDF, n-grams, etc., essentially differing in feature term segmentation and weight calculation methods, and the document's vector dimension is the number of elements in the feature term set [23–25].

3.3   Topics

Topic modeling is a text modeling approach that statistically analyzes the relationships among text, topics, and words at the vocabulary level, thereby facilitating the representation and analysis of text. The idea behind topic modeling mirrors the process of writing and reading articles. During writing, an author must determine the central idea, then organize material and vocabulary around various topics, constructing sentences according to grammatical rules. Topic models can be applied to various natural language processing tasks, such as text classification and summarisation [26–28].

By introducing a "topic" as a latent variable, topic models extend the bag-of-words approach, viewing a document as composed of multiple topics, each constituted by several words. Topic models map words or phrases sharing the same topic onto the same dimension. The likelihood that two different words belong to the same topic depends on the probability that they appear together in the same document or, given a topic, that they are generated with higher probability than other words. Topic models are a specialized type of probabilistic graphical model (PGM) supported by rigorous mathematical theory, and inference based on Gibbs sampling is simple and effective. Assuming there are K topics (usually manually determined, which may lead to potential issues with the model), a document is represented as a K-dimensional vector, with each dimension representing a topic, and the weight indicating the probability of the document in the corresponding topic. In this way, topic models compute the word distribution across topics within a text corpus and determine the weight of each article for different topics.

The most representative topic model is Latent Dirichlet Allocation (LDA), a probabilistic generative model that assumes every document is made up of multiple topics, each consisting of various words. By applying Bayesian inference to the document-word matrix, LDA yields both document-topic and topic-word distributions [29].

$$p(w, d) = p(d) \times \sum_z p(w \mid z)p(z \mid d) \tag{2}$$

$$p(w \mid d) = p(w \mid t) \times p(t \mid d) \tag{3}$$

The essential components of LDA include the corpus, document d, and word w. For each document within the corpus, LDA's generation process mainly includes three steps: (1) extracting a topic t from the topic distribution for each document; (2) drawing a word w from the word distribution corresponding to the chosen topic t; (3) repeating the above process until every word in the document has been traversed [30,31].

3.4   Graph-Based Methods

The earliest representation of text with graphs was proposed by Schenker et al. [32]. Nodes in the graph are English words after Web text parsing, and the edges are the co-occurrence relationships between nodes. The type of edges is based on 3 tags generated from HTML file parsing. The text graph constructed does not consider edge weight information. Models representing text with a graph structure, commonly referred to as GSM (Graph Space Model), preserve the structural information of the text.

GSM construction mainly includes three steps:

(1) Extracting text features to construct the node set V;
(2) Defining relationships between feature terms to determine the set of edges E;
(3) Quantifying nodes and edges as needed, including node attribute quantification and edge weight quantification.

The nodes can be characters, words, phrases, sentences, or other forms, and the total number of nodes equals the number of distinct feature terms, forming the node set V. Edges are formed by relationships between nodes in V, such as co-occurrence relationships. Text graphs can be directed or undirected, and all edges constitute the edge set E. Node attributes and edge weights are set according to task requirements. Besides co-occurrence graphs, syntactic or semantic relationship graphs can also be constructed [32–34].

3.5   Neural Networks

Neural networks were originally a biological concept. Artificial Intelligence has borrowed the structure and functionality of neural networks to design Artificial Neural Networks (ANNs). They consist of a large number of information processing units, also referred to as neurons, interconnected to form a network model, which abstracts, simplifies, and simulates the organizational structure and information processing mechanism of the human brain. Research in cognitive and neural science has found that the human brain abstracts and extracts semantic information from sensory input in a hierarchical manner. Inspired by this, deep learning based on neural networks captures combination features of external input through multilayer network interconnection and weight computation, extracting higher-level features and gradually learning effective feature representation from massive input data. Traditional machine learning methods usually divide complex tasks into many modules or stages, learning each one separately, such as natural language understanding [35], which often requires sentence splitting, word segmentation, part-of-speech tagging, syntactic analysis, semantic analysis, and reasoning. This approach has its shortcomings, as each module needs individual optimization, and the optimization goal may not align with the overall task objective. Additionally, error propagation between modules can cause a progressive amplification effect. Deep learning based on neural networks adopts end-to-end training or learning, without manual intervention or modular division, unifying optimization of the overall task objective. This approach avoids the pitfalls of traditional machine learning, enhancing efficiency and accuracy in handling complex tasks [36,37].

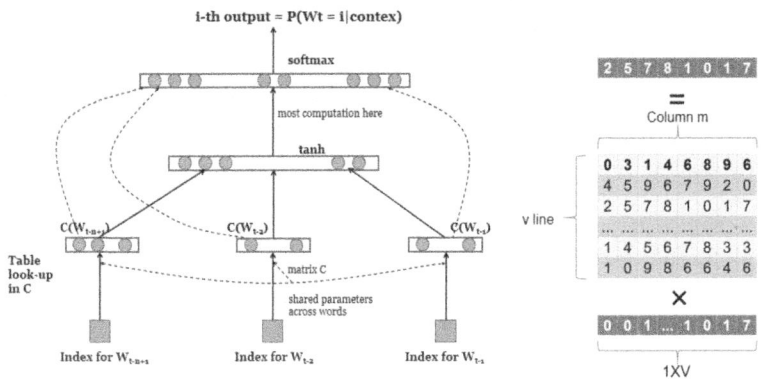

Fig. 2. Neural network language model

Neural networks represent a model for describing complex relationships between data. Deep learning is a method of machine learning that uses neural network models to learn useful features from data. Although often confused, neural networks and deep learning are not exactly the same. Deep learning is based on neural networks, but not all neural networks are deep. Deep neural networks have multiple hidden layers, allowing them to handle high-dimensional and unstructured data, such as speech, images, and videos. In natural language processing (NLP), deep neural networks can automatically extract lexical, syntactic, and semantic features from a large corpus of unlabelled text, facilitating better text understanding and generation. In the field of NLP, word embeddings are a method to convert words into numerical vectors, reflecting their semantic and syntactic information. The generation of word embeddings is typically based on neural network language models, such as NNLM, CBOW, Skip-gram, etc., which learn dense low-dimensional vector representations for each word from extensive text data. The dimensions of word embeddings are usually fixed and cannot be modified or interpreted individually. However, by calculating the similarity or distance between different word vectors, relationships between words, such as synonyms, antonyms, and hypernyms [38], can be discovered. Word embeddings are foundational techniques in NLP and are widely used in text classification, named entity recognition, machine translation, question-answering systems, etc. [39,40]. A basic implementation of NNLM is shown in Fig. 2 [41].

The model consists of an input layer, hidden layer, and output layer, taking the one-hot vector [42] representation of the $t-1$ to $t-n+1$ words w_{t-1}, w_{t-2}, ..., w_{t-n+1}, and predicting and outputting the embedding representation of the t-th word, w_t. The hidden layer maps each input word to a vector C(i) through the parameter matrix $C \in \mathbb{R}^{|V| \times m}$, where $C(i) \in \mathbb{R}^m$ represents the vector corresponding to the i-th word in the vocabulary, $|V|$ represents the number of words in the vocabulary, and m is the dimension of the vector.

4   Neural Model Structures

In the following sections, we will introduce several common neural network language model structures, including early feed-forward neural network language models [42], recurrent neural network language models, and language models under the Transformer framework.

4.1   Feed-Forward Neural Network (FNN) Models

Feed-forward Neural Network (FNN) Language Models (LMs) [43–46] were among the earlier neural network language models. The FNN LM takes historical context as input and outputs a probability distribution over words, as shown in Fig. 3. It includes three layers: input, hidden, and output. The input layer transforms each word in the context window into a continuous vector via a projection layer (or embedding matrix), then concatenates these vectors as input. The hidden layer applies nonlinear transformation, and the output layer uses the softmax function to obtain the posterior probability $P(w_j = i|h_j)$, that is, the probability of a word given a specific history. FNN LM uses a fixed window to gather fixed-length context.

Compared with clm, the language model of feedforward neural network structure has several advantages. First, it can effectively reduce the size of the thesaurus by using subword units, thus reducing the number of parameters and computational complexity of the model. Secondly, it can model the complex dependence relationship between words by multi-layer nonlinear transformation, and improve the expressiveness and prediction accuracy of the model [47]. Finally,

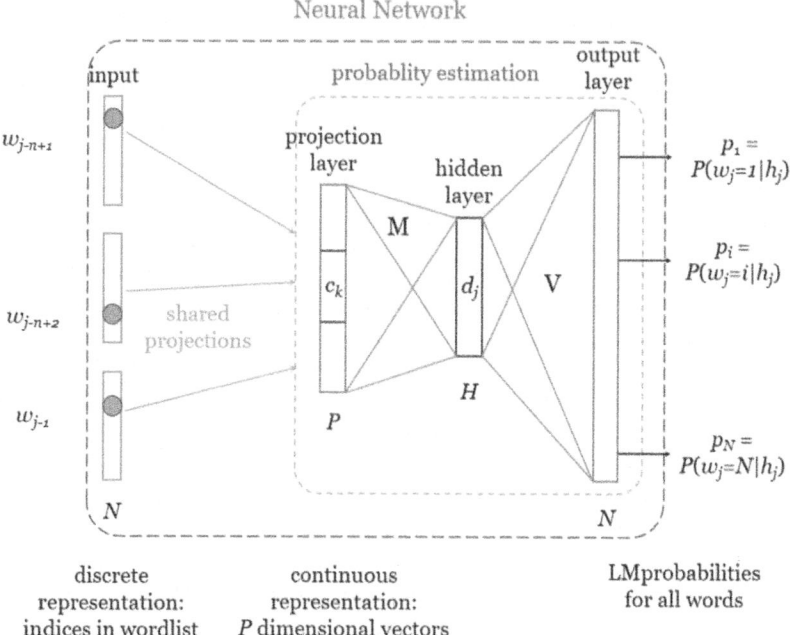

Neural Network

discrete
representation:
indices in wordlist

continuous
representation:
P dimensional vectors

LMprobabilities
for all words

Fig. 3. Neural network language model

it can realize distributed representation by mapping words to low-dimensional continuous Spaces, alleviating the problem of data sparsity and enhancing the generalization ability and robustness of the model.

4.2   Recurrent Neural Network (RNN) Models

Clearly, relying solely on fixed-length context to predict the next word is far from sufficient. In contrast to the finite historical context used in FNN language models, Recurrent Neural Network (RNN) language models [48–52] can utilize context of arbitrary length for predicting the next word.

The structure of an RNN language model is illustrated in Fig. 4. The word at position i, x(i), is first transformed into a one-hot representation, and then the recurrent hidden state h(i + 1) is calculated using the previous hidden state h(i) and the one-hot representation of the word x(i). The expression can be formulated as follows:

$$[h]h(i+1) = f(W \cdot \hat{x}(i) + U \cdot h(i)) \tag{4}$$

Here, $f(\cdot)$ is a nonlinear activation function, W is the weight matrix from the input layer to the hidden layer, and U represents the connections between the previous and current layers. Through iterative computation of hidden states, RNN LMs can encode contexts of varying lengths. Finally, the output layer provides the conditional probability of the word $y(t) = g(V \cdot h(t))$, where V is the weight matrix connecting the hidden and output layers, and $g(\cdot)$ is the softmax activation function.

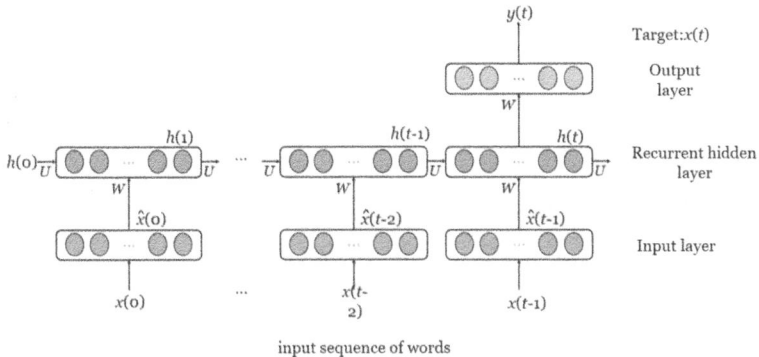

Fig. 4. Neural network language model

Theoretically, RNN LMs do not need the Markov Assumption [53]. They can use all previous history to predict the next word. However, the inherent gradient vanishing problem in RNNs hampers the model's learning [54]. Since the gradient may become extremely small over long distances, in practice, the model weights are actually updated by the nearby context. Generally speaking, RNN LMs struggle to learn dependencies between the current word and its distant historical context, although attention mechanisms can be introduced to alleviate this problem [55,56]. The intrinsic sequential nature of RNNs renders them less powerful compared to Transformer-based LMs equipped with self-attention mechanisms [57].

4.3   Transformer/Self-attention Models

The Transformer architecture [58] is a neural network model based on the self-attention mechanism, and it can efficiently process sequential data such as text, speech, and images. Unlike RNNs, which rely on sequential computation, the Transformer captures long-range dependencies and local features within the sequence using multi-head attention layers. One advantage of the Transformer is that it allows parallelized training and inference, thereby enhancing efficiency and performance. The Transformer's main components are the encoder and decoder. The encoder transforms the input text sequence into a set of vectors representing the context information for each token, while the decoder generates the target text sequence based on the encoder's output and its own input. Both encoder and decoder consist of several identical sub-layers, including multi-head attention layers, feed-forward network layers, residual connection layers, and layer normalization layers. Before entering the encoder, the text sequence must be embedded, and positional embeddings are added to preserve sequence order information. A special sub-layer in the decoder is the masked multi-head attention layer, which prevents the decoder from seeing future tokens, thereby maintaining its autoregressive property.

(a) BERT: Random tokens are replaced with masks, and the document is encoded bidirectionally. Missing tokens are predicted independently, so BERT cannot easily be used for generation.

(b) GPT: Tokens are predicted auto-regressively, meaning GPT can be used for generation. However words can only condition on leftward context, so it cannot learn bidirectional interactions.

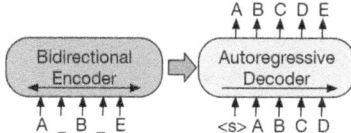

(c) BART: Inputs to the encoder need not be aligned with decoder outputs, allowing arbitary noise transformations. Here, a document has been corrupted by replacing spans of text with a mask symbols. The corrupted document (left) is encoded with a bidirectional model, and then the likelihood of the original document (right) is calculated with an autoregressive decoder. For fine-tuning, an uncorrupted document is input to both the encoder and decoder, and we use representations from the final hidden state of the decoder.

Fig. 5. Illustration of different transformer models, where BERT is the encoder-only model, GPT is the decoder-only model, and BART is the encoder-decoder model

Depending on their use, there are three variations of the Transformer model based on the encoder, decoder, or encoder-decoder structure, as shown in Fig. 5. Encoder-only models can access all positions of a given input and utilize bidirectional context to predict words. They are suitable for tasks requiring complete sentence understanding, such as text classification. Transformer decoder models can only use previous words to predict the current word (i.e., they are autoregressive models) and excel at text generation tasks. Transformer encoder-decoder models have access to all words in the encoding phase and words prior to the current word in the decoding phase, making them suitable for sequence-to-sequence tasks like translation and summarization.

5   Pre-training-Then-Fine-Tuning Paradigm

In the era of deep learning, the traditional training-testing paradigm starts from scratch to learn a model for a specific task. While training task-specific models is an effective approach in NLP, custom-designed models are not extensible to other tasks, and they also require a sufficient amount of labeled data for training. Additionally, the problem of overfitting has hindered the construction of large models.

To address this issue, the pre-training-then-fine-tuning paradigm [59] has been introduced, reshaping the NLP field. This paradigm involves initially training a universal language model on large-scale text and subsequently fine-tuning the model parameters for different downstream tasks. The significance of this approach lies in leveraging the language knowledge learned from the pre-trained models to enhance the performance of downstream tasks, saving computational resources and time.

Pre-training is a vital technique in deep learning. Its aim is to utilize large-scale unlabeled or labeled data to train a universal model with robust expressive power. This model then serves as the foundation for fine-tuning, tailored to various tasks or domains, to achieve improved performance and generalization capabilities.

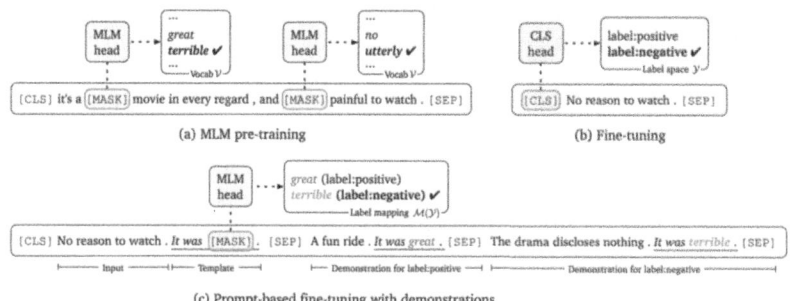

(a) MLM pre-training

(b) Fine-tuning

(c) Prompt-based fine-tuning with demonstrations

Fig. 6. An illustration of (a) LM pre-training, (b) standard fine-tuning, and (c) prompt-based fine-tuning (or prompt-tuning)

Pre-training has multiple advantages. Firstly, it can make full use of massive data resources, learning general features and knowledge, thus enhancing the model's expressive and generalization abilities. Secondly, it alleviates the issue of insufficient labeled data, reducing the amount of data required for specific tasks or domains, and saving manpower and time costs. Lastly, it can enhance the model's transferability across different tasks or domains, simplifying the adaptation process for new tasks, and increasing development efficiency.

Pre-training has widespread applications in the field of Natural Language Processing (NLP). Models trained on extensive text data can accomplish a variety of tasks such as text summarization, translation, sentiment analysis, and more. Based on the pre-training tasks, large models can be categorized into three types: autoencoding modeling, autoregressive language modeling, and seq2seq modeling. In the following sections, we will introduce these three types of language models and their respective pre-training tasks.

5.1   Autoencoding Modeling

Autoencoding modeling is an unsupervised learning method capable of learning word representations from a large amount of unlabeled text. The fundamental idea behind this approach allows the model to generate its own input, thus known as autoencoding. However, if the model were to simply copy the input, it would not learn any useful information. To avoid this, it may be necessary to consider the semantics and logic of the entire sentence. To make full use of the context, autoencoding modeling sometimes corrupts parts of the text, requiring the model to recover the damaged text. This technique can generate bidirectional word representations, taking into account both left and right contextual information. Large models based on autoencoding have become predominant in natural language understanding, with BERT being a milestone model.

BERT utilizes a bidirectional Transformer encoder to generate word representations. To implement autoencoding modeling, BERT introduced the objective of Masked Language Modeling (MLM). Essentially, MLM randomly masks a word in a sentence and attempts to predict the masked word. For example, in the sentence "my dog is hairy," the word "hairy" may be masked with the [MASK] token, and the model is asked to predict this word from the context. BERT selects 15% of the words in the corpus for prediction.

However, one major problem with MLM is that the specific [MASK] token does not appear during fine-tuning. To prevent the model from focusing too much on the [MASK] token, BERT replaces the target word with [MASK] 80% of the time, with random words 10% of the time, and leaves it unchanged 10% of the time. Unlike traditional language models that only predict the next word from left to right or right to left, MLM can fully utilize bidirectional contextual information. Thus, the introduction of MLM in BERT allows the model to capture both global semantics and local syntax of the text, enhancing its expressiveness and generalization ability.

To overcome the problem of vocabulary out-of-bound issues, BERT represents token inputs by summing three types of embeddings:

Word Embeddings: BERT employs WordPiece to divide words into subwords and utilizes word embeddings to represent the context of each subword. This segmentation strategy helps BERT to adaptively deal with various word forms and nuances in different contexts.

Segment Embeddings: In scenarios where there is more than one sentence, such as in Natural Language Inference (NLI) and Question Answering (QA), BERT uses segment embeddings to distinguish between them. This allows for a better understanding of sentence boundaries and relations in multi-sentence tasks.

Position Embeddings: Consistent with traditional Transformers, BERT adds position embeddings to encode the position information of words within a sequence. This spatial information enriches the model's comprehension of the syntax and relationships between words in a sentence.

Though powerful in capturing semantic and syntactic information of text, BERT is not without limitations and deficiencies. To address these issues, many researchers have proposed a series of improvements based on BERT, including four representative BERT variants: RoBERTa [60], DistilBERT [61], ALBERT [62], and KEPLER [63].

RoBERTa, one of the most influential BERT variants, stands for Robustly Optimized BERT Pretraining Approach. By adopting various training techniques and more extensive data, RoBERTa demonstrated that BERT was significantly undertrained. Examples of improvements include larger batch sizes, longer training duration, more datasets, and higher learning rates. RoBERTa outperformed BERT in almost all benchmark tests, despite having the same architecture.

For improving pretraining and inference efficiency, DistilBERT, ALBERT, and TinyBERT [64] have made notable contributions in reducing parameters and accelerating processes. These models significantly lower resource consumption while minimizing performance loss. DistilBERT is a knowledge distillation method, compressing a large model (such as BERT) into a smaller one (like DistilBERT), retaining 95% of language understanding capabilities with only 60% of the parameters. ALBERT is a parameter-sharing method, which shares parameters across all or intra-group layers, thereby reducing the number of parameters and memory usage. TinyBERT is a two-stage learning method, involving pretraining on general data and fine-tuning on task-specific data, utilizing attention mechanisms between teacher and student models to transfer knowledge.

To incorporate external knowledge into BERT, efforts like ERNIE [65] and KnowBERT [66] have utilized embeddings of entities and relationships from knowledge bases to attain rich representations. These models can handle tasks requiring common or domain knowledge, such as entity linking, relationship extraction, etc. KEPLER further co-models knowledge and language through Wikidata descriptions to produce enhanced knowledge embeddings and text representations. The core idea of KEPLER is to transform knowledge base triplets (Entity-Relationship-Entity) into natural language sentences and co-train them with original text.

In summary, these four BERT variants have improved and extended BERT from different angles, demonstrating the vast potential and application value of pretrained models in the natural language processing field (Fig. 7).

5.2 Autoregressive Modeling

Auto Regressive (AR) modeling is a sequential probability-based language modeling method that predicts the probability distribution of the current word based on preceding words. This method aligns well with the nature of language modeling. It is worth noting that AR modeling differs from recurrent modeling, as previous words are provided to the model as input, not as hidden states, thus involving only feedforward networks. Compared to Autoencoding (AE) pretraining, AR pretraining preserves the original form of natural language without using masked tokens, thus reducing the divergence between pretraining and fine-tuning. Moreover, the conditional independence of masked words does not affect AR pretraining. However, AR modeling's unidirectional encoding may not fully utilize bidirectional context information, possibly hindering language understanding.

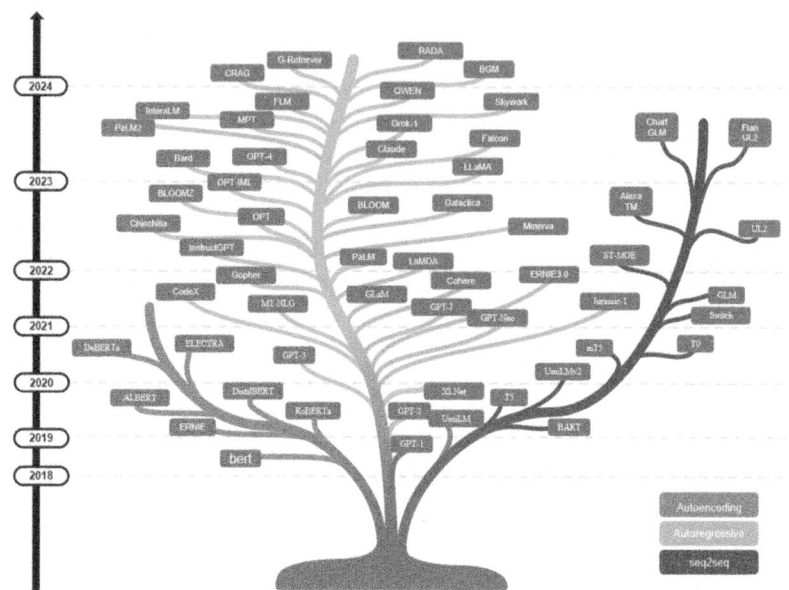

Fig. 7. An evolutionary graph of the research work conducted on LLM

GPT is a typical AR modeling approach and one of the earliest large-scale language models. Unlike BERT, GPT employs a Transformer decoder instead of an encoder to perform the task of predicting the next word. GPT sets a window size of k, asking the model to predict the current word given the preceding k words. GPT's token input consists of token embeddings and position embeddings, with no need for segment embeddings, as GPT can process multiple input paragraphs repeatedly for tasks like multiple-choice and sentence similarity. GPT has achieved excellent results on both large-scale and small-scale downstream datasets. In particular, due to its robust generative capabilities, GPT has even demonstrated remarkable performance in some zero-shot tests.

Subsequent developments introduced GPT-2 [67] and GPT-3 [68], highlighting the immense potential of larger parameter scales and pre-trained corpora. In text generation, GPT models have achieved impressive success, yet the loss of control over generated content may introduce potential risks. Further AR models like CTRL [69] address this issue by estimating the domain of the unsupervised pre-training corpus and controlling the style of the output. To break the constraint of fixed content length and capture long-term dependencies in text, Transformer-XL [70] employs segment-level recurrence and relative positional encoding to resolve context segmentation challenges. However, efficiency tends to decline as models grow larger and input text becomes longer. The Reformer [71] reduces complexity by introducing LSH attention modules and the RevNet framework.

OpenAI has achieved two significant milestones in the field of language models: ChatGPT and GPT-4 [72]. ChatGPT is a conversational model based on the GPT architecture (including GPT-3.5 and GPT-4), specifically optimized for dialogue capability. It excels in communication with humans, possessing rich knowledge, mathematical reasoning abilities, and the ability to accurately understand the context of multi-turn conversations. Later, ChatGPT also supported a plugin mechanism, further enhancing its integration capability with existing tools or applications, making it considered one of the most powerful chatbots to date. GPT-4 extends text input to multimodal signals, performing even better in tackling complex tasks compared to previous models. Through six months of iterative calibration, GPT-4's responses to malicious or provocative queries are also more secure [73]. Additionally, OpenAI has introduced a series of intervention strategies to mitigate potential issues such as hallucinations, privacy concerns, and overreliance, ensuring the safety and reliability of GPT-4. These two milestones have profound significance in advancing future artificial intelligence research, providing important insights for exploring human-like artificial intelligence systems.

Such models typically have billions of parameters. Apart from LLaMA [74] and LLaMA2 [75] (with a maximum of 70B parameters), NLLB [76] (with a maximum of 54.5B parameters), and Falcon [77] (with a maximum of 40B parameters), the parameter range of these models mostly falls between 10B to 20B. Other models within this range include mT5 [78], PanGu-α [79], T0 [80], gpt-neo-20b [81], CodeGen [82], UL2 [83]and mT0 [84]. Among them, FlanT5 (11B version) is the preferred model for instruction tuning research, exploring three aspects: increasing the number of tasks, scaling model size, and fine-tuning with chain of thought prompt data. CodeGen (11B version) is a self-regressive language model designed for code generation, introducing a new benchmark, MTPB, specifically for multi-turn program synthesis, containing 115 expert-generated questions. Recently, CodeGen2 has been released for exploring the impact of choices in model architecture, learning algorithms, and data distribution on the model. StarCoder [85] is another model specifically designed to study coding abilities, and it has achieved excellent results. For multilingual tasks, mT0 (13B version) may be a good candidate model, as it has been fine-tuned for multilingual tasks with multilingual prompts. Additionally, PanGu-α developed based on the deep learning framework MindSpore [86] performs well in zero-shot or few-shot settings for downstream tasks in Chinese. It is worth noting that PanGu-α contains multiple versions of models, with the largest publicly available version having 13B parameters. LLaMA (65B version), as a popular large-scale language model, has demonstrated superior performance in instruction-following tasks. Compared to LLaMA, LLaMA2 has conducted further exploration in reinforcement learning based on human feedback and has developed a chat-oriented version called LLaMA-chat [74]. This version generally outperforms existing open-source models across a range of utility and safety benchmarks.

5.3   Sequence-to-Sequence Modeling

Sequence-to-sequence (seq2seq) modeling is a method utilizing an encoder-decoder framework to carry out text transformation. The encoder encodes an input sequence into a vector, and the decoder predicts the next output word based on this vector and previously generated words. Seq2seq models can handle discrepancies in length and nature between input and output sequences, such as in machine translation and text summarization. However, challenges exist in seq2seq models, such as exposure bias and training efficiency. Techniques like scheduled sampling and beam search have been proposed to alleviate these issues by introducing randomness and diversity during training.

Traditional seq2seq models often rely on RNNs, which face difficulties in parallel computation. The Transformer introduces an elegant attention mechanism, allowing parallel computation in the encoder. It has not only become the backbone for subsequent seq2seq LLMs but also a staple in popular AE and AR models. BART [87], a large Transformer-based seq2seq model, can execute both BERT and GPT functionalities. Its encoder is bidirectional, akin to BERT, while its decoder operates left-to-right, similar to GPT. During pre-training, BART employs an innovative denoising strategy by randomly shuffling the original sentence order and replacing text segments with a mask token. Unlike BERT, BART's corruption methods are more flexible and diverse. BART excels in text generation tasks and is suitable for comprehension tasks, achieving performance comparable to RoBERTa on GLUE and SQuAD, and setting new benchmarks on various summarization, dialogue, and question-answering tasks, with a 6% point increase in ROUGE scores.

Pegasus [88] shares a similar structure with BART, but includes two additional objectives concerning MLM and gap sentence generation. T5 [78] is another representative large seq2seq model where position embeddings are replaced by relative position embeddings. It unifies various tasks in a generative form. In addition to BERT-style masked prediction tasks, T5 pre-training also employs supervised tasks differentiated by prefixes. T0 [80] is a multi-task training language model that enhances generalization to unseen tasks by training on various NLP tasks. Switch is a language model by Google with one trillion parameters, which utilizes Expert-Mixture-of-Experts (MoE) [89] routing technology to improve sample efficiency and training speed. GLM [90] is a general language model that improves pretraining through autoregressive blank filling, achieving excellent performance on various NLP tasks. ST-MoE [91] focuses on designing stable and transferable sparse expert models, achieving state-of-the-art performance in multiple natural language processing tasks. UL2 [83] is a unified pretraining framework that combines various denoising pretraining objectives and introduces the concept of pattern switching to adapt to different downstream fine-tuning schemes. AlexaTM [92] is a multi-language sequence-to-sequence language model by Amazon with 2 billion parameters, demonstrating state-of-the-art performance on multilingual zero-shot tasks.

5.4   Fine-Tuning and Prompt-Tuning

After pre-training, LLMs can attain general capabilities for various tasks. However, increasing research shows that LLM abilities can be further customized for specific goals. Two primary methods for adapting pre-trained LLMs are prompt tuning and alignment tuning. The former aims to enhance (or unlock) LLM capabilities, while the latter aligns LLM behavior with human values or preferences.

Fine-Tuning. In the fine-tuning process, there are several common techniques, with prompt learning and instruction tuning being particularly important.

Prompt Tuning: Prompt tuning is a method based on natural language prompts. A short natural language description serves as a prompt, guiding the model to generate specific outputs. The model is then fine-tuned based on the discrepancy between the generated outputs and the actual targets.

Instruction Tuning: Instruction tuning is a fine-tuning approach [93] founded on instructions. Here, the model is provided with an instruction sequence describing the task it needs to complete. Fine-tuning is then performed based on the model's performance in executing these instructions. The advantage of instruction tuning lies in its ability to help the model adapt to new tasks more swiftly. By providing instructions, we can clearly guide the model's focus, thus enhancing learning efficiency. Moreover, instruction tuning can improve model generalization, allowing better performance with limited data.

Human Feedback Reinforcement Learning (HFRL). Human-Feedback Reinforcement Learning (HFRL) is an algorithm [94] combining human feedback and reinforcement learning aimed at enhancing an agent's learning performance in complex environments. In HFRL, human users guide the learning process of the intelligent agent by providing real-time feedback, such as rewards or punishments.

HFRL consists of several key components:

Agent: The core part of HFRL, responsible for performing tasks in the environment. The agent takes action according to the current state and learns from rewards or punishments signals from the environment.

Environment: An interactive scenario where the agent interacts and performs tasks. The environment can provide real-time feedback on the agent's actions, such as rewards or punishments.

Human Feedback: Human users guide the agent's learning process by offering real-time feedback. The feedback may be rewards (such as praise or encouragement) or punishments (such as criticism or penalties), representing the agent's impact on the environment.

Reinforcement Learning Algorithm: Responsible for training the agent based on its performance in the environment. Common algorithms include Q-learning, Deep Q-Networks (DQNs), etc.

The primary advantage of HFRL lies in its ability to help agents adapt more quickly and achieve better performance in complex environments. By integrating human feedback, agents can better understand human needs and goals, thus improving learning efficiency. Additionally, HFRL can enhance an agent's generalization ability, enabling it to adapt better to new tasks and environments.

However, HFRL also presents challenges, such as designing effective human feedback mechanisms and handling potential noise and uncertainties in human feedback. In practical applications, HFRL algorithms must be designed and adjusted according to the specific characteristics of the task and environment.

Prompt-Tuning. A prompt is a text designed by humans, which is concatenated with the original text to form the new model input. Figure 6 [95] illustrates the pre-training, fine-tuning, and prompt-tuning processes of MLMs. During pre-training, MLMs are trained to predict masked words, assuming that the downstream task involves sentiment analysis of movie reviews. In standard fine-tuning, a new head is trained on top of a PLM to predict sentiment labels. The original input, appended with a designed prompt such as 'It was,' is sent to the PLM. The PLM must assign probabilities to predefined answers, such as 'great' or 'terrible'. If the probability of 'great' is higher, the input label will be positive, and vice versa [96]. Thus, prompt-tuning transforms a distinct downstream task into a word prediction task, bridging the gap between the pre-training and fine-tuning stages.

6   Model Evaluation

Evaluating models is critically important for the advancement of natural language processing (NLP). Unlike the evaluation of traditional models, the assessment of large-scale language models is unique and often challenging to capture with a single metric.

6.1   Intrinsic Evaluation

A good language model should assign higher probabilities to natural text sequences and lower probabilities to unrealistic or random sequences. Perplexity (PPL) is a commonly used metric [97] that evaluates the performance of language models by measuring the inverse probability of a given text sequence, normalized by the number of words [98]. Lower perplexity signifies that the model will be more accurate in text generation tasks.

However, in bidirectional language models, merely maximizing the joint probability of the entire text sequence is insufficient. Therefore, other intrinsic evaluation metrics are needed. A common metric is BLEU (Bilingual Evaluation Understudy) [99], which measures the matching degree by calculating the N-gram overlap between generated text and reference answers.

Specifically, for each N-gram (from 1 to N), we first transform the reference answers and generated text into N-gram sequences, then calculate the number of identical N-grams in both. Next, we calculate the precision (P) and recall (R) for each N-gram, i.e., the ratio of identical N-grams to the total N-grams in the generated text and reference answers. From these values, we can compute the F1 score for each N-gram. Finally, we weight the F1 scores by the N-gram weights and the length of the generated text to obtain the BLEU score. The specific formula is as follows:

$$P = \frac{\text{Number of identical N-grams in generated text and reference}}{\text{Total N-grams in generated text}} \tag{5}$$

$$R = \frac{\text{Number of identical N-grams in generated text and reference}}{\text{Total N-grams in reference}} \tag{6}$$

$$F1 = \frac{2 \cdot P \cdot R}{P + R} \tag{7}$$

$$BLEU = (w_1 \cdot F_1 + w_2 \cdot F_2 + \ldots + w_n \cdot F_n) \cdot \exp\left(1 - \frac{c}{r}\right) \tag{8}$$

where w_i is the N-gram weight, F_n is the F1 score for the N-gram, c is the length of the generated text, and r is the maximum overlap in the reference N-grams.

By incorporating the BLEU metric, we can more comprehensively assess the quality and degree of match in the generated text by bidirectional language models.

6.2   Extrinsic Evaluation

Any downstream task of language models (LMs) can be used for extrinsic evaluation. To comprehensively evaluate the performance of LMs, downstream tasks are often used as benchmarks for extrinsic evaluation. In this regard, GLUE (General Language Understanding Evaluation) [100] and SuperGLUE [101] are two widely adopted benchmarks.

GLUE is a natural language understanding benchmark designed to test model performance across various tasks. It encompasses multiple single-sentence tasks, such as sentiment classification, semantic similarity, and sentence matching, as well as reasoning tasks like natural language inference and multi-sentence inference. GLUE offers a unified set of metrics for assessing model performance on these tasks. By evaluating on the GLUE dataset, one can intuitively understand and compare model performance across different tasks.

SuperGLUE is an enhanced version of GLUE, aiming to provide more challenging language understanding tasks and diverse task formats. SuperGLUE includes new tasks like multiple-choice reading comprehension, natural language inference, and causal reasoning. These tasks cover more complex language understanding and reasoning capabilities, demanding higher generalization and common sense reasoning abilities from models. Furthermore, SuperGLUE improves resources and public leaderboards, making performance comparisons between different models more accurate and reliable.

Through benchmarks like GLUE and SuperGLUE, we can holistically evaluate language models on various tasks and compare and rank models. These benchmarks provide a standardized evaluation framework and metrics, helping researchers and developers better understand and gauge the practical application abilities of language models.

6.3   Relationship Between Intrinsic and Extrinsic Evaluation

Does a lower perplexity in an LM necessarily imply that it will perform well in downstream tasks? In other words, is there a correlation between pre-training tasks (based on word prediction) and subsequent tasks? This question has been explored in various empirical studies, but theoretical research is limited.

Empirical Research. Researchers have designed experiments to understand what knowledge LMs learn from pre-training tasks. For instance [102–107], they employ part-of-speech tagging, constituent tagging, and dependency tagging to measure the degree of syntactic knowledge learned, and named entity tagging, semantic role tagging, and semantic prototype role tagging to test semantic knowledge. Empirical studies have demonstrated that pre-training tasks facilitate LMs in learning linguistic knowledge such as syntax [105] and semantic roles [107]. However, these experimental results only serve as evidence that word prediction tasks are beneficial for downstream tasks, without explaining the underlying mechanisms.

Theoretical Research. Some scholars have attempted to establish mathematical connections between an LM's perplexity and its performance on downstream tasks. Literature [108] has explored text classification tasks, initially positing and then validating that text classification can be reformulated as a sentence completion task. Since the LM's pre-training task is essentially sentence completion, it indeed aids the downstream text classification task. In [109], a generative underlying model was used to exhibit the relationship between pre-training and downstream tasks. Current theoretical studies are confined to considering specific downstream tasks (such as text classification) and are proven valid under certain conditions.

7   Open Issues and Challenges

In just a few short years, large-scale pre-trained language models (LLMs) have rapidly evolved from non-existent technology to ubiquitous entities in the field of machine learning. They demonstrate human-like text generation capabilities, sparking significant interest and applications in many domains. However, the sudden rise of these technologies has also brought about many challenges and concerns.

7.1   Data Complexity and Scale

In the era of large models, the complexity and scale of training data pose a significant challenge. These models typically rely on large-scale text data from the internet for training, which is not only vast in quantity but also complex and diverse in content, filled with metaphors, slang, technical terms, and cultural differences, making comprehensive understanding and review extremely difficult [110]. Training these large language models is a computationally intensive task that requires a significant investment of computational resources, including high-performance hardware and corresponding energy supply. This enormous demand for resources not only increases research and development costs but also raises concerns about environmental impact due to energy consumption and carbon emissions during hardware manufacturing processes [111].

7.2   Bias and Undesirable Output

In the research and application of large language models (LLMs), issues of bias and unintended outputs are increasingly prominent. These problems often stem from inherent biases in the training datasets, which can manifest in the model's outputs once learned [112]. The manifestations of such biases can vary and may include unpleasant, discriminatory, or harmful content. Addressing and mitigating these issues are crucial to ensuring responsible deployment of artificial intelligence. Furthermore, ethical concerns regarding the use of large language models have not been fully resolved. Challenges remain in content filtering, auditing, and accountability mechanisms, as pre-trained models may generate misinformation, hate speech, and biased content [113]. Therefore, ongoing research and development efforts are necessary to ensure that model outputs are fair, unbiased, and compliant with ethical and societal standards.

7.3   Generalization and Few-Shot Learning

Despite achieving satisfactory performance with ample data support, large language models exhibit limited generalization capabilities when dealing with small samples or tasks requiring deep domain knowledge. This phenomenon reveals the models' vulnerability when facing novel or low-resource environments. Consequently, researchers are actively working to enhance the models' learning and adaptation capabilities in limited data contexts, a research direction crucial for advancing the reliability and effectiveness of models in practical applications [114].

7.4   Multimodal Integration

As artificial intelligence applications continue to expand, multimodal integration has become an increasingly important research area. Current trends demand that models possess capabilities not only to understand and generate textual information but also to handle images, audio, and other types of media [115]. This demand for multimodal integration drives model design towards higher levels of complexity, including the acquisition, synchronized processing, and comprehensive evaluation of diverse modal data [116]. Integrating multiple modalities of data into a single model framework not only raises higher demands on the quality and diversity of data but also presents new challenges for model training algorithms and evaluation mechanisms.

7.5   Evaluation Complexity

Evaluating the performance and limitations of large language models (LLMs) is a highly challenging task. In the current evaluation framework, many metrics face difficulties in capturing subtle performance differences of models, leading to widespread questioning of the effectiveness of evaluation methods [117]. Furthermore, concerns about the manipulability and potential deception of evaluation metrics further exacerbate worries about the accuracy of model performance evaluation. To obtain accurate assessments of LLM performance, it is essential to develop and adopt more robust and trustworthy evaluation strategies.

8   Future Research Directions

Recent developments in Large Language Models (LLMs) have positioned them as a pivotal research area, prompting exploration into key focuses and directions aimed at addressing prevalent challenges and unresolved issues. By tackling these issues, there is potential to fully harness the capabilities of LLMs while ensuring their responsible and ethical integration within the AI landscape.

8.1   Advancing Bias Mitigation

Researchers are dedicated to optimizing training data to minimize algorithmic biases to the maximum extent. They are developing effective technologies to eliminate these biases and formulating a series of responsible AI development guidelines [118]. Furthermore, to ensure fairness and justice in the system, they also plan to integrate continuous monitoring and auditing mechanisms into the AI development process. This commitment to reducing bias not only drives advancements in the capabilities of large language models (LLMs) but also ensures adherence to their ethical standards. Researchers are also working hard to establish ethical and legal regulatory frameworks to address evolving regulatory challenges in the field of artificial intelligence. These frameworks will serve as guiding principles for the use of LLMs, ensuring compliance with data protection and privacy regulations. They effectively address previous concerns regarding ethical AI deployment, laying a solid foundation for future development [119].

8.2   Efficiency Optimization

In the rapid development of artificial intelligence, enhancing the training efficiency of large language models (LLMs) has become a central research topic. Researchers are actively exploring and implementing efficient training techniques to address challenges related to resource consumption and environmental impact. Federated learning, as an innovative distributed training method, allows models to be trained across multiple decentralized data sources [120], effectively improving data utilization rates. Additionally, knowledge distillation techniques applied to model compression open new avenues for reducing the computational burden and environmental costs of LLMs [121]. Furthermore, researchers are exploring pruning techniques for structural optimization to reduce the complexity and computational costs of LLMs. These technological advancements not only drive technical innovations in LLMs but also pave the way for building more sustainable and efficient artificial intelligence systems.

8.3   Interpretable Models

Despite the dominance of deep learning based LMs in NLP, they are inherently non-transparent blackbox approaches. Their interpretability warrants attention. There have been efforts to explain blackbox LMs. As mentioned in Sect. 6.3, we conduct empirical studies to understand what PLMs learn through experimental design. This approach can reveal some inner workings of PLMs, but also has limitations. For example, empirical studies often rely on specific tasks, datasets, and evaluation metrics, so may not generalize to other settings. Moreover, empirical studies cannot fully explain the mathematical principles and theoretical foundations of PLMs either. Therefore, we need more mathematical tools and methods to analyze and understand PLM behaviors and performance.

However, progress in this direction may provide some insights but not a satisfactory and unambiguous answer. Providing theoretical explanations or developing interpretable LMs remains a challenging and open problem. One direction for interpretability is to design an interpretable learning model from scratch. For instance, we can incorporate KGs with LMs. In many reasoning tasks like information retrieval [122] and recommender systems [123], KGs have proven powerful. KGs can provide logical paths to justify each prediction, making predictions from LMs more interpretable.

8.4   Multimodal LLMs

Researchers are at the forefront of developing multimodal large language models (MM-LLMs), which integrate information from text, vision, and other modalities. MM-LLMs can extract information from images, videos, and audio to generate relevant text [124], significantly advancing new directions in artificial intelligence applications and effectively meeting the demand for multisensory understanding. MM-LLMs have made significant progress in transitioning from single-modal input/output to versatile "any-to-any" functionalities. These models not only retain the inherent reasoning and decision-making capabilities of LLMs but also support various multimodal tasks such as image captioning and visual question answering [125]. Furthermore, research has found significant improvements in alignment, instruction adaptation, and conversational abilities. These advancements indicate that MM-LLMs will become a pathway towards achieving general artificial intelligence. With ongoing technological advancements, we can expect MM-LLMs to offer more powerful and user-friendly interfaces in the future, allowing users to input and communicate information in a more flexible manner through support for multimodal inputs [126].

8.5   Dynamic Evaluation Metrics and Relevant Benchmarks

The SuperBench comprehensive evaluation framework for large models, jointly developed by Tsinghua University's Basic Model Research Center and Zhongguancun Laboratory, aims to provide objective and scientific evaluation standards. Additionally, the TrustLLM framework comprehensively analyzes the credibility of LLMs, including principles for trustworthy LLMs across different dimensions and a new testing benchmark. Researchers are actively exploring dynamic evaluation metrics and related benchmarks to adapt to continuous changes in language and environment, ensuring accurate assessment of large language model (LLM) performance [127]. They are searching for appropriate measurement standards and developing new benchmarks to overcome limitations in early artificial intelligence capability assessments.

9   Conclusion

This paper outlines the development of language models, analyzes the connection between natural language representation and machine learning, especially deep learning. Statistical natural language processing systems generally consist of training data and statistical models, and traditional machine learning methods face challenges such as high data annotation costs, low quality, and data sparsity, as well as difficulties in feature extraction. Although deep learning still has problems such as insufficient theoretical foundation and poor interpretability, its neural network structure based on multi-layer nonlinear transformations can automatically learn feature representations. Deep learning transforms natural language processing problems from discrete symbol spaces to continuous numerical spaces, changing the definition and solution of problems, and promoting the progress of natural language processing. The current technical trend is to use larger-scale models and more unsupervised text to acquire human knowledge. Finally, this paper points out the future research directions, emphasizing the need for interpretable, trustworthy, and domain-specific language models.

Acknowledgments. This work is supported in part by the "14th Five-Year Plan" Civil Aerospace Pre-Research Project of China under Grant No. D020101, the Natural Science Foundation of China No. 62302213, Innovation Funding of Key Laboratory of Intelligent Decision and Digital Operations No. NJ2023027, Ministry of Industrial and Information Technology Project of Hebei Key Laboratory of Software Engineering, No. 22567637H, the Natural Science Foundation of Jiangsu Province under Grant No. BK20210280.

References

1. Han, X., et al.: Pre-trained models: Past, present and future. AI Open **2**, 225–250 (2021)
2. Ray, J., Johnny, O., Trovati, M., Sotiriadis, S., Bessis, N.: The rise of big data science: a survey of techniques, methods and approaches in the field of natural language processing and network theory. Big Data Cogn. Comput. **2**(3), 22 (2018)
3. Liu, J., Lin, L., Ren, H., Gu, M., Wang, J., Youn, G., Kim, J.-U.: Building neural network language model with POS-based negative sampling and stochastic conjugate gradient descent. Soft. Comput. **22**, 6705–6717 (2018)
4. Vaswani, A., et al.: Attention is all you need. In: Advances in Neural Information Processing Systems, vol. 30 (2017)
5. Devlin, J., Chang, M.-W., Lee, K., Toutanova, K.: BERT: pre-training of deep bidirectional transformers for language understanding. arXiv preprint arXiv:1810.04805 (2018)
6. Raffel, C., et al.: Exploring the limits of transfer learning with a unified text-to-text transformer. J. Mach. Learn. Res. **21**(1), 5485–5551 (2020)
7. Bendersky, M., Croft, W.B.: Modeling higher-order term dependencies in information retrieval using query hypergraphs. In: Proceedings of the 35th International ACM SIGIR Conference on Research and Development in Information Retrieval, pp. 941–950 (2012)
8. Sidorov, G.: Syntactic n-Grams in Computational Linguistics. Springer, Heidelberg (2019). https://doi.org/10.1007/978-3-030-14771-6
9. Mikolov, T., Karafiát, M., Burget, L., Cernocký, J., Khudanpur, S.: Recurrent neural network based language model. In: Interspeech, vol. 2, pp. 1045–1048. Makuhari (2010)
10. Waswani, A., et al.: Attention is all you need. In: NIPS (2017)
11. Devlin, J., Kenton, M.-W.C., Toutanova, L.K.: Bert: Pre-training of deep bidirectional transformers for language understanding. In: Proceedings of naacL-HLT, vol. 1, p. 2 (2019)
12. Roberts, A., et al.: Exploring the limits of transfer learning with a unified text-to-text transformer (2019)
13. Brown, T., et al.: Language models are few-shot learners. In: Advances in Neural Information Processing Systems, vol. 33, pp. 1877–1901 (2020)
14. Granados, A.: Analysis and study on text representation to improve the accuracy of the normalized compression distance. arXiv preprint arXiv:1205.6376 (2012)
15. Dourado, Í.C., Galante, R., Gonçalves, M.A., da Silva Torres, R.: Bag of textual graphs (BoTG): a general graph-based text representation model. J. Assoc. Inf. Sci. Technol. **70**(8), 817–829 (2019)
16. Lin, J.-J.: Applying a co-occurrence matrix to automatic inspection of weaving density for woven fabrics. Text. Res. J. **72**(6), 486–490 (2002)
17. Lek, S., Delacoste, M., Baran, P., Dimopoulos, I., Lauga, J., Aulagnier, S.: Application of neural networks to modelling nonlinear relationships in ecology. Ecol. Model. **90**(1), 39–52 (1996)

18. Ding, Z., Qiu, X., Zhang, Q., Huang, X.: Learning topical translation model for microblog hashtag suggestion. In: Twenty-Third International Joint Conference on Artificial Intelligence (2013)
19. Mikolov, T., Sutskever, I., Chen, K., Corrado, G.S., Dean, J.: Distributed representations of words and phrases and their compositionality. In: Advances in Neural Information Processing Systems, vol. 26 (2013)
20. Nielsen, M.A.: Neural Networks And Deep Learning, vol. 25. Determination Press, San Francisco (2015)
21. Salton, G., Wong, A., Yang, C.-S.: A vector space model for automatic indexing. Commun. ACM **18**(11), 613–620 (1975)
22. Wong, S.K.M., Ziarko, W., Wong, P.C.N.: Generalized vector spaces model in information retrieval. In: Proceedings of the 8th Annual International ACM SIGIR Conference on Research and Development in Information Retrieval, pp. 18–25 (1985)
23. Chew, P.A., Bader, B.W., Helmreich, S., Abdelali, A., Verzi, S.J.: An information-theoretic, vector-space-model approach to cross-language information retrieval. Nat. Lang. Eng. **17**(1), 37–70 (2011)
24. Tsatsaronis, G., Panagiotopoulou, V.: A generalized vector space model for text retrieval based on semantic relatedness. In: 2009 Proceedings of the Student Research Workshop at EACL, pp. 70–78 (2009)
25. Dong, R.F., Liu, C.A., Yang, G.T.: TF-IDF based loop closure detection algorithm for SLAM. J. Southeast Univ. **2**, 251–258 (2019)
26. Hajjem, M., Latiri, C.: Combining IR and LDA topic modeling for filtering microblogs. Procedia Comput. Sci. **112**, 761–770 (2017)
27. Liu, Z., Huang, W., Zheng, Y., Sun, M.: Automatic keyphrase extraction via topic decomposition. In: Proceedings of the 2010 conference on empirical methods in natural language processing, pp. 366–376 (2010)
28. Li, Y., Liu, T., Jiang, J., Zhang, L.: Hashtag recommendation with topical attention-based LSTM. In: Proceedings of COLING 2016, the 26th International Conference on Computational Linguistics: Technical Papers, pp. 3019–3029 (2016)
29. Blei, D.M., Ng, A.Y., Jordan, M.I.: Latent Dirichlet allocation. J. Mach. Learn. Res. **3**(Jan), 993–1022 (2003)
30. Pu, X., Jin, R., Wu, G., Han, D., Xue, G.R.: Topic modeling in semantic space with keywords. In: Proceedings of the 24th ACM International on Conference on Information and Knowledge Management, pp. 1141–1150 (2015)
31. Siu, M., Gish, H., Chan, A., Belfield, W., Lowe, S.: Unsupervised training of an hmm-based self-organizing unit recognizer with applications to topic classification and keyword discovery. Comput. Speech Lang. **28**(1), 210–223 (2014)
32. Schenker, A., Last, M., Bunke, H., Kandel, A.: Graph representations for web document clustering. In: Perales, F.J., Campilho, A.J.C., de la Blanca, N.P., Sanfeliu, A. (eds.) IbPRIA 2003. LNCS, vol. 2652, pp. 935–942. Springer, Heidelberg (2003). https://doi.org/10.1007/978-3-540-44871-6_108
33. Sonawane, S.S., Kulkarni, P.A.: Graph based representation and analysis of text document: a survey of techniques. Int. J. Comput. Appl. **96**(19) (2014)
34. Chen, Y., Lu, H., Qiu, J., Wang, L.: A tutorial of graph representation. In: Sun, X., Pan, Z., Bertino, E. (eds.) ICAIS 2019, Part I. LNCS, vol. 11632, pp. 368–378. Springer, Cham (2019). https://doi.org/10.1007/978-3-030-24274-9_33
35. Allen, J.: Natural Language Understanding. Benjamin-Cummings Publishing Co., Inc. (1995)
36. Jing, K., Xu, J.: A survey on neural network language models. arXiv preprint arXiv:1906.03591 (2019)

37. Liu, H., Zhang, Y., Wang, Y., Lin, Z., Chen, Y.: Joint character-level word embedding and adversarial stability training to defend adversarial text. In: Proceedings of the AAAI Conference on Artificial Intelligence , vol. 34, pp. 8384–8391 (2020)
38. Bengtson, E., Roth, D.: Understanding the value of features for coreference resolution. In: Proceedings of the 2008 Conference on Empirical Methods in Natural Language Processing, pp. 294–303 (2008)
39. Hinton, G.E., et al.: Learning distributed representations of concepts. In: Proceedings of the Eighth Annual Conference of the Cognitive Science Society, vol. 1, p. 12. Amherst, MA (1986)
40. Levy, O., Goldberg, Y.: Neural word embedding as implicit matrix factorization. In: Advances in Neural Information Processing Systems, vol. 27 (2014)
41. Bengio, Y., Ducharme, R., Vincent, P.: A neural probabilistic language model. Advances in Neural Information Processing Systems, vol. 13 (2000)
42. Rodríguez, P., Bautista, M.A., Gonzalez, J., Escalera, S.: Beyond one-hot encoding: lower dimensional target embedding. Image Vis. Comput. **75**, 21–31 (2018)
43. Bengio, Y., Senécal, J.-S.: Adaptive importance sampling to accelerate training of a neural probabilistic language model. IEEE Trans. Neural Netw. **19**(4), 713–722 (2008)
44. Schwenk, H., Gauvain, J.-L.: Training neural network language models on very large corpora. In Proceedings of Human Language Technology Conference and Conference on Empirical Methods in Natural Language Processing, pp. 201–208 (2005)
45. Schwenk, H.: Continuous space language models. Comput. Speech Lang. **21**(3), 492–518 (2007)
46. Arisoy, E., Sainath, T.N., Kingsbury, B., Ramabhadran, B.: Deep neural network language models. In: Proceedings of the NAACL-HLT 2012 Workshop: Will We Ever Really Replace the N-gram Model? On the Future of Language Modeling for HLT, pp. 20–28 (2012)
47. Idrissi, N., Zellou, A.: A systematic literature review of sparsity issues in recommender systems. Soc. Netw. Anal. Min. **10**, 1–23 (2020)
48. Kombrink, S., Mikolov, T., Karafiát, M., Burget, L.: Recurrent neural network based language modeling in meeting recognition. In: Interspeech, vol. 11, pp. 2877–2880 (2011)
49. Mikolov, T., Kombrink, S., Burget, L., Černocký, J., Khudanpur, S.: Extensions of recurrent neural network language model. In: 2011 IEEE international conference on acoustics, speech and signal processing (ICASSP), pp. 5528–5531. IEEE (2011)
50. Chen, X., Ragni, A., Liu, X., Gales, M.J.F.: Investigating bidirectional recurrent neural network language models for speech recognition. In: Proceedings of Interspeech 2017, pp. 269–273. International Speech Communication Association (ISCA) (2017)
51. Sundermeyer, M., Schlüter, R., Ney, H.: LSTM neural networks for language modeling. In: Thirteenth Annual Conference of the International Speech Communication Association (2012)
52. Yang, Z., Dai, Z., Salakhutdinov, R., Cohen, W.W.: Breaking the softmax bottleneck: a high-rank RNN language model. arXiv preprint arXiv:1711.03953 (2017)
53. Kalbfleisch, J.D., Lawless, J.F.: The analysis of panel data under a Markov assumption. J. Am. Stat. Assoc. **80**(392), 863–871 (1985)
54. Hochreiter, S.: The vanishing gradient problem during learning recurrent neural nets and problem solutions. Int. J. Uncertain. Fuzziness Knowl.-Based Syst. **6**(02), 107–116 (1998)

55. Bahdanau, D., Cho, K., Bengio, Y.: Neural machine translation by jointly learning to align and translate. arXiv preprint arXiv:1409.0473 (2014)
56. Deng, H., Zhang, L., Wang, L.: Global context-dependent recurrent neural network language model with sparse feature learning. Neural Comput. Appl. **31**, 999–1011 (2019)
57. Edelman, B.L., Goel, S., Kakade, S., Zhang, C.: Inductive biases and variable creation in self-attention mechanisms. In: International Conference on Machine Learning, pp. 5793–5831. PMLR (2022)
58. Subakan, C., Ravanelli, M., Cornell, S., Bronzi, M., Zhong, J.: Attention is all you need in speech separation. In: ICASSP 2021-2021 IEEE International Conference on Acoustics, Speech and Signal Processing (ICASSP), pp. 21–25. IEEE (2021)
59. Wang, W., et al.: StructBERT: incorporating language structures into pre-training for deep language understanding. arXiv preprint arXiv:1908.04577 (2019)
60. Liu, Y., et al.: RoBERTa: a robustly optimized BERT pretraining approach. arXiv preprint arXiv:1907.11692 (2019)
61. Sanh, V., Debut, L., Chaumond, J., Wolf, T.: DistilBERT, a distilled version of BERT: smaller, faster, cheaper and lighter. arXiv preprint arXiv:1910.01108 (2019)
62. Lan, Z., Chen, M., Goodman, S., Gimpel, K., Sharma, P., Soricut, R.: ALBERT: a lite BERT for self-supervised learning of language representations. arXiv preprint arXiv:1909.11942 (2019)
63. Wang, X., et al.: KEPLER: a unified model for knowledge embedding and pre-trained language representation. Trans. Assoc. Comput. Linguist. **9**, 176–194 (2021)
64. Jiao, X., et al.: TinyBERT: distilling BERT for natural language understanding. arXiv preprint arXiv:1909.10351 (2019)
65. Zhang, Z., Han, X., Liu, Z., Jiang, X., Sun, M., Liu, Q.: ERNIE: enhanced language representation with informative entities. arXiv preprint arXiv:1905.07129 (2019)
66. Peters, M.E., et al.: Knowledge enhanced contextual word representations. arXiv preprint arXiv:1909.04164 (2019)
67. Radford, A., et al.: Language models are unsupervised multitask learners. OpenAI blog **1**(8), 9 (2019)
68. Gao, T., Fisch, A., Chen, D.: Making pre-trained language models better few-shot learners. arXiv preprint arXiv:2012.15723 (2020)
69. Keskar, N.S., McCann, B., Varshney, L.R., Xiong, C., Socher, R: CTRL: a conditional transformer language model for controllable generation. arXiv preprint arXiv:1909.05858 (2019)
70. Dai, Z., Yang, Z., Yang, Y., Carbonell, J., Le, Q.V., Salakhutdinov, R.: Transformer-XL: attentive language models beyond a fixed-length context. arXiv preprint arXiv:1901.02860 (2019)
71. Kitaev, N., Kaiser, Ł., Levskaya, A.: Reformer: the efficient transformer. arXiv preprint arXiv:2001.04451 (2020)
72. OpenAI, R.: GPT-4 technical report. arXiv:2303.08774 (2023). View in Article, 2(5)
73. Ganguli, D., et al. Red teaming language models to reduce harms: methods, scaling behaviors, and lessons learned. arXiv preprint arXiv:2209.07858 (2022)
74. Touvron, H., et al. LLaMA: open and efficient foundation language models. arXiv preprint arXiv:2302.13971 (2023)
75. Touvron, H., et al. LLaMA 2: open foundation and fine-tuned chat models. arXiv preprint arXiv:2307.09288 (2023)

76. Costa-jussà, M.R., et al.: No language left behind: Scaling human-centered machine translation. arXiv preprint arXiv:2207.04672 (2022)
77. Almazrouei, E., et al.: Falcon-40B: an open large language model with state-of-the-art performance (2023). https://falconllmtii.ae. 2022
78. Xue, L., et al.: mT5: a massively multilingual pre-trained text-to-text transformer. arXiv preprint arXiv:2010.11934 (2020)
79. Zeng, W., et al.: PanGu-α: large-scale autoregressive pretrained chinese language models with auto-parallel computation. arXiv preprint arXiv:2104.12369 (2021)
80. Sanh, V., et al.: Multitask prompted training enables zero-shot task generalization. arXiv preprint arXiv:2110.08207 (2021)
81. Black, S., et al.: GPT-NeoX-20B: An open-source autoregressive language model. arXiv preprint arXiv:2204.06745 (2022)
82. Nijkamp, E., et al.: CodeGen: an open large language model for code with multi-turn program synthesis. arXiv preprint arXiv:2203.13474 (2022)
83. Tay, Y., et al.: UL2: unifying language learning paradigms. arXiv preprint arXiv:2205.05131 (2022)
84. Muennighoff, N., et al.: Crosslingual generalization through multitask finetuning. arXiv preprint arXiv:2211.01786 (2022)
85. Li, R., et al.: StarCoder: may the source be with you! arXiv preprint arXiv:2305.06161 (2023)
86. Huawei Technologies Co., Ltd.: Huawei MindSpore AI development framework. In: Huawei Technologies Co., Ltd. (eds.) Artificial Intelligence Technology, pp. 137–162. Springer, Singapore (2022). https://doi.org/10.1007/978-981-19-2879-6_5
87. Lewis, M., et al.: BART: denoising sequence-to-sequence pre-training for natural language generation, translation, and comprehension. arXiv preprint arXiv:1910.13461 (2019)
88. Zhang, J., Zhao, Y., Saleh, M., Liu, P.: PEGASUS: pre-training with extracted gap-sentences for abstractive summarization. In: International Conference on Machine Learning, pp. 11328–11339. PMLR (2020)
89. Shazeer, N., et al.: Outrageously large neural networks: the sparsely-gated mixture-of-experts layer. arXiv preprint arXiv:1701.06538 (2017)
90. Du, Z., et al.: GLM: general language model pretraining with autoregressive blank infilling. arXiv preprint arXiv:2103.10360 (2021)
91. Zoph, B., et al.: ST-MoE: designing stable and transferable sparse expert models. arXiv preprint arXiv:2202.08906 (2022)
92. Soltan, S., et al.: AlexaTM 20B: few-shot learning using a large-scale multilingual seq2seq model. arXiv preprint arXiv:2208.01448 (2022)
93. Liu, H., Li, C., Wu, Q., Lee, Y.J.: Visual instruction tuning. arXiv preprint arXiv:2304.08485 (2023)
94. Griffith, S., Subramanian, K., Scholz, J., Isbell, C.L., Thomaz, A.L.: Policy shaping: integrating human feedback with reinforcement learning. In: Advances in Neural Information Processing Systems, vol. 26 (2013)
95. Gao, T., Fisch, A., Chen, D.: Making pre-trained language models better few-shot learners. In: Zong, C., Xia, F., Li, W., Navigli, R., (eds.) Proceedings of the 59th Annual Meeting of the Association for Computational Linguistics and the 11th International Joint Conference on Natural Language Processing (Volume 1: Long Papers), pp. 3816–3830. Association for Computational Linguistics (2021)
96. Li, Z., Song, M., Zhu, Y., Zhang, L.: Chinese nested named entity recognition based on boundary prompt. In: Yuan, L., Yang, S., Li, R., Kanoulas, E., Zhao, X.

(eds.) WISA 2023. LNCS, vol. 14094, pp. 331–343. Springer, Singapore (2023). https://doi.org/10.1007/978-981-99-6222-8_28

97. Hao, Y., Mendelsohn, S., Sterneck, R., Martinez, R., Frank, R.: Probabilistic predictions of people perusing: evaluating metrics of language model performance for psycholinguistic modeling. arXiv preprint arXiv:2009.03954 (2020)

98. Mnih, A., Teh, Y.W.: A fast and simple algorithm for training neural probabilistic language models. arXiv preprint arXiv:1206.6426 (2012)

99. Papineni, K., Roukos, S., Ward, T., Zhu, W.-J.: BLEU: a method for automatic evaluation of machine translation. In: Proceedings of the 40th Annual Meeting of the Association for Computational Linguistics, pp. 311–318 (2002)

100. Wang, A., Singh, A., Michael, J., Hill, F., Levy, O., Bowman, S.R.: GLUE: a multi-task benchmark and analysis platform for natural language understanding. arXiv preprint arXiv:1804.07461 (2018)

101. Wang, A., et al.: SuperGLUE: a stickier benchmark for general-purpose language understanding systems. In: Advances in Neural Information Processing Systems, vol. 32 (2019)

102. Tenney, I., et al.: What do you learn from context? Probing for sentence structure in contextualized word representations. arXiv preprint arXiv:1905.06316 (2019)

103. Giulianelli, M., Harding, J., Mohnert, F., Hupkes, D., Zuidema, W.: Under the hood: using diagnostic classifiers to investigate and improve how language models track agreement information. arXiv preprint arXiv:1808.08079 (2018)

104. Tenney, I., Das, D., Pavlick, E.: BERT rediscovers the classical NLP pipeline. arXiv preprint arXiv:1905.05950 (2019)

105. Kim, T., Choi, J., Edmiston, D., Lee, S.: Are pre-trained language models aware of phrases? Simple but strong baselines for grammar induction. arXiv preprint arXiv:2002.00737 (2020)

106. Hewitt, J., Manning, C.D.: A structural probe for finding syntax in word representations. In: Proceedings of the 2019 Conference of the North American Chapter of the Association for Computational Linguistics: Human Language Technologies, Volume 1 (Long and Short Papers), pp. 4129–4138 (2019)

107. Rogers, A., Kovaleva, O., Rumshisky, A.: A primer in bertology: what we know about how BERT works. Trans. Assoc. Comput. Linguist. **8**, 842–866 (2021)

108. Saunshi, N., Malladi, S., Arora, S.: A mathematical exploration of why language models help solve downstream tasks. arXiv preprint arXiv:2010.03648 (2020)

109. Wei, C., Xie, S.M., Ma, T.: Why do pretrained language models help in downstream tasks? an analysis of head and prompt tuning. In: Advances in Neural Information Processing Systems, vol. 34, pp. 16158–16170 (2021)

110. Fahad, N.M., Sakib, S., Raiaan, M.A.K., Mukta, M.S.H.: SkinNet-8: an efficient CNN architecture for classifying skin cancer on an imbalanced dataset. In: 2023 International Conference on Electrical, Computer and Communication Engineering (ECCE), pp. 1–6. IEEE (2023)

111. Zhu, X., Li, J., Liu, Y., Ma, C., Wang, W.: A survey on model compression for large language models. arXiv preprint arXiv:2308.07633 (2023)

112. Motoki, F., Neto, V.P., Rodrigues, V.: More human than human: measuring ChatGPT political bias. Public Choice **198**(1), 3–23 (2024)

113. Zhu, L., Xu, X., Lu, Q., Governatori, G., Whittle, J.: AI and ethics—operationalizing responsible AI. In: Chen, F., Zhou, J. (eds.) Humanity Driven AI, pp. 15–33. Springer, Cham (2022). https://doi.org/10.1007/978-3-030-72188-6_2

114. Meng, Y., Michalski, M., Huang, J., Zhang, Y., Abdelzaher, T., Han, J.: Tuning language models as training data generators for augmentation-enhanced few-shot learning. In: International Conference on Machine Learning, pp. 24457–24477. PMLR (2023)

115. Molenaar, I., de Mooij, S., Azevedo, R., Bannert, M., Järvelä, S., Gašević, D.: Measuring self-regulated learning and the role of AI: five years of research using multimodal multichannel data. Comput. Hum. Behav. **139**, 107540 (2023)

116. Azevedo, R., Gašević, D.: Analyzing multimodal multichannel data about self-regulated learning with advanced learning technologies: issues and challenges (2019)

117. He, C., et al.: UltraEval: a lightweight platform for flexible and comprehensive evaluation for LLMs. arXiv preprint arXiv:2404.07584 (2024)

118. Werder, K., Ramesh, B., Zhang, R.: Establishing data provenance for responsible artificial intelligence systems. ACM Trans. Manage. Inf. Syst. (TMIS) **13**(2), 1–23 (2022)

119. Iqbal, U., Kohno, T., Roesner, F.: LLM platform security: applying a systematic evaluation framework to OpenAI's ChatGPT plugins. arXiv preprint arXiv:2309.10254 (2023)

120. Jiang, J., Liu, X., Fan, C.: Low-parameter federated learning with large language models. arXiv preprint arXiv:2307.13896 (2023)

121. Sun, S., Cheng, Y., Gan, Z., Liu, J.: Patient knowledge distillation for BERT model compression. arXiv preprint arXiv:1908.09355 (2019)

122. Dietz, L., Xiong, C., Dalton, J., Meij, E.: The second workshop on knowledge graphs and semantics for text retrieval, analysis, and understanding (KG4IR). In: The 41st International ACM SIGIR Conference on Research & Development in Information Retrieval, pp. 1423–1426 (2018)

123. Yang, Y., Huang, C., Xia, L., Li, C.: Knowledge graph contrastive learning for recommendation. In: Proceedings of the 45th International ACM SIGIR Conference on Research and Development in Information Retrieval, pp. 1434–1443 (2022)

124. Zhang, Z., Zhang, A., Li, M., Zhao, H., Karypis, G., Smola, A.: Multimodal chain-of-thought reasoning in language models. arXiv preprint arXiv:2302.00923 (2023)

125. Zheng, G., Yang, B., Tang, J., Zhou, H.-Y., Yang, S.: DDCoT: duty-distinct chain-of-thought prompting for multimodal reasoning in language models. In: Advances in Neural Information Processing Systems, vol. 36, pp. 5168–5191 (2023)

126. Pan, L., et al.: Learn to explain: multimodal reasoning via thought chains for science question answering. In: Advances in Neural Information Processing Systems, vol. 35, pp. 2507–2521 (2022)

127. Liu, Z., Zhang, Y., Li, P., Liu, Y., Yang, D.: Dynamic LLM-agent network: an LLM-agent collaboration framework with agent team optimization. arXiv preprint arXiv:2310.02170 (2023)

Instruction Tuning Large Language Models for Multimodal Relation Extraction Using LoRA

Zou Li, Ning Pang, and Xiang Zhao[✉]

Laboratory for Big Data and Decision, National University of Defense Technology,
Changsha, China
{pangning14,xiangzhao}@nudt.edu.cn

Abstract. The rapid proliferation of multimodal data on social platforms, particularly text-image pairs, has necessitated advanced methodologies for Multimodal Relation Extraction (MRE) to accurately identify semantic relations between annotated entities. Traditional approaches to MRE often suffer from limited generalizability across different datasets. To address the challenge, we propose to Instruction-Tune Large Language Models for MRE (ITMRE), which further utilizes Low-Rank Adaptation (LoRA) to tailor the Multimodal Large Language Models (MLLMs) for MRE. Our approach simplifies the extraction process using a two-stage multiple-choice question-answer template to first identify the types of two annotated entities in the text and then infer their relations. This structured methodology, combined with targeted LoRA tuning, allows for efficient model optimization without extensive retraining. We demonstrate the effectiveness of ITMRE through comprehensive experiments on the MNRE dataset, where it significantly outperforms existing methods and leading-edge language models such as GPT-3.5 and GPT-4 in terms of precision, recall, and F1 score, with an improvement in F1 scores by over 8%.

Keywords: Multimodal Relation Extraction · Instruction Tuning · Multimodal Large Language Models

1 Introduction

Multimodal relation extraction (MRE) combines data from multiple modalities, such as text and images, to identify the semantic relation between two annotated entities. Over recent years, data on social platforms often appears in the form of text accompanied by images, providing ample data sources for MRE. Therefore, as a useful tool for knowledge structuring, MRE attracts increasing interest and serves a series of applications, including question answering systems, information retrieval, etc.

Current MRE methods primarily rely on deep learning technologies, focusing on the fusion and alignment of image and text representations. Figure 1 shows an

C. Jin et al. (Eds.): WISA 2024, LNCS 14883, pp. 364–376, 2024.
https://doi.org/10.1007/978-981-97-7707-5_30

example of MRE, as well as the basic pipeline of the existing MRE methods and the method proposed in this paper. Chen et al. [1] have experimented with vision-prefix guided fusion methods to utilize hierarchical multiscale visual features. Zhao et al. [2] have tried a tiered multimodal fusion framework to minimize the impact of less relevant visual objects in images on the relation extraction task, achieving state-of-the-art results.

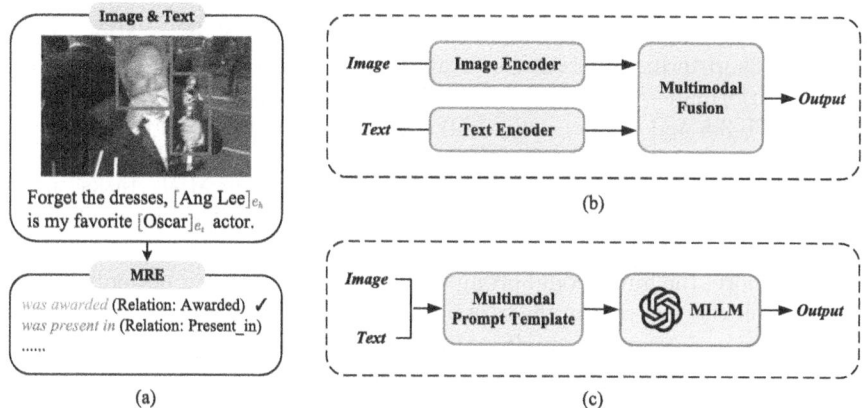

Fig. 1. (a) An example from MNRE dataset. (b) Conventional MRE pipelines. (c) Our proposed method.

However, existing MRE methods focus on custom-designed models for specific datasets or tasks. These models are complex to develop and optimize, and even worse, they exhibit poor generalizability, limiting their transferability to new contexts and rapid deployment. Moreover, these methods struggle to acquire necessary auxiliary knowledge when an external knowledge is involved to enhance textual understanding. Conversely, some other methods retrieve excessively redundant knowledge from external knowledge bases, which can mislead the models.

The advance of large language models (LLMs) offers new perspectives for MRE tasks. Owing to extensive external knowledge which is acquired during the pre-training, LLMs can provide more comprehensive and in-depth information understanding and analysis. Additionally, when these models are trained on data specific to a particular task, they significantly improve their adaptability and performance on that task.

Therefore, in this paper, we propose to a novel method namely ITMRE, which instruction-tunes multimodal large language models (MLLMs) using LoRA for the task of MRE. By leveraging the powerful generalization and understanding capabilities of MLLMs, we can improve the performance of MRE models in extracting relation from multimodal data.

Specifically, we designed a multimodal prompt template that formulates the relation extraction problem as a two-stage multiple-choice task. The first stage identifies the type of entities, also known as Entity Typing, and the second stage uses the type information of the two entities to infer most possible relation type between them by selecting from pre-defined multiple options. After transforming a specific multimodal relation extraction dataset into this prompt template format, we then use a kind of parameter-efficient fine-tuning method, i.e., LoRA, to instructionally optimize the MLLMs. This fine-tuning process allows the model to adapt better to the data of MRE, thereby enhancing the relation extraction performance. Experiments on the public dataset MNRE show that our method significantly outperforms existing multimodal relation extraction methods in terms of precision, recall, and F1 score, and also demonstrates superior performance compared to the latest advanced large language models such as GPT-3.5 and GPT-4. Our contributions are summarized as follows:

- We introduce an innovative application of LLMs for MRE tasks, utilizing instruction-based fine-tuning to improve task-specific adaptability and decrease complexity in both training and deployment.
- Furthermore, the multimodal prompt template we developed not only reduces the complexity inherent in the task but also significantly enhances the model's proficiency in interpreting multimodal data and conducting relational reasoning. The template demonstrates broad applicability, proving effective with the MNRE dataset and easily adaptable to additional datasets.
- Comprehensive experimentation using the MNRE dataset has corroborated the efficacy of our approach. Our method outperforms existing MRE techniques and demonstrates superiority over contemporary advanced large language models, including GPT-3.5 and GPT-4, with a 5%-8% improvement in F1 scores, thus affirming its practical feasibility and advantages.

2   Related Work

2.1   Multi-modal Large Language Models

Recent advances in deep learning have significantly bolstered the field of multimodal intelligence, giving rise to Multimodal Large Language Models (MLLMs) that effectively bridge the gap between diverse data modalities such as text and images. Pioneering works such as CLIP [3,4] and BLIP [5], demonstrate the power of LLMs to enhance multimodal reasoning by aligning visual and textual data through contrastive learning [6]. These models have been pre-trained on extensive datasets to establish robust connections between different modalities. Following this trend, Liu et al. [7] developed LLaVA, further integrating language and visual content, showing remarkable zero-shot capabilities, and extending the utility of LLMs in conversational contexts. Our work extends these foundational models, leveraging their strong reasoning capabilities and instruction tuning techniques, as seen in applications such as Science QA by Zhang et al. [8], to address the challenges of multimodal relation extraction.

2.2 Multi-modal Relation Extraction

Leveraging images to enhance the performance of relation extraction has been receiving increasing attention in information extraction. Zheng et al. [9] released a multimodal relation extraction dataset, MNRE, to address the scarcity of datasets in this field and subsequently proposed a multimodal relation extraction method, MEGA [10], which utilizes visual information and attention mechanisms. Chen et al. [1] designed a visual prefix-guided fusion method, HVP-Net, which incorporates object-level visual information and utilizes hierarchical multi-scale visual features. Li et al. [11] proposed a fine-grained multimodal alignment method using a Transformer to align visual and textual objects in the representation space. Although these methods have achieved commendable results in multimodal entity relation extraction tasks, there are still issues such as insufficient model generalization and transfer capabilities, and a lack of deep understanding of multimodal data.

3 Preliminary

3.1 Task Definition

The goal of multimodal relation extraction (MRE) is to infer the relation between two entities based on the text and its corresponding image. Specifically, each instance consists of a piece of text T and a related image I, where the head entity e_h and the tail entity e_t in T are annotated. Given a relational instance $x = (T, I, e_h, e_t)$, the model needs to combine the auxiliary information from the text T and image I to identify the relation by selecting from a set of predefined relation types.

3.2 Selection of MLLMs

In selecting multimodal large language models (MLLMs) for our study on relation extraction, we prioritized open-source models like LLaVA [12], CogVLM [13], and Qwen-VL [14] due to their flexibility for fine-tuning, notable achievements on various leaderboards, and comparable parameter counts. These characteristics not only facilitate a fair comparison across models but also ensure transparency and adaptability in academic research settings.

LLaVA. It employs a pretrained CLIP-ViT-L/14 for visual encoding and a Vicuna for language model, as well as uses a simple linear projection to align image and text features. It leverages GPT-4-generated datasets for fine-tuning, allowing for advanced instruction-based training to enhance its multimodal capabilities.

CogVLM. Featuring an EVA2-CLIP-E visual encoder and a Vicuna1.5-7B language model, it integrates visual expert modules within each Transformer block for enhanced visual-language interactions. It is pretrained on extensive multimodal datasets and further fine-tuned for precise visual-language tasks.

Qwen-VL. Based on the Qwen7B model, it uses a vision-language adapter with cross-attention to process multimodal inputs. It undergoes multistage training, including high-resolution visual tasks, to enhance its ability to follow complex instructions and understand multimodal data.

4   Methodology

The framework of our proposed ITMRE methodology is illustrated in Fig. 2. Initially, multiple-choice question-answer (MQA) templates classify entities from combined text and image inputs, identifying categories such as person or location. Subsequent questions prompt the model to determine relations between these entities based on predefined options. This streamlined two-stage approach allows for targeted, efficient instruction tuning of MLLMs using LoRA, focusing on the integration of visual and language components without full model retraining. In this section, we elaborate on our proposed approach.

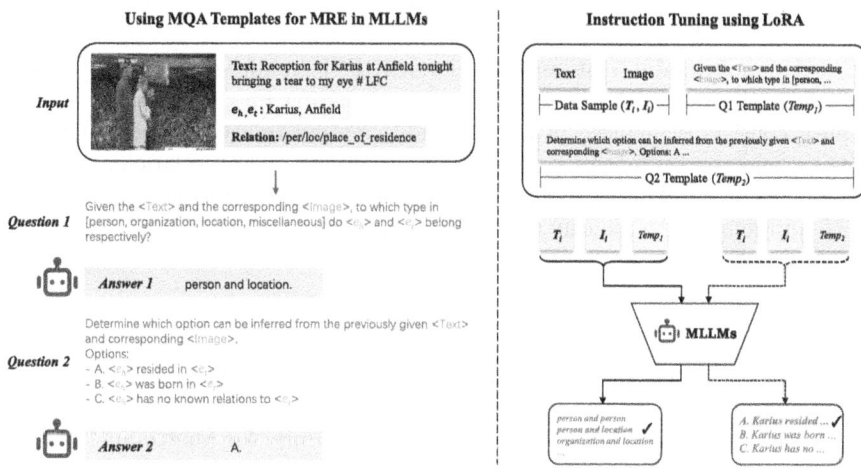

Fig. 2. An overview of our proposed ITMRE framework.

4.1   Multimodal Prompt Design

We transform the MRE task into a MQA problem. This approach leverages structured prompts to guide the relation extraction process, ensuring clarity and focus in model responses.

Stage I: Entity Typing. Initially, the task focuses on identifying the types of entities involved. By integrating head entity e_h and tail entity e_t of a relational instance x from the dataset into our system, we generate the first question by asking the model to determine the entity types based on provided contextual texts and corresponding images. The answer to this question establishes the groundwork for the next stage by confirming the types of each entity.

Stage II: Relation Type Inference. Utilizing the entity types identified in the first stage, we then craft a second question that involves selecting the most appropriate relation between these entities. According to the types of head and tail entities, we can filter out all possible relations. Afterward, we employ pre-defined relation templates to construct multiple choices that have been successfully used in previous studies [15,16]. For a "person" head entity and a "location" tail entity, we construct the choice for a possible relation "/per/loc/place_of_birth" as "$[e_h]$ was born in $[e_t]$", where $[e_h]$ and $[e_t]$ are filled with the specific entity mentions in the original text. By constructing such templates for all possible relations, relation extraction is converted into a multiple-choice option task. This approach aids in maintaining consistency with past research and enhances the precision of the model's output by aligning it with established relational patterns.

Algorithm 1 Generation of Two-Stage Questions for MRE

Require: Text T, Images I, Entities (e_h, e_t),
     List of entity types $\mathcal{T}$, Set of relation templates $\mathcal{R}$
Ensure: Set of questions $Q = \{(q_1, O_1), (q_2, O_2)\}$
 1: Initialize question set $Q \leftarrow \emptyset$
 2: **for** each sample (T, I, e_h, e_t) in the dataset **do**
 3:   $q_1 \leftarrow$ "Determine the entity types for e_h and e_t."
 4:   $O_1 \leftarrow (t_{e_h}, t_{e_t})$
 5:   Initialize options for question 2, $O_2 \leftarrow \emptyset$
 6:   **for** each (t_h, t_t, r) in $\mathcal{R}$ **do**
 7:    **if** $t_{e_h} = t_h \wedge t_{e_t} = t_t$ **then**
 8:     $q_2 \leftarrow q_2 \cup \{\text{fill } T(r) \text{ with } e_h, e_t\}$
 9:    **end if**
10:   **end for**
11:   $q_2 \leftarrow$ "Determine which option..." + "Options: " + q_2
12:   $O_2 \leftarrow \text{MLLM}(q_2)$ *or* Correct Answer(q_2)
13:   $Q \leftarrow Q \cup \{((q_1, O_1), (q_2, O_2))\}$
14: **end for**
15: **return** Q

Following the stages of entity typing and relation type inference detailed earlier, our methodology employs Algorithm 1 to automate the generation of two-stage questions. Here a list of entity types $\mathcal{T} = \{t_1, \ldots, t_m\}$ is defined, where

m is the number of entity types, including "per" (person), "loc" (location), etc. We can define the template $T(r)$ for each relation in the pre-defined collection of relations $\mathcal{R} = \{r_1, \ldots, r_n\}$, where n is the cardinality of $\mathcal{R}$. Each relation $r \in \mathcal{R}$ corresponding to a template can be abstracted as (t_h, t_t, r), where t_h and t_t denote the entity types of head and tail entities participated in relation r. This algorithm systematically processes each dataset sample, first establishing entity types for the provided entities (e_h and e_t) and then using these types to formulate a second, more complex question. This question challenges the model to identify the most appropriate relation by selecting from options that are dynamically generated based on predefined relation templates appropriate to the ascertained entity types.

The reformulation of relation extraction into a two-stage MQA task enables our model to effectively utilize its capabilities in language comprehension and decision-making.

4.2   Dataset Transformation for Instruction Tuning

To effectively apply instructional tuning techniques such as LoRA to our multi-modal relation extraction model, it is essential to transform the existing dataset into a format that supports this kind of fine-tuning. This involves converting the dataset into a series of instruction-response pairs, which are used to guide the model in understanding and executing the tasks more accurately.

The transformation process begins by reformatting each instance in the dataset into two distinct MQA pairs that align with the two-stage approach described in the Sect. 4.1. To better illustrate this transformation, we present an example in Table 1. In the Q&A-1, the model is asked to select the types corresponding to the two entities in the text, $type_h$ and $type_t$, from a given set of entity types. In the Q&A-2, the model is asked to make a judgment regarding the relation between the two entities. The options for the question include all relations corresponding to both $type_h$ and $type_t$ as well as none, and are generated from a predefined template. Table 1 provides a clear example of how instruction-response pairs are structured, where the parts that need to be replaced according to the actual situation are marked in green.

This structured transformation approach not only facilitates the fine-tuning process but also ensures that the model's learning is deeply aligned with the specific requirements of multimodal relation extraction tasks. By generating these precise instruction-response pairs, we enable the model to focus on relevant details, thereby enhancing its ability to infer correct relations based on multimodal inputs.

4.3   LoRA-Based Instruction Tuning

To optimize our multimodal relation extraction model, we employ the Low-Rank Adaptation (LoRA) technique, a method known for its parameter efficiency and effectiveness in adapting large language models to specific tasks. LoRA modifies the model in a way that is both resource-efficient and capable of achieving

Table 1. Example of Instruction Tuning Dataset Construction

Q&A-1	**Question:** Given the text and the corresponding image, to which type in [List L_a] do [c_a] and [c_t] belong respectively? Text: [Text] / Image: [Image]
	Answer: [$type_a$] and [$type_t$]
Q&A-2	**Question:** Determine which option can be inferred from the previously given text and corresponding image. Options: - A. [Entities in Relation 1 Template] - B. [Entities in Relation 2 Template] ... - N. [c_a] has no known relations to [c_t]
	Answer: [Option]

substantial performance improvements without extensive retraining of the entire network.

Concept of LoRA. LoRA introduces small low-rank matrices for efficient fine-tuning by decomposing updates $\Delta\mathbf{W}$ to the pre-trained weight matrix $\mathbf{W}_0$ into trainable matrices $\mathbf{B}$ and $\mathbf{A}$. This is represented as:

$$\mathbf{W}' = \mathbf{W}_0 + \mathbf{BA}, \tag{1}$$

where $\mathbf{B} \in \mathbb{R}^{d \times r}$ and $\mathbf{A} \in \mathbb{R}^{r \times k}$, with r much smaller than d and k, minimizing computational overhead and preserving the model's structure. During training, only $\mathbf{B}$ and $\mathbf{A}$ are adjusted, while $\mathbf{W}_0$ remains unchanged. This selective updating focuses on task-specific adaptations, enhancing performance in multimodal relation extraction without compromising pre-trained capabilities.

Implementation in Instruction Tuning. The MLLMs we employed are generally structured into three core components: a vision encoder, a bridge module, and an LLM backbone. The vision encoder processes images into visual feature vectors, transformed by the bridge module into textual feature spaces, integrating with language queries for the LLM backbone. Both the bridge module and the LLM backbone are fine-tuned using LoRA, refining the model's multimodal data processing without comprehensive retraining, enhancing MRE performance by modifying key layers. The vision encoder remains frozen to preserve feature extraction stability. LoRA fine-tunes the model efficiently with minimal parameters, avoiding overfitting and maintaining computational efficiency.

5 Experiment

5.1 Dataset and Evaluation Metrics

Dataset. Currently, there are fewer publicly available datasets in the MRE field. Our study utilizes the widely-used MNRE dataset, which is developed by Zhang et al. [9,10]. Derived from multimodal named entity recognition datasets Twitter15 and Twitter17, and enhanced with real-time Twitter data. Each sample in the MNRE dataset comprises an image paired with a text snippet, where each text is annotated with two entities and their types, as well as the relation between them. For samples where the relation is not annotated as "None", each constitutes an entity-relation tuple, with both the entity and relation types being predefined. This dataset has two versions: MNRE-1 with 31 relation types and MNRE-2, an improved version with 23 clearer relation types. Statistical details of both versions are shown in Table 2.

Table 2. Statistical Information of the MNRE Dataset

	# Images	# Words	# Sentences	# Entities	# Relations	# Instances
MNRE-1	10,089	172k	14,796	20,178	31	10,089
MNRE-2	9,201	258k	9,201	30,970	23	15,485

Note: # denotes the count.

Evaluation Metrics. For our experiments on the MNRE dataset, we evaluate model performance using precision (*Pre.*), recall (*Rec.*), and Macro-F1 scores (*F1*), metrics that are widely used in relational extraction. The Macro-F1 Score, which averages the F1 scores across all classes, is particularly important for measuring performance amid class imbalance, providing a holistic view of model capabilities.

5.2 Model Settings and Baselines

Model Settings. For the implementation of MLLMs, we use three specific models: "llava-v1.5-7b" (LLaVA), "cogvlm-chat-v1.1" (CogVLM), and "Qwen-VL-Chat" (Qwen-VL). The MNRE dataset is transformed into MQA format following the methodologies described previously and split approximately into an 8:2 ratio for training and testing sets, respectively. The training set is used for instruction tuning, while the testing set, with hidden answers, is used to evaluate extraction accuracy based on the model's responses. The training parameters for instruction tuning using LoRA were meticulously selected based on prior experimental insights and model presets, tailored to optimize performance. Key settings include training for 3 epochs with a per-device batch size of 2 or 4, 2 gradient accumulation steps, and a learning rate of 5e-6. For the LoRA-specific parameters, the rank (r) of the low-rank matrices was set between 16 or 32, with a scaling factor (alpha) of 16 and a dropout rate of 0.05.

Baselines. Our experimental evaluation includes a variety of models to benchmark against our approach. Baseline models encompass traditional single-modal models like PCNN [17] and MTB [18], alongside advanced multimodal models such as UMT [19], UMGF [20], BERT+SG [10], MEGA [9], and MoRe [21]. Additionally, we utilize OpenAI's GPT models: GPT-3.5 (gpt-3.5-turbo-1106) and GPT-4 (gpt-4-0613), which are known for their enhanced reasoning capabilities. Due to API limitations, 300 sample instances were randomly selected when using GPT models, and only text messages were supported. The prompting templates were designed using zero-shot prompting (i.e., they were directly asked to answer the relationships between entities).

5.3   Comparison with SOTA Methods

Table 3 demonstrates the comparison of the experimental results of each baseline model with MLLMs after instruction tuning on the MNRE dataset, highlighting the effectiveness of our proposed ITMRE methodology.

Table 3. Performance (%) of baselines and our methods on the MNRE dataset

Method		MNRE-1			MNRE-2		
		Pre.	Rec.	F1	Pre.	Rec.	F1
Text-based	PCNN	**67.22**	45.03	53.93	62.85	49.69	55.50
	MTB	61.17	57.03	59.03	64.46	57.81	60.95
	GPT-3.5	44.90	45.16	45.03	45.25	47.36	46.28
	GPT-4	63.57	63.24	63.41	64.73	65.09	64.90
Multimodal	UMT	62.74	61.52	62.12	62.93	63.88	63.40
	UMGF	\	\	\	64.38	66.23	65.29
	BERT+SG	64.23	61.59	62.88	62.95	62.65	62.80
	MEGA	\	\	\	64.51	68.44	66.42
	MoRe	64.57	67.58	66.04	65.25	67.32	66.27
ITMRE (Ours)	LLaVA	65.51	**70.33**	**67.84**	66.88	70.29	68.54
	CogVLM	64.12	68.54	66.25	64.68	68.51	66.54
	Qwen-VL	66.60	68.57	67.57	**68.62**	**71.32**	**69.94**

The experimental results show that, except for the GPT model, the multimodal models basically outperform the single-text model on both datasets, especially MEGA and MoRe on the MNRE-2 dataset, with F1 score of 66.42% and 66.27%, respectively. This indicates that image information plays a greater role in helping relation extraction for short, context-less text information in the MNRE dataset. And the performance of the fine-tuned MLLMs on the MNRE dataset basically outperforms all the baseline models, with LLaVA reaching 67.84% F1 score on the MNRE-1 dataset, and Qwen-VL reaching 69.94% F1 score on the

MNRE-2 dataset. This indicates that the MLLMs have better generalization and inference abilities on the MRE task.

In addition, the introduction of the GPT model in our experiments provides a comparative perspective, showing that even state-of-the-art language models with significantly larger parameter counts, such as GPT-3.5 and GPT-4, do not have the same level of performance as MLLMs using the ITMRE methodology, reflecting the superiority of the ITMRE methodology. The incorporation of MQA templates in the ITMRE approach enabled precise relation extraction by sequentially addressing entity type identification followed by relation inference.

5.4   Analysis of Instruction Tuning Using LoRA

Table 4 presents the average F1 scores before and after instruction tuning using LoRA, highlighting the significant improvement in performance for all models.

Table 4. Average pre- and post-tuning F1 scores (%) on MNRE datasets

	LLaVA	CogVLM	Qwen-VL
ITMRE (Pre-Tuning)	42.85	37.01	47.90
ITMRE (Post-Tuning)	68.18	66.39	68.75
Improvement	**+25.33**	**+29.38**	**+20.85**

The demonstrated performance improvements validate the efficacy of ITMRE using LoRA, emphasizing not only the importance of accurate multimodal fusion but also the adaptability and precision in instruction tuning. These results indicate that instruction tuning of the models using LoRA resulted in significant enhancements in handling multimodal data used for relation extraction. Overall, the ITMRE methodology leverages LoRA and MQA templates, creating a robust framework that significantly boosts MLLM performance in MRE tasks. This methodology demonstrates substantial improvements in precision and recall and highlights potential for scalable and efficient multimodal learning applications.

6   Conclusion

Our study confirms the efficacy of the ITMRE methodology, which integrates MQA templates with LoRA instruction-tune to optimize multimodal relation extraction tasks. This approach significantly simplifies the relation extraction process while improving the adaptability and performance of MLLMs on complex multimodal datasets. By achieving superior results on the MNRE dataset in terms of precision, recall, and F1 scores, ITMRE sets a new standard for MRE tasks, offering a scalable and efficient solution that is adaptable to diverse applications. This breakthrough paves the way for future advancements in multimodal learning, fostering further exploration into the integration of varied data modalities for enhanced computational understanding.

Acknowledgement. This work was partially supported by National Key R&D Program of China (No. 2022YFB3103600), NSFC (Nos. U23A20296, 62272469), and The Science and Technology Innovation Program of Hunan Province (No. 2023RC1007).

References

1. Chen, X., et al.: Good visual guidance makes a better extractor: hierarchical visual prefix for multimodal entity and relation extraction. arXiv preprint arXiv:2205.03521 (2022)

2. Zhao, Q., Gao, T., Guo, N.: TSVFN: two-stage visual fusion network for multimodal relation extraction. Inf. Process. Manage. **60**(3), 103264 (2023)

3. Radford, A., et al.: Learning transferable visual models from natural language supervision. In: International conference on machine learning. PMLR (2021)

4. Kong, Z., Li, W., Zhang, H., Yuan, X.: FocusCap: object-focused image captioning with clip-guided language model. In: Yuan, L., Yang, S., Li, R., Kanoulas, E., Zhao, X. (eds.) WISA 2023. LNCS, vol. 14094, pp. 319–330. Springer, Singapore (2023). https://doi.org/10.1007/978-981-99-6222-8_27

5. Li, J., Li, D., Xiong, C., Hoi, S.: BLIP: bootstrapping language-image pre-training for unified vision-language understanding and generation. In: International Conference on Machine Learning, pp. 12888–12900. PMLR (2022)

6. Zhang, S., et al.: Instruction tuning for large language models: a survey. arXiv preprint arXiv:2308.10792 (2023)

7. Liu, H., Li, C., Li, Y., Lee, Y.J.: Improved baselines with visual instruction tuning. arXiv preprint arXiv:2310.03744 (2023)

8. Lu, P., et al.: Learn to explain: multimodal reasoning via thought chains for science question answering. Adv. Neural. Inf. Process. Syst. **35**, 2507–2521 (2022)

9. Zheng, C., Wu, Z., Feng, J., Fu, Z., Cai, Y.: MNRE: a challenge multimodal dataset for neural relation extraction with visual evidence in social media posts. In: 2021 IEEE International Conference on Multimedia and Expo (ICME). IEEE (2021)

10. Zheng, C., Feng, J., Fu, Z., Cai, Y., Li, Q., Wang, T.: Multimodal relation extraction with efficient graph alignment. In: Proceedings of the 29th ACM International Conference on Multimedia, pp. 5298–5306 (2021)

11. Li, L., Chen, X., Qiao, S., Xiong, F., et al.: On analyzing the role of image for visual-enhanced relation extraction. arXiv preprint arXiv:2211.07504 (2022)

12. Liu, H., Li, C., Wu, Q., Lee, Y.J.: Visual instruction tuning. Adv. Neural Inf. Process. Syst. **36** (2024)

13. Wang, W., Lv, Q., Yu, W., Hong, W., Qi, X., et al.: CogVLM: visual expert for pretrained language models. arXiv preprint arXiv:2311.03079 (2023)

14. Bai, J., et al.: Qwen-VL: a versatile vision-language model for understanding, localization, text reading, and beyond. arXiv preprint arXiv:2308.12966 (2023)

15. Zhang, K., Gutiérrez, B.J.: Aligning instruction tasks unlocks large language models as zero-shot relation extractors. arXiv preprint arXiv:2305.11159 (2023)

16. Sun, Y., Zhang, K., Su, Y.: Multimodal question answering for unified information extraction. arXiv preprint arXiv:2310.03017 (2023)

17. Zeng, D., Liu, K., Chen, Y., Zhao, J.: Distant supervision for relation extraction via piecewise convolutional neural networks. In: Proceedings of the 2015 Conference on Empirical Methods in Natural Language Processing, pp. 1753–1762 (2015)

18. Soares, L.B., FitzGerald, N., Ling, J., Kwiatkowski, T.: Matching the blanks: distributional similarity for relation learning. arXiv preprint arXiv:1906.03158 (2019)

19. Yu, J., Jiang, J., Yang, L., Xia, R.: Improving multimodal named entity recognition via entity span detection with unified multimodal transformer. In: Proceedings of the 58th Annual Meeting of the Association for Computational Linguistics (2020)
20. Zhang, D., Wei, S., Li, S., Wu, H., Zhu, Q., Zhou, G.: Multi-modal graph fusion for named entity recognition with targeted visual guidance. In: Proceedings of the AAAI Conference on Artificial Intelligence, vol. 35, pp. 14347–14355 (2021)
21. Wang, X., Cai, J., Jiang, Y., Xie, P., Tu, K., Lu, W.: Named entity and relation extraction with multi-modal retrieval. arXiv preprint arXiv:2212.01612 (2022)

Instruction Tuning with LLMs
for Programming Exercise Generation

Guolong Zeng, Qinchen Xue, and Xuesong Lu[✉]

East China Normal University, Shanghai, China
{glzeng,10205501413}@stu.ecnu.edu.cn, xslu@dase.ecnu.edu.cn

Abstract. Large language models (LLMs) have been applied to help programming education on aspects such as question answering and program repair. While they make students learn more efficiently, how to use LLMs to help increase teaching efficiency is rarely explored. In this paper, we focus on harnessing LLMs to automatically generate programming exercises with the goal of alleviating teachers' workload and enhancing teaching efficiency. We first evaluate the performance of seven open-source LLMs using prompts, and then fine-tune two winning LLMs using instructions constructed with the Evol-Instruct and the ACES algorithms, respectively. Experimental results demonstrate the improved performance on the two LLMs after the instruction tuning. Additionally, our contribution encompasses the formulation of evaluation metrics and the exploration of various prompt methods.

Keywords: Programming exercise generation · Open-source LLMs · Instruction Tuning

1 Introduction

Programming education at scale is of great significance to the digital transformation of today's society since the demand for programming skills is ubiquitous in various industries. Hence previous studies have investigated AI-assisted approaches for programming knowledge tracing [14], program repairs [4], code summary [16] and so on, and shown that these intelligent tools can effectively support large-scale self-study and increase learning efficiency. While these tools are more student-centered, teachers also need intelligent means to improve teaching efficiency for large-scale students, such as answering questions and creating exercises, which are much more challenging to solve using traditional AI technologies.

The emergence of large language models (LLMs) provides new alternatives for developing intelligent teaching tools. In this work, we particularly investigate the problem of automatically generating programming exercises for programming courses using LLMs. The problem is important because teachers are often asked to create new exercises and manually creating programming exercises is time-consuming. Moreover, to enable personalized learning for large-scale students,

C. Jin et al. (Eds.): WISA 2024, LNCS 14883, pp. 377–385, 2024.
https://doi.org/10.1007/978-981-97-7707-5_31

a huge number of exercises are needed to cater for students of diverse coding abilities and knowledge status. Recently, lots of open-source LLMs [1,17] have been developed and shown comparable performance with commercial models. In educational scenarios, open-source LLMs are often preferred over commercial LLMs because they are much cheaper to use and can avoid privacy issues with local deployment. Hence, in this study we evaluate representative open-source LLMs and attempt to improve their performance via instruction tuning.

Specifically, we first evaluate seven open-source LLMs on the AlpacaEval Leaderboard[1] [6] by prompting them to generate programming exercises. Then, we pick two LLMs with the best performance and construct instructions to further fine-tune them. The instructions are constructed using the Evol-Instruct algorithm [15] and the ACES algorithm [9], respectively. Finally, we explore different prompt methods to guide the two tuned LLMs for programming exercise generation and show the improved performance over the base variants.

2   Related Work

Many studies have explored the use of LLMs in programming education. Finnie-Ansle et al. [3] evaluate the performance of Codex in solving several common student programming tasks and find that Codex can solve the tasks in many cases and provide diverse solutions. Zhang et al. [19] use Codex to repair bugs in student programming assignments. Phung et al. [7] develop a method based on Codex to provide high-precision feedback on syntax errors in Python code. Also, Phung et al. [8] evaluate the performance of ChatGPT and GPT-4 in solving different programming tasks, including program repair, hint generation, grading feedback, pair programming, contextualized explanation and task synthesis.

For the task of programming exercise generation, several studies have explored the ability of OpenAI's models. For instance, Sarsa et al. [10] evaluate Codex for programming exercise generation and code explanation. They simulate a complete process of programming assignments, including problem generation, student training and feedback. Phung et al. [8] evaluate ChatGPT and GPT-4 on a slightly different task, which generates an exercise based on a student's buggy program to help the student consolidate the knowledge. Pourcel et al. [9] design the ACES algorithm based on the P3 dataset [11] and construct a Python puzzle dataset through prompting GPT-4. They find that the dataset constructed by ACES is more diverse and innovative. Our work differs from them on two aspects. First, we evaluate the ability of open-source LLMs for programming exercise generation. Second, we adopt instruction-tuning to improve the performance of open-source LLMs.

---

[1] https://tatsu-lab.github.io/alpaca_eval.

3 The Methodology

3.1 The Candidate Open-Source LLMs

The AlpacaEval [6] Leaderboard is one of the representative leaderboards that evaluate the ability of LLMs to follow general user instructions. To evaluate the ability on programming exercise generation, we select seven models from the leaderboard: Alpaca-7B [12], Vicuna-13B [2], ChatGLM-6B [18], OpenChatV2-W-13B [13], WizardLM-13B [15], Yi-34B-Chat [17], Aligner-2B+Qwen1.5-32B-Chat [1,5].

Then, we prompt the seven LLMs to generate programming exercises as shown in Fig. 1.

> You are a Python programming teacher and need to create programming exercises.
>
> Please create a *low* difficulty exercise which tests students' mastery level on the concepts of *branch statement and function*.
>
> Your Response:

Fig. 1. A prompt for generating programming exercises.

The words in bold italics indicate the required difficulty and the inspected concepts of the exercise. The difficulty levels include low, medium and high. The concepts include data type, operator, branch statement, loop statement, list, dictionary, function, class, string, sorting, searching, recursion and dynamic programming. For each prompt, we randomly select a difficulty level and one or two concepts to form the prompt.

Table 1. The performance of the seven candidate LLMs and GPT-4.

Model	Absolutely Right	Barely Right	Not Acceptable
Alpaca-7B	3	5	42
Vicuna-13B	3	9	38
ChatGLM-6B	6	12	32
OpenChatV2-W-13B	13	12	25
WizardLM-13B	16	20	14
Yi-34B-Chat	**16**	**24**	**10**
Aligner-2B+Qwen1.5-32B-Chat	**23**	**17**	**10**
GPT-4	*47*	*3*	*0*

We generate 50 prompts using the aforementioned method and let each LLM generate 50 programming exercises. We manually examine the generated exercises and divide them into three categories: absolutely right, barely right and not acceptable. Absolutely right means that the exercise has no error and can be used directly by teachers. Barely right means that the exercise has clear content with a few errors, and can be used by teachers subject to necessary corrections. Not acceptable means that the content of the exercise is hard to understand so that it cannot be used. The results are reported in Table 1. We observe that GPT-4 performs significantly better than the open-source LLMs, which generates 47 absolutely right exercises and 3 barely right exercises. No exercise is not acceptable. Among the open-source LLMs, Aligner-2B+Qwen1.5-32B-Chat and Yi-34B-Chat outperform other models. Based on the results, we select the two LLMs for fine-tuning.

3.2   Fine-Tuning the Open-Source LLMs

We apply instruction tuning [20] on Aligner-2B+Qwen1.5-32B-Chat and Yi-34B-Chat. Particularly, we adopt two algorithms to construct the instructions, namely, Evol-Instruct [15] and ACES (Autotelic Code Exploration via Semantic descriptors) [9].

The Evol-Instruct algorithm [15] starts with an initial set of instructions and rewrites them using a powerful LLM step by step into more complex instructions. The rewriting includes In-depth Evolving, which enhances instructions by making them more complex and difficult through prompts, and In-breadth Evolving, which enhances topic or skill coverage and overall diversity of instructions. In our task, we prompt GPT-4 to rewrite instructions. We select 80 programming exercises from a university's online judge system, which cover all the three levels of difficulty and the aforementioned concepts. For each exercise, we manually write the corresponding prompt like Fig. 1 and append the exercise as the response to form an instruction. Hence, the initial set contains 80 instructions. The In-depth Evolving includes refining inspected concepts, promoting difficulty levels and refining application scenarios. The In-breadth Evolving includes adding or changing inspected concepts or application scenarios and changing difficulty levels.

Using the method, in each iteration we evolve each existing instruction either in depth or in breadth with equal probability and double the number of instructions. The iteration terminates once we obtain the required number of instructions.

The ACES algorithm [9] is a sampling-based method to generate diverse instructions with LLMs. In each iteration, it samples semantic features from a pre-defined semantic space and uses the features to retrieve closest examples from an existing instruction dataset. The retrieved examples are then used as few-shot examples to guide a powerful LLM to generate a new instruction, which is added into the existing dataset. In our task, the dimensions of the semantic space are defined as the unique difficulty levels and concepts, that is, [low, medium, high, data type, operator, branch statement, loop statement, list, dictionary, function, class, string, sorting, searching, recursion, dynamic programming]. Each instruction in the existing dataset is associated with a multi-hot vector in the space, where each dimension has value 1 or 0, indicating whether the exercise of the instruction has the corresponding difficulty level or concept. Then we run the ACES algorithm as follows. In each iteration, we sample at random a multi-hot vector from the semantic space and retrieve from the existing dataset the closest vectors to the sampled vector based on the Hamming distance. Based on the vectors, we further retrieve three instructions, where the instruction corresponding to a vector closer to the sampled vector has the higher priority to be retrieved. Finally, we use the three exercises in the retrieved instructions as few-shot examples to prompt GPT-4 to generate a new exercise, and further use the new exercise to guide GPT-4 to generate the corresponding prompt. The new prompt and new exercise form a new instruction, which is added into the existing dataset. The iteration terminates once we obtain the required number of instructions.

4   Performace Evaluation

4.1   The Datasets

We use GPT-4 to generate instruction data as follows. We first adopt Evol-Instruct to generate 920 instructions so that we have in total 1000 instructions after combining with the initial set of 80 instructions. Then we manually examine the generated instructions and discard those with low quality. Eventually we obtain 722 instructions out of 1000. For fair comparison, we also construct 722 instructions using the ACES algorithm. We examine each generated instruction manually before adding it into the existing dataset, until we obtain 722 high-quality instructions.

4.2   Hyperparameter and Evaluation Metrics

The training epoch is set to 10. The maximum input and output lengths are set to 2048. The batch size is set to 8. The learning rate is set to 2e-5. We use AdamW as the optimizer. The two LLMs are fine-tuned on an NVIDIA A800 GPU. When prompting, the temperature is set to 0.

We consider four metrics: sensibleness, novelty, readiness and accuracy. The first three are defined in [10]. Sensibleness measures whether the exercise describes a practical problem for students to solve. Novelty measures whether the exercise is a copy of or similar to an existing exercise in the training set or searched via Google and GitHub. Readiness measures whether the exercise can be used by teachers with minor modifications. In addition, we define accuracy, which measures whether the exercise meets the requirements of the prompt.

4.3 The Results

We compare the performance of the two LLMs before and after fine-tuning, and use GPT-4 as reference. To verify the effectiveness of in-context learning, we adopt zero-shot, one-shot and three-shot to prompt each LLM to generate 50 programming exercises with varying difficulty levels, concepts and scenarios. The results are shown in Table 2. In the tables, "random" means we randomly select existing exercises as the context, "selected" means we manually select exercises that have the similar difficulty levels, concepts and scenarios with the content in the prompt, and "similar" means we select exercises that are the most similar to the content of the prompt. The results show there is a clear gap between the two open-source LLMs and GPT-4.

The Impact of Instruction-Tuning. First, for sensibleness and novelty, we observe apparent performance improvements after instruction-tuning on both LLMs. The improvements for sensibleness are around 3~4 exercises by using either Evol-Instruct or ACES. For novelty, using ACES brings much more improvements than using Evol-Instruct on both LLMs, which is as expected because ACES focuses more on generating diverse instructions. Note that all LLMs including GPT-4 perform poor for novelty compared to other metrics, indicating that the LLMs essentially memorize existing content in real world. Second, for readiness we observe apparent improvements on Yi-34B-Chat, and observe no improvement on Aligner-2B+Qwen1.5-32B-Chat. This may indicate Aligner-2B+Qwen1.5-32B-Chat already has relatively good ability at generating ready-for-use content before fine-tuning. Finally, we observe no improvement for accuracy on both LLMs. This may indicate the LLMs are already good at understanding the requirements of the prompt and the comprehension is hard to improve via instruction-tuning.

The Impact of In-Context Learning. First, we can observe consistent improvements by using one-shot and three-shot in-context learning over zero-shot prompts for readiness and accuracy on both LLMs. For sensibleness and novelty, no consistent improvement is observed. This is reasonable because in-context learning mainly forces an LLM to generate content in the form similar to the examples. Among the four metrics, readiness and accuracy are much easier to learn from the examples compared to sensibleness and novelty. Second, we do not observe consistent improvements by using three-shot in-context learning over one-shot for readiness and accuracy. This may indicate that the quality of in-context learning examples is more important than the quantity. Finally, we observe that in most cases, using selected and similar examples performs better than using random examples, for both one-shot and three-shot in-context learning. This indicates the importance of careful selection of in-context learning examples.

Table 2. Main Results.

LLM	Zero-Shot	One-Shot			Three-Shot		
		random	selected	similar	random	selected	similar
Results for sensibleness							
Qwen	39	40	40	41	40	40	41
Qwen+Evol-Instruct	**42**	**43**	**45**	**45**	**43**	**44**	**43**
Qwen+ACES	**42**	**43**	**43**	**44**	40	**42**	**42**
Yi	31	29	31	31	28	29	29
Yi+Evol-Instruct	**33**	**33**	**35**	**33**	**31**	**34**	**35**
Yi+ACES	**36**	**33**	**35**	**36**	**33**	**36**	**35**
GPT-4	48	49	49	49	49	49	49
Results for novelty							
Qwen	11	11	11	10	10	13	14
Qwen+Evol-Instruct	**15**	**13**	**16**	**13**	**13**	**13**	12
Qwen+ACES	**17**	**18**	**17**	**17**	**16**	**17**	**16**
Yi	4	6	7	7	5	5	5
Yi+Evol-Instruct	**10**	**9**	**9**	**7**	**8**	**10**	**10**
Yi+ACES	**16**	**16**	**18**	**16**	**17**	**18**	**19**
GPT-4	18	20	20	21	19	20	19
Results for readiness							
Qwen	23	27	27	28	30	29	31
Qwen+Evol-Instruct	24	25	27	27	24	30	28
Qwen+ACES	22	27	29	28	30	30	29
Yi	16	18	18	18	20	19	19
Yi+Evol-Instruct	**24**	**26**	**24**	**24**	**24**	**24**	**25**
Yi+ACES	**23**	**23**	**24**	**24**	**23**	**24**	**24**
GPT-4	44	47	47	48	48	48	48
Results for accuracy							
Qwen	30	31	33	34	34	34	33
Qwen+Evol-Instruct	30	30	31	33	30	34	34
Qwen+ACES	27	28	30	32	31	32	33
Yi	17	22	24	24	21	24	25
Yi+Evol-Instruct	21	21	25	26	22	23	23
Yi+ACES	21	22	24	25	22	22	22
GPT-4	44	46	46	47	47	46	48

5 Conclusion

In this work, we study the problem of programming exercise generation with open-source large language models. We first evaluate seven open-source LLMs listed on the AlpacaEval Leaderboard and identify two winning LLMs. Then we construct two instruction sets using Evol-Instruct and ACES, respectively, and fine-tune the two LLMs. Finally, we adopt different prompt methods to guide the LLMs for exercise generation. The results show that the tuned LLMs have improved performance for sensibleness, novelty and readiness, and have no improvement for accuracy.

References

1. Bai, J., et al.: Qwen technical report. arXiv preprint arXiv:2309.16609, 2023
2. Chiang, W.-L., et al.: Vicuna: an open-source chatbot impressing GPT-4 with 90%* chatgpt quality. See https://vicunalmsys.org. Accessed 14 Apr 2023. 2(3):6, 2023
3. Finnie-Ansley, J., Denny, P., Becker, B.A., Luxton-Reilly, A., Prather, J.: The robots are coming: exploring the implications of OpenAI codex on introductory programming. In: Proceedings of the 24th Australasian Computing Education Conference, pp. 10–19 (2022)
4. Han, S., Wang, Y., Lu, X.: Errorclr: semantic error classification, localization and repair for introductory programming assignments. In: Proceedings of the 46th International ACM SIGIR Conference on Research and Development in Information Retrieval, pp. 1345–1354 (2023)
5. Ji, J., et al.: Aligner: achieving efficient alignment through weak-to-strong correction. arXiv preprint arXiv:2402.02416, 2024
6. Li, X., et al.: Alpacaeval: an automatic evaluator of instruction-following models (2023)
7. Phung, T., et al.: Generating high-precision feedback for programming syntax errors using large language models. arXiv:2302.04662, 2023
8. Phung, T., et al.: Generative AI for programming education: benchmarking Chat-GPT, GPT-4, and human tutors. In: Proceedings of the 2023 ACM Conference on International Computing Education Research, vol. 2, pp. 41–42 (2023)
9. Pourcel, J., Colas, C., Oudeyer, P.Y., Teodorescu, L.: ACES: generating diverse programming puzzles with autotelic language models and semantic descriptors. arXiv preprint arXiv:2310.10692, 2023
10. Sarsa, S., Denny, P., Hellas, A., Leinonen, J.: Automatic generation of programming exercises and code explanations using large language models. In: Proceedings of the 2022 ACM Conference on International Computing Education Research, vol. 1, pp. 27–43 (2022)
11. Schuster, T., Kalyan, A., Polozov, O., Kalai, A.T.: Programming puzzles. arXiv preprint arXiv:2106.05784, 2021
12. Taori, R., et al.: Stanford alpaca: an instruction-following llama model (2023)
13. Wang, G., Cheng, S., Zhan, X., Li, X., Song, S., Liu, Y.: Openchat: advancing open-source language models with mixed-quality data. arXiv preprint arXiv:2309.11235, 2023

14. Wang, L., Sy, A., Liu, L., Piech, C.: Deep knowledge tracing on programming exercises. In: Proceedings of the Fourth 2017 ACM Conference on Learning@ Scale, pp. 201–204 (2017)
15. Xu, C., et al.: Wizardlm: empowering large language models to follow complex instructions. arXiv preprint arXiv:2304.12244, 2023
16. Yang, C., Wu, J.: An approach of code summary generation using multi-feature fusion based on transformer. In: Yuan, L., Yang, S., Li, R., Kanoulas, E., Zhao, X. (eds.) Web Information Systems and Applications. WISA 2023. LNCS, vol. 14094, pp. 271–283. Springer, Singapore (2023). https://doi.org/10.1007/978-981-99-6222-8_23
17. Young, A., et al.: Yi: open foundation models by 01. AI. arXiv preprint arXiv:2403.04652, 2024
18. Zeng, A., et al.: GLM-130B: an open bilingual pre-trained model. arXiv preprint arXiv:2210.02414, 2022
19. Zhang, J., et al.: Repairing bugs in python assignments using large language models. arXiv preprint arXiv:2209.14876, 2022
20. Zhang, S., et al.: Instruction tuning for large language models: a survey. arXiv preprint arXiv:2308.10792, 2023

Security

A Blockchain-Based Dynamic Symmetric Searchable Encryption Scheme for Sharing Elderly Health Data

Zhongyi Yu, Xiaomei Dong$^{(\boxtimes)}$, Ziqiang Wen, and Tiezheng Nie

School of Computer Science and Engineering, Northeastern University, Shenyang 110169, China
xmdong@mail.neu.edu.cn

Abstract. In recent years, the smart healthcare and elderly care has emerged as a new elderly care model, with elderly health data playing a significant part in this model. Distributed storage based on blockchain technology facilitates the storage and sharing of elderly health data due to its decentralization, immutability, and traceability. Nevertheless, this approach encounters challenges, including high storage costs on the blockchain and potential privacy leakage during data sharing. To address these issues, we propose a dynamic searchable symmetric encryption (DSSE) scheme for elderly health data sharing based on blockchain technology called BEDSSE. Our scheme utilizes a hybrid on-chain and off-chain storage approach combining blockchain with IPFS, which reduces the storage overhead on the blockchain. Tailored to the storage characteristics of blockchain, we optimize the storage structure of the fish-bone chain, which ensures both forward and backward security in DSSE. We also design a tiered authorization mechanism that allows institutions to access elderly health data when the elderly encounter special circumstances. Then we introduce a supervision chain, maintained by supervision institutions, where health data query records are stored. During the data access process in our scheme, user identity information is anonymized to protect user privacy while ensuring traceability. Finally, the effectiveness of our scheme is demonstrated through relevant experiments.

Keywords: Health data sharing · Blockchain · IPFS · DSSE · Forward security · Backward security

1 Introduction

In this era of rapid development of internet technology, scholars have proposed a new model called smart healthcare and elderly care. It extends traditional integrated elderly care and medical services online, and facilitates resource sharing and collaboration between medical institutions [1]. This model enhances service efficiency and improves the quality of life for the elderly.

Blockchain technology was first proposed by Satoshi Nakamoto in 2008 [2]. Fundamentally, blockchain is a type of distributed shared database. Its decentralization, immutability, and traceability make data sharing more secure and efficient. With the

© The Author(s), under exclusive license to Springer Nature Singapore Pte Ltd. 2024
C. Jin et al. (Eds.): WISA 2024, LNCS 14883, pp. 389–402, 2024.
https://doi.org/10.1007/978-981-97-7707-5_32

rise of blockchain technology, its applications in medical data sharing are increasingly widespread [3]. However, the storage capacity of blocks and the block generation speed severely limit the practical applications of blockchain.

Outsourcing data has become prevalent, and Searchable Symmetric Encryption (SSE) has arisen widespread interest as a method to protect outsourced data [4]. While enabling encrypted data querying, SSE faces challenges like data updates and information leakage. Overall, there are four major challenges that still need to be overcome:

- The storage of a large amount of elderly health data on the blockchain increases the storage overhead and leads to decreased blockchain performance.
- To ensure the security of elderly health data, it is necessary to encrypt the data. However, this also makes searching and updating the encrypted data challenging.
- Currently, the sharing of health data relies on the active authorization of the elderly. This mechanism does not consider the characteristics and issues faced by the elderly.
- Although blockchain provides a certain level of security for users, the issues with privacy leakage and lack of regulation still exist.

Based on the issues mentioned above, we propose a scheme called BEDSSE. The main contributions are as follows:

- We utilize IPFS to store encrypted elderly health data, blockchain stores file indexes. This approach significantly alleviates the storage overhead on blockchain.
- We encrypt elderly health data using DSSE, effectively shortening the maximum fish-bone chain length while ensuring forward and backward security during search and update processes.
- We designed a tiered authorization mechanism using Shamir secret sharing technique, allowing institutional users to access elderly health data through cooperation between institutions when the elderly are unable to grant permission.
- We anonymized user identities and designed a supervision chain that preserves user privacy while also enabling the accountability tracing of users who leak information.

The rest of the paper is organized as follows: In Sect. 2, we discuss the related work on DSSE and blockchain. In Sect. 3, We provide an introduction to our scheme in detail. In Sect. 4, we analyze the correctness and security of our scheme and conduct experimental evaluations. In Sect. 5, we summarize our work in this paper.

2 Related Work

Since Song et al. [5] proposed Symmetric Searchable Encryption (SSE), the field has attracted significant attention. Kamara et al. [6] proposed a sublinear Dynamic Symmetric Searchable Encryption (DSSE), that updates data but leaks keyword hashes. Stefanov et al. [7] and Bost [8] developed DSSE schemes that satisfy forward security, but Bost's scheme [8] requires multiple exponentiations during updates, which leads to lower efficiency. Chamani et al. [9] proposed a DSSE scheme using an inverted index structure that ensures both forward and backward security, but this approach requires maintaining an index list on the client side, which lowers the update efficiency. He et al. [10] proposed a scheme for forward and backward security using a fish-bone chain, but the fixed length of the chain impedes token generation and decreases efficiency.

Most existing searchable encryption schemes assume that the cloud server is honest but curious. However, server failures or business interests might result in incomplete or incorrect results [11]. Blockchain technology can address this issue. Xu et al. [12] proposed a DSSE scheme based on consortium blockchain, which fails to consider DSSE's forward and backward security, risking sensitive keyword leakage. Tang et al. [13] introduced a DSSE scheme for medical data sharing based on Ethereum, achieving forward security but with high gas costs during updating. Liu et al. [14] developed an attribute-based SSE scheme using consortium blockchain, but it fails to implement data updating. Although, Du et al. [15] proposed a medical data sharing scheme based on blockchain and IPFS with hybrid encryption to reduce the overhead of encryption, however, it still does not support data updating.

In this paper, we propose a DSSE scheme for sharing elderly health data using blockchain and IPFS for hybrid storage. We encrypt health data and ensure forward and backward security in updating and searching through using fish-bone chain. Our tiered authorization mechanism provides a new way to access elderly health data when the elderly are unable to grant permission. To enhance the oversight of data queries, we introduce a supervision chain and anonymize users to protect their privacy.

3 Scheme Design

In this section, we describe our proposed scheme BEDSSE in detail. First, we present the corresponding system model (see Sect. 3.1). Then we describe in detail the five phases of our scheme (see Sect. 3.2). We also explain how the blockchain stores the index structure, and how we optimize the fish-bone chain (see Sect. 3.3).

3.1 System Model

In our scheme, there are eight entities involved: data owner, data user, InterPlanetary File System, blockchain, third-party institution, key generation center, supervision chain, and supervision institution. The system model is shown in Fig. 1. The roles of each entity are described as follows:

1. Data Owner (DO): mainly consists of the elderly. DO can encrypt and upload their health data to IPFS. They can extract keywords and encrypt UID, the CID from IPFS, the symmetric key AK, the file operation type op, then send them and the set of keywords to the key generation center.
2. Data User (DU): mainly consists of the users from the medical, elderly care, and volunteer institutions. DU can obtain encrypted file information of health data through uploading search tokens. After passing the tiered authorization mechanism, DU can decrypt the ciphertext to access the elderly health data.
3. InterPlanetary File System (IPFS): stores the encrypted health data of the elderly and returns the file's CID according to DO's requirements.

4. Blockchain (BC): stores the encrypted index of the elderly health data, and updates the index according to update requests. The blockchain returns results based on query requests, and sends query records to the supervision chain.
5. Third-party Institution (TI): joins BC as a full node. It maintains complete information of BC and shares the encryption key with the DO for the update token.
6. Key Generation Center (KGC): generates system key MK and symmetric keys AK for encrypting health data, along with public keys PK, secret keys SK, subkeys SK_{OS}, and re-encryption keys RK for users. Additionally, it generates update tokens for DO and manages key-recovery and re-encryption tasks.
7. Supervision Chain (SC): stores query records from BC and is maintained by the supervision institution.
8. Supervision Institution (SI): mainly consists of government agencies, it keeps a local record of the real and encrypted identities of DU. The accountability can be traced based on the contents that are recorded in SC.

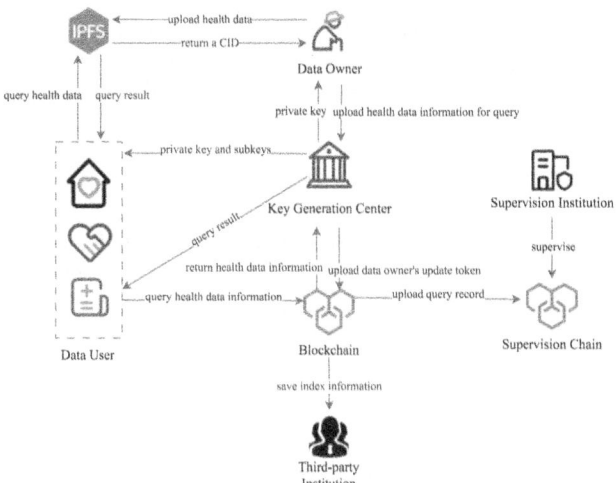

Fig. 1. System model

3.2 Description of Proposed Scheme

There are five phases with twelve algorithms in our scheme: system initialization, data encryption, data update, data retrieval and data decryption. The symbols involved in the algorithms are shown in Table 1.

System Initialization

1. $Setup(1^\lambda) \rightarrow (F, H, ct, MK)$. KGC inputs a security parameter λ, defines a pseudo-random function $F : \{0, 1\}^\lambda \times \{0, 1\}^* \rightarrow \{0, 1\}^\lambda$, and then selects a collision-resistant

hash function $H : \{0, 1\}^* \rightarrow \{0, 1\}^\lambda$. Then, KGC declares the system constant *Flen*, sets the global counter $ct = Flen$, and generates system key $MK = (K_1, K_2, K_3)$, where $K_i \xleftarrow{\$} \{0, 1\}^\lambda (i = 1, 2, 3)$.

2. *KeyGen(MK, F, t, n, p)* $\rightarrow$ $(AK, PK_O, SK_O, PK_U, SK_U, SK_{OS})$. KGC inputs the master key *MK*, pseudo-random function F to generate AK, (PK_O, SK_O) for DO and (PK_U, SK_U) for DU. Then according to the Shamir secret sharing technique, KGC chooses SK_O as secret. KGC selects a large prime $p \in GF(p)$, and defines a polynomial of degree t-1 as the Eq. (1). The SK_{OS} is generated according to the Eq. (2) by using K_3 to obtain a random x, and distributed to various institutions.

$$F(x_i) = SK_O + a_1x_1 + a_2x_2^2 \cdots + a_{t-1}x_i^{t-1}, s_i = F(x_i)(i = 1, 2, \cdots, n) \qquad (1)$$

$$SK_{OS_i} = \{(x_1, s_1), (x_2, s_2), \cdots, (x_n, s_n)\}(i = 1, 2, \cdots, n) \qquad (2)$$

Table 1. Symbols used in this scheme

Symbols	Meaning of the symbols
MK	The system master key
ct	The global counter
(PK_O, SK_O)	The secret key and public key pair for DO
(PK_U, SK_U)	The secret key and public key pair for DU
SK_{OS}	The subkey of DO's secret key
AK	The AES symmetric key
RK	The proxy re-encryption conversion key
T_{wj}	The keyword update token
CT_{wj}	The ciphertext of keyword update token
$EX(w) = \{C_{CID}, C_{AK}, UID\}$	The data receiving set of file information
$EX_R(w) = \{C_{RECID}, C_{REAK}\}$	The packet of re-encrypted $EX(w)$
$RC = \{Result, C_{UUID}\}$	The search result and DU's encrypted identity
$C_{mi} = \{C_{CID}, C_{AK}, C_{UID}, C_{op}\}$	The ciphertext set of IPFS CID, symmetric key, DO's encrypted identity and operation type

Data Encryption

1. *HDEncrypt(HD, AK)* $\rightarrow$ (C_{HD}). DO encrypts the elderly heah data *HD*, using the AES encryption algorithm to generate the ciphertext C_{HD}.
2. *HDUpload(C_{HD}, AK, PK_O, op, UID)* $\rightarrow$ (Msg). The health data ciphertext C_{HD} is uploaded to IPFS which generates a unique hash *CID*, serving as the identifier

for files, and returns it to DO. Subsequently, DO uses RSA encryption algorithm to encrypt *CID*, *UID* and *AK*.The operation type *op*, taken from the set $\{1, 0\}$, where '1' represents the addition and '0' represents the deletion, then DO encrypts *op* to C_{op}. DO extracts the keyword set W_{Si} from the health data, containing the keyword $w_j \in W_{Si}$, $(j = 1, 2, \cdots, |W_{Si}|)$. Finally, the encrypted data C_{mi} is packaged, consisting of C_{CID}, C_{AK}, C_{UID}, C_{op}. DO sends $Msg = \{C_{mi}, W_{Si}\}$ to KGC.

Data Update

1. *UpdateTokenGen*$(Msg, F, H, MK, ct) \rightarrow (TS, ct\prime)$. After receiving Msg, KGC selects a keyword w_j from the keyword set W_{Si}. It generates a keyword token $C_{wj} = F(K_1, w_j)$. Then, using a hash function as the Eq. (3) to generate a keyword update token $T_{wj} = H^{ct}(C_{wj})$. The token T_{wj} is encrypted to produce CT_{wj} by K_2.. The set $TS = \{CT_{wj}, C_{mi}\}$ is transmitted to the blockchain. Then, update the global counter $ct' = ct - 1$.

$$H^k(C_{wj}) = \begin{cases} H(C_{wj}), & k = 1 \\ H(H^{k-1}(C_{wj})), & k \geq 2 \end{cases} \tag{3}$$

2. *UpdateIndex*$(TS, MK, H, WSD) \rightarrow (WSD')$. Upon receiving TS, the blockchain issues an update transaction that utilizes CT_{wj} to update the keyword index chain in the local world state database (*WSD*) of TI. The data index chain is updated using CT_{wj} as key and C_{mi} as valueSince TI also holds K_2, TI decrypts CT_{wj} and stores it in the local world state database. The detail is shown in Algorithm 1.

Algorithm 1. $UpdateIndex(TS, MK, H, WSD) \rightarrow (WSD')$

Blockchain:

1:    **while** $TS \neq \emptyset$ **do**

2:      $ts \xleftarrow{\$} TS$

3:      $TS \leftarrow TS \backslash \{ts\}$

4:      Parse TS as $\{CT_{wj}, C_{mi}\}$

5:      $T_{wj} \leftarrow dec(CT_{wj}, K_2)$

6:      **if** $T_{wj} = \perp$ **then**

7:        $dic_{wj} = \emptyset$

8:        $key \leftarrow H(T_{wj} || 0)$

9:        $value \leftarrow search(key, dic_{wj})$

10:       **if** $value = \perp$ **then**

11:         $value \leftarrow H(T_{wj} || 1) \oplus (C_{mi} || \perp)$

12:         $dic_{wj} \leftarrow dic_{wj} \cup \{key, value\}$

13:       **else**

14:         $dic_{wj} \leftarrow dic_{wj} \backslash \{key, value\}$

15:         $rk \xleftarrow{\$} \{0, 1\}^{\lambda}$

16:         $dic_{wj} \leftarrow dic_{wj} \cup \{key, \ H(T_{wj} || 1) \oplus (C_{mi} || rk)\}$

17:         $dic_{wj} \leftarrow dic_{wj} \cup \{H(rk || 0), \ H(rk || 1) \oplus H(T_{wj} || 1) \oplus value\}$

18:      $WSD \leftarrow WSD \cup dic_{wj}$

Data Retrieval

1. $SearchTokenGen(w, F, H, ct) \rightarrow (T_w, C_{UUID})$. DU provides a keyword w and generates $C_w = F(K_1, w)$. Using a hash function to generate a keyword search token $T_w = H^{ct}(C_w)$, then DU sends T_w and its encrypted identity C_{UUID} to blockchain.
2. $DataRetrieval(T_w, H, WSD) \rightarrow (EX(w))$. After blockchain receives the token T_w, TI retrieves the data index chain from the head node to the end based on T_w, then sends $Result$ to KGC. KGC traverses the $Result$ and establishes $EX(w)$ to receive all undeleted file information. The detail is shown in Algorithm 2.

Algorithm 2. $DataRetrieval(T_w, H, WSD) \rightarrow \left(EX(w)\right)$

Blockchain:
1:    $Result \leftarrow \emptyset$
2:    **while** $T_w \neq H^{FLen}(C_w)$ **do**
3:        $key \leftarrow H(T_w||0)$
4:        $value \leftarrow search(key, WSD)$
5:        **if** $value \neq \perp$ **then**
6:            $(C_{mi}||rk) \leftarrow value \oplus H(T_w||1)$
7:            **while** $rk \neq \perp$ **do**
8:                $Result \leftarrow Result \cup C_{mi}$
9:                $value \leftarrow search(H(rk||0), WSD)$
10:               $(C_{mi}||rk) \leftarrow value \oplus H(rk||1)$
11:            $Result \leftarrow Result \cup C_{mi}$
12:        $T_w \leftarrow H(T_w)$
13: Send $Result$ to KGC

KGC:
1:    $EX(w) \leftarrow \emptyset$
2:    **while** $Result \neq \emptyset$ **do**
3:        $\{C_{CID}, C_{AK}, C_{UID}, op\} \leftarrow dec(C_{mi}, SK_O)$
4:        **if** $op = 1$ **then**
5:            $EX(w) \leftarrow EX(w) \cup \{C_{CID}, C_{AK}, UID\}$
6:        **else**
7:            $EX(w) \leftarrow EX(w) \backslash \{C_{CID}, C_{AK}, UID\}$
8: Send $EX(w)$ to DU

---

3. $QueryRecordUpload(Result, C_{UUID}) \rightarrow (RC)$. After DU executing the query, the blockchain sends $RC = \{Result, C_{UUID}\}$ to SC, which includes $Result$ and the DU's encrypted identity information. Then SI saves the corresponding relationship between $UUID$ and C_{UUID} to facilitate the tracking of query information leakage.

Data Decryption

1. $Authorization(SK_{OS}, H, PK_U, SK_U) \rightarrow (SK_O)$. Before generating the search token, DU needs to request data access permission from DO. If DO agrees, DO send SK_O to KGC. If DO does not agree, a rejection is returned to DU. If DO does not respond, DU uses the RSA signature algorithm to sign the subkey request Q and sends it to other institutions. Upon verifying the signature, they send their subkey to KGC, and KGC recovers the complete key based on the Shamir key recovery algorithm as:

$$SK_O = f(0) = \sum_{i=1}^{t} f(x_i) \prod_{j=1, j \neq i}^{t} \frac{-x_j}{x_i - x_j} (\bmod p) \tag{4}$$

2. $ReEncrypt(SK_O, PK_U, EX(w)) \rightarrow (RK, EX_R(w))$. KGC generates the re-encryption key RK through the SK_O and PK_U. By comparing the UID of the DO that authorized DU with the UID in the $Result$, KGC re-encrypts the right elements in $EX(w)$ to establish a re-encrypted data receiving set $EX_R(w)$, which is sent to DU.

3. $Decrypt(EX_R(w), SK_U) \rightarrow (HD, UID)$. DU decrypts the ciphertexts in $EX_R(w)$ based on its own secret key SK_U. Find the corresponding C_{HD} from IPFS based on the CID, and use AK to decrypt the C_{HD}. Obtain the elderly health information HD.

3.3 Fish-Bone Chain Storage Structure in Blockchain

In our scheme, blockchain serves as a storage medium with indexes stored in the world state database. We use TI as a full node, and the block body include transactions for updating indexes, as shown in Fig. 2. Since blockchain transactions are public, storing update tokens directly would reveal their relationships. Therefore, these tokens are encrypted using symmetric encryption algorithm AES. TI decrypts these tokens when storing them into the local world state database. Moreover, because of the immutability of the blockchain, the index structure stored in blockchain is completely trustworthy.

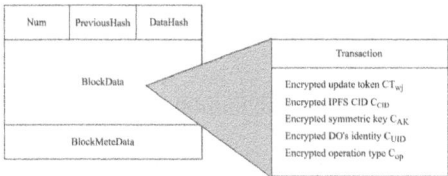

Fig. 2. Block data structure

Since the fish-bone chain sets a maximum chain length from the beginning, each update reduces the available length. We propose a method, where the fish-bone chain reaches its maximum length, a sub-chain is added to the tail node of the keyword index chain. The sub-chain retains all the structures of the fish-bone chain, thus allowing for the storage of a large number of update requests. The index structure described above is shown in Fig. 3.

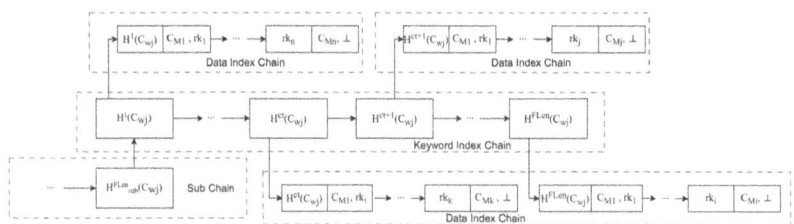

Fig. 3. Fish-bone chain structure

4 Security Analysis and Performance Evaluation

In this section, we analyze and evaluate correctness (see Sect. 4.1), security (see Sect. 4.2) and performance (see Sect. 4.3) of our scheme.

4.1 Correctness Analysis

There is one place to prove of Sect. 3.2.

Correctness Analysis of Data Retrieval. We ensure the uniqueness of the search token T_w and the head node of the data index chain by utilizing the anti-collision property of the hash function H. Then, we ensure the accuracy of retrieval as follows:

$$
\begin{aligned}
Result &= \left\{ (value_h \oplus H(T_w||1)) \bigcup\nolimits_{i=1}^{n} (value_s \oplus H(rk_i||1)) \right\} \\
&= \left\{ \begin{array}{c} (H(T_w||1) \oplus (C_{m0}||rk_0) \oplus H(T_w||1)) \\ \bigcup_{i=1}^{n} (H(rk_i||1) \oplus (C_{mi}||rk_i) \oplus H(rk_i||1)) \end{array} \right\} \\
&= \left\{ (C_{m0}||rk_0) \bigcup\nolimits_{i=1}^{n} (C_{mi}||rk_i) \right\} \\
&= \left\{ \bigcup\nolimits_{i=0}^{n} C_{mi} \right\}, \ (rk_n = \perp)
\end{aligned} \tag{5}
$$

4.2 Security Analysis

In this section, we provide the security definition and proof of Sect. 3.2.

Definition 1. For the dynamic searchable symmetric encryption scheme $\prod$ to be L-adaptively-secure. We define a leakage function $L = (L_{setup}, L_{update}, L_{search})$, for any PPT adversary A, there is a simulator S such that:

$$
\left| Pr\left[Real_A^{\prod}(\lambda) = 1 \right] - Pr\left[Ideal_{A,S,L}^{\prod}(\lambda) = 1 \right] \right| \leq negl(\lambda) \tag{6}
$$

Definition 2. Forward security means that addition operations do not leak whether a previously searched keyword is in the added files. Backward security means that the search queries on keyword w do not reveal the files which have been deleted.

Theorem 1. F is a pseudo-random function. H is a hash function. $|DB(w)|$ is the number of keyword/file pairs. $sp(w) = \{ct|(ct, w) \in Q\}$, which is the search mode of keyword w under counter ct. If the BEDSSE scheme is L-adaptively-secure, and is both forward and backward secure, then the leakage functions of our scheme are as follows:

$$
\begin{cases} L_{setup}(\lambda) = \lambda \\ L_{update}(op, DOC) = \left(\sum_{w \in W_s} |DB(w)| \right) \\ L_{search}(w) = (sp(w)) \end{cases} \tag{7}
$$

Prove. In order to prove the security of our scheme, we build a simulator S that takes the leakage function as input to simulate the real world called ideal world. We build two efficient adversary B_1, $Adv_{F,B_1}^{prf}(\lambda)$, B_2, $Adv_{H,B_2}^{hash}(\lambda)$, that distinguish between F,H and true random function. Our scheme satisfies Eq. (8), the theorem is proved.

$$
Pr\left[Real_A^{BEDSSE}(\lambda) = 1 \right] - Pr\left[Ideal_{A,S,L}^{BEDSSE}(\lambda) = 1 \right] \leq Adv_{F,B_1}^{prf}(\lambda) + Adv_{H,B_2}^{hash}(\lambda) \tag{8}
$$

4.3 Evaluation of Performance

In this section, we evaluate the efficiency of our proposed scheme. The experimental environment is based on Intel(R) Core (TM) i7–97000 CPU @ 3.00 GHz with 16 GB RAM, run on the Ubuntu 20.04.6 LTS operating system. Primarily using the Hyperledger Fabric framework, writing SDK and smart contracts in Golang.

We compare our scheme with others in terms of security and functionality. Our scheme utilizes IPFS and blockchain hybrid storage to reduce blockchain storage costs. It employs dynamic searchable encryption based on the fish-bone chain for updating and searching ciphertexts, ensuring both forward and backward security. We also optimized the storage and structure of the fish-bone chain. We implement a tiered authorization mechanism using Shamir secret sharing technique, and anonymize user identities to protect privacy. Comparison results are shown in Table 2.

Table 2. Comparison with other schemes

	MITRA [9]	CLOSE-FB [10]	PCIB-MIS [15]	Our scheme
Identity anonymity	✗	✗	✓	✓
Data update	✓	✓	✗	✓
Forward security	✓	✓	✗	✓
Backward security	✓	✓	✗	✓
Low client cost	✗	✓	✓	✓
Flexible authorization	✗	✗	✗	✓
Blockchain	✗	✗	✓	✓

As shown in Fig. 4, we compare our scheme with CLOSE-FB [10]. Our experimental data keyword count is between 1000 and 10000. Both schemes show linear increases in keyword token generation time with increasing keyword numbers, but our scheme has a time advantage. This is because the maximum length of the fish-bone chain in the CLOSE-FB [10] is fixed, with the initial value of global counter being the maximum chain length, affecting token generation time. We reduce the maximum length of the fish-bone chain by adding sub-chain at the end of the fish-bone chain, which reduces our token generation time.

As shown in Fig. 5, we compare our scheme with CLOSE-FB [10] and MITRA [9]. Our experimental data uses the number of files from 100 to 550. In the experiment, update time of all schemes increases with the number of files. Our scheme initially has lower efficiency due to using blockchain, which requires a consensus stage during data updating. However, as the number of files and keywords increases, the time consumed by blockchain consensus becomes less than update time. Our scheme and the CLOSE-FB [10] scheme use the fish-bone chain as the update index, ensuring fixed local storage costs but CLOSE-FB [10] requires more rounds of hash computations. In contrast, MITRA [9] uses an inverted index structure, where the user needs to generate a list of indexes, and the storage and update costs increase with the number of files. Thus, our scheme update time is much better compared to others.

As shown in Fig. 6, we compared the search time of three schemes as the number of matching entities. We can see that CLOSE-FB [10] with its fixed chain length, requires many rounds of hashing, resulting in lower search efficiency. Our scheme and MITRA [9] perform better in searching. However, as the number of keywords increases, the search efficiency of MITRA [9] drops because it needs to retrieve the local keyword list. Therefore, our search performance is better than other schemes.

As shown in Fig. 7, Starting from 10 blocks, we compared the number of files that could be stored on-chain and off-chain from 10 to 100 blocks. In our experiments, each block is 1MB, elderly health data is 500KB, and the information of encrypted health data is 1KB. It is evident that the hybrid approach allows for the storage of more data under the same number of blocks, significantly saving on blockchain space.

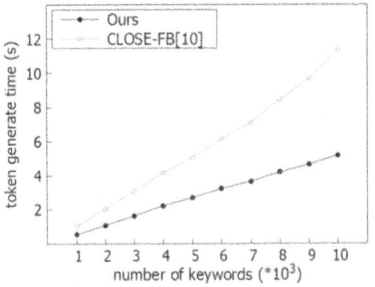

Fig. 4. Time overhead of token generation

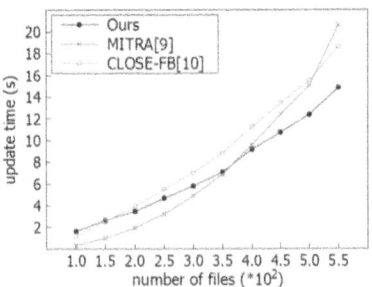

Fig. 5. Time overhead of update

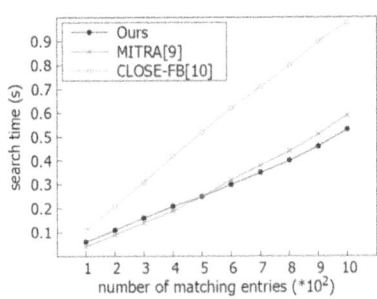

Fig. 6. Time overhead of search

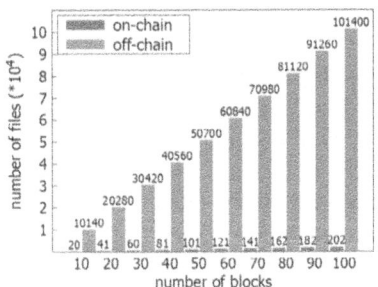

Fig. 7. Storage capacity: on-chain vs off-chain

5   Conclusion

In this paper, we propose BEDSSE, and utilize blockchain and IPFS hybrid storage to reduce blockchain overhead. We optimize the fish-bone chain structure to reduce token generation and update time, ensuring forward and backward security while enabling files updating and searching. We design a tiered authorization mechanism that allows institutional users to access elderly health data through cooperation when the elderly can't grant permission. Additionally, we introduce a supervision chain that stores user query requests and protects user privacy through identity anonymization. Finally, through security analysis and performance evaluation, we demonstrate the efficiency of our scheme. However, our scheme also has some limitations, such as reliance on iterative hashing for token generation. To implement our proposed scheme, we need to overcome the network overhead of real blockchain system and design a user-friendly interface for the elderly. In the future, we plan to optimize the limitations in our scheme.

Acknowledgments. This work is supported by the Major projects of National Social Science Foundation of China (No. 21&ZD124): research on community home care model and quality safety system based on blockchain. We also thank anonymous reviewers for the helpful reports.

References

1. Wang, C., Shen, Z., Sun, X.: A theoretical framework of elderly users' demand for smart senior care and health care. Inf. Stud. Theory Appl. **43**(11), 71–78 (2020). (in Chinese)
2. Nakamoto, S.: Bitcoin: a peer-to-peer electronic cash system (2008)
3. Xu, X., Dong, X., Li, X., He, G., Xu, S.: Patient-friendly medical data security sharing scheme based on blockchain and proxy re-encryption. In: Zhao, X., Yang, S., Wang, X., Li, J. (eds.) Web Information Systems and Applications. WISA 2022. LNCS, vol. 13579, pp. 615–626. Springer, Cham (2022). https://doi.org/10.1007/978-3-031-20309-1_54
4. Li, F., Ma, J., Miao, Y., Liu, X., Ning, J., Deng, R.H.: A survey on searchable symmetric encryption. ACM Comput. Surv. **56**(5), 1–42 (2023)
5. Song, D.X., Wagner, D., Perrig, A.: Practical techniques for searches on encrypted data. In: Proceeding 2000 IEEE Symposium on Security and Privacy (S&P), pp. 44–55 (2000). https://doi.org/10.1109/SECPRI.2000.848445
6. Kamara, S., Papamanthou, C., Roeder, T.: Dynamic searchable symmetric encryption. In: Proceedings of the 2012 ACM Conference on Computer and Communications Security, pp. 965–976 (2012). https://doi.org/10.1145/2382196.2382298
7. Stefanov, E., Papamanthou, C., Shi, E.: Practical dynamic searchable encryption with small leakage. Network & Distributed System Security Symposium (2014). https://doi.org/10.14722/ndss.2014.23298
8. Bost, R.: $\sum$ οφος: forward secure searchable encryption. In: Proceedings of the 2016 ACM SIGSAC Conference on Computer and Communications Security, pp. 1143–1154 (2016). https://doi.org/10.1145/2976749.2978303
9. Ghareh Chamani, J., Papadopoulos, D., Papamanthou, C., Jalili, R.: New constructions for forward and backward private symmetric searchable encryption. In: Proceedings of the 2018 ACM SIGSAC Conference on Computer and Communications Security, pp.1038–1055 (2018). https://doi.org/10.1145/3243734.3243833
10. He, K., Chen, J., Zhou, Q., Du, R., Xiang, Y.: Secure dynamic searchable symmetric encryption with constant client storage cost. IEEE Trans. Inf. Forensics Secur. **16**, 1538–1549 (2021). https://doi.org/10.1109/TIFS.2020.3033412
11. Sun, G., Wang, Y., Li, Z., Han, R., Wan, M., Yuan, T.: Survey of searchable encryption based on blockchain. J. Nanjing Univ. Posts Telecommun. (Nat. Sci. Edition) 65–78 (2024). (in Chinese)

12. Xu, C., Yu, L., Zhu, L., Zhang, C.: A blockchain-based dynamic searchable symmetric encryption scheme under multiple clouds. Peer-to-Peer Netw. Appl. **14**, 3647–3659 (2021). https://doi.org/10.1007/s12083-021-01202-6

13. Tang, X., Guo, C., Choo, K.K.R., Liu, Y., Li, L.: A secure and trustworthy medical record sharing scheme based on searchable encryption and blockchain. Comput. Netw. **200**, 108540 (2021)

14. Liu, J., Wu, M., Sun, R., Du, X., Guizani, M.: BMDS: a blockchain-based medical data sharing scheme with attribute-based searchable encryption. In: ICC 2021-IEEE International Conference on Communications, pp. 1–6 (2021). https://doi.org/10.1109/ICC42927.2021.9500966

15. Du, X., Liu, S., Han, Z., Huo, Z., Wang, Y.: Patient-centric medical information sharing scheme based on IPFS and blockchain. J. Comput. Appl. 1001–9081 (2024). (in Chinese)

Enhancing Medical Data Sharing with an Attribute-Based Dynamic Verifiable Searchable Encryption Scheme Using Blockchain

Shuai Hao, Xiaomei Dong$^{(\boxtimes)}$, Ziqiang Wen, and Tiezheng Nie

School of Computer Science and Engineering, Northeastern University, Shenyang 110169, China
xmdong@mail.neu.edu.cn

Abstract. Searchable encryption (SE) can retrieve ciphertext without decrypting it, effectively meeting the privacy protection needs for medical data sharing in cloud environments. However, SE applications in medical data sharing also face issues related to data security and the trustworthiness of results. To ensure that unauthorized users cannot access medical data, fine-grained access control is necessary. Additionally, since cloud servers may return tampered results to data users, the verification of returned results is also required. To this end, we propose an attribute-based dynamic verifiable searchable encryption scheme that utilizes attribute-based encryption for fine-grained access control, and guarantees forward security during the update process by maintaining the update state locally for keyword ciphertext updating. Our scheme leverages the tamper-proof feature of blockchain to verify search results by designing a smart contract, thus ensuring the trustworthiness of the results and the fairness of the verification. Combined with the designed authentication tag with cumulative and updating properties, our scheme can be applied to the dynamic data update scenarios. Finally, we formally prove the security of the proposed scheme, and experimental results show that the scheme has better efficiency in keyword encrypt and keyword search phases.

Keywords: Blockchain · Searchable encryption · Attribute-based encryption · Results verification · Forward security

1 Introduction

Traditional cloud-based medical data sharing is mainly based on plaintext data, and sensitive private information in patients' medical data may be stolen maliciously. Searchable encryption (SE) [1], as an encryption primitive, can retrieve ciphertext without decrypting it, which can effectively support medical data sharing in the cloud environment and meet users' privacy protection needs.

However, existing SE schemes also have security and functional limitations. First, in one-to-many medical data sharing scenarios, it is necessary to ensure that only authorized users can search for ciphertext based on the keywords, and many searchable encryption

C. Jin et al. (Eds.): WISA 2024, LNCS 14883, pp. 403–414, 2024.
https://doi.org/10.1007/978-981-97-7707-5_33

schemes allow users performing search tasks to retrieve ciphertext using arbitrary keywords, so that the data owner cannot control the access to specific data. Second, many existing searchable encryption schemes only support static data, and the privacy of updated data cannot be guaranteed during dynamic updating of data. Finally, in the face of the possibility that cloud servers may return maliciously tampered results, many current searchable encryption schemes overly rely on trusted entities to verify the returned results, which cannot guarantee the fairness of the verification and reliability in practice, and in addition, the verification of results in many schemes are not friendly enough to support the dynamic data update scenarios.

To this end, in order to realize effective data privacy protection in one-to-many medical data sharing scenarios and guarantee trustworthy search results, we propose an attribute-based dynamic verifiable searchable encryption scheme using blockchain, and the main contributions include:

- Fine-grained access control: combined with ciphertext-policy attribute-based encryption (CP-ABE) [2] based on linear secret sharing scheme (LSSS), our scheme embeds access policy into keyword indexes, supports fine-grained access control to ensure that only authorized users can search successfully and decrypt data.
- Forward security: our scheme supports dynamic keyword index update by maintaining the update state locally, which can ensure forward security during data update.
- Utilizing the blockchain's tamper-proof feature, our scheme supports the verification of the integrity of the search results returned from the cloud server by the blockchain smart contract, which ensures the trustworthiness of the search results and the fairness of the verification. In addition, we also design an authentication tag with cumulative and updating properties, enabling effective update during the data updating process.

The rest of this paper is organized as follows. In Sect. 2, we discuss the existing related works on searchable encryption. In Sect. 3, we describe our scheme in detail. In Sect. 4, we give the correctness and security analysis of our scheme. In Sect. 5, we provide a comparative functional analysis and experimental evaluation of our scheme, and in Sect. 6, we give the conclusion.

2 Related Works

In scenarios such as telemedicine and smart healthcare, there exists a need for one-to-many medical data sharing, and how to control user's access rights when sharing medical data to avoid users having unlimited search capabilities is an issue that requires special attention. Attribute-based encryption [3] enables users to specify access policies to realize fine-grained access control. [4] and [5] introduced attribute-based encryption in cloud environment to control access of data. Cai et al. [6], Niu et al. [7], and Guo et al. [8] implemented SE schemes in combination with ciphertext policy attribute-based encryption. By embedding an access policy on the secure index, these schemes allow searching only if user attributes are satisfied, realizing fine-grained keyword search control. However, none of these schemes consider forward security in the case of data update.

In the case of data outsourcing, there is a possibility that cloud servers may return tampered search results, and thus the search results need to be verified to ensure the integrity of the results. Chai et al. [9] first proposed a verifiable symmetric SE scheme, where the user can verify the results based on the hash sequences of the retrieval paths. Ge et al. [10] proposed a searchable encryption scheme based on symmetric encryption that supports the verification of the search results, and Wu et al. [11] proposed a verifiable forward-secure SE scheme in a multi-user environment that utilizes multiset hashing to verify data. However, these schemes still have limitations in practice, and in the actual verification process, it is often necessary to verify the search results by a trusted entity such as a data user [9, 10, 12] or an auditor [13], which cannot guarantee the fairness and reliability of the verification. Li et al. [14] applied searchable encryption to decentralized storage by using blockchain to achieve a fair search between user and cloud servers, but the verification in this scheme is static and cannot effectively support dynamic data update. Du et al. [15] proposed a blockchain-based verifiable searchable encryption scheme, supporting integrity verification of data returned from cloud servers in dynamic scenarios, but the search operation of this scheme is performed by a blockchain smart contract, which leads to lower search efficiency.

In this paper, we proposed a SE scheme that satisfies forward security by combining attribute-based encryption, which provides fine-grained control of data access rights and effectively protects privacy in scenarios of dynamic data updating. In addition, our scheme introduces blockchain to achieve fair verification of search results, and with the designed authentication tag, our scheme can realize effective verification in scenarios of dynamic data updating.

3   Scheme Design

In this section, we present the system model (see Sect. 3.1), algorithm description (see Sect. 3.2), and smart contract design of the proposed scheme (see Sect. 3.3).

3.1   System Model

The system consists of four entities: data owner (DO), data user (DU), blockchain and cloud server (CS). The system model is shown in Fig. 1.

- Data owner (DO): DO is considered as a fully trusted entity in the system. DO generates public parameters of the system, maintains its own private key, defines access policies, extracts keywords from files, generates keyword ciphertexts to construct indexes, and uploads the ciphertexts of the files to the cloud server for storage. In addition, DO also generates the private keys for the data user (DU) and delivery parameters for DU.
- Data user (DU): DU registers with DO when joining the system, generates search token related to the query keyword, deploys the verification smart contract, and sends search token to the cloud server for searching. The final search result is obtained after verification by the blockchain smart contract.
- Blockchain: the blockchain stores the authentication tag of encrypted files, and verifies the integrity of the search results returned by the cloud server by deploying the verification smart contract to ensure the trustworthiness of the search results and fair transactions.

- Cloud server (CS): CS provides data storage service in the system, and after DU submits the search token, CS performs the keyword search. In the process of keyword search, CS will verify whether DU's attributes satisfy the access policy at the same time, and when the keyword search is successful, the cloud server will return the search results to the blockchain smart contract for public verification.

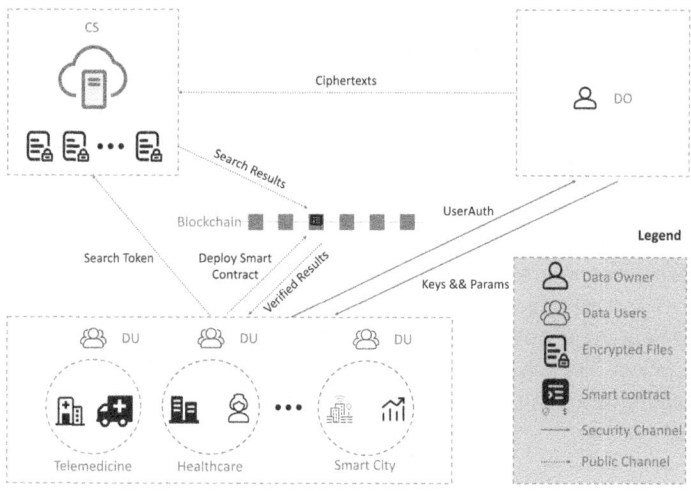

Fig. 1. System Model

3.2 Algorithm Description

The scheme proposed in this paper consists of seven probabilistic polynomial algorithms, which are described as follows:

- $Setup(1^k, U)$: DO executes the algorithm by taking the security parameter k and attribute set U as inputs, the algorithm first generates two multiplicative cyclic groups (G and G_T) of prime order p, where g is the generator of group G. After that, the algorithm initializes a bilinear mapping function $e : G \times G \to G_T$, defines collision-resistant hash function and pseudo-random functions: $H_1 : \{0, 1\}^* \to Z_p, F : \{0, 1\}^k \times \{0, 1\}^* \to \{0, 1\}^l, P : \{0, 1\}^k \times \{0, 1\}^* \to \{0, 1\}^l$, and then generates the pseudo-random function keys: $K_F \leftarrow \{0, 1\}^k, K_P \leftarrow \{0, 1\}^k$. For each attribute $att_i \in U$, the algorithm randomly selects $h_i \in_R G$ to obtain the attribute value set $U_v = \{h_1, h_2, \cdots, h_{|U|}\}$. Finally, the algorithm randomly selects the parameters $\alpha, \beta, \gamma \in Z_p$, and outputs the system public parameters and the private key of DO: $PP = \{g, e(g, g)^\alpha, g^\beta, U_v, H_1, F, P\}, SK = \{g^\alpha, \gamma, K_F, K_P\}$.
- $Encrypt(PP, SK, F, (M, \rho)) \to CT, AVT$: DO executes this algorithm and performs the following processes respectively:
- DO extracts the keyword set W from the file set F, defines a LSSS access policy $(M, \rho), M$ denotes the access matrix with l rows and n columns, and the function

ρ maps the ith row of the matrix $M_{l \times n}$ to the corresponding attribute value. DO randomly selects the vector $\vec{v} = (s, y_2, \cdots, y_n)^T \in Z_p, r_1, \cdots, r_l \in_R Z_p^*$, where s is the secret value to be shared, and computes $\lambda_i = M_i \cdot \vec{v}, i \in [1, l]$. For each keyword $w_j \in W, j = 1, \cdots, |W|$, DO randomly selects $\sigma_j \in_R Z_p^*$ as the state value, and maintains a dictionary with $(keyword, statevalue)$ as key-value pairs in the form of $\Sigma[w_j] = \sigma_j$ locally, and computes the keyword ciphertext I_w as:

$$C_{w_j} = e\left(g^{\sigma_j H_1(w_j)}, g\right)e(g, g)^{\gamma s}, C' = g^s, C_i' = g^{\beta \lambda_i}h_{\rho(i)}^{-r_i}, D_i' = g^{r_i} \qquad (1)$$

2. DO generates a symmetric key k_j for the keyword w_j and encrypts each file f_i (assuming $i = 1, \cdots, m$) containing the keyword w_j using a secure symmetric encryption algorithm (such as AES), and gets the encrypted ciphertexts ct as follows:

$$E_i = Enc_{k_j}(f_i), C_k = k_j \cdot e(g, g)^{\alpha s} \qquad (2)$$

DO then constructs the authentication tag AVT on the encrypted ciphertexts E_i as follows:

$$AVT_j = F\left(w_j\right) + F\left(\sigma_j\right) + \sum_{i=1}^{m} P(E_i) \qquad (3)$$

After getting the authentication tag, DO sends it to the blockchain for storage. Finally, DO uploads the ciphertext $CT = (I_w, ct)$ to the CS for storage.

- *UserAuth*$(PP, SK, S_{uid}) \rightarrow UK, \Sigma, K_F, K_P$: When DU joins the system, DU submits the attribute set S_{uid} to the DO, and DO selects $t \in Z_p$ randomly for each registered user, computes DU's attribute key $K = g^\alpha g^{\beta t}, L = g^t, K_x = h_x^t, x \in S_{uid}$, the trapdoor key $K_T = g^\gamma g^{\beta t}$. Finally, DO will return the private key set $UK = \{K, L, K_x, K_T\}$, the state Σ, the pseudo-random function keys K_F and K_P to DU.
- *Trapdoor*$(PP, UK, w_j) \rightarrow T_w$: DU randomly chooses $r_a \in Z_p$, computes the blinded keys $L' = g^{tr_a}, K_x' = h_x^{tr_a}, K_T' = g^\gamma g^{\beta tr_a}$, querying the state Σ to compute the trapdoor $T = g^{\sigma_j H_1(w_j)}$ based on the search keyword w_j, and finally gets the search token $T_w = \{L', K_x', K_T', T\}$.
- *Search*$(CT, T_w) \rightarrow R$: DU sends the search token T_w to the CS, CS calculates $Q_1 = e(C', K_T'), Q_2 = \prod_{i \in S}\left(e(C_i', L')e(D_i', K_i')\right)^{\omega_i}$, then CS will determine whether the following equation holds:

$$\frac{Q_1}{Q_2}e(T, g) = C_{w_j} \qquad (4)$$

If the equation holds, search is successful and indicates that DU's attributes satisfy the access policy, after which CS constructs a blockchain transaction and sends the search results $R = \{E_i, C_k, C', C_i', D_i'\}$ to the smart contract for verification.

- *Decrypt*$(R, UK, Proof) \rightarrow f_i$: When receiving the search result, DU will first judge the *Proof* value of the verification result returned by the blockchain smart contract, if *Proof* $= 0$, it means the verification fails, DU rejects the search result; if *Proof* $= 1$, it indicates that the results returned by the cloud server have not been tampered with, then DU accepts the search result. Using the attribute key, DU performs the following computations to obtain the symmetric key and finally decrypts ciphertext to obtain the plaintext of the file.

$$k_j = \frac{C_k \prod_{i=1}^{S} \left(e\left(C_i', L\right) e\left(D_i', K_i\right)\right)^{\omega_i}}{e(c_0, K)}, f_i = Dec_{k_j}(E_i) \tag{5}$$

- *Update*$(PP, SK, \Sigma, f_i') \rightarrow CT\prime$: DO executes this algorithm for keyword index update when data update exists. For the keyword $w_j \in W$ that needs to be updated, DO randomly selects $\sigma_j' \in_R Z_p^*$ as the new state value, sets $\Sigma[w_j] = \sigma_j'$, and computes the new ciphertext $CT\prime$ as:

$$C_{w_j}' = e\left(g^{\sigma_j' H_1(w_j)}, g\right) e(g, g)^{\gamma s}, E_i' = Enc_{k_j}(f_i') \tag{6}$$

DO then re-uploads the ciphertexts $CT\prime$ to the CS. Utilizing cumulative and updating properties of AVT, the authentication tag on the blockchain is updated as:

$$AVT_j' = AVT_j - F(\sigma_j) + F(\sigma_j') - \sum_{i=1}^{M} P(E_i) + \sum_{i=1}^{m'} P(E_i') \tag{7}$$

3.3   Concrete Design of Contract

In this subsection, we will present the design of smart contract deployed on the blockchain.

Result Verification Contract (RVC): DU can deploy RVC to verify the integrity of the results returned by CS. RVC works as follows: DU and CS need to agree on the search fees in advance, after which DU deploys the RVC to the blockchain. After completing the search, RVC accepts the verification token sent by DU and the search result returned by CS as input to verify the integrity of the results. The pseudo-code of RVC is shown below, following the computation process mentioned in the algorithm.

Smart Contract RVC

Input: $VerifyToken, R$
Output: $R, Proof$

1: **function** $Verify(VerifyToken, R)$
2:     $proof \leftarrow 0; Value \leftarrow \{0\}^l;$
3:     $F(w_j), F(\sigma_j) \leftarrow VerifyToken;$
4:     **for** $E_i \in R$ **do**
5:         $Value \leftarrow Value + P(E_i)$
6:     **end for**
7:     $AVT = F(w_j) + F(\sigma_j) + Value$
8:     **if** $(AVT = AVT_j) proof = 1$
9:     **else** $proof = 0$
10:    **return** $(R, Proof)$
11: **end function**

The verification process is executed by the decentralized blockchain smart contract. Assuming that CS and DU are honestly executing the search process, CS can obtain the corresponding search fees while DU can receive the correct search results, which ensures the fairness of the verification.

4 Security Analysis

In this section, we give proofs of the correctness and security of the scheme.

4.1 Correctness of Search

$$Q_1 = e\left(C', K_T'\right) = e\left(g^s, g^\gamma g^{\beta tr_a}\right) = e(g, g)^{\beta tr_a s} e(g, g)^{\gamma s} \tag{8}$$

$$
\begin{aligned}
Q_2 &= \prod_{i \in S} \left(e\left(C_i', L'\right) e\left(D_i', K_i'\right)\right)^{\omega_i} \\
&= \prod_{i \in S} \left(e\left(g^{\beta \lambda_i} h_i^{-r_i}, g^{tr_a}\right) e\left(g^{r_i}, h_i^{tr_a}\right)\right)^{\omega_i} \\
&= \prod_{i \in S} \left(e\left(g^{\beta \lambda_i}, g^{tr_a}\right) e\left(h_i^{-r_i}, g^{tr_a}\right) e\left(g^{r_i}, h_i^{tr_a}\right)\right)^{\omega_i} \\
&= e(g, g)^{\beta tr_a \sum_{i \in S} \omega_i \lambda_i} \\
&= e(g, g)^{\beta tr_a s}
\end{aligned}
\tag{9}
$$

Based on the Eqs. (8) and (9), we can continue to get:

$$\frac{Q_1}{Q_2} = \frac{e(g, g)^{\beta tr_a s} e(g, g)^{\gamma s}}{e(g, g)^{\beta tr_a s}} = e(g, g)^{\gamma s} \tag{10}$$

CS verifies whether the equation $\frac{Q_1}{Q_2} e\left(g^{\sigma_j H_1(w_j)}, g\right) = C_{w_j}$ holds, if the equation holds, it implies that the attributes of DU satisfy the access policy and the search algorithm is executed successfully. At this point, CS can send the search results R to smart contract for verification.

4.2 Security Proof

Security of Keyword
Theorem 1: (q-BDHE assumption): let G, G_T be groups of prime order p, g be a generator of G, and $e : G \times G \to G_T$ be a bilinear map. If an adversary A is given

$$
\begin{aligned}
\vec{y} &= \{g, g^s, g^a, \cdots, g^{a^q}, \cdots, g^{a^{2q}}, \\
&\forall_{1 \le j \le q} g^{sb_j}, g^{a/b_j}, \cdots, g^{a^q/b_j}, \cdots, g^{a^{2q}/b_j}, \\
&\forall_{1 \le j,k \le q, k \ne j} g^{asb_k/b_j}, \cdots, g^{a^q sb_k/b_j}\},
\end{aligned}
\tag{11}
$$

where $a, s, b_1, \cdots, b_q \in_R Z_p^*$, $Z \in_R G_T$. The challenger B tries to determine whether $Z = e(g, g)^{a^{q+1} s}$ holds. Let $Adv_{q\text{-}BDHE}^B(1^k)$ be the probability of probabilistic polynomial time (PPT) algorithm B to solve the q-BDHE problem, and q-BDHE assumption is holds if $Adv_{q\text{-}BDHE}^B(1^k)$ is negligible under the security parameter k.

Proof: In this proof, we will construct Algorithm B which will utilize adversary A to solve the q-BDHE problem. Thus, Algorithm B will simulate and play a game with adversary A based on the q-BDHE problem. If adversary A is able to win the keyword attack game with a non-negligible advantage, then challenger B will be able to solve the q-BDHE difficulty problem with a non-negligible advantage. This game consists of the following phases:

- **Initialization phase**: Given the keyword space W, challenger B takes the q-BDHE challenge $\vec{y}, Z$ as input. The adversary A outputs the challenge access structure (M^*, ρ^*), where M^* has l^* rows and n^* columns.

The challenger B chooses random $\alpha', \gamma' \in Z_p$ and sets $\alpha = \alpha' + a^{q+1}$ and $\gamma = \gamma' + a^{q+1}$ implicitly by constructing the equations: $e(g, g)^\alpha = e(g^a, g^{a^q})e(g, g)^{\alpha'}$ and $e(g, g)^\gamma = e(g^a, g^{a^q})e(g, g)^{\gamma'}$. For each x where $x \in U$, a random value z_x is chosen from Z_p. X represents the set of indices i for $\rho^*(i) = x$, then let

$$h_x = g^{z_x} \prod_{i \in X} g^{aM^*_{i,1}/b_i} \cdot g^{a^2 M^*_{i,2}/b_i} \cdots g^{a^{n^*} M^*_{i,n^*}/b_i} \tag{12}$$

Otherwise, if $X = \emptyset$, let $h_x = g^{z_x}$.

- **Query Phase I**: In this phase, the challenger B responds to adversary's private key queries for attribute set S that does not satisfy M^*.

The challenger first randomly chooses a value $r \in Z_p$. Then it finds a vector $\vec{\omega} = (\omega_1, \cdots, \omega_{n^*}) \in Z_p^*$ with the property that $\omega_1 = -1$ and for all i where $\rho^*(i) \in S$, the condition $\vec{\omega} \cdot M_i^* = 0$ holds. By the definition of LSSS, such a vector must exist. The challenger then implicitly defines t as follows:

$$r + \omega_1 a^q + \omega_2 a^{q-1} + \cdots + \omega_{n^*} a^{q-n^*+1} \tag{13}$$

The above calculation is performed by setting L as $L = g^r \prod_{i=1,\cdots,n^*} (g^{a^{q+1-i}})^{\omega_i} = g^t$. Then the challenger computes K and K_T as:

$$K = g^{\alpha'} g^{ar} \prod_{i=2,\cdots,n^*} (g^{a^{q+2-i}})^{\omega_i}, \quad K_T = g^{\gamma'} g^{ar} \prod_{i=2,\cdots,n^*} (g^{a^{q+2-i}})^{\omega_i}, \tag{14}$$

For attributes $x \in S$ for which there is no i that satisfies $\rho^*(i) = x$, the challenger can simply compute $K_x = L^{z_x}$. As for attribute x used in the access structure, the challenger constructs K_x as follows:

$$K_x = L^{z_x} \prod_{i \in X} \prod_{j=1,\cdots,n^*} (g^{(a^j/b_i)r} \prod_{k=1,\cdots,n^*,k \neq j} \left(g^{a^{q+1+j-k/b_i}}\right)^{\omega_k})^{M^*_{i,j}} \tag{15}$$

- **Challenge**: The adversary gives two keywords $w_0, w_1 \in W$ to the challenger. The challenger then flips a coin μ, chooses random value $\sigma_\mu \in_R Z_p^*$, and creates the challenge ciphertext $C_{w_\mu} = e\left(g^{\sigma_\mu H(w_\mu)}, g\right) Z e\left(g^s, g^{\gamma'}\right)$ and $C' = g^s$. Next, the challenger will choose random values $y'_2, \cdots, y'_{n^*} \in Z_p$ and divide the secret using the vector as follows:

$$\vec{v} = \left(s, sa + y'_2, sa^2 + y'_3, \cdots, sa^{n-1} + y'_{n^*}\right)^T \in Z_p^{n^*} \tag{16}$$

For $i = 1, \cdots, n^*$, defining R_i as the set that satisfies $\rho^*(i) = \rho^*(k)(k \neq i)$ for all k, other challenge ciphertext components can be then generated as follows:

$$C_i' = h_{\rho^*(i)}^{r_i'} \left(\prod_{j=2,\cdots,n^*} (g^a)^{M_{i,j}^* y_j'} \right)(g^{b_i s})^{-z_{\rho^*(i)}} \left(\prod_{k \in R_i} \prod_{j=1,\cdots,n^*} (g^{a^j s(b_i/b_k)M_{k,j}^*}) \right)$$

(17)

$$D_i' = g^{-r_i'} g^{-sb_i}$$

(18)

- **Query Phase II**: Repeat as **Phase I**.
- **Guess**: A outputs a guess μ' for μ, if $\mu = \mu'$, then A wins, showing that A can win the keyword attack game with a non-negligible advantage, then B can solve the q-BDHE difficulty problem with a non-negligible advantage, but this contradicts the q-BDHE difficulty assumption, so A cannot recover the keyword information from the keyword ciphertext if the q-BDHE assumption holds.

Forward Security

Our scheme sends the encrypted keyword ciphertext using the new state to the CS during the updating process, which is indistinguishable from a random value under the guarantees of the hard problem assumption, any external adversary cannot compute it in possession of only public information. Since the keyword is encrypted using the new state, the previous trapdoor can no longer be used to search for the updated data. Therefore, our scheme ensures forward security.

Security of Authentication Tag

The authentication tag in our scheme ensures unforgeability. Any external adversary who can forge a tag that can pass the verification means that the adversary can give a valid pseudo-random function output without knowing the key K_P of the pseudo-random function p, which contradicts the security of the pseudo-random function. In addition, due to the introduction of the state value $F(\sigma_j)$ in the authentication tag, it can effectively resist replay attacks during the data update process.

5 Performance Analysis and Evaluation

5.1 Function Comparison

We compare the functionality of our scheme with some attribute-based searchable encryption schemes proposed in recent years, the comparison results are shown in Table 1. The comparison results show that our scheme has certain advantages in functional characteristics, ensures forward security by updating the keyword index during data updating, and also supports the verification of search results utilizing the authentication tag with updating and accumulating characteristics. Additionally, our scheme utilizes the tamper-proof feature of the blockchain and the designed smart contract to perform the results verification, which can ensure the fairness of verification. In summary, our scheme provides comprehensive functionality compared with other schemes.

Table 1. Function Comparison

	Ref [7]	Ref [8]	Ref [12]	Ours
Access Structure	Access Tree	LSSS	Access Tree	LSSS
Searchable	✓	✓	✓	✓
Forward security	✗	✗	✗	✓
Blockchain	✓	✗	✓	✓
Verifiable	✗	✗	✓	✓
Fair verification	✗	✗	✗	✓

5.2 Experiment Evaluation

We implement our scheme with two other schemes and compare the performance of Encrypt (keyword encryption process) and Search algorithms of the different schemes. Our experimental environment is under Ubuntu 20.04 with Intel(R) Core (TM) i7-8750H CPU @ 2.20 GHz and 4 GB RAM, using the bilinear pairing package based on 512-bit elliptic curve group. The results of the experiments are shown in Figs. 2, 3, 4, and 5.

Figure 2 shows the performance of the Encrypt algorithm with the number of keywords kept constant and the number of attributes gradually increased. It can be seen that our scheme and scheme [7] are much better than scheme [8] in terms of efficiency performance, and as the number of attributes increases, the time overhead curve of our scheme basically stays the same and much lower, with higher efficiency.

Figure 3 shows the performance of the Search algorithm when the number of keywords is kept constant and the number of attributes is gradually increased. The search time of all the schemes grows linearly with the increase of the number of attributes. The search time of our scheme is higher than scheme [7] at first, but scheme [7] needs to perform recursive operations when executing the keyword search due to the use of an access tree as the access policy, and the search time will be much higher when the recursive reaches a certain depth. Our scheme has slower growth with better performance.

Figure 4 shows the performance of Encrypt algorithm in the case that the number of attributes remains unchanged and the number of keywords gradually increases. It can be seen that under the above conditions, the keyword encryption time overhead of all the schemes grows linearly with the increase of the number of keywords, but the time growth of our scheme and scheme [7] is smaller, thus, our scheme can be better applied to the sharing of medical data.

Figure 5 shows the performance of Search algorithm when the number of attributes is kept constant and the number of keywords is gradually increased. Although the time overhead of scheme [7], scheme [8] and our scheme rises with the increase of the number of keywords, but our scheme performs better and is more suitable for large-scale data sharing scenarios.

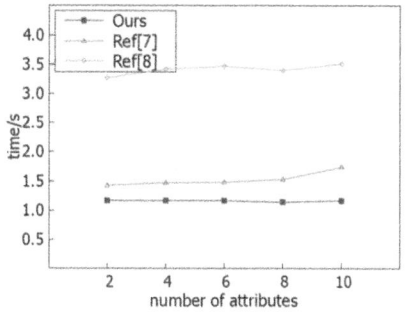

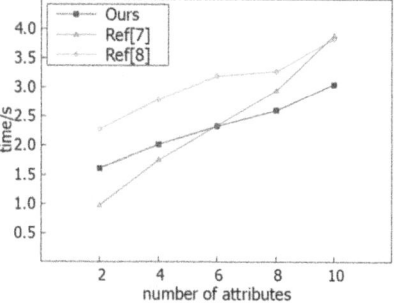

Fig. 2. Encrypt time of fixed keywords is 500   **Fig. 3.** Search time of fixed keywords is 500

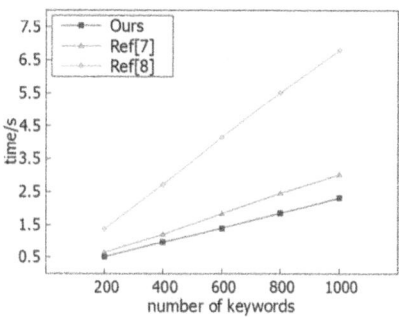

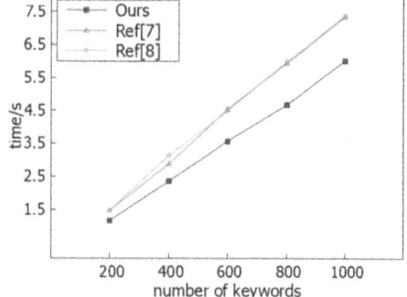

Fig. 4. Encrypt time of fixed attributes is 10   **Fig. 5.** Search time of fixed attributes is 10

6 Conclusion

In this paper, we focus on the privacy protection problem during medical data sharing and propose a blockchain-based attribute-based dynamic verifiable searchable encryption scheme. Our scheme enables fine-grained access control of user rights during data access, ensuring forward security by updating the keyword ciphertext through updating the state value when the data update occurs. By designing a secure and reliable blockchain smart contract to replace the traditional trusted entity to verify the results, the trustworthiness of the search results and the fairness of the verification are ensured. Additionally, we design an authentication tag with cumulative and updating properties, it is more friendly to support dynamic data updating scenarios. In the future, considering access policy attribute hiding and data traceability will be our main improvement direction.

Acknowledgments. This work is supported by the Major projects of National Social Science Foundation of China (No. 21&ZD124): research on community home care model and quality safety system based on blockchain. We also thank anonymous reviewers for the helpful reports.

References

1. Song, D.X., Wagner, D., Perrig, A.: Practical techniques for searches on encrypted data. In: Proceeding 2000 IEEE Symposium on Security and Privacy (S&P), pp. 44–55 (2000). https://doi.org/10.1109/SECPRI.2000.848445

2. Bethencourt, J., Sahai, A., Waters, B.: Ciphertext-policy attribute-based encryption. In: IEEE Symposium on Security and Privacy (S&P), pp. 321–334 (2007). https://doi.org/10.1109/SP.2007.11

3. Sahai, A., Waters, B.: Fuzzy identity-based encryption. In: Cramer, R. (eds.) Advances in Cryptology – EUROCRYPT 2005. EUROCRYPT 2005. LNCS, vol. 3494, pp. 457–473. Springer, Berlin, Heidelberg (2005). https://doi.org/10.1007/11426639_27

4. Wang, S., Zhang, D., Zhang, Y., Liu, L.: Efficiently revocable and searchable attribute-based encryption scheme for mobile cloud storage. IEEE Access **6**, 30444–30457 (2018)

5. Zhang, N., Dong, X., Wen, Z., Yang, C.: Secure mutual aid service scheme based on blockchain and attribute-based encryption in time bank. In: Yuan, L., Yang, S., Li, R., Kanoulas, E., Zhao, X. (eds.) Web Information Systems and Applications. WISA 2023. LNCS, vol. 14094, pp. 403–414. Springer, Singapore (2023). https://doi.org/10.1007/978-981-99-6222-8_34

6. Cai, C., Weng, J., Yuan, X., Wang, C.: Enabling reliable keyword search in encrypted decentralized storage with fairness. IEEE Trans. Dependable Secure Comput. **18**(1), 131–144 (2018)

7. Niu, S.F., Xie, Y.Y., Yang, P.P., Du, X.: Cloud-assisted attribute-based searchable encryption scheme on blockchain. J. Comput. Res. Dev. **58**(4), 811–821 (2021). (in Chinese)

8. Guo, L., Li, Z., Yau, W.C., Tan, S.Y.: A decryptable attribute-based keyword search scheme on ehealth cloud in Internet of Things platforms. IEEE Access **8**, 26107–26118 (2020)

9. Chai, Q., Gong, G.: Verifiable symmetric searchable encryption for semi-honest-but-curious cloud servers. In: 2012 IEEE international conference on communications (ICC), pp. 917–922 (2012). https://doi.org/10.1109/ICC.2012.6364125

10. Ge, X., et al.: Towards achieving keyword search over dynamic encrypted cloud data with symmetric-key based verification. IEEE Trans. Dependable Secure Comput. **18**(1), 490–504 (2019)

11. Wu, A., Yang, A., Luo, W., Jinghang, W.: Enabling traceable and verifiable multi-user forward secure searchable encryption in hybrid cloud. IEEE Trans. Cloud Comput. (2022)

12. Zheng, Q., Xu, S., Ateniese, G.: VABKS: verifiable attribute-based keyword search over outsourced encrypted data. In: IEEE INFOCOM 2014-IEEE Conference on Computer Communications, pp. 522–530 (2014). https://doi.org/10.1109/INFOCOM.2014.6847976

13. Soleimanian, A., Khazaei, S.: Publicly verifiable searchable symmetric encryption based on efficient cryptographic components. Des. Codes Crypt. **87**(1), 123–147 (2019)

14. Li, H., Tian, H., Zhang, F., He, J.: Blockchain-based searchable symmetric encryption scheme. Comput. Electr. Eng. **73**, 32–45 (2019)

15. Du, R., Wang, Y.: Verifiable blockchain-based searchable encryption with forward and backward privacy. In: 2020 16th International Conference on Mobility, Sensing and Networking (MSN), pp. 630–635 (2020). https://doi.org/10.1109/MSN50589.2020.00105

Principal Component Analysis Scheme Based on Homomorphic Encryption in a Distributed Environment

Dong Wang[1,2]([✉]), Ming cheng Ma[1,2], Lingli Liu[1,2], Xiongxiong Du[1,2], Xiaoruo Li[1,2], and Bingnan Zhu[1,2]

[1] Henan International Joint Laboratory of Intelligent Network Theory and Key Technology, Henan University, Kaifeng 475000, China
juliawdd@henu.edu.cn
[2] School of Software, Henan University, Kaifeng 475000, China

Abstract. Principal component analysis (PCA) is a widely used technique in the field of machine learning and one of the main dimensionality reduction methods. PCA can convert high-dimensional data into lower-dimensional representations, thereby helping to extract important information from data, and is an effective tool for data analysis and pattern recognition. In this paper, we propose a scheme for executing PCA on encrypted data based on homomorphic encryption in a distributed environment. In a distributed setting, computing nodes do not need to perform matrix operations such as matrix multiplication or addition, which reduces the computational burden on the nodes. To ensure the security of the data, we employ the CKKS homomorphic encryption scheme, which allows for approximate calculations on real numbers, meeting the requirements of machine learning. Additionally, we horizontally partition the data to facilitate its application in distributed computing. Furthermore, we optimize the detailed computation of HPCA (Homomorphic PCA), yielding promising results on various datasets.

Keywords: Homomorphic Encryption · Distributed Computing · Principal Component Analysis (PCA) · CKKS scheme · Goldschmidt's Algorithm

1 Introduction

Principal Component Analysis (PCA) [1], as a classic multivariate data analysis method, has a long history and wide-ranging applications. Its basic principles date back to the early 20th century and were proposed by statisticians such as Karl Pearson. However, its widespread application in the fields of statistics and data analysis gradually became mainstream in the late 20th century and the end of the 20th century. Since the 80s of the last century, with the rapid development of computer technology and the popularization of multivariate data analysis methods, principal component analysis (PCA) has attracted more and more attention. During this period, there was a growing awareness of the urgent need for solutions to problems such as data dimensionality reduction, feature extraction, and data visualization. PCA, as a simple and effective method, gradually became an important tool for addressing these issues. With the improvement of computer hardware performance and the development of computer software, the computation of PCA has become more efficient and feasible. This has provided broader application space for PCA in multiple fields, including statistics, pattern recognition, image processing, signal processing, and data mining. In these fields, PCA is widely used for tasks such as data dimensionality reduction, feature extraction, data compression, data visualization, and pattern recognition, providing important theoretical foundations and practical methods for data analysis and processing. Principal component analysis (PCA) is one of the most popular dimensionality reduction methods in the field of machine learning. The main idea of PCA is to convert the original n-dimensional features into a lower-dimensional space (usually k-dimensional), where these k-dimensional features are entirely new orthogonal features, also known as principal components. These principal components are obtained by reconstructing the original n-dimensional features to form a k-dimensional feature space.

PCA has a wide range of applications in a variety of fields, from biomedical science to financial quantitative analysis. One of the most common applications is facial recognition. However, the widespread use of PCA has raised some privacy concerns and issues. With the increasing prevalence of these privacy concerns, the need for performing PCA on encrypted data has become a current requirement. Existing research techniques to address these issues include homomorphic encryption (Hardy et al., 2017) [2] and secure multiparty computation (Mohassel and Zhang, 2017) [3]. However, these approaches present certain challenges. For some homomorphic encryption schemes (Privacy Preserving PCA for Multiparty Modeling) [4], a trusted third party is required for data decryption and re-encryption. Secure multiparty computation schemes involve repeated communication, increasing the number of communication rounds and posing risks of privacy leakage for both approaches.

Several studies [5–7, 20], and [8] have confirmed the application of the CKKS scheme in various machine learning techniques. This paper focuses on the principal component analysis (PCA) scheme based on CKKS homomorphic encryption scheme in a distributed environment. Previously, there were few endeavors to conduct Principal Component Analysis (PCA) on encrypted data. However, Lu et al. [9] and Rathee et al. [10] utilized the BGV scheme for PCA. However, they only applied PCA to datasets with relatively few attributes (≤ 20). Panda [11] first proposed large-scale data PCA based on CKKS, which can handle datasets with more attributes. However, during encrypted computation, there were occasional instances of non-convergence of some data, leading to significant computational errors.

2 Related Work

2.1 CKKS Homomorphic Encryption Scheme

The CKKS (Cheon-Kim-Kim-Song) [12] scheme is a hierarchical homomorphic encryption scheme, and its security is based on the difficulty of the RLWE (Ring Learning With Errors) problem. The RLWE problem is a challenging mathematical problem in cryptography, and by solving the RLWE problem, the CKKS scheme can provide secure encryption and homomorphic computing functions. Compared to other homomorphic encryption schemes, the CKKS algorithm has unique characteristics: it performs approximate computations and supports encrypted computations of real and complex numbers in ciphertext environments. This feature aligns well with the computation methods commonly used in machine learning, as most computations in machine learning are performed in an approximate manner.

For an integer power M of 2, let $\varnothing$-$M.(X)$ be the M-th cyclotomic polynomial of degree $N = \emptyset M$, $\emptyset Mx = x - \xi...x - \xi j...(x - \xi M - 1)$, where $\xi = e^{\frac{2\pi i}{M}}$. The ring $R = \mathbb{Z} x/\emptyset M$ x and the key distribution X key, the noise distribution X $error$, the required distribution X enc for encryption, and the integer M and P modulus qL are selected. The CKKS homomorphic encryption scheme contains the following algorithms:

Key generation: KeyGen(1^λ) Sampling $s = Xkey$, $y = RqL$, $e = Xerror$, generate private key $SK = (1, s)$, public key $pk = (x, y)$, where $x = -ys + e$ modq L. Sampling $y' = RqL2$, $e' = Xerror$. Let $evk = (x', y')$, where $x' = -y's + e' + Ps^2$.

Encrypt $Encpk(m)$: For the plaintext polynomial $m \in R$, sampling $a = Xenc$, $e0 = e1 = Xerror$, which outputs a password $c = a * pk + m + e0,e1$.

Decrypt $Decsk(s)$: Enter a ciphertext at the l layer x, calculate $m' = m + e = (s,sk)(modqL)$.

Addition $addition(s1,s2)$: Input the ciphertext $s_1, s_2 \in R^2_{qL}$, out put $S = s1, + s2(modqL)$.

Multiplication $multiplication(s 1,s 2)$: Input the ciphertext $s 1, s 2 \in Rq L2$, let $z0, z1, z2 = (x1x2, y1x2 + y2x1, y1y2)(modqL)$, out put $S = z0,z1 + P - 1 * d2 * evk(modqL)$.

Revolve $revolve_{gk}(s, r)$: Input the ciphertext $s \in R^2_{qL}$, and spin convert the key gk, the plaintext vector of output s is rotated r bits Ciphertext s'.

Rescale $Rsl \to l'(s)$: Input the ciphertext $s \in Rq L2$, and out put $s' = (ql'ql * s)(mod ql')$.

2.2 Principal Component Analysis (PCA)

PCA is a widely used data processing technology, and its main role is to reduce the dimensionality of data and extract the features of matrices in data analysis. Its core idea is to transform high-dimensional data into a lower-dimensional space while identifying the main features that preserve the maximum variance of the original data. Specifically, given a vector x, PCA aims to find a projection direction u such that the variance of the vector x projected along this direction is maximized. In other words, PCA seeks to identify a dimension u that retains as much of the original data's variability as possible after projection. Therefore, PCA can be described as a problem of maximizing variance.

$$\max_{\mu} \frac{1}{N} \sum_{i=1}^{N} \left(u^T x_i - u^T \mu \right)^2$$

$$\max_{\mu} \frac{1}{N} \mu^T \sum \mu$$

$$||u|| = 1$$

where μ is the mean vector, and $\Sigma = i = 1 n x i - \mu x i - \mu T$ is the covariance matrix.

Through the above formula, we can derive that finding the maximum eigenvector of the matrix corresponds to maximizing the variance problem. Solving the problem of maximizing variance can typically be achieved through iterative methods like the power iteration. In the power iteration, we repeatedly multiply a random vector u by the matrix A to find the leading eigenvector of A. With continuous iteration, the matrix product result Au will converge towards the main eigenvector.

In the First Principal Component algorithm, we eschew the use of the covariance matrix Σ, opting instead for the sum of outer products of xi. This approach allows us to store just a single vector rather than the entire covariance matrix Σ, which can be more efficient in a homomorphic setting. Furthermore, we circumvent the need for any matrix operations. If we seek subsequent eigenvectors (e.g., the 2nd, 3rd, or L-th largest), we employ an eigenvalue shifting process. By combining this process with power iteration, we can obtain the L largest eigenvectors of the covariance matrix Σ, representing the top L principal components of the given data matrix X.

Algorithm   First principal component

Input: s : Row vector data matrix
Output: y,q: The maximum eigenvalue of X and the corresponding eigenvector.
for i=1 to t do:
  $z = \sum_{s \in S} S^t(s * w)$
  $y = ||z||$
  $q = \frac{z}{||z||}$
end for
return y,q

---

3 The Proposed Method

In this scheme, distributed computing is employed, involving a data owner, users, and multiple computation nodes. The data owner performs initial operations on the data, including data partitioning and encryption. Computation nodes receive encrypted data from the data owner for computation and return the final results to the data owner. The data owner decrypts and integrates the data.

In data encryption, we adopt the CKKS homomorphic encryption scheme and utilize ciphertext packing technique, ensuring that the ciphertext consists of vectors with approximately equal lengths. The advantage lies in the ability to directly operate on polynomials of the ciphertext's sub-ciphertext lengths in subsequent computations, rather than polynomials of the ciphertext lengths. Moreover, in our proposed scheme, during Principal Component Analysis (PCA) calculations under ciphertext state, there is no need for matrix multiplication and division operations. This significantly reduces the computational time overhead. Thus, we ensure data security while enhancing computational efficiency, providing a feasible solution for data privacy protection and efficient data processing.

3.1 Vector Operations

We treat all vectors as row vectors and focus only on row-filled vectors in the ciphertext. We ensure that the number of zeros in each partition is the same by evenly distributing the number of ciphertext slots across all vectors. We assume that each vector is d-dimensional and the total number of ciphertext slots is N. At this time, the number of partitions in the ciphertext is N/k, where k is a factor of N greater than or equal to d. In this scenario, we assume that N is an integer power of 2. Therefore, calculating the value of k is equivalent to calculating the nearest power of 2 that is greater than or equal to d. We can use the binary search method, which can ensure that the entire process is completed efficiently in $O(\log\log(N))$ steps.

Element-wise summation: Considering that we treat the size of the ciphertext as a power of 2, to perform homomorphic addition on all elements in the ciphertext, we need to achieve this by rotating the ciphertext by powers of 2 and adding it to itself.

```
Algorithm    Sum (X)

Input: x : Ciphertext
Output:  y : Sum of all the elements in ciphertext
z=Ciphertext()
y = x
t = log(N)
for i=1 to t do:
   z = levorotation(y, 2^{i-1})
   y=y+z
end for
return y
```

```
Algorithm    Partial sum (X)

Input: x : Ciphertext
Output: y : sum of all subciphertexts given values
W = ciphertext(1,1,1,1…||0,0,0,0…)
y = ciphertext(x)
t = log(d) - 1
for i= t to 0 do:
   z = dextrorotation(w, 2^l)
   X1 = multiplication(w,y)
   X2 = multiplication(z,y)
   X2 = levorotation(X2, 2^l)
   Y = X1 + X2
   if i>0 then:
      z = levorotation(z, 2^l + 2^{l-1})
      w = multiplication(z,w)
   end if
end for
l = log(d)
for i=0 to l do:
   z= dextrorotation(y, 2^l)
   y=y+z
end for
return y
```

The partial sum of the ciphertext refers to the sum of all elements within each sub-ciphertext of the given ciphertext. Let $x = [u'_1, u'_2, u'_3, \ldots, u'_i,]$ be a ciphertext, where $u'_i = (u_i \| 0,0, \ldots, 0)$ is a sub-ciphertext. The partial sum of x is $y = [Su_1, Su_2, Su_3, \ldots, Su_i]$, where $Su_i = \left(\sum_{j=1}^{k} u_{ij}, \sum_{j=1}^{k} u_{ij}, \ldots, \sum_{j=1}^{k} u_{ij}, \right)$

Inner product: The inner product refers to the operation on two ciphertexts, where one ciphertext x contains all sub-ciphertexts, and the other ciphertext y contains i sub-ciphertexts. We can calculate the inner product between these two ciphertexts by multiplying each corresponding element of them and then summing the resulting partial ciphertexts.

```
Algorithm    Inner product (x,y)
```

```
Input: x : Ciphertext, y : Vector packed Ciphertext.

Output: z : Inner product vectors j and y

z= multiplication (x,y)

z=partialsum(z)

return z
```

3.2 Homomorhpic Goldschmidt's Algorithm

Goldschmidt's algorithm [13] is an iterative algorithm designed to be used to calculate the value of a fraction. The algorithm gradually approximates the exact value of the score through a series of iterative steps. An improved version of the Goldschmidt algorithm can iteratively compute the square root of a given value as well as the reciprocal of the square root. The Goldschmidt algorithm is faster than directly using Newton's iteration method [14], but it requires an approximate value of $\frac{1}{\sqrt{x}}$. In this paper, an approximation algorithm (Algorithm 1) is used to obtain this approximate value, and the Goldschmidt algorithm is improved to conform to the ciphertext operations of the CKKS homomorphic encryption.

```
Algorithm 1    Approximation of 1/√x
```

```
Input: x , d , [a , b] , µ , f , g , ε

Output: guess : Approximation of 1/√x

Using pivot-tangent method to find x2,P,x1

Calculate β(P, x) and y0 = f(x)

Count Newton's iterations d times to get yd

Return guess
```

To approximate the value of $\frac{1}{\sqrt{x}}$, Algorithm 1 can compute an approximate value for $\frac{1}{\sqrt{x}}$ given the value of x. It utilizes the secant method to compute the reciprocal square root and is adapted from Panda [15]. Initially, two straight lines are used to approximate the curve $\frac{1}{\sqrt{x}}$. Within a larger interval, there is a line that approximates the curve and captures a slow decline in the value trend, while the other line represents a rapid increase in the value trend of the curve. The intersection of these two lines is known as the pivot point. Here, l and r denote the upper and lower bounds of the initial guess, respectively, where:

$$\beta(x) = \varphi\left(\frac{P}{r-l}, \frac{x}{r-l}\right)$$

$$\varphi(a, b) = \frac{1 + \mathrm{sgn}(a - b)}{2}$$

$$f(x) = \beta(x) \cdot L1(x) + (1 - \beta(x)) \cdot L2(x).$$

With the Goldschmidt algorithm, we are able to simultaneously compute the square root and its reciprocal. The enhanced homomorphic Goldschmidt algorithm described in Algorithm 2 is a variant of the Goldschmidt algorithm [16], which integrates both multiplication and addition operations:

Algorithm 2   Homomorphic_Goldschmidt algorithm

```
Input: d: Ciphertext
Output: z: Norm of X, t: Inverse of norm of X
Minus = Ciphertext(-1.0)
Zero-five = Ciphertext(0.5)
w = Re-encrypt (PartialSum(square(s)))
t = multiplication (w, Zero-five)
w1 = Inv_Norm(w)
y = Re-encrypt (w1)
z = multiplication (w, y)
for i=1 to do:
exchange = multiplication (multiplication (s, t), Minus)
r = Ciphertext_Plus (exchange, Zero-five)
z = Ciphertext_Plus (w, multiplication (z, r))
t = Ciphertext_Plus (h, multiplication (t, r))
if depth(w) < 3:
    z = Re-encrypt (z)
    t = Re-encrypt (t)
end if
end for
t=2*t
return t , z
```

3.3 Homomorphic PCA

We utilize the homomorphic CKKS encryption scheme to implement homomorphic Principal Com-ponent Analysis (PCA). The specific implementation details are described in Algorithm 3.

In the algorithm3 Homomorphic PCA, we employ the homomorphic power method for homo-morphic computations, it is used to calculate the first l principal components on the encryption matrix X. First, we initiate the process by finding the maximum eigenvalue of the covariance matrix. Then, we iteratively find the subsequent eigenvectors using the eigenvalue shifting method. To facilitate this process, we also employ Algorithm 2 to compute norms and the inverse of norms.

```
Algorithm 3     Homomorphic PCA
```

```
Input: s: Ciphertext
Output:  y: The first L principal components of s
Minus ← Ciphertext(-1.0)
for i=1 to L do:
   for j = 1 to d do:
```
$\quad\quad$ x1 $= \sum_{s\in S} multiplication(s, innerproduct(s,r))$
```
      x1 = SubSum(x1)
      x1 = Re-encrypt (x1)
```
$\quad\quad$ x2 $= \sum_{\lambda\in\Lambda, w\in W} multiplication(\lambda, multiplication(w, innerproduct(w,r)))$
```
      x2 = Re-encrypt (x2)
      x2 = multiplication (x2, Minus)
      s = Ciphertext_Plus (x1, x2)
      inv ,val = Homomorphic_Goldschmidt(s)
      r = Re-encrypt (multiplication (inv, s))
   end for
y←y∪r
Λ←Λ ∪ val
return y
```

We use algorithm 4 to perform PCA in a homomorphic manner in a distributed environment. In Algorithm 4, the data owner first divides the data horizontally and then calculates the average vector for each subset and subtracts the average vector from all other vectors so that the data matrix is cen-tered on its average. Then, by performing PCA calculations on the encrypted data, algorithm 3 is used to obtain the principal component of the original data from the computing node. Finally, the principal components are multiplied by the data owner to obtain a low-dimensional representation of the origi-nal data.

```
Algorithm 4    DHPC
```

```
Input: x: raw data
Output: y: data after dimensionality reduction
Data Owner:
Horizontally divide the dataset x
Calculate the average vector
Encrypted data is distributed to compute nodes
Receive compute node data for decryption, multiplication, and inte-
gration
Compute nodes:
    Receive ciphertext sent by the data owner
    A PCA calculation is performed and the result is returned to
the data owner
```

---

4 Implementation Details and Results

We have implemented the DHPC algorithm described in this article using Python. We utilize SEAL-Python [17] (the Python binding of the Microsoft SEAL library [18]) to implement the CKKS scheme on python. All experiments were conducted on a machine equipped with an Intel(R) Core(TM) i5-8250U CPU @ 1.60 GHz 1.80 GHz, running on a 64-bit Windows 10 environment with PyCharm Community edition, 15.9 GB of RAM, and 921 GB of disk capacity. We conducted experiments on the Yale Face Database [19], computing the top principal components of the dataset and validating the accuracy of our algorithm by calculating the R2 score. The efficiency of our algorithm is represented by the runtime of the principal component analysis.

The Yale Face Database has been appropriately scaled to ensure that its eigenvalues are sufficiently small for proper data handling. Originally composed of three channels, the database was subsequently transformed into a single channel. Subsequently, the images, initially sized at 195×231 pixels, were resized to 16×16 pixels using bicubic interpolation.

We have re-encrypted some parameters to eliminate noise. Ideally, re-encryption can be replaced by bootstrapping without altering any part of the algorithm. We ensure that no ciphertext in the input needs to be re-encrypted, making the frequency of re-encryption independent of the dataset. To achieve this, we observe that the maximum depth required for each ciphertext in Algorithm 3 is only $\log(k) + 2$. To further reduce the frequency of re-encryption, we also limit the number of re-encryptions for feature vectors. Because of this, we set the maximum re-encryption count to $\log(k) + 3$ in Algorithm 3.

We initially compute the first few principal components of the dataset and utilize the R2 score of the reconstructed data as our evaluation metric. We also validate the efficiency of our algorithm by comparing the results obtained with Panda's HPCA method. Typically, R2 scores between 0.3 and 0.7 are considered very suitable for original data.

The specific results are presented in Table 1, where s denotes the dimensionality of each vector, k represents a factor of N greater than or equal to d, N denotes the total number of ciphertext slots, x signifies the sample size, l indicates the number of retained maximum eigenvalues, c denotes the number of participating parties, D represents the depth of computation, R2 signifies the calculated score, and Time indicates the runtime of principal component analysis in minutes.

Table 1. Experimental comparison of HPCA and DHPCA

scheme	s	k	N	x	l	c	D	R2	Time
HPCA	256	256	32768	165	4	1	11	0.529	9.75
	256	256	32768	165	5	1	11	0.573	13.82
	256	256	32768	165	6	1	11	0.579	16.24
DHPC	256	256	32768	165	4	5	11	0.626	5.15
	256	256	32768	165	5	5	11	0.642	7.22
	256	256	32768	165	6	5	11	0.649	9.61

From Table 1 Experimental comparison of HPCA and DHPCA, we observe that our distributed PCA algorithm performs better on encrypted data compared to Panda's solution, with an 18% improvement in R2 score and a 1.89x increase in computation time.

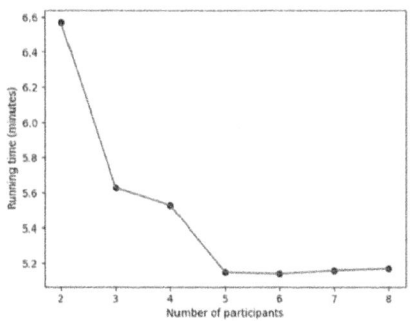

Fig. 1. Number of compute nodes

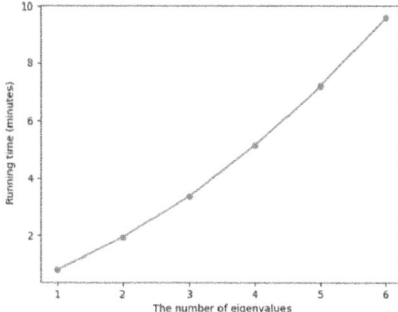

Fig. 2. Number of eigenvalues

Figure 1 shows the running time of different compute nodes in the distributed state, and it can be found that the running time tends to be stable when there are more than five nodes. Figure 2 shows the running time when different amounts of principal components are retained at five compute nodes, and it can be found that the more principal component eigenvalues are retained, the longer the running time.

5   Conclusion

This paper introduces a novel distributed Principal Component Analysis (PCA) scheme based on CKKS fully homomorphic encryption in a distributed environment. Initially, the data owner horizontally partitions the data and encrypts it using CKKS before distribution. Subsequently, each node receives the ciphertext and performs computations, returning the results to the data owner. Finally, the data owner decrypts the computations and integrates the data. While our approach ensures data security, the computational overhead in the ciphertext state remains substantial. Future work will focus on streamlining computations to enhance efficiency.

References

1. Pearson, K.: Principal components analysis. Lond. Edinb. Dublin Philos. Mag. J. Sci. **6**(2), 559 (1901)
2. Hardy, S., Henecka, W., Ivey-Law, H., et al.: Private federated learning on vertically partitioned data via entity resolution and additively homomorphic encryption. arXiv preprint arXiv:1711. 10677, 2017
3. Mohassel, P., Zhang, Y.: Secureml: a system for scalable privacy-preserving machine learning. In: 2017 IEEE Symposium on Security and Privacy (SP), pp. 19–38. IEEE (2017)
4. Liu, Y., Chen, C., Zheng, L., et al.: Privacy preserving PCA for multiparty modeling. arXiv preprint arXiv:2002.02091, 2020
5. Crockett, E.: A low-depth homomorphic circuit for logistic regression model training. Cryptology ePrint Archive, 2020
6. Boura, C., Gama, N., Georgieva, M., et al.: Chimera: combining ring-LWE-based fully homomorphic encryption schemes. J. Math. Cryptol. **14**(1), 316–338 (2020)
7. Boura, C., Gama, N., Georgieva, M., Jetchev, D.: Simulating homomorphic evaluation of deep learning predictions. In: Dolev, S., Hendler, D., Lodha, S., Yung, M. (eds.) Cyber Security Cryptography and Machine Learning. CSCML 2019. LNCS, vol. 11527, pp. 212–230. Springer, Cham (2019). https://doi.org/10.1007/978-3-030-20951-3_20
8. Kim, M., Song, Y., Wang, S., et al.: Secure logistic regression based on homomorphic encryption: design and evaluation. JMIR Med. Inform. **6**(2), e8805 (2018)
9. Lu, W., Kawasaki, S., Sakuma, J.: Using fully homomorphic encryption for statistical analysis of categorical, ordinal and numerical data. Cryptology ePrint Archive, 2016
10. Rathee, D., Mishra, P.K., Yasuda, M.: Faster PCA and linear regression through hypercubes in HElib. In: Proceedings of the 2018 Workshop on Privacy in the Electronic Society, pp. 42–53 (2018)
11. Panda, S.: Principal component analysis using CKKS homomorphic scheme. In: Dolev, S., Margalit, O., Pinkas, B., Schwarzmann, A. (eds.) Cyber Security Cryptography and Machine Learning. CSCML 2021. LNCS, vol. 12716, pp. 52–70. Springer, Cham (2021). https://doi. org/10.1007/978-3-030-78086-9_4
12. Cheon, J.H., Kim, A., Kim, M., Song, Y.: Homomorphic encryption for arithmetic of approximate numbers. In: Takagi, T., Peyrin, T. (eds.) Advances in Cryptology – ASIACRYPT 2017. ASIACRYPT 2017. LNCS, vol. 10624, pp. 409–437. Springer, Cham (2017). https://doi.org/ 10.1007/978-3-319-70694-8_15
13. Markstein, P.: Software division and square root using Goldschmidt's algorithms. In: Proceedings of the 6th Conference on Real Numbers and Computers (RNC'6), vol. 123, pp. 146–157 (2004)

14. Sutherland, S.: Finding roots of complex polynomials with Newton's method. Boston University, 1989

15. Panda, S.: Polynomial approximation of inverse SQRT function for FHE. In: Dolev, S., Katz, J., Meisels, A. (eds.) Cyber Security, Cryptology, and Machine Learning. CSCML 2022. LNCS, vol. 13301, pp. 366–376. Springer, Cham (2022). https://doi.org/10.1007/978-3-031-07689-3_27

16. Wikipedia contributors: Goldschmidt's algorithm — Wikipedia, the free encyclopedia (2021). https://en.wikipedia.org/wiki/Methodsofcomputingsquareroots

17. Python binding for the Microsoft SEAL library. https://github.com/Huelse/SEAL-Python

18. Microsoft SEAL (release 4.0.0), microsoft Research, Redmond, WA. https://github.com/Microsoft/SEAL

19. Belhumeur, P.N., Hespanha, J.P., Kriegman, D.J.: Eigenfaces vs. fisherfaces: recognition using class specific linear projection. IEEE Trans. Pattern Anal. Mach. Intell. **19**(7), 711–720 (1997)

20. Wang, D., Jin, C., Li, H., Perkowski, M.: Proof of activity consensus algorithm based on credit reward mechanism. In: Wang, G., Lin, X., Hendler, J., Song, W., Xu, Z., Liu, G. (eds.) Web Information Systems and Applications. WISA 2020. LNCS, vol. 12432, pp. 618–628. Springer, Cham (2020). https://doi.org/10.1007/978-3-030-60029-7_55

Secure Reinsurance Data Sharing Scheme Based on Blockchain and Multi-level Attribute-Based Encryption

Xiaolin Yue[1], Ziqiang Ma[2(✉)], Juanyang Zhang[3], Yajie Lan[1], and Jiali Chen[1]

[1] School of Information Engineering, Ningxia University, Yinchuan 750021, China
{xiaoliny,yajielan,jialic}@stu.nxu.edu.cn
[2] Ningxia Key Laboratory of Artificial Intelligence and Information Security for Channeling Computing Resources from the East to the West, Yinchuan 750021, China
maziqiang@nxu.edu.cn
[3] Collaborative Innovation Center for Ningxia Big Data and Artificial Intelligence Co-founded by Ningxia Municipality and Ministry of Education, Yinchuan 750021, China
jyzhang@nxu.edu.cn

Abstract. As an important component of the insurance sector, reinsurance allows an insurer to protect itself from the risk of large claims by passing on some of its insurance liabilities to other insurers. Sharing reinsurance data between different institutions can simplify the reinsurance procedure. However, data sharing between different institutions poses challenges regarding privacy protection and access control. To address these issues, we propose ReinChain, a blockchain-based secure data sharing scheme that realizes access control of hierarchical data. In access control, we design ML-ABE to manage various types of data access control and enable access to data of different sensitivity levels. Finally, we have simulated the proposed scheme on Hyperledger Fabric. The security analysis and simulation experiments show that the proposed scheme is secure and robust in encryption and decryption.

Keywords: Reinsurance · Blockchain · Attribute-Based Encryption

1 Introduction

Reinsurance plays a pivotal role in mitigating risk. Through reinsurance agreements, insurers transfer some of their risks to other entities, diversifying their exposure [4]. However, the reinsurance process can be pretty complex because it involves third parties (brokers, insurers, reinsurers) [2]. Also, contracts can be split among multiple reinsurers to cut risk. As a result, traditional reinsurance systems face challenges in streamlining processes, protecting privacy, and implementing access controls.

Blockchain technology is a digital ledger that securely records transactions across a network of computers [5]. It enables secure, transparent, and immutable

C. Jin et al. (Eds.): WISA 2024, LNCS 14883, pp. 428–435, 2024.
https://doi.org/10.1007/978-981-97-7707-5_35

transactions when used to share data between parties, while also preserving privacy and enhancing trust [17]. In such scenarios, implementing appropriate authorization and encryption techniques is imperative to safeguard data security and privacy [13]. Given the sensitivity of policy, establishing access control mechanisms for various data types becomes essential to prevent misuse or leakage [12]. This is particularly important in reinsurance transactions, which involve multiple parties and a substantial volume of policies [3].

To address the challenges above, we propose ReinChain, a blockchain-based reinsurance data sharing scheme that provides secure access control for varies data types. In conclusion, we make the following contributions:

- We present a novel reinsurance data sharing framework, using consortium blockchain to store encrypted personal policies. This approach facilitates the establishment of dataflows among insurers, reinsurers, and policyholders, while using ML-ABE to ensure data security.
- We design a Multi-Level Attribute-Based Encryption (ML-ABE) scheme that integrates with blockchain to handle different data types in reinsurance workflows. By establishing a parent-child relationship between insurance and reinsurance data, as well as implementing an attribute-based encryption-verification mechanism, our solution expedites the processing of different data types and is better aligned with real-world business requirements.

This paper is organized as: Sect. 2 reviews related work, Sect. 3 details our ML-ABE scheme, Sect. 4 analyzes security and experiments, and Sect. 5 summarizes our work.

2   Related Work

2.1   Reinsurance

Reinsurance is a risk management strategy that allows companies to transfer parts of risk to other insurers, protecting them from catastrophic losses. The process involves various elements, such as risk layering, premium negotiations, and the involvement of reinsurance brokers, who facilitate communication between the parties [11]. Additionally, there has been a notable increase in the utilization of digital tools to enhance the efficiency and security of reinsurance operations. For instance, Yadav et al. [14] proposed a blockchain framework for vehicle insurance, using smart contracts for claims automation.

2.2   CP-ABE

Cipher Policy Attribute-Based Encryption(CP-ABE) is an encryption technique that encrypts data and associates access policies with the ciphertext [1]. It enables data owners to define policies based on attributes, allowing only authorized users with matching attributes to decrypt the data. In terms of access control, Zhang et al. [16] introduced a CP-ABE scheme for time bank in multi-authority systems. Guo et al. [7] proposed an extended CP-ABE scheme to overcome existing limitations in secure data storage.

2.3 Hyperledger Fabric

Hyperledger Fabric is an open-source, cross-industry blockchain technology. Fabric offers three node types, including peer nodes, orderer nodes, and client applications. Peer nodes host ledgers and smart contracts, enabling them to execute transactions and maintain ledger states. Orderer nodes are responsible for ordering transactions into blocks and distributing them to peer nodes for validation and commitment. Client applications interact with the network, submitting transactions and querying ledger data.

3 Design of ReinChain

In this section, we present an overview of ReinChain for reinsurance (see Sect. 3.1), then we introduce a ML-ABE scheme(see Sect. 3.2).

3.1 System Model

In our model, ML-ABE encrypts reinsurance contract data, ensuring access only to those meeting specific attribute criteria. Furthermore, only authorized data requesters with the right attributes can verify and decrypt re-encrypted data using their private keys. There are six entities in the proposed framework. Here are the functional descriptions for each entity.

1. Certificate Authority(CA): Responsible for managing the system's keys.
2. Data Owners(DO): Includes policyholders and owners insurance data.
3. Data Producers(DP): Insurance companies generating insurance data.
4. Data Requesters(DR): Companies utilizing insurance data.
5. Server(Srv): Responsible for storing permission verification data.
6. Blockchain(BC): Stores encrypted data.

The system model, shown in Fig. 1, comprises three data processing flows.

1. Reinsurance Contract (RC): After authorization, the DP encrypt data packet $v_0 = (RC\|seed)$ with ML-ABE and upload it to the BC.
2. Personal Policy (m_1): As insureds register on the consortium blockchain, the DP generate policy data, derive a key k, encrypt m_1 with k, and uploads it to the BC. Subsequently, the data packet $v_1 = (GID\|salt)$ is encrypted using the ML-ABE and uploaded to the server. If DR request m_1, they need to decrypt data packets v_0 and v_1 first, derive the key using the key derivation algorithm, and then decrypt m_1. Note that the encryption of m_1 uses symmetric encryption and the GID stands for the Global Identifier.
3. Claims Data (m_2): When a claim arises, the insured's data is encrypted with their public key and uploaded. The DR generates verification data, and upon approval, the insured generates a proxy key, which is then used by the server to re-encrypt the original ciphertext and send it to the DR.

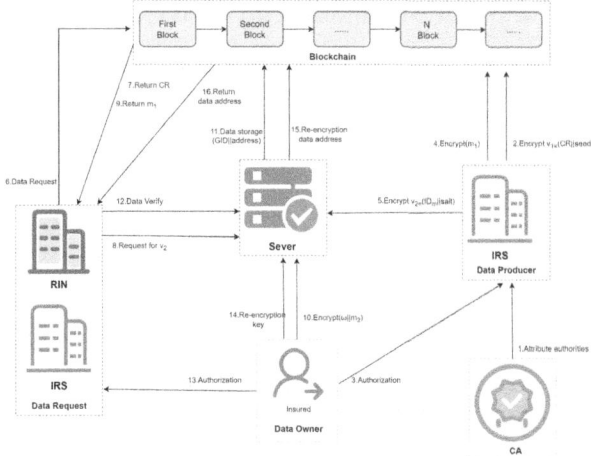

Fig. 1. System model.

3.2 Multi-level Attribute-Based Encryption(ML-ABE)

In ML-ABE, an access structure enables granting access to two data types, with the DO being responsible for administering permissions for both data types. Upon successful verification, it generate a proxy re-encryption key, allowing the DR to access higher-sensitivity data using their private key after re-encryption. In our scheme, m_1 is accessible to users whose attributes include "buss" representing traditional business, whereas m_2 is accessible to users with attributes including "clm" representing claims business. Only users involved in claims processing have access to m_2. The ML-ABE scheme consists of the following steps:

Step1 System Initialization. The CA initially executes the *Setup* algorithm to create global parameters.

$Setup(1^\lambda) \rightarrow (MSK, PK)$. It is performed by CA to get the system master key MSK and public key PK, it chooses $(p, g, G_0, \alpha, \beta)$, where G_0 is a bilinear group of prime order p and $\alpha, \beta \in Z_p$. Let g be a generator of G_0, the calculation is as follows:

$$PK = (G_0, g, h = g^\beta, f = g^{\frac{1}{\beta}}, e(g, g)^\alpha), MK = (\beta, g^\alpha) \qquad (1)$$

Step2 Data Encryption and Uploading. The DP randomly choose a seed and encrypt $(RC \| seed)$ using the *Encrypt* algorithm. Then DP encrypts m_1 with a symmetric key derived from the *seed* and *salt* using a key derivation function (KDF).

$Encrypt(PK, M, A) \rightarrow CT$. This algorithm takes a system public key PK, plaintext M and θ, and an access tree A as the attribute-based access structure as input. It then encrypts the plaintext M to get the ciphertext CT.

$$T, \tilde{C} = Me(g,g)^{\alpha s}, C = h^s$$

$$CT = \begin{pmatrix} \overline{C} = \theta e(g,g)^{\alpha s_i}, \overline{C}' = h^{s_i} \\ \forall y \in Y : C_y = g^{q_y(0)}, C_y' = H(att(y))^{q_y(0)} \\ C_{last} = g^{q_{last}(0)}, C_{last}' = H(att(last))^{q_{last}(0)} \end{pmatrix} \tag{2}$$

Step3 Key Generation. In the key generation phase, the CA employs the *KeyGen* algorithm to generate a SK for ABE.

$KeyGen(PK, SK, A) \rightarrow SK$. CA runs the algorithm to generate a user's private key for ML-ABE. The algorithm begins by choosing a random number $r \in Z_p$. Each attribute $j \in S$ randomly selects a value $r_j \in Z_p$. Finally, it computes the SK as follows:

$$SK = \begin{pmatrix} D = g^{\frac{\alpha+r}{\beta}}, \forall j \in S : D_j = g^r \cdot H(j)^{r_j}, D_j' = g^{r_j} \\ D_{last} = g^{q_{last}(0)}, C_{last}' = H(att(last))^{q_{last}(0)} \end{pmatrix} \tag{3}$$

Step4 Data Decryption. The DR initially submit a request to BC for data access. Once the verification is successful, it can decrypt the ciphertext.

$Decrypt(PK, CT, SK) \rightarrow (M, \theta)$. If S meets the access structure A, they can decrypt the ciphertext correctly and retrieve the plaintext M [1]. Additionally, if specific attributes are met, they can obtain θ.

$$DecryptTheta(CT, SK) = \frac{e(D_{last}, C_{last})}{e(D_{last}', C_{last}')}, \quad \frac{\overline{C}}{\frac{e(\overline{C}', D)}{A'}} = \theta \tag{4}$$

Step5 Higher Sensitivity Data Authorization. In this phase, DR seeking access to the higher sensitivity data m_2 needs to generate a verified data token ω using the *VerGen* algorithm.

1. $VerGen(M, \theta) \rightarrow \omega$. This algorithm inputs the plaintext M and θ acquired from the previous decryption step and returns verification data ω.
2. $Verify(\omega) \rightarrow (0,1)$. If the verification is successful, DR will be granted the corresponding operational permissions.

4   Security Analysis and Performance Evaluation

4.1   Security Analysis

Data Privacy Preservation. Our scheme employs a hybrid encryption approach, using ABE for reinsurance contracts and symmetric encryption for individual contracts. This ensures data security both during transmission and storage. Additionally, the claim data encrypts with the owner's public key, guaranteeing access only to authorized users.

Access Control. With ML-ABE, we control data access based on specific attributes, ensuring only necessary individuals can see reinsurance details. Sensitive claim data gets encrypted using a re-encryption key, limiting access to authorized parties or those with the right decryption key.

4.2   Comparisons of Security Properties

We compared our scheme's security properties with others (Table 1). While Guo's study [7] encrypted files with a hierarchical tree, it neglected data relationships. Blockchain-based approaches [6, 15] addressed access control but not hierarchical data issues. [9, 10] tackled hierarchical data access, yet used diverse methods. Our work extends beyond by implementing access control across different data types based on a unified access structure.

Table 1. Comparisons of security properties.

Relevant literature	Encryption scheme	Blockchain based	Access control	Hierarchical data type	Policy expressiveness	Nested attribute support
[7]	CP-ABE	N	Y	Y	Y	Y
[10]	CP-ABE	Y	Y	Y	N	N
[15]	CP-ABE	Y	Y	N	N	N
[9]	CP-ABE	Y	Y	Y	N	N
[8]	CP-ABE	Y	Y	N	N	N
[6]	RBAC	Y	Y	N	N	N
Ours	ML-ABE	Y	Y	Y	Y	Y

4.3   Performance Evaluation

To evaluate the proposed method, we tested the average latency and throughput of each function. Then we compared ML-ABE and BSW-ABE [1] in the key generation, encryption, and decryption times as the number of attributes varied. Testing was conducted on an Intel Core i7-6700 CPU @3.40 GHz with 8 GB of memory running Ubuntu 22.04 LTS.

Performance of Smart Contract. We simulated a blockchain network using Hyperledger Fabric v2.4 and Java for smart contracts. Tests with Caliper v0.5 revealed that $EncCR$ and $DecCR$ functions, handling attribute encryption/decryption, had higher latency. As shown in Fig. 2, apart from $EncCR$ and $DecCR$, all other functions are complete within 3 s or less. The experiment results illustrate that the extra overhead caused by the verification is very small. As shown in Fig. 3, the system demonstrates robust processing capabilities in encryption and decryption. Additionally, the throughput of other processes ranges between 14 to 17 transactions per second, indicating moderate performance levels.

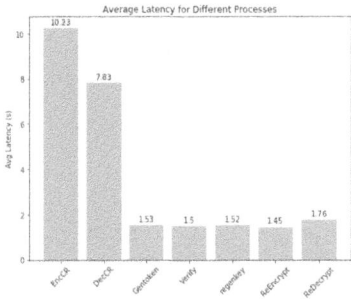
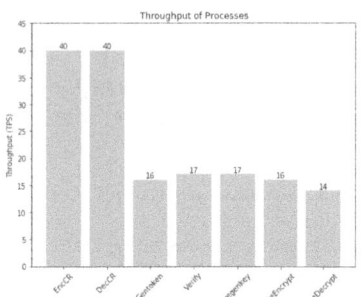

Fig. 2. Average Latency   **Fig. 3.** Throughput

ABE Computation Overhead. We simulate our ML-ABE scheme using JPBC with Java, comparing key generation, encryption, and decryption overhead with [1] for 4–20 attributes. Our enhanced scheme enables a single access structure for multiple data types, resulting in higher time costs (Fig. 4). This complexity supports advanced access control and security features, offering stronger data protection in practical applications (Figs. 5 and 6).

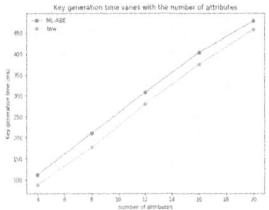
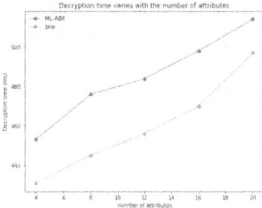

Fig. 4. Key generation   **Fig. 5.** Encryption   **Fig. 6.** Decryption

5   Conclusion

In this paper, we propose an ML-ABE scheme based on blockchain to achieve secure data sharing and access control in reinsurance. To accommodate the scenario of multi-level data sharing in reinsurance, we design a multi-level attribute-based encryption approach based on ABE, which enables an access structure to access two types of data with different sensitivity levels. The security and experimental results demonstrate that our proposed scheme can meet the designed security objectives.

Disclosure of Interests. The authors declare that they have no competing interests that are relevant to the content of this article.

References

1. Bethencourt, J., Sahai, A., Waters, B.: Ciphertext-policy attribute-based encryption. In: 2007 IEEE Symposium on Security and Privacy, pp. 321–334. IEEE (2007)
2. Brophy, R.: Blockchain and insurance: a review for operations and regulation. J. Financ. Regul. Compliance **28**(2), 215–234 (2020)
3. Cheng, X., Chen, F., Xie, D., Sun, H., Huang, C.: Design of a secure medical data sharing scheme based on blockchain. J. Med. Syst. **44**(2), 52 (2020)
4. Cummins, J.D., Dionne, G., Gagné, R., Nouira, A.: The costs and benefits of reinsurance. Geneva Pap. Risk Insur. Issues Pract. **46**, 177–199 (2021)
5. Eyal, I., Gencer, A.E., Sirer, E.G., Van Renesse, R.: {Bitcoin-NG}: a scalable blockchain protocol, pp. 45–59 (2016)
6. Gai, K., She, Y., Zhu, L., Choo, K.K.R., Wan, Z.: A blockchain-based access control scheme for zero trust cross-organizational data sharing. ACM Trans. Internet Technol. **23**(3), 1–25 (2023)
7. Guo, R., Li, X., Zheng, D., Zhang, Y.: An attribute-based encryption scheme with multiple authorities on hierarchical personal health record in cloud. J. Supercomput. **76**(7), 4884–4903 (2020)
8. Li, Y., Chen, R., Rahmani, R.: Secure data sharing in internet of vehicles based on blockchain and attribute-based encryption. In: 2023 IEEE International Conference on Smart Internet of Things, pp. 56–63. IEEE (2023)
9. Malamas, V., Kotzanikolaou, P., Dasaklis, T.K., Burmester, M.: A hierarchical multi blockchain for fine grained access to medical data. IEEE Access **8** (2020)
10. Mittal, S., Ghosh, M.: A novel two-level secure access control approach for blockchain platform in healthcare. Int. J. Inf. Secur. **22**(4), 799–817 (2023)
11. Sayegh, K., Desoky, M.: Blockchain application in insurance and reinsurance. Skema Business School, France (2019)
12. Trivedi, S., Malik, R.: Blockchain technology as an emerging technology in the insurance market (2022)
13. Wang, C., Wang, S., Cheng, X., He, Y., Xiao, K., Fan, S.: A privacy and efficiency-oriented data sharing mechanism for IoTs. IEEE Trans. Big Data **9**(1), 174–185 (2022)
14. Yadav, A.S., Charles, V., Pandey, D.K., Gupta, S., Gherman, T., Kushwaha, D.S.: Blockchain-based secure privacy-preserving vehicle accident and insurance registration. Expert Syst. Appl. **230**, 120651 (2023)
15. Zhang, D., Wang, S., Zhang, Y., Zhang, Q., Zhang, Y., et al.: A secure and privacy-preserving medical data sharing via consortium blockchain. Secur. Commun. Netw. **2022** (2022)
16. Zhang, N., Dong, X., Wen, Z., Yang, C.: Secure mutual aid service scheme based on blockchain and attribute-based encryption in time bank. In: Yuan, L., Yang, S., Li, R., Kanoulas, E., Zhao, X. (eds.) Web Information Systems and Applications. WISA 2023. LNCS, vol. 14094, pp. 403–414. Springer, Singapore (2023). https://doi.org/10.1007/978-981-99-6222-8_34
17. Yang, C., Dong, X., Zhang, N., Wen, Z.: Will data sharing scheme based on blockchain and weighted attribute-based encryption. In: Yuan, L., Yang, S., Li, R., Kanoulas, E., Zhao, X. (eds.) Web Information Systems and Applications. WISA 2023. LNCS, vol. 14094, pp. 391–402. Springer, Singapore (2023). https://doi.org/10.1007/978-981-99-6222-8_33

Information System Applications

Prediction Method of Type 2 Diabetes Mellitus Based on a Combination of Hybrid Feature Selection and Random Forest

Yunming Wang[1], Jiangang Hu[1(✉)], Xinru Fan[2], Xiue Gao[3], and Changzheng Liu[4]

[1] College of Automation and Electrical Engineering, Dalian Jiaotong University, Dalian 116028, China
1490314979@qq.com
[2] Hebei University, Baoding 071002, China
[3] College of Computer Science and Intelligent Education, Lingnan Normal University, Zhanjiang 524048, China
[4] College of Information Science and Technology, Shihezi University, Shihezi 832003, China

Abstract. Type 2 diabetes mellitus(T2DM) has become a major social problem threatening the health of the population; the ability to predict its prevalence can help in prevention and early treatment. Existing prediction methods face difficult discovery of predictive T2DM risk factor features and low accuracy. To address these shortcomings, we propose a T2DM prediction method based on a combination of hybrid feature selection and random forest. First, an algebraic combination of original features is used to construct a candidate feature set of risk factors. Second, the RReliefF method is used to screen the maximum relevant features to obtain the maximum relevant feature set, and the mRMR algorithm is used to eliminate redundant features to obtain the maximum relevant minimum redundant feature set (important feature set). Again, the key feature set is obtained by causal replacement of the important feature set. Finally, a diabetes prediction model is constructed using the key feature set and the random forest algorithm. Experimental results show that the model helps to screen the features with predictive contribution and effectively improves the accuracy of diabetes prediction, which can provide some reference for T2DM prevention and treatment research.

Keywords: Type 2 diabetes mellitus · Prediction model · Hybrid feature selection · Random forest · Causal substitution

1 Introduction

T2DM is a complex chronic disease with different etiologies, which has now become an epidemic disease that seriously affects human health [1–3], and effectively preventing and treating it has become an urgent problem. Analysis of the

C. Jin et al. (Eds.): WISA 2024, LNCS 14883, pp. 439–450, 2024.
https://doi.org/10.1007/978-981-97-7707-5_36

relationship between risk factors and T2DM and developing diabetes prevalence prediction models are the keys to revealing the pathogenesis of diabetes and also effectively help prevent and treat diabetes [4–7].

In recent years, researchers have preferred machine learning and data mining methods to construct nonparametric prediction models for diabetes mellitus. Intelligent algorithms used to construct prediction models generally include Neural Networks, Naive Bayes, Decision Trees, Support Vector Machines, and Random Forests. Soltani [8] et al. used probabilistic neural networks to predict T2DM and showed that the method outperformed logistic regression and linear perceptron. Kandhasamy [9] et al. compared and analyzed J48, random forest, K-nearest neighbor, and support vector machine algorithms on a Pima Indian dataset, in which random forest achieved the highest prediction accuracy. Mohebbi [10] et al. used traditional neural networks and multilayer perceptron neural networks to detect T2DM and showed that the prediction accuracy of the former was higher than that of the latter. Kumari [11] et al. constructed a prediction model with a fusion of Naive Bayes, random forest, and logistic regression, and the experimental results showed that the prediction accuracy of the integrated approach is higher than that of the other underlying classifiers, but with this the complexity of the model increases considerably.

Evidently, intelligent algorithms are well suited for building diabetes prediction models, easily handle high-dimensional data and nonlinear problems, and can improve the classification prediction accuracy to varying degrees, but the prediction accuracy is constrained by the quality of the data [12–14] and dataset preprocessing is usually required before training. None of the above smart algorithm models performs data feature extraction, which limits the prediction accuracy. The diabetes prediction model of Yuvraj [15] et al. selected significant features using an information gain selection method, and the results showed improved model prediction accuracy. Mercaldo [16] et al. used two feature selection algorithms, greedy stepwise and best-first, to select significant features to improve the performance of the classifier. Rakshit [17] et al. used correlated feature selection methods for feature extraction, which improved the model prediction accuracy. Choubey [18] et al. used principal component analysis and linear discriminant for feature selection and the prediction accuracy after feature extraction was significantly improved in both cases, but there was a problem of insufficient features for model training due to data limitations. Negi [19] et al. used both wrapper and ranker methods to process the dataset separately, but the prediction accuracy after feature selection was not improved, mainly because feature selection reduced the dataset dimensionality, and the selected features did not have sufficient prediction contribution.

Intelligent prediction methods can easily deal with high-dimensional data and nonlinear problems, but the quality of the data directly affects the prediction accuracy of the model. These methods are usually combined with feature selection methods to build prediction models. The feature selection method can tap the features of diabetes risk factors with predictive contribution and effectively improve the prediction accuracy of diabetes. However, the problem of too few dimensions of the input data or insufficient predictive contribution of the selected features can lead to the inability to further improve the prediction accuracy.

To address the shortcomings of existing T2DM prediction methods that there are few diabetes risk factor features with prediction contribution, the difficulty in extracting them, and that the accuracy is low, we propose a T2DM prediction method based on a combination of hybrid feature selection and random forest. The algebraic combination of risk factors is used to construct more risk factor features without prior knowledge, which solves the problem of fewer features with prediction contribution; the RReilfF algorithm and mRMR algorithm are used for feature selection to solve the problem of difficult extraction of important features. The causal discovery method is used for feature set construction to solve the problem of insufficient feature prediction contribution after feature selection. Finally, combined with the random forest algorithm, a prediction model that classifies T2DM is established.

2   Materials and Methods

2.1   Risk Factor Candidate Feature Set

T2DM is a chronic disease affected by multiple risk factors. It is extremely difficult to obtain data on many risk factors and their contributions to model prediction; furthermore, these risk factors are often interrelated and interact with each other. It is clearly important to construct a risk factor candidate feature set.

To this end, in this paper, the number of candidate risk factors are increased by algebraically combining them with the existing risk factors. It is assumed that $\{X_1, X_2, \ldots, X_N\}$ is the set of variables of diabetes risk factors, where is the number of risk factor variables. The algebraic combination of (1) is chosen as the risk factor candidate feature set.

$$\left\{X_i, X_i^2, 1/X_i^2, 1/X_i, X_i X_j, X_i^2 X_j\right\} \tag{1}$$

This allows the selection of features with a more predictive contribution, which in turn improves the prediction accuracy. However, the dimensionality of the risk factor candidate feature increases significantly, and a reasonable feature selection algorithm needs to be designed.

2.2   Hybrid Feature Selection Algorithm

Although there are features with a more predictive contribution to the risk factor candidate feature set, too many features will certainly lead to an inefficient prediction and an insignificant improvement in the prediction accuracy. Screening the features with a high predictive contribution is required. For this reason, a hybrid feature selection algorithm is proposed in this paper. The principle of the algorithm is briefly described as follows:

First, feature relevance is considered. The RReilfF algorithm is used to filter features with high relevance to diabetes and obtain the maximum relevant feature set and remove irrelevant features.

Second, feature redundancy is considered. The mRMR algorithm is used to remove redundant features to obtain the mRMR feature set.

Finally, feature causality is considered. A feature causal substitution is applied using the IFCL risk factor causal discovery method to obtain the key feature set, which in turn is used to create the predictive model.

The principle of the hybrid feature selection algorithm is shown in Fig. 1. The algorithm involves RReilfF, mRMR, and IFCL algorithms, which are briefly described as follows.

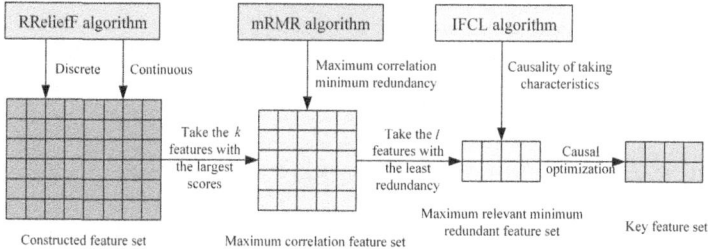

Fig. 1. Schematic diagram of hybrid feature selection algorithm

RReliefF Algorithm. The RReliefF algorithm [20] is a filtered feature selection method that determines the relevance of the features based on their ability to discriminate between nearest neighbor samples and removes irrelevant features based on the relevance value. The features are assigned weight values as evaluation values $W[A]$, based on the features that are most relevant to the target class filtered.

The feature weights $W[A]$ are then calculated as follows:

$$w[A] = \frac{N_{dC\&dA}[A]}{N_{dC}} - \frac{(N_{dA}[A] - N_{dC\&dA}[A])}{m - N_{dC}} \tag{2}$$

where N_{dC} denotes the weights conditional on different predicted values, $N_{dA}[A]$ denotes the weights under different feature conditions, $N_{dC\&dA}[A]$ denotes the set of weights at different predicted values under different feature conditions. The specific parameters are solved in [21]

mRMR Algorithm. The mRMR algorithm [22] is a correlation measurement algorithm based on mutual feature information, which is commonly used to remove redundant features by finding the set of features in the original set of features that are most relevant to the final output result but with the least relevance between each other.

The maximum relevancy D can be expressed as follows:

$$maxD(S,c), D = \frac{1}{|S|} \sum_{x_i \in S} I(x_i; c) \tag{3}$$

where $I(x_i; c)$ is the mutual information between the feature parameter x_i and the target category c. The minimum redundancy R can be expressed as follows:

$$minR(S,c), R = \frac{1}{|S|^2} \sum_{x_i, x_j \in S} I(x_i; x_j) \tag{4}$$

where $I(x_i; x_j)$ is the mutual information between the feature parameter x_i and the feature parameter x_j.

Combining the maximum correlation D and the minimum redundancy R. the optimal algorithm used to obtain the difference criterion is as follows:

$$\max \left(\frac{1}{|s|} \sum_{x_i \in S} I(x_i; c) - \frac{1}{|s|^2} \sum_{x_i, x_j \in S} I(x_i; x_j) \right) \tag{5}$$

IFCL Algorithm. The IFCL algorithm [23] is used to discover the existence of causal relationships between diabetes risk factors and achieve a causal substitution of the risk factors to obtain the key feature set, thereby achieving a further reduction in the dimensionality of the risk factor candidate feature set.

Given the risk factor observation data $O = \{\vec{o_1}, \vec{o_2}, \cdots, \vec{o_j}, \cdots, \vec{o_m}\}$, where $\vec{o_j}$ is an n dimensional vector, that is $\vec{O_j} = (o_{j,1}, o_{j,2}, \cdots, o_{j,n}), 1 \leq j \leq m$. Let o_{j,p_i} denote the observation value that contains X_{P_i}. According to the joint distribution $P(X)$ and the causal structure G, the log-likelihood of the observations is as follows:

$$L(G; O) = \sum_{j=1}^{m} \sum_{i=1}^{n} \log \left(P(X_i = o_{j,i} | X_{P_i} = o_{j,P_i}) \right) \tag{6}$$

The causal structure obtained by maximizing (6) is not necessarily correct because it is possible that there are Markov equivalence classes of causal structures with the same maximum likelihood but different structures that require this type of causal structure. It is necessary to distinguish such causal structures. Using the additive noise model $X_i = F_i(X_{P_i}) + E_i$ as the causal relationship generation mechanism, F_i is the causal function of X_i.

$$\begin{aligned} P(X_i = o_{j,i} | X_{P_i} = o_{j,P_i}) \\ = P(E_i = o_{j,i} - F_i(o_{j,P_i}) | X_{P_i}) \\ = P(E_i = o_{j,i} - F_i(o_{j,P_i})) \end{aligned} \tag{7}$$

From (6) and (7), the likelihood of the observed data is equal to the likelihood of the noise in the observed data. Letting $S = \langle G, F \rangle$ be its causal structure, the likelihood of the noise of the observed data is given as follows:

$$L\left(S; O\right) = \sum_{j=1}^{m} \sum_{i=1}^{n} \log\left(P\left(E_i = o_{j,i} - F_i\left(o_{i,P_i}\right)\right)\right) \tag{8}$$

Equation (8) is the transformed objective function. To avoid excessive redundancy and false causality edges, the Bayesian information criterion is added, and the threshold is adjusted to obtain the IFCL algorithm model as follows:

$$\overline{L_B}\left(S; O\right) = \sum_{i=1}^{n} \left(\sum_{j=1}^{m} \log\left(P\left(E_i = o_{j,i} - F_i\left(o_{j,P_i}\right)\right)\right) - \frac{d_i \log\left(m\right)}{2} + \alpha \right) \tag{9}$$

where d_i is the coefficient used to estimate X_i, and α is the adjusted threshold.

3 Experimental Process and Analysis

3.1 Experimental Process

RFs [24–26] have strong operational and generalization capabilities. In [12], it was shown that RFs achieve satisfactory results for diabetes prediction. Therefore, this paper proposes a T2DM prediction method based on the combination of hybrid feature selection and RF, and a flowchart of the algorithm applied is shown in Fig. 2.

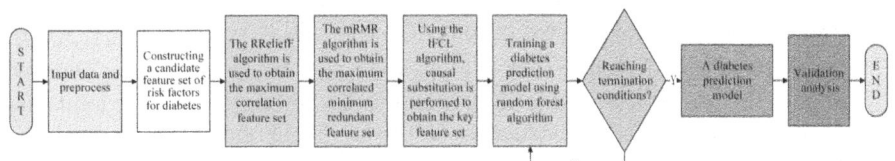

Fig. 2. Flow chart of T2DM prediction algorithm based on mixed feature selection and random forest

3.2 Experimental Data

The experiment data(Pima Indian Diabetes data set) were obtained from the Kaggle platform and placed in two diabetes datasets with sample sizes of 768 [27] and 2000 [28]. The datasets included 9 variables: number of pregnancies X_1, 2-h plasma glucose concentration in the oral glucose tolerance test X_2, diastolic blood pressure (mm Hg) X_3, thickness of the triceps skinfold (mm) X_4, 2-h serum insulin (mu U/ml) X_5, body mass index(BMI) X_6, diabetic lineage function X_7, age X_8, and diabetes diagnosis variables, where the diabetes lineage function contained genetic information on diabetes in the families of the subjects.

3.3   Experiment Results and Analysis

Analysis of Feature Sets Obtained by Different Feature Selection Algorithms. To better analyze the advantages of the hybrid feature selection algorithm proposed in this paper in terms of prediction contribution, we had two samples of size 768 and 2000. From the sample of size 768, three feature sets were selected: the original feature set was defined as Feature set 1, Feature set 2 (important feature set) was obtained based on the screening of RReliefF algorithm and mRMR algorithm, Feature set 3 (key feature set) was obtained based on the causal discovery algorithm. The same procedure was applied to the sample of size 2000 with the respective Feature sets numbered 4–6. T2DM prediction Models 1 to 6 were based on the combination of Feature sets 1 to 6 respectively.

Table 1. Feature structure of Pima Indian feature set

Sample size	Feature set	Feature set structure	Prediction model
768	1	$X_1, X_2, X_3, X_4, X_5, X_6, X_7, X_8$	Model 1
	2	$X_2X_6, X_1X_8^2, X_3X_6^2, X_2^3, X_1X_7^2, X_4X_8^2, X_2X_3^2, X_4X_6^2$	Model 2
	3	$X_2X_6, X_8^3, X_3X_6^2, X_2^3, X_8X_7^2, X_6X_8^2, X_2X_3^2, X_6^3$	Model 3
2000	4	$X_1, X_2, X_3, X_4, X_5, X_6, X_7, X_8$	Model 4
	5	$X_2X_8, X_6X_7, X_3X_6^2, X_2^3, X_2X_6^2, X_1X_4^2, X_2X_3^2, X_7X_8^2$	Model 5
	6	$X_2X_8, X_6X_7, X_3X_6^2, X_2^3, X_2X_6^2, X_8X_4^2, X_2X_3^2, X_7X_8^2$	Model 6

To better compare the performance of different feature selection algorithms, the number of features of Feature sets 1–6 is shown in Table 1. As can be seen from Table 1, Feature set 3 performs causal substitution for four features $X_1X_8^2$, $X_1X_7^2, X_4X_8^2, X_4X_6^2$ in Feature set 2; Feature set 6 performed causal substitution for one feature $X_1X_4^2$ in Feature set 5.

Training and Prediction of the T2DM Prediction Models. RF is used to train Models 1–6, and their ROC curves are shown in Fig. 3. The red line represents the RF ROC curve for the original data, the blue line represents the RF ROC curve for the important feature data, and the green line represents the RF ROC curve for the key feature data.

As shown in Fig. 3, the AUC evaluation index of the model with the sample size of 768 is much smaller than that of the model with the sample size of 2000. This indicates that sample size can significantly improve the model training accuracy. Meanwhile, the ROC curves of each of the three feature sets with the same sample set are nearly the same and the AUC evaluation indexes are all the same, which indicates that the feature sets generated by feature extraction and causal replacement do not degrade the performance of the training model.

The training and prediction accuracies of Models 1–6 are shown in Table 2.

As can be seen from Table 2, Under the sample size of 768, the three models have the same training accuracy, the prediction accuracy of Model 2 is 0.8%

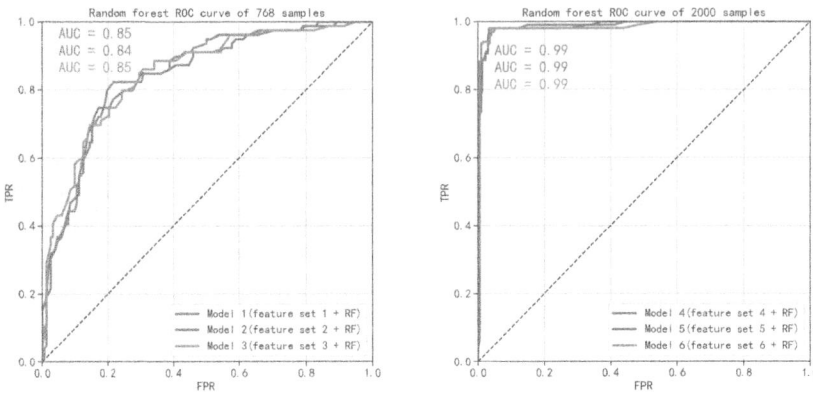

(a) ROC curve with sample size of 768   (b) ROC curve with sample size of 2000

Fig. 3. Random forest ROC curve under PIMA data set

higher than that of Model 1, and the prediction accuracy of Model 3 is 2.2% higher than that of Model 2; also under the sample size of 2000, the three models The comparison models have the same training accuracy, the prediction accuracy of Model 5 is 0.5% higher than that of Model 4, and the prediction accuracy of Model 6 is 1.2% higher than that of Model 5. This shows that a good feature extraction method is of great help to improve the prediction accuracy of the model. Comparing the prediction accuracy under two sample sizes, we found that the prediction accuracy under the sample size of 2000 was much higher than that under the sample size of 768, which is because with the increase of sample size, the misclassification rate of RF training is significantly reduced, and the generalization ability of the model is improved. This is consistent with the fact that the AUC evaluation index under the sample size of 2000 is much higher than that under the sample size of 768.

Table 2. Summary of training and prediction results of each model

Sample size	Model	Training accuracy	Prediction accuracy
768	Model 1	80.3%	77.1%
	Model 2	80.3%	77.9%
	Model 3	80.3%	80.1%
2000	Model 4	99.9%	95.3%
	Model 5	99.9%	95.8%
	Model 6	99.9%	97.0%

Feature set 3 was obtained from Feature set 2 by causal substitution and Feature set 6 was obtained from Feature set 5 by causal substitution. Thus, we can conclude that the causal discovery method is very helpful in diabetes risk factor feature extraction, even all disease risk factor feature extraction, and causal substitution obtain features with higher prediction contribution, the prediction accuracy of the diabetes prediction model also improves with them.

Model Feature Contribution Analysis. The contributions of all features to the prediction of each model under the sample sizes of 768 are shown in Fig. 4(a-c).

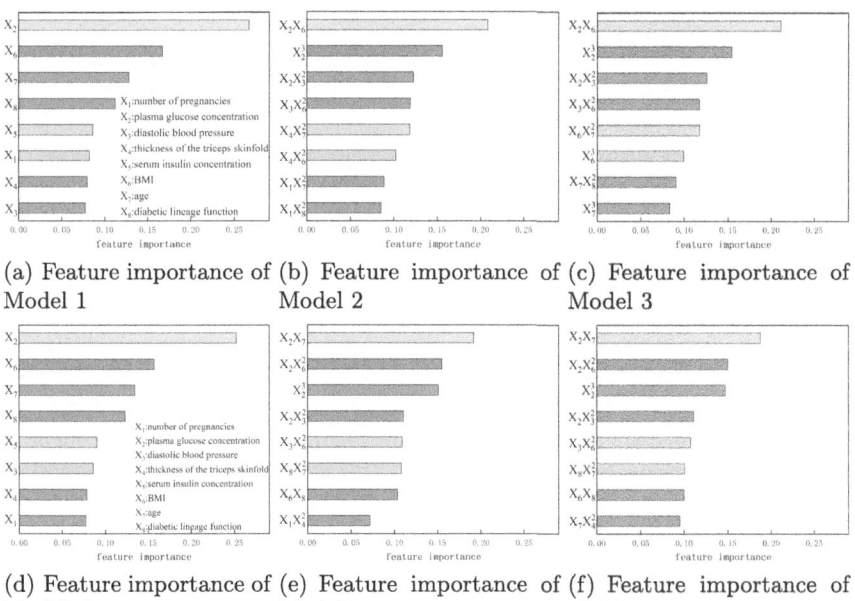

(a) Feature importance of Model 1

(b) Feature importance of Model 2

(c) Feature importance of Model 3

(d) Feature importance of Model 4

(e) Feature importance of Model 5

(f) Feature importance of Model 6

Fig. 4. Feature importance of the model

It can be seen from Fig. 4(a-c) that under the condition of 768 sample size, the contribution values of features X_2 and X_6 with their constructs X_2X_6 and X_2^3 ranked at the top, which indicates that these features have great contribution to the prediction of Models 1–3, that is, blood glucose, BMI, the product of blood glucose and BMI, and the cube of blood glucose have great effect on the prediction of diabetes. Meanwhile, the comparisons of Fig. 4(a-c) show that the distribution of feature contributions is more balanced in Fig. 4(b,c), and the prediction accuracy of Model 2 and Model 3 are also improved.

It can be seen from Fig. 4(d-f) that under the condition of the sample size of 2000, the feature contribution of X_2, X_6, X_8 and its constructors $X_2X_6^2$, X_2^3 and

$X_2 X_8$ rank at the top, which indicates that these features contribute greatly to the prediction of Models 4–6. Under the sample size of 2000, the feature contribution distribution of the model is more noticeable. The feature contribution distributions of Models 5 and 6 are more balanced than that of Model 4. Compared with Fig. 4(e,f), $X_8 X_4^2$ was obtained after causal replacement of $X_1 X_4^2$, and the contribution of feature $X_8 X_4^2$ is greater. The feature contribution of Model 6 is more balanced than that of Model 5. Therefore, we draw the following conclusions: The method based on hybrid feature selection can make the feature contribution distribution of model prediction more balanced, which increases the model prediction accuracy.

4   Discuss

Through Fig. 3, it is found that the hybrid feature selection method proposed in this paper improves the prediction accuracy of the RF-based diabetes prediction model to a certain extent with nearly the same ROC curve under different datasets. At the same time, after selecting feature variables that have more contributory power using the feature selection method, the accuracy of the prediction model can be improved with the same computational burden as the original data. In particular, since all samples of the pima Indian dataset were female.

From Fig. 4, we can see that the contribution of the characteristic variables X_2, X_6, X_8 and the construction of these three characteristic variables are all at the top of the list, where X_2 is the 2-h plasma glucose concentration in the oral glucose tolerance test, X_6 is the body mass index, and X_8 is the age. We know that these three characteristic variables are by far the factors most associated with diabetes. There is a direct correlation between plasma glucose concentration and HBA1c, which is closely related to diabetes mellitus and its complications. BMI is a commonly used index to measure the degree of body fat and thinness. Obesity and diabetes belong to metabolic diseases, so obesity and diabetes are closely related. Studies [29] found that the incidence of diabetes increases with age. According to the analysis, the hybrid feature selection method in this paper is reliable, and the prediction results obtained by RF also have theoretical bases.

The main reasons for the predictive advantages of our method are: (1) There are many types of T2DM risk factors and they are interrelated, so constructing a risk factor candidate feature set can expand the predictive features and help screen the features with more predictive contribution; (2) Due to the large number of candidate features and the existence of many redundant features, the hybrid feature selection algorithm can effectively screen the maximum relevant minimum redundant features, and causal substitution can effectively compensate for the missing information.

5   Conclusions

First, to address the problem of too few existing T2DM prediction features, the original features are algebraically combined to effectively find features with higher prediction contribution; second, to address the feature extraction problem of the candidate feature set of T2DM risk factors, a hybrid feature selection algorithm is proposed and the key feature set is obtained; finally, based on the T2DM prediction method combining hybrid feature selection and RF, relevant validation experiments are designed. Experimental analysis shows that the model has a strong generalization ability, and its method has better prediction advantages.

Acknowledgments. This research was supported by the Fund Project of XJPCC (2023AB018-10, 2022ZD077).

References

1. Strelitz, J., Ahern, A.L., Long, G.H., et al.: Moderate weight change following diabetes diagnosis and 10 year incidence of cardiovascular disease and mortality. Diabetologia **62**(1), 1391–1402 (2019)
2. Guidelines for the prevention and treatment of type 2 diabetes in China (2017). Chin. Diabetes Soc. **38**(04), 292–344 (2018)
3. Saxena, R., Sharma, S.K., Gupta, M., Sampada, G.C.: A comprehensive review of various diabetic prediction models: a literature survey. J. Healthc. Eng. **2022**, 8100697 (2022)
4. Yuvaraj, N., SriPreethaa, K.R.: Diabetes prediction in healthcare systems using machine learning algorithms on Hadoop cluster. Clust. Comput. **22**(1), 1–9 (2019)
5. Jaiswal, V., Negi, A., Pal, T.: A review on current advances in machine learning based diabetes prediction. Prim. Care Diabetes **15**(3), 435–443 (2021)
6. VijiyaKumar, K., Lavanya, B., Nirmala, I., et al.: Random forest algorithm for the prediction of diabetes. In: IEEE International Conference on System, Computation, Automation and Networking, pp. 1–5 (2019)
7. Ashiquzzaman, A., Tushar, A.K., Islam, M.R., Kim, J.-M.: Reduction of Overfitting in Diabetes Prediction Using Deep Learning Neural Network. ArXiv 2017, abs/1707.08386
8. Soltani, Z., Jafarian, A.: A new artificial neural networks approach for diagnosing diabetes disease type II. Int. J. Adv. Comput. Sci. Appl. **7** (2016)
9. Kandhasamy, J.P., Balamurali, S.: Performance analysis of classifier models to predict diabetes mellitus. Procedia Comput. Sci. **47**, 45–51 (2015)
10. Mohebbi, A., Aradottir, T.B., Johansen, A.R., et al.: A deep learning approach to adherence detection for type 2 diabetics. In: 2017 39th Annual International Conference of the IEEE Engineering in Medicine and Biology Society (EMBC), pp. 2896–2899 (2017)
11. Kumari, S.V., Kumar, D., Mittal, M.: An ensemble approach for classification and prediction of diabetes mellitus using soft voting classifier (2021)
12. Kushan, D.S., Joanne, C.E., Christopher, A.B., et al.: Use and performance of machine learning models for type 2 diabetes prediction in community settings: a systematic review and meta-analysis. Int. J. Med. Inform. **143**, 104268 (2020)
13. Deberneh, H.M., Kim, I.: Prediction of type 2 diabetes based on machine learning algorithm. Int. J. Environ. Res. Public Health **18**(6), 3317 (2021)
14. Diao, Y., Sun, Z., Zhou, Y.: A multi-label imbalanced data classification method based on label partition integration. In: Yuan, L., Yang, S., Li, R., Kanoulas, E., Zhao, X. (eds.) Web Information Systems and Applications. WISA 2023. LNCS, vol. 14094, pp. 14–25. Springer, Singapore (2023). https://doi.org/10.1007/978-981-99-6222-8_2

15. Yuvaraj, N., SriPreethaa, K.R.: Diabetes prediction in healthcare systems using machine learning algorithms on Hadoop cluster. Clust. Comput. 1–9 (2017)
16. Mercaldo, F., Nardone, V., Santone, A.: Diabetes mellitus affected patients classification and diagnosis through machine learning techniques. In: International Conference on Knowledge-Based Intelligent Information & Engineering Systems (2017)
17. Rakshit, S., et al.: Prediction of diabetes Type-II using a two-class neural network. In: International Conference on Computational Intelligence, Communications, and Business Analytics (2017)
18. Choubey, D.K., Kumar, M., Shukla, V., Tripathi, S., Dhandhania, V.K.: Comparative analysis of classification methods with PCA and LDA for diabetes. Curr. Diabetes Rev. **16**(8), 833–850 (2020)
19. Negi, A., Jaiswal, V.: A first attempt to develop a diabetes prediction method based on different global datasets. In: 2016 Fourth International Conference on Parallel, Distributed and Grid Computing (PDGC), pp. 237–241 (2016)
20. Deberneh, H.M., Kim, I.: Hyperparameter importance analysis based on N-RRReliefF algorithm. Int. J. Environ. Res. Public Health **18**(6), 3317 (2021)
21. Robnik-Sikonja, M., Kononenko, I.: Theoretical and empirical analysis of ReliefF and RReliefF. Mach. Learn. **53**, 23–69 (2003)
22. Peng, H., Long, F., Ding, C.: Feature selection based on mutual information: criteria of max-dependency, max-relevance, and min-redundancy. IEEE Trans. Pattern Anal. Mach. Intell. **27**(8), 1226–1238 (2005)
23. Gao, X., Xie, W., Wang, Z., Chen, B., Zhou, S.: Improved functional causal likelihood-based causal discovery method for diabetes risk factors. Comput. Math. Methods Med. **2021**, 5552085 (2021)
24. Zhang, Z.L., Sun, Y., Tuo, X.Q., et al.: A study on the predictive value of random forest algorithm on the risk of diabetes in a medical examination population. Chin. Gen. Pract. **22**(9), 1021–1026 (2019)
25. Hasanah, W., Munggaran, L.C.: Comparison of Naïve Bayes and random forest methods for diabetes prediction. Int. J. Comput. Appl. **174**, 13–18 (2021)
26. Shin, Y.S., Lee, N., Chigon, H.: A research on the key factors for classification of diabetes based on random forest (2020)
27. Kaggle, Pima Indians Diabetes Database. https://www.kaggle.com/datasets/uci ml/pima-indians-diabetes-database
28. Kaggle, Pima Indians Diabetes Database. https://www.kaggle.com/code/chirag 9073/diabetes-using-deep-learning/input
29. Wang, T., Zhao, Z., Wang, G., et al.: Age-related disparities in diabetes risk attributable to modifiable risk factor profiles in Chinese adults: a nationwide, population-based, cohort study. Lancet Healthy Longev. (2021). published online Sept 7

Named Entity Recognition Using EHealth-BiLSTM-CRF Combine with Multi-head Self-attention for Chinese Medical Information

Bin Wang and Fangjiao Jiang[✉]

School of Physics and Electronic Engineering, Jiangsu Normal University, Xuzhou 221000, China
jiangfj@jsnu.edu.cn

Abstract. This research investigates how to extract crucial information from large amounts of medical data in an effective and precise manner. The BERT pre-training language model has become a popular tool for named entity recognition techniques. However, the method of randomly masking a single token used in the BERT pre-training strategy does not fully utilize the training data's lexical, syntactic, and semantic structure for modeling. As a result, BERT is not a suitable solution for Chinese entity identification, so this research employs Ernie-Health (a Chinese language representation model pre-trained over large-scale biomedical text corpora) as the model's embedding layer, adds a multi-head self-attention layer, and presents an EHealth-BiLSTM-Attention-CRF model that focuses on medical entity recognition. The F1 value of the model provided in this study for the three datasets CMeEE, Yidu-S4K, and cMedQANER is 78.34%, 90.24%, and 86.62%, respectively. The experimental results reveal that the proposed method outperformed other traditional methods in the task of named entity recognition in the Chinese medical field.

Keywords: Named Entity Recognition · Ernie-Health · Self-Attention

1 Introduction

With the rapid advancement of artificial intelligence technology, people from many areas of life are actively researching and utilizing machine learning and deep learning technologies. Natural language processing is a major study area with substantial applications in text and sentiment analysis, named entity recognition, question answering systems, knowledge graphs, and so on. Named Entity Recognition (NER) is a critical problem in natural language processing that seeks to detect things with specified meaning in text and classify them into predetermined categories. The medical field, particularly the Chinese medical field [1], has a large variety of professional medical terminology with complex meanings. Furthermore, in this era of data explosion, a great number of electronic medical records have been integrated into the medical process. A significant technical challenge is determining how to extract useful information from large amounts of medical data and reliably identify medical entities.

Traditional naming recognition algorithms include rule-based and statistical model-based approaches. The rule-based method has the advantages of robust interpretability, a broad application range, and ease of use, but it is dependent on manually given rules. Statistical model-based methods turn named entity recognition into sequence labeling. However, with the advancement of deep learning, particularly the introduction of models like as BiLSTM [2], CNN [3], and BERT [4], the challenges of data sparsity and context dependence in named entity recognition have been overcome.

In this paper, we propose a deep learning model dedicated to Chinese medical named entity recognition. Specifically, the model consists of four data processing layers. They are the medical pre-trained model word embedding layer, which is used to process the input text into word vectors. The memory network layer is used to capture the semantic information of the context. The self-attention layer is used to calculate the correlation between each element in the current input sequence. The output classification layer returns the conditional probability associated with the matching label and completes the classification using end-to-end training. The major contributions are as follows:

- We build a new named entity recognition model that encodes the input language sequence containing medical text using a pre-trained language model that is more specialized in the medical field, and then uses the multi-head self-attention mechanism to dynamically process the input's context information.
- We conduct experiments on three official launched datasets to demonstrate the effectiveness and accuracy of our proposed model. Experimental results show that the proposed model has a significant improvement over the baseline model.

2   Related Work

Named Entity Recognition (NER) is an essential problem in natural language processing that aims to recognize meaningful entities in texts, such as the names of people, locations, and organizations. With the development of deep learning technology, NER has found widespread application in information extraction, question answering systems, machine translation, and other domains. Erik [5] first proposed a named entity recognition method based on conditional random field CRF. Hamilton [6] et al. first used LSTM neural network for named entity recognition task. Peng and Mark [7] proposed a model combining LSTM and CRF to jointly train NER tasks, which improved the F-value by 5% compared to previous related results.

With the proposal of the concept of "smart healthcare", an increasing number of deep learning models are being applied to the task of named entity recognition in electronic medical records, and the accuracy of identifying entities from medical texts is improving. Ji [8] et al. used the BiLSTM-CRF model to identify meaningful named items in Chinese electronic medical records, and the F1 value for this technique on the test dataset was 87.68%. Li [9] et al. incorporated the attention mechanism into the neural network and proposed an enhanced clinical named entity recognition technique BiLSTM-Att-CRF. Li [10] et al. pre-trained the BERT model on Chinese clinical records and used an LSTM network and CRF to categorize clinical terms in electronic medical records.

3   The Proposed Model

The structure diagram of named entity recognition model we introduced is shown in Fig. 1. Firstly, the input text sequence containing medical content is encoded into a word vector sequence through the EHealth pre-training model, and the output vector sequence is sent to the BiLSTM network to obtain the feature information of context semantics. Then, the Attention layer dynamically calculated the attention weight of each position according to the current obtained feature information, so that the model could selectively focus on different parts of the sequence to better capture key information. Finally, the sequence processed by the attention mechanism will be input into the CRF layer for decoding. In the following, we will briefly introduce each part of the model.

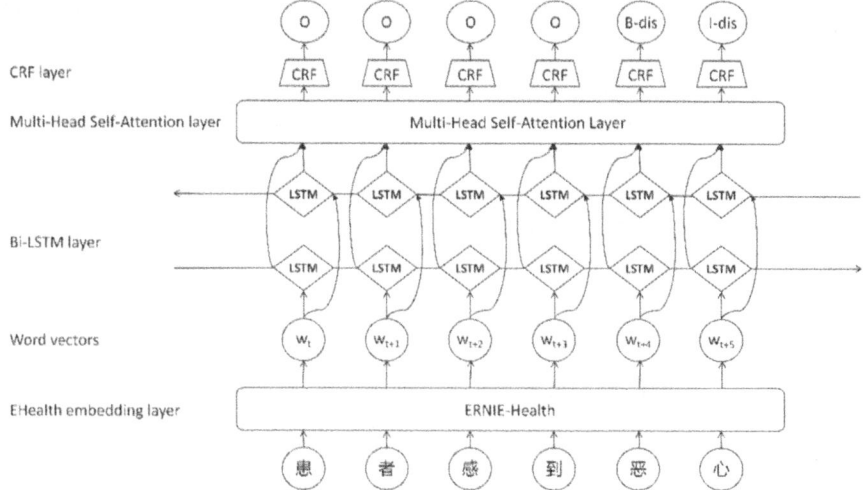

Fig. 1. The model structure diagram

3.1   Ernie-Health (EHealth) Embedding Layer

Google's BERT model has performed well in a variety of NLP downstream tasks by randomly masking 15% of words or words and using Transformer's multi-layer self-attention bidirectional modeling capability. However, the BERT model focuses primarily on cloze learning at the character or English word level and does not fully utilize the lexical structure, grammatical structure, and semantic information in the training data to learn and model.

ERNIE [11] (Enhanced Representation through Knowledge Integration) is a model proposed by Baidu in 2019. The main improvement over BERT is the improvement of the masking mechanism in the pre-training stage. Its mask is not the basic word piece mask, but consists of three levels of masks. They are basic-level masking (word piece) + phrase level masking (WWM style) + entity level masking. As shown in Fig. 2.

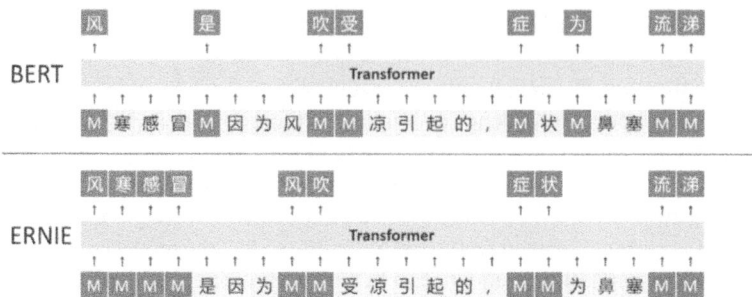

Fig. 2. The different masking strategy between BERT and ERNIE

ERNIE achieves state-of-the-art results on Chinese NLP tasks by modeling lexical structure, grammatical structure, and semantic information in the training data in a unified manner. Ernie-health [12] is a Chinese medical model built based on ERNIE pre-trained language model. It further learns the massive 126.9 GB medical text data through medical knowledge enhancement technology, which significantly increases the understanding and modeling ability of medical professional knowledge and accurately grasped professional medical knowledge. In addition, Ernie-Health investigates a multi-level semantic discrimination pre-training job to increase the model's learning efficiency for medical knowledge.

3.2 Multi-head Self-attention Layer

The multi-head self-attention [13] mechanism gathers various levels of information in the sequence using several concurrent self-attention layers. Each self-attention layer focuses on various local information before fusing it to gain more comprehensive information, boosting the model's expressive capabilities.Longer texts make it difficult for the encoder and decoder to reach full memory, and not all words can be used to recognize medical entities, therefore we introduce a multi-head self-attention layer. At the same time, it addresses the complexities of Chinese medical text nomenclature and the uneven distribution of entities.

The scaled dot product is usually used as the attention score function of the self-attention mechanism, and the calculation formula is shown in (1).

$$Attention(Q, K, V) = SoftMax\left(\frac{QK^T}{\sqrt{d_k}}\right)V \qquad (1)$$

Among them, Q, K and V are three matrices created using the same input but different parameters. First, the correlation between Q and K was calculated by multiplying them, and the attention matrix was created following Softmax normalization. Then divided by $\sqrt{d_k}$, the purpose is to reduce the weight variance between characters and avoid the problem of gradient disappearance caused by small weights in the training process. Finally, a vector representation with mixed relations is constructed by weighted summing of V using the acquired weights. Their structure is shown in Fig. 3.

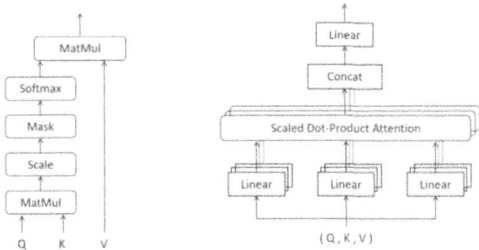

Fig. 3. Self-attention structure and Multi-head Self-attention structure

3.3 BiLSTM Layer

In the task of NER, compared with Recurrent Neural Network [14] (RNN), Long short-term memory network solves its difficulty in capturing long-distance dependencies in text. In contrast, LSTM is better able to capture long-term dependencies in text sequences through its unique gating mechanism, especially the forget gate, input gate and output gate. This leads to an improvement in named entity recognition's robustness and accuracy. Its cell structure diagram is shown in Fig. 4.

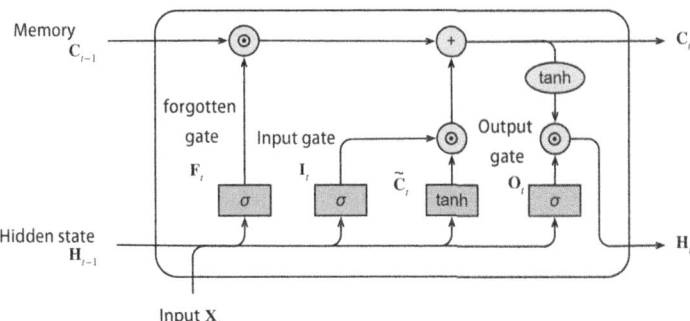

Fig. 4. Long short-term Memory network unit structure diagram

Assume there are h hidden units, a batch size of n, and d inputs. Thus, the input is $X_t \in \mathbb{R}^{n \times d}$, and the concealed state from the previous time step is $H_{t-1} \in \mathbb{R}^{n \times h}$. The specific calculation method is as follows:

$$\begin{cases} I_t = \sigma(X_t W_{xi} + H_{t-1} W_{hi} + b_i), \\ F_t = \sigma(X_t W_{xf} + H_{t-1} W_{hf} + b_f), \\ O_t = \sigma(X_t W_{xo} + H_{t-1} W_{ho} + b_o), \\ \tilde{C}_t = \tanh(X_t W_{xc} + H_{t-1} W_{hc} + b_c), \\ C_t = F_t \odot C_{t-1} + I_t \odot \tilde{C}_t, \\ H_t = O_t \odot \tanh(C_t). \end{cases} \quad (2)$$

where the input gate is $I_t \in \mathbb{R}^{n \times h}$, the forget gate is $F_t \in \mathbb{R}^{n \times h}$, and the output gate is $O_t \in \mathbb{R}^{n \times h}$. σ is the sigmoid activation function used as the output function of the fully connected layer. W_{xi}, W_{xf}, $W_{xo} \in \mathbb{R}^{d \times h}$ and W_{hi}, W_{hf}, $W_{ho} \in \mathbb{R}^{h \times h}$ are the weight parameters, b_i, b_f, $b_o \in \mathbb{R}^{1 \times h}$ is the bias parameter. $\tilde{C}_t \in \mathbb{R}^{n \times h}$ is the candidate memento, The memory cell $C_t \in \mathbb{R}^{n \times h}$ represents the state of the cell at the current time. Finally, according to the state of the output gate and the memory cell, the final output of LSTM, namely the hidden state $\mathbf{H}_t \in \mathbb{R}^{n \times h}$, can be obtained.

BiLSTM is a bidirectional long short-term memory neural network architecture that can process both forward and reverse input sequences. It contains two LSTM layers, forward and reverse, which are processed along the two directions of the sequence, respectively. In this way, bidirectional context information can be captured at each time step, which helps to better understand long-term dependencies in sequence data.

3.4 CRF Layer

Linear chain conditional random field are also known as conditional random field (CRF) were introduced by Lafferty et al. (2001). A conditional random field [15] is a probabilistic graphical model that generates a conditional probability distribution $P(Y|X)$ over an output sequence $Y = (Y_1, Y_2, \cdots Y_n)$ of random variables based on an input sequence $X = (X_1, X_2, \cdots X_n)$.

$$P(Y_i|X, Y_1, \cdots, Y_{i-1}, Y_{i+1}, \cdots, Y_n) = P(Y_i|X, Y_{i-1}, Y_{i+1}) \tag{3}$$

Conditional random fields are widely used in sequence labeling tasks, where a global feature function can be used to describe the dependence between the input and output sequences, and the feature function's weights are learned using the maximum entropy principle or maximum likelihood estimation. Let P be the weight matrix of the decoding layer output; then its scoring function is $Score(X, y)$:

$$Score(X, y) = \sum_{i=0}^{n} A_{y_i, y_{i+1}} + \sum_{i=1}^{n} P_{i, y_i} \tag{4}$$

Among these, A is the transition matrix, n represents the length of the series, $A_{y_i, y_{i+1}}$ represents the transition probability from y_i to y_{i+1}, and P_{i, y_i} represents the probability that the i-th word is categorized as y_i. In this method, some rule limitations can be introduced to the decoding layer's output to prevent the aforementioned inappropriate predicted labels, while also taking into account the entity context relationship. Finally, using the maximum likelihood approach for maximum a posterior probability $P(y|x)$, obtain the loss function values of the model.

$$\log P(y|x) = S(x, y) - \sum_{i=0}^{n} S(x, y_i) \tag{5}$$

4   Experiment and Result Analysis

4.1   Experimental Dataset

In this experiment, three data sets were used to evaluate the model's performance for Chinese medical entity recognition. The first dataset is Yidu-S4K, derived from one of the CCKS 2019 evaluation tasks and manually edited according to the real medical record distribution. The other dataset is CMeEE [16], which is the dataset of the "Chinese Medical Named Entity Recognition" sub-task in Chinese Medical Information Processing Benchmark (CBLUE). In this dataset, the words of the article are automatically segmented before labeling, and all medical entities have been correctly segmented. The final dataset is the Chinese community medical question answering cMedQANER [17] sub-dataset in ChineseBlue, which was launched by Alibaba Company. Entity nesting is not permitted in this experiment.

Table 1. Statistics of Yidu-S4K,CMeEE and cMedQANER

Yidu-S4K		CMeEE		cMedQANER	
Entity	Numbers	Entity	Numbers	Entity	Numbers
DISEASE	5535	dis	42553	physiology	461
DRUG	2307	sym	25523	test	604
EXAMINATIONS	1317	pro	20704	disease	4653
TEST	1785	equ	1800	time	262
TREATMENT	1191	dru	9850	drug	646
BODY	11520	ite	8455	symptom	2636
		bod	47509	body	2880
		dep	735	department	167
		mic	5086	crowd	861
				feature	366
				treatment	1318

There are three primary annotation approaches for named entity recognition tasks: BIO, BMES, and BIOES. This experiment using the BIO annotation method to process the JSON data set format. Label B represents the identified entity's first word, label I represents its non-first word, and label O represents the unnamed entity. Yidu-S4k dataset contains six labels such as name of diseaes (DISEASE), name of drugs (DRUG) and medical examinations (EXAMINATIONS). The cMeEE dataset contains nine labels such as diseases name (dis), symptoms name (sym), and Medical equipments (equ). The cMedQANER dataset contains 11 kinds of labels such as name of body parts (body), Diagnostic departments (department), and Type of patients (crowd). The entity types and number of these three datasets are shown in Table 1.

4.2 Experimental Environment and Parameter Settings

The experiment in this section was created using the Python programming language and the Pytorch deep learning development framework. The specific software and hardware parameters of the experiment are shown in Table 2.

In order to ensure the fairness of the experiment, except for some special experimental parameters, such as the number of Transformer layers, the number of hidden layers, and

Table 2. Experimental setting

Environment	Configuration
Operating system	Windows10
RAM	191GB
GPU	8 * NVIDIA® Tesla T4
CPU	Intel(R) Xeon(R) Gold 5218 CPU
Python version	Python 3.9
Pytorch version	Pytorch 2.1.2

the vocabulary size of different pre-trained models, all parameters are kept as consistent as possible. The detailed experimental model parameters are as follows: the length of the maximum allowed input text is 128 characters, the hidden layer dimension of the LSTM is 512, and the dropout is 0.5. The Adam optimizer was used to train the model, with a 5e-5 training learning rate and a 5e-5 weight decay. The multi-head self-attention layer consists of 12 self-attention heads, whose hidden state dimension is 128, and dropout is also 0.5. In the experiments, the embedding size of the five pre-trained models is 768, which is determined by the design architecture of the models. They all have 12 Transformer layers and 12 attention heads in each layer. The batch size is 64 for all three datasets.

4.3 Experimental Results and Analysis

In this paper, we propose a named entity recognition model called EHealth-BiLSTM-Attention-CRF. The model is based on EHealth, a pretrained language model with fixed parameters focused on medical information processing, together with multi-head self-attention layers. In order to prove the effectiveness of this model, we choose six deep neural network models related to this model to compare with the model proposed in this paper and each model is trained 10 times, and the results obtained are averaged. The six comparison models are as follows.

1. Bert-BiLSTM-CRF, the most commonly used deep neural network model in named entity recognition tasks, consists of the pre-trained language model Bert as the word embedding layer, combined with a long short-term memory network and a conditional random field layer.
2. Bert-BiLSTM-CRF(WWM), the Bert model upgraded by whole word masking(wwm), combined with the model composed of BiLSTM-CRF.
3. RoBert-BiLSTM-CRF(WWM), compared with Bert, RoBert removes the Next Sentence Prediction (NSP) task, uses a larger dataset, more stringent data processing and pre-training Settings, and improves compared with Bert on related tasks.
4. Ernie-BiLSTM-CRF, which combines the Enhanced Representation through Knowledge Integration (ERNIE) pre-trained model with the BiLSTM-CRF network, The difference with Bert model was stated above.

5. McBert-BiLSTM-CRF, McBert is based on Bert and trains a Bert-like Chinese pre-trained model for medical treatment.
6. EHealth-BiLSTM-CRF, the baseline model of this paper, uses a deep neural network model combining the EHealth pre-trained language model with the BiLSTM-CRF network.

The evaluation metrics in this experiment are the same as most named entity recognition tasks: precision, recall, and F1 score. The formulas for the three indices are as follows:

$$P = \frac{TP}{TP + FP} \times 100\% \tag{6}$$

$$R = \frac{TP}{TP + FN} \times 100\% \tag{7}$$

$$F_1 = 2 \times \frac{P \times R}{P + R} \times 100\% \tag{8}$$

In these formulas: TP represents the number of correctly identified entities, FP represents the number of irrelevant entities recognized, and FN represents the number of entities not recognized. The F1 score is the harmonic mean of precision and recall, with higher values indicating better performance. Table 3 shows the performance of the seven models on Yidu-S4K dataset, and Table 4 and Table 5 show the performance of the models on CMeEE and cMedQANER datasets.

Table 3. Experimental comparison results on Yidu-S4K

Model	Precision	Recall	F1-Score
Bert-BiLSTM-CRF	85.36%	89.37%	87.31%
Bert-BiLSTM-CRF (WWM)	86.60%	91.03%	88.76%
RoBert-BiLSTM-CRF (WWM)	87.54%	90.25%	88.88%
Ernie-BiLSTM-CRF	88.97%	90.46%	89.22%
McBert-BiLSTM-CRF	87.78%	90.80%	89.25%
EHealth-BiLSTM-CRF	88.98%	90.79%	89.89%
EHealth-BiLSTM-Attention-CRF	89.89%	90.58%	90.24%

Through the comparative experimental results of these seven models on three data sets, it can be seen that in the general field model, Ernie's effect of recognizing Chinese entities is better than Bert and the improved model of Bert. Some pre-trained language models in biomedical field, such as McBert, achieve 1.94%, 3.18%, 3.41% higher F1 score than Bert on three datasets; The F1 value of EHealth is 0.67%, 1.12%, 1.45% higher than that of Ernie. This shows that the pre-trained language model in the biomedical domain is more effective in understanding the text information in the related domain than the model in the general domain. The EHealth BiLSTM-CRF deep neural network

Table 4. Experimental comparison results on CMeEE

Model	Precision	Recall	F1-Score
Bert-BiLSTM-CRF	70.58%	77.70%	73.97%
Bert-BiLSTM-CRF (WWM)	71.76%	79.32%	75.35%
RoBert-BiLSTM-CRF (WWM)	72.33%	79.58%	75.78%
Ernie-BiLSTM-CRF	72.71%	79.75%	76.07%
McBert-BiLSTM-CRF	75.27%	79.13%	77.15%
EHealth-BiLSTM-CRF	75.07%	79.42%	77.19%
EHealth-BiLSTM-Attention-CRF	77.27%	79.43%	78.34%

Table 5. Experimental comparison results on cMedQANER

Model	Precision	Recall	F1-Score
Bert-BiLSTM-CRF	79.96%	84.01%	81.94%
Bert-BiLSTM-CRF (WWM)	80.72%	84.51%	82.57%
RoBert-BiLSTM-CRF (WWM)	82.53%	85.43%	83.96%
Ernie-BiLSTM-CRF	83.21%	85.06%	84.13%
McBert-BiLSTM-CRF	83.56%	87.22%	85.35%
EHealth-BiLSTM-CRF	83.54%	87.71%	85.58%
EHealth-BiLSTM-Attention-CRF	85.15%	88.15%	86.62%

with multi-head self-attention mechanism achieves an F1 score 0.35% higher than the baseline model on the Yidu S4K dataset, which may not be very high. However, the improvement on CMeEE and cMedQANER datasets is 1.15% and 1.04% respectively. This is because, as the entity information in the dataset rises and becomes congested, the self-attention mechanism can dynamically calculate the attention weights of different entity information, allowing the model to selectively focus on different entities and capture entity information more effectively. The experimental results show that our proposed named entity model is superior to other general domain or biomedical domain models in the task of Chinese biological named entity recognition.

5   Conclusion

Aiming at the complexity of named entity recognition in the Chinese biomedical field and the uneven distribution of entities between experimental datasets, we apply a Chinese biomedical language model EHealth and combine multi-head self-attention layers to form a deep neural network model EHealth-BiLSTM-Attention-CRF. We conduct extensive experiments with several different biomedical annotation datasets and several different models. Through this experiment, it can be concluded that a domain-specific model is more suitable for domain-specific natural language understanding tasks, and the self-attention mechanism can selectively focus on different parts of the sequence to better capture critical information. Through many comparative experiments, the results show that the proposed model is superior to other models in recognizing diseases, drugs, body and other biomedical terms. In the future, we will improve named entity recognition in the Chinese medical area using various methodologies, as well as investigate other fields to solve diverse entity recognition difficulties.

References

1. Ren, Q., Li, K., Yang, D., et al.: TCM function multi-classification approach using deep learning models. In: Yuan, L., et al. (eds.) Web Information Systems and Applications. WISA 2023. LNCS, vol. 14094, pp. 246–258. Springer, Singapore (2023). https://doi.org/10.1007/978-981-99-6222-8_21
2. Siami-Namini, S., Tavakoli, N., Namin, A.S.: The performance of LSTM and BiLSTM in forecasting time series. In: 2019 IEEE International Conference on Big Data, pp. 3285–3292. IEEE (2019)
3. LeCun, Y., Bottou, L., Bengio, Y., et al.: Gradient-based learning applied to document recognition. Proc. IEEE **86**(11), 2278–2324 (1998)
4. Devlin, J., Chang, M.W., Lee, K., et al.: Bert: Pre-training of deep bidirectional transformers for language understanding. In: 2019 Association for Computational Linguistics, pp. 4171–4186. Association for Computational Linguistics (2019)
5. Sang, E.F., De Meulder, F.: Introduction to the CoNLL-2003 shared task: language-independent named entity recognition. In: Proceedings of the Seventh Conference on Natural Language Learning at HLT-NAACL, pp. 142–147. Association for Computational Linguistics (2003)
6. Hammerton, J.: Named entity recognition with long short-term memory. In: Proceedings of the Seventh Conference on Natural Language Learning at HLT-NAACL, pp. 172–175. Association for Computational Linguistics (2003)
7. Peng, N., Dredze, M.: Improving named entity recognition for chinese social media with word segmentation representation learning. In: Proceedings of the 54th Annual Meeting of the Association for Computational Linguistics, pp. 149–155. Association for Computational Linguistics (2016)
8. Ji, B., Liu, R., Li, S., et al.: A BILSTM-CRF method to Chinese electronic medical record named entity recognition. In: Proceedings of the International Conference on Algorithms, Computing and Artificial Intelligence, pp. 166–169. IEEE (2019)
9. Li, L., Zhao, J., Hou, L., et al.: An attention-based deep learning model for clinical named entity recognition of Chinese electronic medical records. BMC Med. Inform. Decis. Mak. **19**(5), 1–22 (2019)
10. Li, X., Zhang, H., Zhou, X.H.: Chinese clinical named entity recognition with variant neural structures based on BERT methods. J. Biomed. Inform. **107**, 103422 (2020)
11. Zhang, Z., Han, X., Liu, Z., et al.: ERNIE: Enhanced language representation with informative entities. In: Proceedings of the 57th Annual Meeting of the Association for Computational Linguistics, pp. 1441–1451. Association for Computational Linguistics (2019)
12. Wang, Q., Dai, S., Xu, B., et al.: Building chinese biomedical language models via multi-level text discrimination. arXiv preprint arXiv:2110.07244 (2021)
13. Vaswani, A., Shazeer, N., Parmar, N., et al.: Attention is all you need. In: Proceedings of the 31st International Conference on Neural Information Processing Systems (NIPS'17), pp. 6000–6010. Curran Associates Inc (2017)

14. Zaremba, W., Sutskever, I., Vinyals, O.: Recurrent neural network regularization. arXiv preprint arXiv:1409.2329 (2014)
15. Xu, L., Li, S., Wang, Y., Xu, L.: Named entity recognition of BERT-BiLSTM-CRF combined with self-attention. In: Xing, C., Fu, X., Zhang, Y., Zhang, G., Borjigin, C. (eds.) Web Information Systems and Applications. WISA 2021. LNCS, vol. 12999, pp. 556–564. Springer, Cham (2021). https://doi.org/10.1007/978-3-030-87571-8_48
16. Zhang, Z., Chen, M., et al.: CBLUE: a Chinese biomedical language understanding evaluation benchmark. In: Proceedings of the 60th Annual Meeting of the Association for Computational Linguistics, pp. 7888–7915. Association for Computational Linguistics (2022)
17. Zhang, N., Jia, Q., Yin, K., et al.: Conceptualized Representation Learning for Chinese Biomedical Text Mining. arXiv preprint arXiv:2008.10813 (2020)

AGNE: Attentional Graph Convolutional Network Embedding for Knowledge Concept Recommendation in MOOCs

Jiahui Chen[1], Dan Meng[2], Xiangyun Gao[1], Liping Zhang[1], and Chao Kong[1(✉)]

[1] School of Computer and Information, Anhui Polytechnic University, Wuhu, China
{2230911107,2220920103}@stu.ahpu.edu.cn, {zhanglp,kongchao}@ahpu.edu.cn
[2] OPPO Research Institute, Shenzhen, China

Abstract. The knowledge concept recommendation aims to identify and suggest specific knowledge concepts that a student needs to master on Massive Open Online Courses (MOOCs) platforms, addressing the information overload issue and offering a tailored educational experience. Existing studies have constructed heterogeneous information networks to capture accurate representations of both learners and concepts, thereby mitigating the challenges posed by data sparsity. However, these approaches have limits in their ability to adequately synthesize and portray node-related data and nearby connections within the graph structure. Furthermore, they underutilize the potential of heterogeneous network data and fail to adapt or customize it in response to the dynamic nature of learning processes. To address these issues, we propose a new method named AGNE, short for *Attentional Graph Convolutional Network Embedding*, consists of three components. Firstly, we construct a heterogeneous information network that includes knowledge concepts, students, and other entities like videos, courses, and instructors. We employ meta-path embedding techniques to generate node embeddings enriched with semantic content and utilize Graph Convolutional Networks to integrate contextual information of these nodes. Secondly, we utilize a graph attention network to augment the efficacy of information dissemination among the entities. Lastly, we implement matrix factorization to refine and enhance the recommendation algorithm. These components are systematically amalgamated to produce more precise recommendation outcomes. Extensive experiments have been conducted on the large public MOOCCubeX dataset, demonstrating the superiority of AGNE against several state-of-the-art methods.

Keywords: Online Learning · Knowledge Concept Recommendation · Graph Convolutional Network · Heterogeneous Information Network

1 Introduction

In recent years, online education platforms have witnessed a surge in student and teacher participation due to their wealth of resources. Massive Open Online

Courses (MOOCs) offer students access to a vast array of learning materials and track their learning behaviors on the platform [1]. XuetangX [2] stands as one of China's largest MOOC platforms, offering comprehensive and impactful courses that enable students to quickly grasp the essence of their studies.

However, accompanied by the rise of such learning platforms is a high dropout rate. According to statistics [3], the majority of students who register for courses from the beginning ultimately fail to complete their studies.

Therefore, research on recommending knowledge concepts to learners has begun to emerge. These studies mostly utilize heterogeneous information networks with graph neural networks to learn the features of users and concepts. However, these methods often fail to fully utilize the contextual information in HIN. For instance, embedding techniques like Word2Vec and metapath2vec are widely used, but cannot fully capture the contextual information between nodes. Additionally, existing research has shown limitations in recognizing the contributions of neighbor nodes, causing low efficiency of information dissemination among the entities.

To overcome the limitations of existing methods, this paper proposes an innovative AGNE method. In the first step, to comprehensively integrate heterogeneous graph information, we employ the Meta-Path-Guided Deep Walk (MDK) algorithm for node sampling and embedding. To further enrich the capability of node representation, we then utilize graph convolutional networks (GCN) to aggregate information from contextual nodes encountered during random walks, generating auxiliary embeddings, which are then integrated with the original embeddings in subsequent calculations, thus improving performance. In the second step, to maximize the utilization of information from neighboring nodes, we adopt a strategy based on graph attention networks (GAT). Specifically, we first evaluate the feature contributions between nodes and their neighbors, then weigh these contributions using attention mechanisms, and finally transform these weighted features using nonlinear functions to form updated node embeddings. This method dynamically adjusts the influence weights of neighbors on the current node, providing a more accurate depiction of the relationships between nodes. In the final step, we employ matrix factorization techniques for rating prediction. The main contributions of this study can be summarized as follows.

– We utilize Graph Convolutional Networks (GCN) to integrate contextual information from random walks into auxiliary embeddings, thereby enhancing node representations across heterogeneous graphs.
– We employ Graph Attention Networks (GAT) to optimize the embeddings of nodes and their neighbors, effectively enhancing our ability and precision in handling complex heterogeneous graph data.
– We perform extensive experiments on the large public MOOCCubeX dataset to illustrate the effectiveness and rationality of AGNE.

The remainder of the paper is organized as follows. We first review the related work in Sect. 2. We formulate the problem in Sect. 3, before delving into details of the proposed method in Sect. 4. We perform extensive empirical studies in Sect. 5 and conclude the paper in Sect. 6.

2   Related Work

Our work is related to the following research directions.

2.1   Recommender System in MOOCs

Recommender systems in MOOCs have become a research hotspot nowadays. Some research attempts to introduce traditional recommendation methods into the field of smart education, such as the e-commerce sector methods. However, course selection behavior is fundamentally different from commodity purchasing behavior. Learning a course generally requires a long period and continuously consumes students' attention and energy. This difference leads to more severe data sparsity issues for recommender systems in MOOCs.

To address this dilemma, existing research attempts to combine graph learning to improve performance. For example, KGAN [4] utilizes knowledge graph-enhanced algorithms in course recommendation to automatically and iteratively estimate learners' potential interests, HGNN [5] models the relationships between learners through hypergraphs and simultaneously considers both the long-term and short-term sequential relationships of courses.

With further exploration of online learning, researchers have found that directly recommending courses may overlook users' need for an in-depth understanding of concepts. As a result, work related to **knowledge concept recommendation** has begun to attract attention. To the best of our knowledge, the earliest work is ACKRec [6], which combines heterogeneous information networks, graph convolutional neural networks, and attention mechanisms. Wang argued that ACKRec did not consider the characteristics of nodes. Therefore, they proposed Multi-HIN [7], which dynamically assigns contextual information to each node. MOOCIR [8] adopts different attention mechanisms and changes the objective function in ACKRec from Pointwise to Pairwise, thereby improving the recommendation effectiveness. Some approaches frame the knowledge concept recommendation problem as a reinforcement learning task, such as HinCRec-RL [9], as an improvement proposed by the original authors of ACK-Rec, and the method combining reinforcement learning with knowledge graphs by Jiang et al. [10]. CERec-ME [11] designed a novel loss function that simultaneously considers community structure information and node neighborhood information. ConceptGCN [12] uses a pre-trained transformer language model encoder (SBERT) to provide personalized and transparent recommendations. CL-KCRec [13] modeled the explicit and implicit relations in the heterogeneous information network of MOOCs and employed contrastive learning to balance their contributions to the final recommendations.

Most of these studies introduce heterogeneous information networks(HIN) to alleviate sparsity. However, they fail to fully leverage the high-order neighborhood information of nodes in the HIN, which is a limitation. Inspired by graph neural networks, this study utilizes GCN and GAT to fully explore the contextual information of nodes to optimize embedding effectiveness.

2.2 Heterogeneous Information Network Embedding

The purpose of Heterogeneous Graph embedding is to learn representations in a lower-dimensional space while preserving the diverse structures and semantics of heterogeneous graphs, to support downstream tasks such as node or graph classification, node clustering, and link prediction [14]. Metapath2vec [15] uses metapath-based random walks for sampling and learning heterogeneous graph embeddings. PME [16] treats each type of link as a relationship and utilizes a specific matrix for each relationship to transform nodes into distinct metric spaces. HetGNN [17] uses GNN to learn embeddings while simultaneously considering both heterogeneous structural information and heterogeneous content information for each node. PGCN [18] transforms user-item interactions into sub-graphs and employs a HGNN to propagate information across these graphs for collaborative filtering signal capture. DyHNE [19] is an incremental update method based on the theory of matrix perturbation, which considers both the heterogeneity and evolution of heterogeneous graphs when learning node embeddings.

3 Problem Formulation

We first introduce the notations used in this paper and then formalize the knowledge concept recommendation problem to be addressed.

Notations. In our study, MOOCs platform data is modeled as a Heterogeneous Information Network (HIN), represented as $G = \{V, E\}$, with V including sets such as Learners, Concepts, Teachers, Courses, and Videos, and E as the set of links. Both V and E are associated with two mapping functions: $\phi : V \longrightarrow O$ and $\varphi : E \longrightarrow R$. where O and R are sets of object and link types, respectively, with $|O| + |R| > 2$. This HIN underpins a network schema $S = (O, R)$, acting as a meta-template. Meta-paths are derived between entity pairs via the schema, denoted by composite relations $O_1 \xrightarrow{R_1} O_2 \xrightarrow{R_2} \ldots \xrightarrow{R_1} O_{l+1}$.

Problem Definition. The goal of the recommendation system is to predict the most relevant knowledge concepts that a learner is likely to need or be interested in based on their previous interactions and learning history.

Input: The interaction matrices $R \in \mathbb{R}^{n \times m}$, as well as the embeddings of learners and concepts.

Output: A ranking list for all candidate items. The $top - K$ items in the ranking list will be selected for recommendation.

4 Methodology

In this section, we present the proposed AGNE model to address the three major challenges mentioned in Sect. 1 and demonstrate the proposed model framework in Fig. 1.

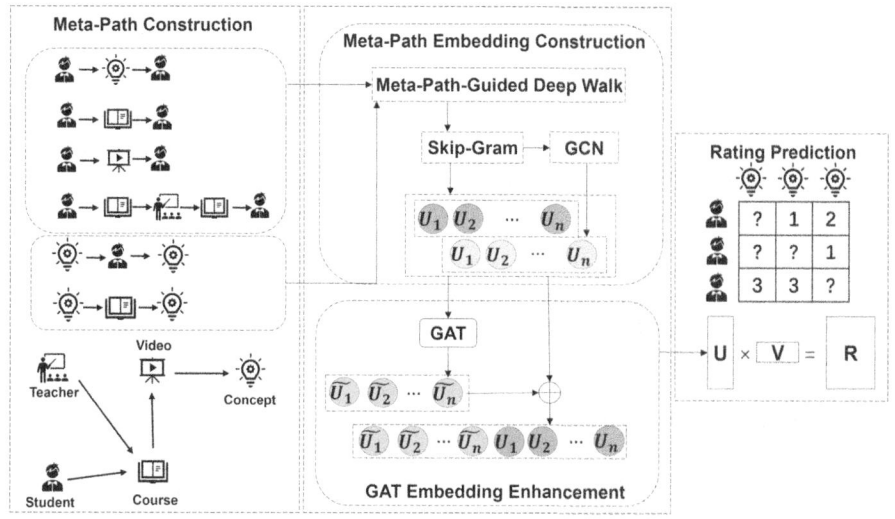

Fig. 1. An overview of AGNE model.

4.1 Meta-path Embedding Construction

Meta-path-Based Deep Walk. Firstly, we perform meta-path-based random walks to generate embeddings, and the walking strategy is as follows:

$$P\left(v^{i+1} \mid v_t^i, P\right) = \begin{cases} \frac{1}{|N_{t+1}(v_t^i)|}, & (v_t^i, v^{i+1}) \in E, \Phi(v^{i+1}) = t+1 \\ 0, & (v_t^i, v^{i+1}) \in E, \Phi(v^{i+1}) \neq t+1 \\ 0, & (v_t^i, v^{i+1}) \notin E \end{cases} \tag{1}$$

In this description, v_t^i represents the node visited at step i in the random walk, where t indicates the type of that node, and v^{i+1} is the next node to be visited. The set $N_{t+1}(v_t^i)$ contains all neighbor nodes connected to v_t^i whose node types are $t+1$. The probability function P is used to calculate the probability of transitioning from v_t^i to v^{i+1}.

Specifically, if there is an edge between v_t^i and v^{i+1}, and the node type of v^{i+1} conforms to the next expected node type $t+1$ according to the network scheme, then the probability of transitioning to v^{i+1} is $\frac{1}{|N_{t+1}(v_t^i)|}$, indicating that all eligible nodes have an equal chance of being selected. If there is no edge between v_t^i and v^{i+1}, or if the node type of v^{i+1} is not $t+1$, then the transition probability is 0. This ensures that the random walk strictly follows the sequence of node types prescribed by the meta-path, helping to accurately capture and represent the structural relationships between nodes.

Contextual Graph Embedding. After obtaining random walk paths, we use the skip-gram algorithm to project the node sequences into a low-dimensional space to learn distributed representations of nodes. Specifically, the optimization objective of this algorithm is $\arg\max_\theta \sum_{v \in V} \sum_{c \in C(v)} \log p(c|\theta(v))$. Here, the v refers to the current node, while $C(v)$ is the context node set obtained through metapath-based random walks, and θ denotes a function that maps each node v to a vector in a d-dimensional space. By applying this function to each node, we can thus obtain the node embedding e^p, p indicates the specific metapath under which the embedding is generated.

Auxiliary Embedding Generator. After obtaining embeddings from all meta-paths, we further process these embeddings using Graph Convolutional Networks (GCN) to generate auxiliary embeddings. Leveraging the context node set from the Meta-path-based Deep Walk, GCN aggregates contextual information to enhance and capture node features more comprehensively.

We can use the adjacency matrix M to describe the connection relationships of context node sequences in the heterogeneous graph G, and then aggregate the contextual neighbor information of nodes through GCN. To account for self-connections, we augment the adjacency matrix M with an identity matrix $M = M + I$. Before proceeding with aggregation, we need to normalize the adjacency matrix M to consider the degrees of each node, ensuring that the influence of different nodes remains balanced during the aggregation process, the formula can be defined as follows $\Delta = \sum_j M_{ij}$ and $\tilde{M} = \Delta^{-\frac{1}{2}} M \Delta^{-\frac{1}{2}}$. Finally, We apply $ReLU$ to introduce non-linearity to enhance model complexity and utilize a $softmax$ function to normalize the outputs into a probability distribution: $\hat{e}^p = softmax(\tilde{M} ReLU(\tilde{M} e^p W_0^p) W_1^p)$, where the matrices W_0^p and W_1^p serve as transformation weights.

4.2   Graph Attention Network Enhancement

We employ the graph attention network to collect the neighboring information of each node. For meta-path l, we compute the attention scores of neighboring nodes towards the target node $e_v^{(l)}$: $a_{v,j}^{(l)} = ReLU(\eta^{(l)T} \cdot [W^{(l)} e_v^{(l)} || W^{(l)} e_j^{(l)}])$, where $\eta^{(l)T}$ represents the attention parameter vector under the l-th meta-path and $W^{(l)}$ is a trainable weight matrix. After calculating the contribution of all neighboring nodes, we can obtain the normalized contribution scores $\tilde{a}_{v,j}^{(l)}$:

$$
\tilde{a}_{v,j}^{(l)} = \frac{exp(t_{vj}^{(l)})}{\sum_{k=1}^{k \in N_v} exp(t_{vk}^{(l)})},
$$

$$
\tilde{e}_v^{(l)}(K) = \overset{K}{\underset{k=1}{||}} \sigma(\sum_{j \in \mathcal{N}_v} [\tilde{a}_{vj}^{(l)}]^k [W^{(l)}]^k e_j^{(l)}).
$$

(2)

After obtaining the normalized contribution scores, we use a multi-head attention mechanism to calculate the new feature of node v, as mentioned in Equation (2). σ is a nonlinear transformation function, $[W^{(l)}]^k$ is a trainable weight matrix for the k-th attention head. After the calculations are finished, the results from all heads are concatenated together. Additionally, to preserve the information from the unprocessed embeddings, we concatenate the embeddings processed by the GAT with the original embeddings $e_v^{(l)} = \left[\tilde{e}_v^{(l)}(K) \| e_v^{(l)}\right]$.

Then, we apply a nonlinear depth fusion technique to merge all various types of embedded information from the meta-path into the targeted embedded information for a single node: $e_v = \sigma\left(W_u^{(l)}\left(\sigma(M^{(l)}e_v^{(l)} + b^{(l)}) + \sigma(M^{(l)}\hat{e}_v^{(l)} + b^{(l)})\right)\right)$, In this definition, $M^{(l)}$ and $b^{(l)}$ represent the conversion matrix and offset vector for meta-path l. $W_u^{(l)}$ denotes the preference weight of user v under meta-path l. It should be noted that $\hat{e}_v^{(l)}$ is the auxiliary embedding of $e_v^{(l)}$ previously generated under path l, which has also been processed by GAT.

4.3   Matrix Factorization-Based Prediction

In the final module, we merge the embedded vectors of users e_u and items e_i into the matrix factorization model to calculate ratings. The equation is defined as: $\hat{r}_{u,i} = u_u^T \cdot v_i + \alpha \cdot e_u^T \cdot \gamma_i + \beta \cdot \gamma_u^T \cdot e_i$, where γ_i and γ_u are two hidden factors, with α and β as adjustable parameters. The ultimate optimization objective is as follows:

$$\ell = \sum_{(u,i,r_{u,i})\in\mathcal{R}} (r_{u,i} - \hat{r}_{u,i})^2 + \sum \lambda(\| \theta \|_2^2). \tag{3}$$

To prevent overfitting, regularization constraints are added to the loss function. θ represents all trainable parameters in the model. During the training process, gradient descent is used, which calculates derivatives and iteratively updates parameters towards the negative gradient direction until the model converges.

5   Empirical Study

To evaluate the proposed model, multiple experiments were conducted on the large public MOOCCubeX dataset. Through experimental analysis, we aim to answer the following research questions.

RQ1: How does the performance of the proposed model compare to the state-of-the-art methods?

RQ2: How can the proposed model improvements in this paper, which involve using GCN for auxiliary embeddings and GAT for aggregating meta-path embeddings with neighbor weights, enhance existing methods' performance?

RQ3: How do the different meta-paths affect the performance of this model?

In what follows, we first introduce the experimental settings and answer the above research questions in turn to demonstrate the rationality of our methods.

Table 1. The details of MOOCCubeX dataset.

Entities	Statistics	Relations	Statistics	Density
users	3,175	user-concept	26,551	0.007265438
concepts	1,151	user-course	13,152	0.034519685
courses	120	user-video	32,949	0.001464940
teachers	188	course-teacher	313	0.013874113
videos	7,084	video-concept	160,943	0.019738685

Dataset. We conducted experiments on a large public MOOCCubeX dataset [1]. MOOCCubeX is maintained by the Knowledge Engineering Group of Tsinghua University and supported by XuetangX, one of the largest MOOC websites in China. We selected a portion of data from early September to the end of December 2020 for analysis. The statistics of our experimented dataset are summarized in Table 1.

Meta-path Definition. We can define various meta-paths based on the dataset.

For the user aspect, we have designed the following four meta paths: User-Knowledge-User (UKU), User-Video-User (UVU), User-Course-User (UCU), and User-Course-Teacher-Course-User (UCTCU). For the knowledge concepts aspect, we designed the following two metapaths: Knowledge-Course-Knowledge (KCK) and Knowledge-User-Knowledge (KUK).

Baselines. We co mpare AGNE with the following baselines:

- BPR [20]: The BPR method utilizes stochastic gradient descent to optimize the ranking preferences of users for items.
- MLP [21]: The MLP method employs multilayer perceptrons to learn non-linear mappings and capture complex patterns in data for improved predictive accuracy.
- DeepFM [22]: DeepFM is a recommendation system model that combines Factorization Machines (FM) and deep neural networks.
- FISM [23]: An item-to-item collaborative filtering algorithm leverages item-based factorization models to capture item-item interactions.
- NAIS [24]: An item-to-item collaborative filtering algorithm utilizes neural attentive item similarity models to capture personalized item interactions.
- metapath2vec [15]: A method utilizes graph embedding techniques to learn representations of nodes in heterogeneous information networks.
- ACKRec [6]: A knowledge concept recommendation combines heterogeneous information networks and attention mechanisms.

---

[1] https://github.com/THU-KEG/MOOCCubeX

- Multi-HIN [7]: Derived from ACKRec, it dynamically assigns contextual information to each node and Learning item representations using the skip-gram model with Gumbel-Softmax.

Evaluation Metrics. We evaluate all methods using widely adopted metrics, including Hit Ratio at $top-K$ items ($HR@K$), Normalized Discounted Cumulative Gain at $top-K$ items ($NDCG@K$), and Mean Reciprocal Rank (MRR). The knowledge concepts with which users last interacted are taken as positive instances, and 99 randomly sampled knowledge concepts that users have never interacted with are taken as negative instances. Metrics are calculated based on these 100 instances.

Table 2. Performance comparison on the MOOCubeX.

Methods	HR@5	HR@10	HR@20	NDCG@5	NDCG@10	NDCG@20	MRR
BPR	0.5514	0.7029	0.7886	0.3983	0.4473	0.4791	0.3810
MLP	0.4984	0.6264	0.7149	0.3720	0.4129	0.4451	0.3501
DeepFM	0.5489	0.6755	0.7743	0.3959	0.4370	0.4670	0.3763
NAIS	0.5344	0.6601	0.7697	0.4427	0.4705	0.4928	0.4047
FISM	0.5923	0.6978	0.8061	0.4409	0.4752	0.5026	0.4155
metapath2vec	0.6050	0.7143	0.8079	0.4533	0.4902	0.5168	0.4245
ACKRec	0.6620	0.7679	0.8375	0.5155	0.5499	0.5648	0.4888
Multi-HIN	0.6453	0.7720	0.8523	0.5013	0.5561	0.5790	0.5015
AGNE	**0.6995***	**0.8031***	**0.8796***	**0.5531***	**0.5853***	**0.6032***	**0.5250***

* indicates that the improvements are statistically significant for $p < 0.01$ judged by paired t-test

5.1 Performance Comparison (RQ1)

To demonstrate the recommendation performance of our model AGNE, we compared it with other state-of-the-art methods. As shown in Table 2, we have the following key observations.

- AGNE significantly outperforms models such as BPR, MLP, and DeepFM, underscoring that graph neural networks more effectively capture the interactions between users and items in MOOCs data. Moreover, AGNE also shows substantial improvements over item-based collaborative filtering algorithms like FISM and NAIS, For instance, compared to FISM, AGNE sees a 26.4% increase in MRR, an 23.2% increase in NDCG@10, and a 15.1% increase in HR@10, highlighting the benefits of integrating heterogeneous information networks in MOOCs recommendation systems.
- AGNE demonstrates considerable advantages in comparison with HIN-based recommendation systems such as metapath2vec, ACKRec, and Multi-HIN.

Compared to Multi-HIN, AGNE shows improvements of 4.7% in MRR, 5.3% in NDCG@10, and 4.0% in HR@10, respectively. This improvement can be attributed to two factors: firstly, the generation of auxiliary embeddings captures more information about the graph structure. Secondly, the introduction of graph attention networks (GAT) enables the model to more effectively utilize information from neighboring nodes, further enhancing the performance of the recommendation system.

Table 3. Performance comparison of AGNE and its variants.

Methods	HR@5	HR@10	HR@20	NDCG@5	NDCG@10	NDCG@20	MRR
$AGNE_{gcn}$	0.6668	0.7753	0.8465	0.5200	0.5649	0.5758	0.4974
$AGNE_{gat}$	0.6894	0.7942	0.8692	0.5395	0.5703	0.5890	0.5123
AGNE	**0.6995***	**0.8031***	**0.8796***	**0.5531***	**0.5853***	**0.6032***	**0.5250***

* indicates that the improvements are statistically significant for $p < 0.01$ judged by paired t-test.

5.2   Ablation Study (RQ2)

To verify whether our model's performance has been enhanced by two significant improvements, we design two model variants: The model $AGNE_{gcn}$, which removes the GAT enhancement, and the model $AGNE_{gat}$, which removes the auxiliary embeddings. Table 3 presents the experimental results of various variants of AGNE, all of which outperform the baseline method metapath2vec in HIN and the state-of-the-art method ACKRec in knowledge concept recommendation. We can observe that AGNE successfully outperformed its two variants on the MOOCCubex dataset, this demonstrates the rationale behind the improvements of utilizing GCN to generate auxiliary embeddings with contextual information and aggregating neighbor information using GAT networks, both of which contributed to significant improvements.

5.3   Meta-path Selection (RQ3)

Figure 2 displays the performance of the AGNE model when different meta-paths are selected. The data indicates that UKU has the most significant effect on enhancing model performance, while UCTCU contributes relatively less to performance improvement. However, overall, each meta-path has a positive impact on performance. When all four meta-paths are used simultaneously, the model demonstrates optimal performance.

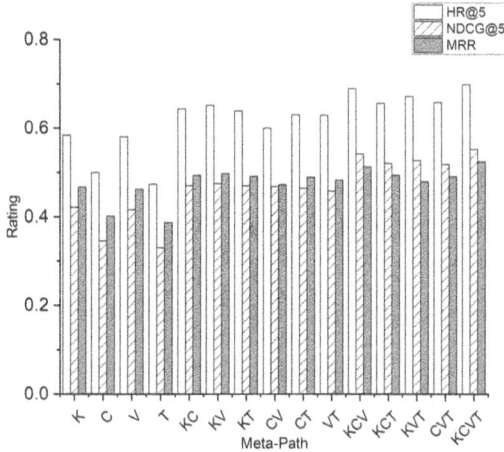

Fig. 2. The performance under different combinations of meta-paths. Here, K stands for the abbreviation of the meta-path UKU, C stands for UCU, V stands for UVU and T stands for UCTCU. KC represents the simultaneous use of UKU and UCU, etc.

6   Conclusions

This paper introduces AGNE, a novel approach designed to enhance knowledge concept recommendations on MOOC platforms. AGNE addresses challenges such as data sparsity and suboptimal use of heterogeneous network data by combining graph convolution techniques with a graph attention network. Testing on the MOOCCubeX dataset shows that AGNE significantly outperforms current methods. Future research should focus on improving accuracy by incorporating more user behavior data and integrating additional information sources like knowledge graphs and social networks to better adapt to changing educational environments and meet individual learner needs more effectively.

Acknowledgments. This work has been supported by the Science Research Project of Anhui Higher Education Institutions (No. 2023AH050914), the National Natural Science Foundation of China Youth Fund (No. 61902001), the Science and Technology Project of Wuhu City under Grant No. 2023pt07 and 2023ly13, the Quality Engineering Project of Anhui Higher Education Institutions (No. 2023zybj018), and the Quality Improvement Program of Anhui Polytechnic University (No. 2022lzyybj02).

References

1. Liang, Z., Mu, L., Chen, J., Xie, Q.: Graph path fusion and reinforcement reasoning for recommendation in MOOCs. Educ. Inf. Technol. **28**(1), 525–545 (2023)
2. Yu, J., et al.: MOOCCubeX: a large knowledge-centered repository for adaptive learning in MOOCs. In: CIKM 2021, pp. 4643–4652 (2021)
3. Wang, R., Cao, J., Xu, Y., Li, Y.: Learning engagement in massive open online courses: a systematic review. Front. Educ. **7**, 1074435 (2022)

4. Zhang, H., Shen, X., Yi, B., Wang, W., Feng, Y.: KGAN: knowledge grouping aggregation network for course recommendation in MOOCs. Expert Syst. Appl. **211**, 118344 (2023)

5. Wang, X., Ma, W., Guo, L., Jiang, H., Liu, F., Xu, C.: HGNN: hyperedge-based graph neural network for MOOC course recommendation. Inf. Process. Manage. **59**(3), 102938 (2022)

6. Gong, J., et al.: Attentional graph convolutional networks for knowledge concept recommendation in MOOCs in a heterogeneous view. In: SIGIR2020, pp. 79–88 (2020)

7. Wang, X., Jia, L., Guo, L., Liu, F.: Multi-aspect heterogeneous information network for MOOC knowledge concept recommendation. Appl. Intell. **53**(10), 11951–11965 (2023)

8. Piao, G.: Recommending knowledge concepts on MOOC platforms with meta-path-based representation learning. In: EDM 2021 (2021)

9. Gong, J., et al.: Reinforced MOOCs concept recommendation in heterogeneous information networks. ACM Trans. Web **17**(3) (2023)

10. Jiang, L., et al.: Reinforced explainable knowledge concept recommendation in MOOCs. ACM Trans. Intell. Syst. Technol. **14**(3), 1–20 (2023)

11. Ye, B., Mao, S., Hao, P., Chen, W., Bai, C.: Community enhanced course concept recommendation in MOOCs with multiple entities. In: KSEM 2021, pp. 279–293 (2021)

12. Alatrash, R., Chatti, M.A., Ain, Q.U., et al.: ConceptGCN: knowledge concept recommendation in MOOCs based on knowledge graph convolutional networks and SBERT. Comput. Educ. Artif. Intell. **6**, 100193 (2024)

13. Gu, H., Duan, Z., Xie, P., Zhou, D.: Modeling balanced explicit and implicit relations with contrastive learning for knowledge concept recommendation in MOOCs. In: Proceedings of the ACM on Web Conference 2024, pp. 3712-3722 (2024)

14. Yu, C., Fei, L., Chen, F., Chen, L., Wang, J.: Heterogeneous graphs embedding learning with metapath instance contexts. In: Yuan, L., Yang, S., Li, R., Kanoulas, E., Zhao, X. (eds.) WISA 2023. LNCS, vol. 14094, pp. 149–161. Springer, Singapore (2023). https://doi.org/10.1007/978-981-99-6222-8_13

15. Dong, Y., Chawla, N.V., Swami, A.: MetaPath2Vec: scalable representation learning for heterogeneous networks. In: SIGKDD 2017, pp. 135–144 (2017)

16. Chen, H., Yin, H., Wang, W., Wang, H., Nguyen, Q.V.H., Li, X.: PME: projected metric embedding on heterogeneous networks for link prediction. In: SIGKDD 2018, pp. 1177–1186 (2018)

17. Zhang, C., Song, D., Huang, C., Swami, A., Chawla, N.V.: Heterogeneous graph neural network. In: SIGKDD 2019, pp. 793–803 (2019)

18. Xu, Y., Zhu, Y., Shen, Y., Yu, J.: Learning shared vertex representation in heterogeneous graphs with convolutional networks for recommendation. In: IJCAI 2019, pp. 4620–4626 (2019)

19. Wang, X., Lu, Y., Shi, C., Wang, R., Cui, P., Mou, S.: Dynamic heterogeneous information network embedding with meta-path based proximity. IEEE Trans. Knowl. Data Eng. **34**(3), 1117–1132 (2022)

20. Rendle, S., Freudenthaler, C., Gantner, Z., Schmidt-Thieme, L.: BPR: Bayesian personalized ranking from implicit feedback. In: UAI 2009 (2009)

21. He, X., Liao, L., Zhang, H., Nie, L., Hu, X., Chua, T.S.: Neural collaborative filtering. In: WWW 2017, pp. 173–182 (2017)

22. Rendle, S.: Factorization machines with libFM. ACM Trans. Intell. Syst. Technol., 1–22 (2012)

23. Kabbur, S., Ning, X., Karypis, G.: FISM: factored item similarity models for top-n recommender systems. In: SIGKDD 2013 (2013)
24. He, X., He, Z., Song, J., Liu, Z., Jiang, Y.G., Chua, T.S.: NAIS: neural attentive item similarity model for recommendation. IEEE Trans. Knowl. Data Eng. 2354–2366 (2018)

Sepsis Mortality Prediction with Electronic Health Records Based on Sequential and Attention-Based Models

Xianbo Liu[1], Kaiyuan Wang[2(✉)], Weifan Wang[3(✉)], Yi Luo[1(✉)], Peng Ren[4(✉)],
Yuhang Hu[1], Zeming Li[1], Xiangkuan Li[5], Zhentao Hu[3], Wenyao Li[6],
and Chunxiao Xing[4]

[1] School of Computer Science and Technology, Beijing Institute of Technology,
Beijing 100081, China
{liuxianbo,luoyi,huyuhang_21,lizeming}@bit.edu.cn
[2] School of Accountancy, Central University of Finance and Economics, Beijing 100081, China
[3] School of Artificial Intelligence, Henan University, Zhengzhou 450046, China
{henuwwf,hzt}@henu.edu.cn
[4] BNRist, DCST, RIIT, Tsinghua University, Beijing 100084, China
{renpeng,xingcx}@tsinghua.edu.cn
[5] Beijing National Day School, Beijing 100049, China
[6] School of Software, Henan University, Kaifeng 475004, China

Abstract. Sepsis is a leading cause of death in the ICU. The mortality prediction driven by medical data is vitally important for sepsis prevention and treatment. However, the task of risk prediction is particularly challenging due to the complexity and heterogeneity of medical data. The past models tend to focus only on sequential models or time embedding. In this paper, we propose a novel model architecture DiagNet, which utilizes sequential and global analysis of patient information to improve the prediction accuracy. For DiagNet, we design a comparison experiment with four existing models Retain, Dipole, RetainEX and HiTANet, and the ablation study with the next-best variant on the MIMIC-IV and eICU datasets. Evaluations showcase that DiagNet outperforms other four models, especially on the MIMIC-IV dataset, achieving both a superior F1-Score and AUC score. Comparing with the next-best variant of the architecture, DiagNet performs better on the most crucial metric, AUC on both datasets. This research contributes to the field by providing an enhanced model architecture for healthcare risk prediction, offering the potential for improved patient care and outcomes.

Keywords: Sepsis · Mortality prediction · Electronic health records · Attention-based

1 Introduction

Sepsis is an extremely serious and life-threatening infection in the ICU. Every year more than 49 million people are infected with sepsis all over the world and nearly one in three of those die [1]. Based on the fact, mortality prediction, a subset of risk prediction in

© The Author(s), under exclusive license to Springer Nature Singapore Pte Ltd. 2024
C. Jin et al. (Eds.): WISA 2024, LNCS 14883, pp. 476–486, 2024.
https://doi.org/10.1007/978-981-97-7707-5_39

healthcare is a key area of research, which uses various statistical and machine learning techniques [2] to predict the likelihood health outcomes for individual patients [3]. This predictive model can effectively anticipate potential health problems and support to intervene early, potentially improving patient outcomes and increasing the efficiency of healthcare delivery.

Due to patient state being sequential in nature, the RNN, introduced by Elman [4], and the LSTM, introduced by Hochreiter et al. [5], have proven effective and versatile when solving a variety of risk prediction tasks. Reddy et al. [6] use a mixture of a RNN and LSTM to predict re-admission rate for lupus patients, beating the classical logistic regression and multilayer perceptron. Besides, the Retain architecture proposed by Choi et al. [7] also utilizes a RNN to predict heart failure, outperforming both a pure RNN and MLP on the task. Through the attention mechanism of Transformer architecture, HiTANet, proposed by Luo et al. [8], goes even further by adding a separate time encoding to a Transformer.

Unfortunately, the methods mentioned above tend to focus only on the characteristics of data sequence or temporal embedding. Therefore, we propose a new model architecture, DiagNet, which improves on the previous approach. By combining the strengths of the current state-of-the-art sequential and attention-based models, DiagNet can demonstrate superior performance in the field of risk prediction.

Our contributions are as follows. To deal with the high degree of sparsity inherent in EHR data, DiagNet initially per forms a linear projection, before then passing on the data to the sequential and global analysis components. To address the irregularly parsed data, DiagNet takes a learnable time-embedding, analogous to the one proposed in HiTANet [8], and adds it to the linear projection prior to passing it into the global analysis component. Finally, to aid in interpreting the model's predictions, DiagNet has two separate mechanisms.

This paper has 6 chapters. In Sect. 1, we introduce the research topic and current approaches. Then we briefly describe our methods and contributions, and illustrate the structure of the paper. Section 2 introduces four related work and their models Retain, Dipole, RetainEX, and HiTANet. Section 3 reviews electronic health records, focusing specifically on the MIMIC-IV and eICU databases, which includes 6660 and 13564 patients with sepsis respectively. Section 4 presents our proposed model, DiagNet, which covers the sequential analysis component, global analysis component and analysis fusion. Then in Sect. 5, we evaluate the performance of DiagNet with other related work, to demonstrate the excellent performance of our models. At last, Sect. 6 discusses the findings, limitations, and potential future work related to the study.

2   Related Work

In order to solve the problem of risk prediction, varying architectures have been proposed. In this section, four models that shows good performance will be discussed. These models are Retain, Dipole, RetainEX and HiTANet.

Created by Choi et al. [7], Retain models patient visits by performing a single learned projection, then applying two RNNs to calculate separate attention weightings. These are then multiplied with the projection to create the final context vector, which is finally mapped to the number of predicting classes.

Ma et al. [9] proposed Dipole, which uses a Bidirectional RNN to model patient visits in a sequential order. The paper discussed three different attention mechanisms and applied the best performing one on top of the model. The final context vector is then calculated by summing the product of the hidden states and the attention.

Similar to Dipole, Kwon et al. [10] also used a Bi-directional RNN as the backbone of their architecture to create their model, RetainEX. In contrast, RetainEX do two projections instead of one initially, to calculate two separate attention values. Finally, the earlier linear projection is multiplied pointwise with these two separate attention values and summed over each time t to calculate the final context vector.

The above three models are all based on sequential paper models, while HiTANet, proposed by Luo et al. [8], uses Transformer as its backbone. The model has two distinct components, the visit level analysis component and the global level analysis component. In the visit level analysis, the model puts the projections added with time encoding into a Transformer, to calculate an attention weighting. Then in the global level, the model mimics the attention proposed by Vaswani et al. [11] to combine the hidden state and the pure time encodings to generate additional attention. At last, it merges these attention weightings and then multiplies with the hidden state to generate its final context vector.

All of these models make good use of the attention mechanism, which greatly enhances the interpretability of the models. However, each of these models has its own disadvantages. For Retain, the use of a single linear projection to represent patient visits simplifies the complexity of the model, making it impossible to fully capture the complexity and diversity of patient information. And for Dipole, though its use of a Bidirectional RNN can help to capture dependencies in both past and future data points, it may still hard to capture long-term dependencies when the relevant data points are far apart. As for RetainEX, its performance depends heavily on the quality of the linear projection and attention mechanisms, which affects the robustness of the model. One potential weakness of HiTANet is its strong focus on time, which may ignore other more important factors, thus leading to sub-optimal performance.

3 Datasets

Electronic Health Records (EHRs) represent a collection of patient health information in a digital format [12]. Designed to capture the complete health information of individual patients across time, EHRs provides a comprehensive, longitudinal record of a patient's health history. EHRs can include a wide range of data, not only personal information like age and weight, but also medical data such as immunization status and vital signs.

In this paper, we choose two typical EHRs to test our model architecture, MIMIC-IV and eICU. In order to make the data meet the requirements of our model, we select sepsis patients and remove all other patients. Each of these datasets is processed, filtered, and missing values are imputed with the help of expert knowledge.

3.1 Mimic-IV

Introduced by Johnson et al. [11], MIMIC-IV is a publicly accessible EHR dataset derived from the Beth Israel Deaconess Medical Center (BIDMC). The MIMIC-IV dataset consists of 6660 total sepsis patient stays. Each patient's stay in the emergency unit is recorded with around 15 different samples. In addition to recording demographic and observation data, each sample also gives multiple different diagnoses of each patient, such as diabetes, hypertension, and cancer.

Furthermore, the dataset features a critical predictor label that determines whether a patient survived or passed away within 90 days' post-emergency unit admission. This binary classification is represented as a Boolean variable in our model. It should be noted that the proportion of classifications is not balanced, with 80.3% of patients surviving and 19.7% patients dying.

3.2 EICU

Introduced by Pollard et al. [13], the eICU Collaborative Research Database is a comprehensive and freely accessible database that facilitates research in critical care. The eICU dataset comprises 13,564 sepsis patient stays. During the patients' stay in the emergency unit, 15 different samples are taken. In addition to recording various monitoring data, the samples also include multiple different diagnoses of each patient, such as SOFA (Sequential Organ Failure Assessment) scores, whether the patient has heart disease and so on.

Like the MIMIC-IV dataset, the eICU dataset also represents patient survival status as a Boolean variable. Similar to MIMIC-IV, there is a class imbalance within the eICU dataset, with "surviving" patients accounting for 83 percent, and the "passed" patients at 17 percent.

4   Model Architecture

In this section, we will introduce our novel model, DiagNet. As shown in Fig. 1, the model contains three components, sequential analysis, global analysis and analysis merging, which mimic the doctor's diagnostic process from several aspects such as patient status and time embedding.

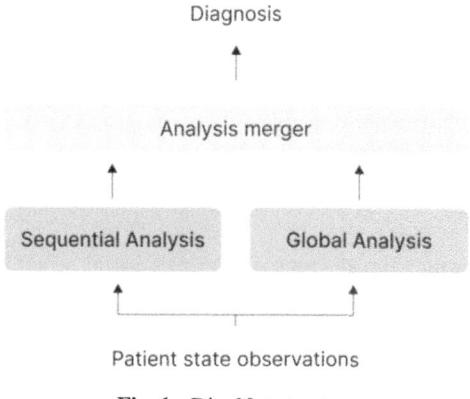

Fig. 1. DiagNet structure

4.1 Sequential Analysis

The sequential analysis mimics the analysis a doctor performs when intermittently checking up on the patient before coming to a decision.

First, the patient state x_t is projected to an embedding vector $e_t \in \mathbb{R}^m$ with a standard hidden layer and a ReLU operation:

$$e_t = \text{RELU}(W_e x_t + b_e) \tag{1}$$

where $W_e \in \mathbb{R}^{m \times l}$ is the weight of the projection layer and $b_e \in \mathbb{R}^m$ is the bias.

Then each embedding is fed into an LSTM one-by-one, getting each subsequent hidden state $h_i \in \mathbb{R}^k$ through the following:

$$h_i = \text{LSTM}(e_{i-1}, h_{i-1}) \tag{2}$$

In this way, each h_i can represent the understanding of the patient up until state i. As the final understanding of the whole sequence, the final hidden state h_n is taken as the output of this component to send to the diagnosis merger.

Figure 2 illustrates the entire process of sequential analysis.

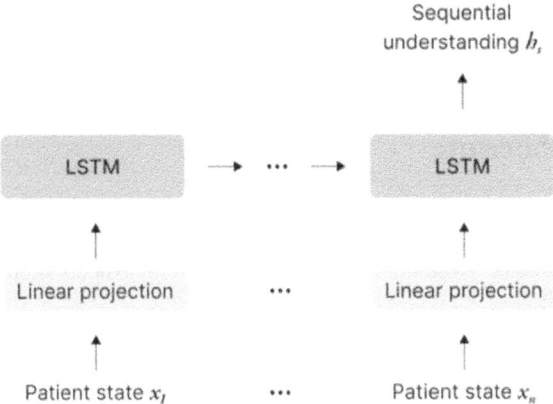

Fig. 2. DiagNet's sequential analysis component

4.2 Global Analysis

The global analysis mimics the analysis a doctor performs when looking at the whole patient sheet in parallel to checking up on the patient.

Similar to the sequential analysis component, the global analysis also begins by projecting each patient state x_i to an embedding vector $e_{t'} \in \mathbb{R}^m$ with a hidden layer and a ReLU operation:

$$e_{t'} = \text{RELU}(W_{e'} x_t + b_{e'}) \tag{3}$$

where $W_{e'} \in \mathbb{R}^{m \times 1}$ is the weight of the projection layer and $b_{e'}$ is the bias.

In order to provide information on the distance between each observation, we use an additional time-embedding, which is introduced by Luo et al. [8]. Through this time-embedding, Transformer will consider the exact distance between each state. To generate it, the time δ_t of each patient state x_t is taken and mapped to an in-between latent representation o_t through the following operation:

$$o_t = 1 - \tanh((W_o \frac{\delta_t}{180} + b_o)^2) \tag{4}$$

where $W_o \in \mathbb{R}^{a \times 1}$ is the weight matrix projecting the time into a latent space, b_o, $o_t \in \mathbb{R}^a$ are the bias vector and the latent representation. In this operation, through point-wise multiplication, tanh activation function and invert, the correlation of patient status increased.

To feed the information into the Transformer, o_t is projected to another latent representation r_t of the same dimension:

$$r_t = \tanh(W_r O_t + b_r) \tag{5}$$

Here, $W_r \in \mathbb{R}^{m \times a}$ is the weight matrix and $b_r \in \mathbb{R}^m$ is the bias.

Then the time-embedding and the patient state embedding are added to get the complete embedding v_t, which is fed into a single Multi-head Transformer. The Transformer learns both longer and short-term dependencies of the data to provide the multiple different hidden states:

$$v_t = e_{t'} + r_t \tag{6}$$

$$[h_1, \ldots, h_n] = \text{Transformer}[v_1, \ldots, v_n] \tag{7}$$

Since only a single hidden state is required to summarize the data, we use the approach proposed by Ma et al. [9]. Each h_t is projected into $\alpha_{t'}$, which is the attention weighting of individual state:

$$\alpha_{t'} = W_\alpha h_t + b_\alpha \tag{8}$$

where $W_\alpha \in \mathbb{R}^{1 \times m}$ is the weight matrix projecting the hidden state to a scalar, and $b_\alpha \in \mathbb{R}$ is the bias.

Since these values are unrestricted, we take the normalization through Softmax operation to obtain α_t. Finally, those weightings are applied as scalar multiplications to each hidden state h_t and summed to retrieve the final global patient understanding h_g.

$$[\alpha_1', \ldots, \alpha_n'] = \text{Softmax}[\alpha_1, \ldots, \alpha_n] \tag{9}$$

$$h_g = \sum_{i=1}^{n} \alpha_i h_i \tag{10}$$

Figure 3 illustrates the entire process of global analysis.

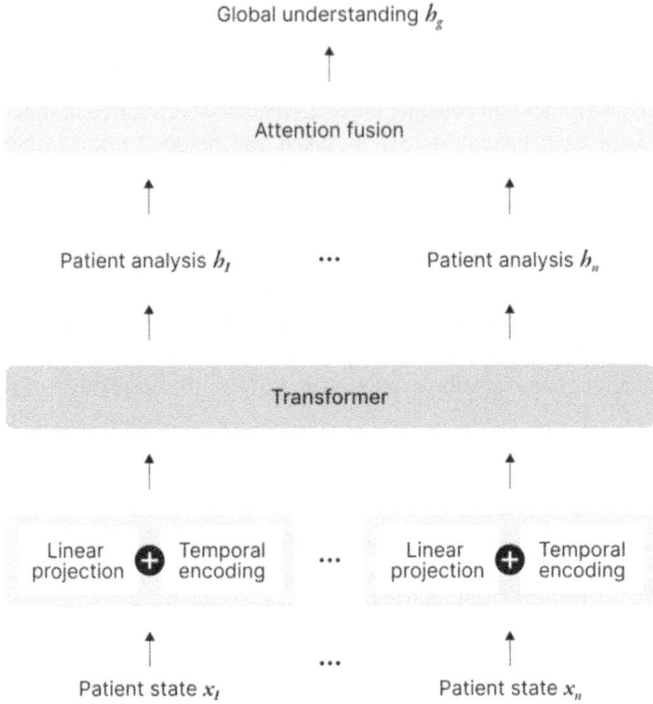

Fig. 3. DiagNet's global analysis component

4.3 Analysis Fusion

For the contextual state vector from the sequential analysis h_r and from the global analysis h_g, we need to add them evenly through a Layer Normalization operation:

$$h = \text{LayerNorm}(h_r + h_g) \tag{11}$$

Which gives equal weighting to each analysis, and normalizes the total values for the final diagnosis step.

Finally, a single linear projection is done and the Sigmoid activation function is applied to obtain the final diagnosis:

$$y' = \text{Sigmoid}(W_f h + b_f) \tag{12}$$

Here, $W_f \in \mathbb{R}^{1 \times m}$ is the weight matrix of the final projection layer, and $b_f \in \mathbb{R}$ is the bias.

5 System Evaluation

In this section, to verify the performance of our model, we conduct a comparison experiment between DiagNet and the related models mentioned above, and conduct an ablation study on DiagNet.

5.1 Experiment and Metric

In order to conduct a comparative analysis of our models, we choose the models introduced earlier, Retain, Dipole, RetainEX and HiTANet. The first three models initially use the RNN as the backbone network. However, in the study of Luo et al. [8], Retain and Retain used the LSTM to conduct experiments. To ensure the fairness of the experiment, we also choose LSTM as the backbone network of Dipole.

For the datasets, we split both the MIMIC-IV and eICU dataset into a train, validation, and test bucket. For each dataset, the ratio is 75% for training, 5% for validation, and 20% for testing respectively.

To measure and compare the performance of the models, we choose two metrics, the main metric AUC and the secondary metric, F1-Score.

AUC. The Receiver Operator Characteristic (ROC) Area Under Curve (AOC) score de scribes the ability of a binary classifier as the threshold between both is altered [14]. The calculation way is by plotting the True positive rate (TPR, see Eq. 13) against the false positive rate (FPR, see Eq. 14), then taking the area under the curve

$$TPR = \frac{TP}{\text{\# of ALL Samples}} \tag{13}$$

$$FPR = \frac{FP}{\text{\# of Negative Samples}} \tag{14}$$

Here, TP means instances correctly identified as positives and FP means instances falsely identified as negatives.

In the ideal case, the AUC is 1, and when the model predicts completely randomly, the AUC is 0. Since the AUC is not dependent on the distribution of the data classes, we choose it as the main metric.

F1-Score. F1-score is a weighted average between both precision and recall, and it can be used as a single indicator of a models performance.

$$F1 - score = \frac{2 * \text{Precision} * \text{Recall}}{\text{Precision} + \text{Recall}} \tag{15}$$

Here, precision measures the fraction of correct predictions over the amount of total pre dictions, and recall measures the fraction of total instances of what we want to measure, over the total number of positive instances.

5.2 Final Comparison

To demonstrate the performance of our model, we compare it to other models on both datasets (Table 1).

As the result shows, DiagNet surpasses all other models on the MIMIC-IV dataset. For the F1-Score, it is 0.004 higher than the second-best model, RetainEX. But on the main metric AUC, it is over 0.017 better than RetainEX, which shows that our model is far ahead of other models on the MIMIC-IV dataset. For the larger eICU dataset, Dipole is the best model. However, DiagNet, as the second-best model, only lags by 0.004 in

Table 1. Final model comparison on eICU & MIMIC-IV

Method	MIMIC-IV		eICU	
	F1-Score	AUC	F1-Score	AUC
Retain	0.787	0.767	0.812	0.782
Dipole	0.794	0.775	0.850	0.809
RetainEX	0.798	0.780	0.825	0.773
HiTANet	0.787	0.757	0.838	0.790
DiagNet	0.802	0.797	0.846	0.804

F1-score and 0.005 in the AUC. At the same time, DiagNet outperforms the remaining three models by a wide margin on the eICU model.

Combining the two datasets, DiagNet boasts a stronger overall performance than the other models, with a remarkable improvement in AUC of 0.017 compared to Dipole and 0.048 compared to RetainEX.

5.3 Ablation Study

For the proposed model architecture, multiple experiments are conducted with each of the different components as a point of comparison. Finally, we choose two models: DiagNet as the final model and DiagNet* as the next-best variant with the difference in the analysis fusion component. DiagNet* uses a trained attention weighting similar to the "Dynamic Attention Fusion" introduced by Luo et al. [8] (Table 2).

Table 2. Ablation study of DiagNet

Model	MIMIC-IV		eICU	
	F1-Score	AUC	F1-Score	AUC
DiagNet*	0.839	0.794	0.841	0.803
DiagNet	0.802	0.797	0.846	0.804

For the MIMIC-IV dataset, DiagNet performs extremely well on the F1-Score, while DiagNet is approximately 0.003 better on the AUC. However, since AUC is the most important metric, the overall selection is still DiagNet.

For the eICU dataset, it is obvious that DiagNet performed better on both metrics. In this dataset, DiagNet is 0.005 better on the F1-Score and 0.001 better on the AUC. Although the difference between the two models is slight, it is enough to prove that the DiagNet is the better one.

All in all, through the ablation study, DiagNet shows better performance on both datasets than the next-best variant.

6   Conclusions

In this paper, we propose a novel model architecture, DiagNet. Through the experiments on the two types of EHR datasets, our model proves its superiority in risk prediction tasks over state-of-the-art models. This work makes several key contributions.

In terms of model architecture, our model introduces three components, sequential analysis, global analysis and analysis merging. On the one hand, we project data linearly, which creates either a denser or a wider representation of the patient state, and then pass it to sequential and global analysis components. In this way, we solve the high degree of sparsity inherent in EHR data. On the other hand, we take a learnable time-embedding and add it to the linear projection before passing it into the global analysis component, which addresses the irregularly parsed data.

In terms of interpretability, DiagNet has two separate mechanisms. Similar to Dipole [9], we add the hidden state output of the global analysis component with an attention head, which increases the interpretability of our model. By investigating the values of the localized attention for each hidden state, medical practitioners can identify which patient observations have a larger influence on the final prediction.

Our model proves its excellent performance in risk prediction, but it still has some limitations. One problem is that though MIMIC-IV and eICU datasets are typical examples of EHRs, they still don't cover all HER data and real-world situations. And another potential shortcoming of DiagNet lies in its resource-heavy training demands. The high complexity of the model necessitates substantial computational power for training and inference. When computational resources or time are limited, it can have a significant impact on model hyperparameter optimization and performance.

In our future work, we will focus on the performance of the model on different sequence lengths and varied, time separate data points. At the same time, we will generalize the model to a larger dataset to evaluate its scalability.

Acknowledgments. This work was supported by the National Natural Science Foundation of China under Grant 62076027.

References

1. Rudd, K.E., et al.: Global, regional, and national sepsis incidence and mortality, 1990–2017: analysis for the global burden of disease study. Lancet **395**(10219), 200–211 (2020)
2. Zhang, Y., et al.: HKGB: an inclusive, extensible, intelligent, semi-auto-constructed knowledge graph framework for healthcare with clinicians' expertise incorporated. Inf. Process. Manage., 102324 (2020). https://doi.org/10.1016/j.ipm.2020.102324
3. Liang, Y., Wang, H., Zhang, W.: A knowledge-guided method for disease prediction based on attention mechanism. In: Zhao, X., Yang, S., Wang, X., Li, J. (eds.) Web Information Systems and Applications, WISA 2022. Lecture Notes in Computer Science, vol. 13579, pp. 329–340. Springer, Cham (2022). https://doi.org/10.1007/978-3-031-20309-1_29
4. Elman, J.L.: Finding structure in time. Cogn. Sci. **14**(2), 179–211 (1990)
5. Hochreiter, S., Schmidhuber, J.: Long short-term memory. Neural Comput. **9**(8), 1735–1780 (1997)

6. Reddy, B.K., Delen, D.: Predicting hospital readmission for lupus patients: an RNN-LSTM-based deep-learning methodology. Comput. Biol. Med. **101**, 199–209 (2018)
7. Choi, E., et al.: Retain: an interpretable predictive model for healthcare using reverse time attention mechanism. In: Advances in neural information processing systems, vol. 29 (2016)
8. Luo, J., et al.: HiTANet: hierarchical time-aware attention networks for risk pre diction on electronic health records. In: Proceedings of the 26th ACM SIGKDD International Conference on Knowledge Discovery & Data Mining, pp. 647–656 (2020)
9. Ma F, et al.: Dipole: diagnosis prediction in healthcare via attention-based bidirectional recurrent neural networks. In: Proceedings of the 23rd ACM SIGKDD International Conference on Knowledge Discovery and Data Mining, pp. 1903–1911 (2017)
10. Kwon, B.C., et al.: Retainvis: Visual analytics with interpretable and interactive recurrent neural networks on electronic medical records. IEEE Trans. Visual Comput. Graph. **25**(1), 299–309 (2018)
11. Johnson, A., et al.: Mimic-iv. PhysioNet (2020). https://physionet.org/content/mimiciv/1.0/. Accessed 23 Aug 2021
12. Zhang, Y., et al.: A heterogeneous multi-modal medical data fusion framework supporting hybrid data exploration. Health Inf. Sci. Syst. **10**, 22 (2022). https://doi.org/10.1007/s13755-022-00183-x
13. Pollard, T.J., et al.: The eICU collaborative research database, a freely available multi-center database for critical care research. Sci. Data **5**(1), 1–13 (2018)
14. McClish, D.K.: Analyzing a portion of the roc curve. Med. Decis. Making **9**(3), 190–195 (1989)

Alleviating Collapsing Problem in Policy Topic Discovery via Soft Clustering-Based Regulation

Yuqi Wang[(✉)] and Chen Liu

North China University of Technology, Beijing, China
1985712426@qq.com, liuchen@ncut.edu.cn

Abstract. Topic modeling aims to extract the core information implicit in natural language texts. To develop an effective topic model, high-quality input are crucial. However, existing policy texts suffer from the long-tail issue, i.e., a large proportion of words are rarely mentioned in the context, which are called long-tail words. As a result, the topic model trained on these texts tends to recommend frequent items, which leads to high similarity and poor usability of the topics obtained by the model, preventing mainstream topic models from being directly applied to policy texts. To address the above challenge in topic modeling of policy texts, in this paper, we proposes the Embedding Soft Clustering-based Regulation Neural Topic Model for policy topic discovery (**ESCRTopic**). ESCRTopic first uses Auto-Encoder to learn topic representation self-supervisely, then processing a soft clustering-based embedding regularization via Sinkhorn algorithm based transport optimization to mitigate the long-tailed problem from the perspective of clustering. The experimental results show that our method outperforms other state-of-the-arts topic models in terms of the classical evaluation indexes of topic models, and significantly improves the diversity of topic generation results.

Keywords: policy topic modeling · topic collapsing · cluster regulation

1 Introduction

Policy texts are a crucial tool for the government to communicate guidelines, policies, and resolutions to the public, which consist of rules and regulations. Nowadays, the number and content of policy texts are increasing rapidly, which makes the importance of quickly obtaining the core of it more crucial. The topics of texts is comprised of keywords that reflect the core of texts. Using topics and corresponding keywords enables quick comprehension and summarization of a vast amount of text. The topic model [1] maps the high-dimensional word space of text vectors to the low-dimensional target topic space, which can mine the hidden structural and semantic information of the document and ultimately generate topics for the document. With the satisfactory results of topic modeling, topic model has begun to be widely used in various applications [7,11].

C. Jin et al. (Eds.): WISA 2024, LNCS 14883, pp. 487–499, 2024.
https://doi.org/10.1007/978-981-97-7707-5_40

However, there is a serious long-tail distribution problem implicit in the policy data [16]: a small number of words appear frequently, occupying the main exposure of the policies, while the majority of words appear rarely. It makes existing topic models being forced to pay too much attention to frequent words and lacking the ability to learn words that appear rarely. The generated policy topic semantics are concentrated, i.e., facing the topic collapsing problem (the semantic similarity between topics is high and topic keywords are highly repetitive). Due to the long-tail problem, existing topic models cannot perform topic extraction on the complete of policy texts. Therefore, the topic collapsing problem in policy topic extraction is an urgent problem that needs to be solved.

To overcome these obstacles, we first analyze the root cause of the topic collapsing problem in policy topic modeling and then propose our Embedding Soft Clustering-based Regularization Neural Topic Model (ESCRTopic). The core of ESCRTopic model is the topic representation learning and the soft clustering-based regularization objective of topic and word embeddings. The topic representation learning part uses auto-encoder for document reconstruction and topic distribution learning. The regularization part analyzes the topic extraction process from the perspective of clustering and explores a new optimization objective for topic and word embeddings to mitigate the topic collapsing problem. To be specific, ESCR first assumes that topic embeddings are cluster centers and word vectors are cluster samples. Then, based on this, the modeling and optimization objective of the topic model is to correctly cluster word vectors into the vicinity of topic embedding. ESCR further enforces the topic embeddings to be the centers of independent word vectors, which means every single word is supposed to be associated with one topic. So that the semantic overlap between words of topics will be as small as possible, and the relative distance between topic representations will be as large as it can be, which will not only reduce the repetition among the extracted topics, but also cover more policy content. We utilize the optimal transport problem to figure this objective out, which is added to the topic representation learning by a regularization loss term. Extensive experiments over the policy dataset demonstrate our ESCRTopic not only alleviates the topic collapsing problem but also improves the topic quality.

The main contributions of this work are as follows:

- We analyze how the long-tail problem in policy data affects topic models, which reveals the reason why topic collapsing occurs in topic models nowadays.
- We propose a novel ESCRTopic framework with soft clustering-based regularization to force the topic embeddings to be the centers of independent word vectors, which effectively alleviates the topic collapsing problem and improves the topic quality.
- We collated policy documents from central and local government agencies between 2020 and 2023 and produced a policy dataset, based on which we conducted an experimental study. Extensive experiments are conducted to validate the effectiveness of our proposal.

2   Related Work

2.1   Probabilistic Topic Models

The probability-based topic model can be traced back to the latent semantic analysis [5](LSA) model based on singular value decomposition proposed by Kontostathisa, which effectively summarizes and extracts the document information while achieving the dimensionality reduction of the document. However, such model faces problems such as "one word with multiple meanings" and "many words with one meaning".

2.2   Neural Topic Models

In recent years, neural network-based topic models have become a hot research topic [3,8,12,13]. They usually utilizes neural networks to process the representation learning of documents that contain potential topic information. Early models, such as the NTM [3] model, compared with the probabilistic model, it doesn't require prior assumptions and can obtain high-quality topic results. [8] first tried to use variation encoders for topic modeling and proposed NVDM. [9] reconstructed the structure of adversarial learning applied to the topic domain and designed a topic-based sampling strategy. [6] proposed the BERTopic model based on pre-trained language models and the category-based TF-IDF method, which can be used without representation training. [15] proposed the NSTM model, which is based on reconstruction loss and auto-encoders. It improves the training speed of neural topic models. Further optimization was carried out by [12], which proposed the WeTe model with the SOTA results.

However, none of the above models has been concerned with the long-tail problem in policy data, therefore, these models produce poor results when faced with policy texts. To solve this problem, our ESCRTopic proposes an optimization regularization objective from the perspective of clustering and utilizes the Sinkhorn algorithm [4] of optimal transport to figure it out. To be specific, it forces topic representations to be as far apart as possible, and word embeddings fit as closely as possible to a single topic. Finally, such objective is integrated into the objective function of the neural topic model as a regularization term, so that the ability to solve long-tailed problem can be carried out together with the learning of topics, and the topic collapsing problem can be effectively alleviation.

3   Problem Analysis

Here, we first introduce the basic notations. Suppose that there are V different words in a given document set X, where each document is represented by x. The topic model needs to mine K topics implicit in the document collection, where the $k - th$ topic can be defined as a distribution (topic-word distribution) $\beta_k \epsilon R^V$ for all words V. The topic-word distribution matrix can be expressed as $\beta = (\beta_1, \beta_2, \ldots, \beta_K) R^{V \times K}$, where each element indicates which words the topic is composed of. For documents, the document-topic distribution θ can be used to represent the probability that the article belongs to the topic and $\theta \epsilon \Delta_k$, where Δ_k is a probability simplex, $\Delta_k = \left\{ \theta \epsilon R_+^k \mid \sum_{k=1}^K \theta_k = 1 \right\}$.

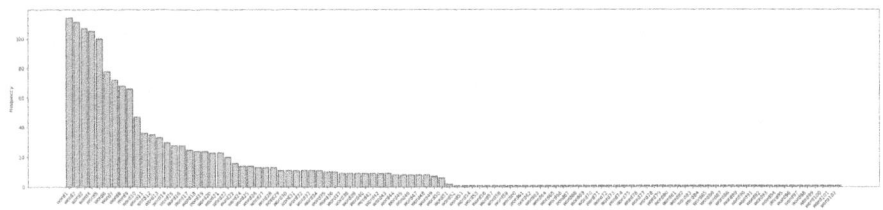

Fig. 1. Visualization of frequency distribution of policy text keywords.

Figure 1 displays the statistics of the word frequency distribution in a sample of the policy dataset. It illustrates that policy words suffer from a severe long-tail problem, i.e., certain frequent words and expressions commonly used in the policy domain appear frequently in most policy documents, while the vast majority of words appear rarely. The main reason for it can be related to the unique style of policy text, which can be summarized with the following three points: First, policy text often contains technical terms and commonly used words. Second, policies are typically structured and expressed in a specific way. Finally, policy text in Chinese follows the objective laws of language, such as Ziff's law [16].

Neural topic models typically utilize reconstruction loss to learn topic information. The main objective is to ensure that the document distribution reconstructed by a neural network model, such as a variation encoders, closely matches the original document distribution. However, when analyzing policy text data, the topic model tend to concentrate on a limited number of frequent words as a result of the long-tail effect of the words. The model found that only fitting these frequent words could effectively reduce the loss. While this approach may bring the reconstructed distribution closer to the original distribution and minimize the reconstruction loss, it may also result in a lack of diversity in the topics identified. It causes the topic vectors learned by the model to be closer to the representation of frequent words in the semantic space, resulting in higher topic homogeneity. Additionally, the topic model does not learn from the rich long-tail text, which means that the extracted topics cannot cover the overall text dataset in terms of semantics. As a result, the model lacks the ability to learn topics with fewer occurrences but independent meanings. In this context, the topics extracted by the topic model may exhibit high repetition and low semantic diversity, making them unsuitable for downstream analysis tasks. This issue is referred to as the topic collapsing problem in this paper.

To better illustrate the topic collapsing problem, we presents visualizations of the topic vectors and word vectors learned by the neural network-based topic model NSTM. As depicted in Fig. 2(a), the topic vectors produced by the neural network-based topic model NSTM are concentrated in the dense word vector region. This suggests that the topics extracted by the NSTM model and the vector space of high-frequency words are closely related, indicating that the topics are concentrated in the semantic vicinity of the high-frequency words and cannot cover all words uniformly.

We further collected the topic results of NSTM and ESCRTopic on the policy dataset. Table 1 shows the sampled topics, where the keywords are listed in descending order via probabilities and the repetitive words are underlined. This table reveals that the topics mined by the NSTM model suffer from an obvious topic collapsing problem, while the topics discovered by our proposed method, ESCRTopic, are more diverse and of higher quality. Therefore, despite the advancement of NSTM, it still has limitations.

Table 1. Visualized of a case study: sampled three discovered topics by NSTM and ESCRTopic. Repetitive words are underlined.

Keyword 1	全国 (National)	卫生院 (Health center)	教育 (Education)	代表大会 (Congress)	乡镇 (Township)	中学 (Middle school)
Keyword 2	代表大会 (Congress)	乡镇 (Township)	中学 (Middle school)	人民 (People)	医院 (Hosptial)	学生 (Student)
Keyword 3	人民 (People)	医院 (Hospital)	学生 (Student)	人大 (National People's Congress)	医疗 (Medical)	教师 (Teacher)
Keyword 4	届 (Session)	医疗 (Medical)	教师 (Teacher)	国务院 (State Council)	药品 (Drug)	教学 (Teaching)
Keyword 5	会 (Conference)	药品 (Drug)	教学 (Teaching)	届 (Session)	养老 (Retirement)	课后 (After school)
Keyword 6	大会 (Conference)	农村 (Urban)	教课 (Teaching)	全国 (National)	器材 (Material)	完善 (Refining)
Topic	Topic 1	Topic 2	Topic 3	Topic 1	Topic 2	Topic 3
Model		NSTM			ESCRTopic	

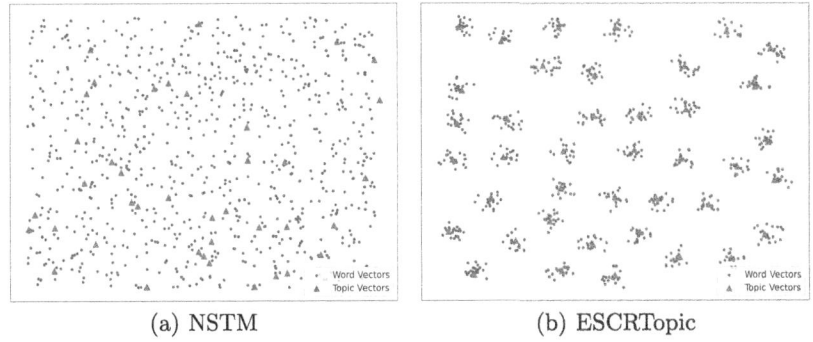

(a) NSTM   (b) ESCRTopic

Fig. 2. Visualization of the topic representations and word embeddings of NSTM and ESCRTopic model.

4 Methodology

4.1 Embedding Soft Clustering-Based Regularization

ESCR assumes topic modeling as a clustering process, where the clustering center denotes the topic representations and the sample represents the word vectors. So our target is to cluster topic representations and their corresponding word vectors into pairs, increasing the distance between topics and ensuring each word belongs to only one topic, i.e., center. Therefore, the main idea of ESCR is to increase the distance between each topic representation and make the distributional relationship between topics and words sparse through regularization terms, which ensures that the topic vectors are semantically distinct from each other. To achieve this goal, two technical challenges must be overcome:

Is there a way to jointly learn the proposed clustering-based objective term and the topic modeling objective? Neural network-based topic models usually optimize parameters by minimizing the reconstruction loss, which is the difference between the reconstructed document distribution and the original document distribution. If the traditional KMeans method is used for clustering and calculating the regularization, It' s not possible to co-optimize this term with the topic reconstruction loss [15]. To address this issue, we utilize probabilistic (soft) clustering to cluster topic and word representations.

Should we need and how should we generate sparse thematic clustering relationships? From the clustering view, it's important for each word to belong to only one topic as much as possible. This enables the clustering algorithm to group individual samples around specific centers and guarantees that the clusters for each topic remain distinct from one another. To accomplish this objective, we approach the issue through the lens of optimal transport, i.e., resolving the soft assignment relationship between topics and words.

Here, the first constraint of ESCR is the restriction on the cluster size of word vectors in the semantic space that correspond to the clustering center, which is used to avoid empty or duplicate topics. Assume we have K topics and V clustering samples, the cluster size of the k_{th} topic is $n_k = s_k * V$, where $(s_1, \ldots, s_k)^T \epsilon \Delta_k$ is the allocation vector of all cluster sizes. Inspired by previous works [12,15], we utilizes a special case of the Dirichlet prior distribution, i.e., uniform distribution, where the proportion of word embedding cluster sizes corresponding to each topic is equal, i.e., $1/K$.

As previously stated, the main objective of ESCR is to establish a sparse soft assignment relationship between topic and word vectors. It aligns perfectly with the optimal transport problem, which aims to generate a sparse and optimal solution. Consequently, we utilizes the way of solving optimal transport problem to evaluate and enhance the clustering matching relationship between word vectors and topic representations. We first give the definitions of the discrete measures (distributions) $\gamma = \sum_{j=1}^{V} \frac{1}{V} \delta_{w_j}$ and $\phi = \sum_{k=1}^{K} \frac{1}{K} \delta_{t_k}$ for the topic t_k and word vector W_j. Using this discrete distribution, we can formulate the optimal transmission problem between topic and words as Eq. 1, which provides the objective expression of the entropy regularized optimal transmission problem.

$$\mathrm{argmin}_{\pi \epsilon R^{V*K}} \mathcal{L}_{OT}(\gamma, \phi), \tag{1}$$

where $\mathcal{L}_{OT}(\gamma, \phi)$ can be represent as:

$$\mathcal{L}_{OT}(\gamma, \phi) = \sum_{j=1}^{V}\sum_{k=1}^{K} ||w_j - t_k||^2 \pi_{jk} + \sum_{j=1}^{V}\sum_{k=1}^{K} \varepsilon \pi_{jk} \left(\log\left(\pi_{jk}\right) - 1\right), \tag{2}$$

where the first term $\sum_{j=1}^{V}\sum_{k=1}^{K} ||w_j - t_k||^2 \pi_{jk}$ is the original optimal transport problem solution, and $C_{i,j} = ||w_j - t_k||^2$ is the Euclidean distance between the topic representations t_k and the word vectors w_j also known as the transport cost. And the second term $\sum_{j=1}^{V}\sum_{k=1}^{K} \varepsilon \pi_{jk} \left(\log\left(\pi_{jk}\right) - 1\right)$ is the entropy regularization. It's the transport weight between the topic representation and the word vector, indicating the relationship between the word and the topic, i.e., whether the clustered samples are attributed to the clustering center. Adding this term can not only make this problem microscopic and back-propagate (trainable), but also smooth out the solution space of the optimal problem to avoid extreme assignment values. The parameters π can be learned by minimizing the $\mathcal{L}_{OT}(\gamma, \phi)$ in Eq. 2. Here, we assume that π_ε^* is the corresponding optimal solution under the parameter value, at this time, π_ε^* is the soft clustering allocation relationship between words and topics, i.e., the allocation relationship between the clustering samples and the clustering center.

So the aim of ESCR is to minimize the total distance between topic representations and word embeddings, which can be achieved by weighting the distance by soft assignment parameter π_ε^*. By doing so, the average distance between topics and words can be reduced, resulting in compact clustering results and high match degree between samples and their centers. The below equation presents the formal formulation of ESCR based regularization objective function:

$$\mathcal{L}_{ESCR} = \sum_{j=1}^{V}\sum_{k=1}^{K} ||w_j - t_k||^2 \pi_{\varepsilon,jk}^*, \tag{3}$$

it should be noted that, in this paper, we use the Sinkhorn algorithm [4] to figure out π_ε^* efficiently, i.e., $\pi_\varepsilon^* = sinkhorn(\varepsilon, \gamma, \phi) = argmin_{\pi \epsilon R^{V*K}} \mathcal{L}_{OT}(\gamma, \phi)$

By incorporating the aforementioned formulas into the topic modeling process, ESCR can ensure that the topic representations are located in the clustering centers of the word vectors. Additionally, the word vectors are uniformly distributed in the vicinity of different clustering centers, which effectively prevents empty or redundant topic representations. Furthermore, ESCR can increase the distance between topic vectors, evenly distributing them throughout the semantic space. This effectively addresses the issues of topic repetition and collapsing.

4.2 Empowered ESCR on Neural Topic Model

Here, we present the ESCRTopic model for policy topic modeling. ESCRTopic jointly optimizes the objective function of the topic model and soft clustering regularization. Figure 3 shows the overview of ESCRTopic.

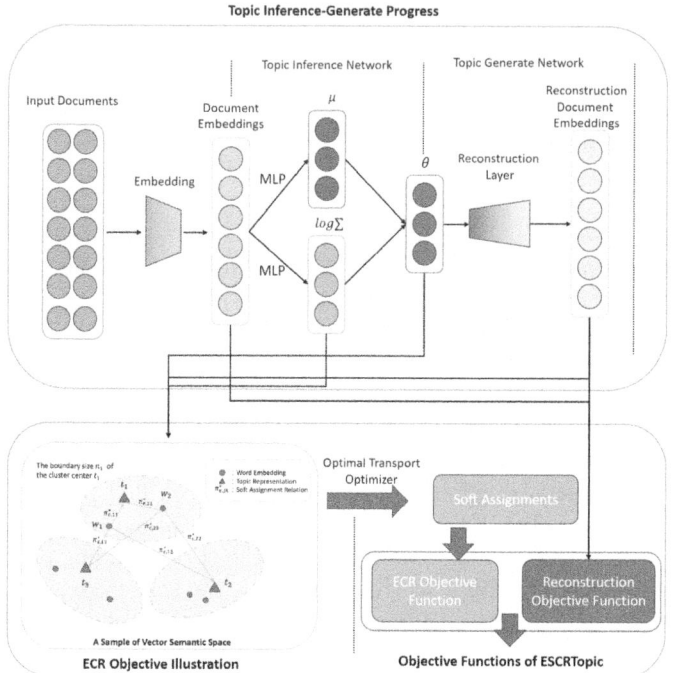

Fig. 3. Visualization of the ESCRTopic model.

We utilizes the autoencoder to reconstruct policy documents and learn relevant information, such as topic representations, by optimizing the gap between the reconstructed and original document distributions. We first use autoencoder to obtain the document-topic distribution $\theta = softmax(r)$, where r is the latent variable. The core parameters of its probability distribution, mean μ and variance Σ, are both obtained by the autoencoder. The neural encoder receives the document embedding inputs x, which is obtained using a pre-trained language model, e.g., GloVe following mainstream work [9]. It then outputs the parameters $\mu = f_\mu(x)$ and $\sum = diag(f_\Sigma(x))$ of the variational distribution. The latent variable r is modeled using the reparameterization technique[6,16], where $r = \mu + \epsilon \sum^{1/2}$ (ϵ follows a standard normal distribution). To reconstruct the document representation, It's necessary to obtain the topics-words distribution β. Previous methods typically model β as $\beta = W^T * T$, which fails to capture the spatial relationship between topics and words. To address this issue, we adopt a modeling approach that incorporates ESCR regularization, which constrains the model to better mine topic representations by clustering the topic and word vector representations. In this paper, we similarly model the distribution relationship between topics and words, as shown in Eq. 4:

$$\beta_{jk} = \frac{e^{-||w_j - t_k||^{2/\tau}}}{\sum_{k'=1}^{K} e^{-||w_j - t_k,||^{2/\tau}}} \tag{4}$$

where τ is the temperature coefficient. Equation 4 models the topic-word distribution, which reflects the relationship between the topic representation and word embeddings. It's important to note that the soft clustering assignment parameter π_ε^* derived in Sect. 4.1 is not utilized here to represent the distribution between topics and words. This is because π_ε^* is supposed to force each word to belong to only one topic, resulting in a very sparse representation of the topic-word distribution. However, in policy data, It's possible for any word to belong to different topics simultaneously, which may help to reconstruct the document distribution better and allow the model to learn a more realistic topic distribution. Furthermore, the regularization term proposed in Sect. 4.1 constrains the clustered sample word vectors to belong to only one clustering center topic vector, which eliminates unreasonable word-topic connections and retains accurate and reasonable topic-word distribution.

Based on above, given a representation $(x^1, x^2, ..., x^N)$ of N documents, the objective function of the topic model proposed can be formulated as follows:

$$\mathcal{L}_{TM} = \frac{1}{N} \sum_{i=1}^{N} -(x^i)^T \log\left(softmax(\beta\theta^i)\right) + KL, \tag{5}$$

where the first term is the document reconstruction loss, and the second term is the KL scatter of the original and reconstructed document distributions. The final objective function of our ESCRTopic can be formulated as:

$$\mathcal{L} = \min_{\Theta, W, T} \mathcal{L}_{TM} + \cdot\lambda_{ESCR} * \mathcal{L}_{ESCR}, \tag{6}$$

where λ_{ESCR} is the hyperparameter. With the objective function in Eq. 6, we enable the topic model to avoid generate repetition topics,i.e., prevent topic collapsing, and enhance the stability and diversity of topics in policies.

Upon analysis of Fig. 2(b), It's evident that ESCRTopic uniformly distributes the learned word vectors in the semantic space, and the topic vectors effectively cover the various clusters of word vectors. This indicates that the model's topic results can better cover the semantic space, i.e., results generated by ESCRTopic proposed are reliable, stable and non-repetitive .

5 Experiments

5.1 Experimental Settings

Datasets. To evaluate the capability of our ESCRTopic model on policy data, 56,091 pieces of data were crawled from public government websites in the country's provinces and cities. To ensure data quality, we conducted a preliminary screening to filter out duplicate policies, policies with empty text, policies with only names, and policies with content that refers to other policies. After screening and de-weighting, the resulting policy dataset in this paper contains a total of 36,572 pieces of data. The dataset includes policy title, main text, release time, release organization, and policy subject information from the original website.

Baselines. We compared ESCRTopic with seven classic topic models. **LDA** [2] is a classic probabilistic topic model. **BTM** [14] is good at capturing local word co-occurrence patterns. **Topic2Vec** [10] can learn topic representations in the same semantic vector space as words. **NTM** [8] is the first neural topic model. **NSTM** [15] is a topic model optimized by reconstruction loss. **BERTopic** [6] is a novel topic model with pre-trained language models and c-TF-IDF. **WeTe** [12] utilizes the optimal transport problem to minimize reconstruction loss.

Metrics and Hyper-parameters. We adopt three popular metrics, i.e., topic coherence (TC), topic diversity (TD), and topic quality (TQ) [1], to evaluate the performance of all models. All baselines were implemented using the code published by the author, and grid search were used for parameter searching while training.

Table 2. Performance comparison of the ESCRTopic model and baseline models on the TC, TD, and TQ for scenarios with 10, 20, and 30 generated topics.

Methods	K=10			K=20			K=30		
	TC	TD	TQ	TC	TD	TQ	TC	TD	TQ
LDA	0.231	0.655	0.151	0.262	0.670	0.176	0.230	0.635	0.146
BTM	0.079	0.630	0.050	0.040	0.590	0.024	0.101	0.629	0.064
Topic2Vec	0.124	0.622	0.077	0.065	0.611	0.040	0.229	0.615	0.141
NTM	0.334	0.798	0.267	0.350	0.801	0.280	0.291	0.650	0.189
NSTM	0.404	0.810	0.327	0.404	0.800	0.323	0.383	0.799	0.306
BERTopic	0.421	0.700	0.295	0.425	0.755	0.321	0.393	0.690	0.271
WeTe	0.451	0.827	0.373	0.410	0.851	0.349	0.400	0.811	0.324
ESCRTopic	**0.492**	**0.900**	**0.443**	**0.481**	**0.950**	**0.457**	**0.431**	**0.925**	**0.399**
%Improve.	9.09%	8.83%	18.72%	13.18%	11.63%	30.97%	7.75%	14.06%	22.90%

5.2   Performance Comparison

Table 2 presents the overall performance comparison of baseline models and ESCRTopic with the number of generated topics of 10, 20, and 30. It shows that NSTM, BERTopic, and WeTe, which are neural topic models, improve on average by almost two times compared to LDA and outperform all the traditional models. This may be because traditional models disregard word order and contextual information, potentially leading to the loss of important semantic information. In contrast, topic models based on deep learning have a stronger representation ability, allowing for a more accurate capture of the text's topic structure. Bertopic, as an no-training-required model, performs pretty well in terms of effectiveness, second only to WeTe, which employs bidirectional conditional transport optimizes and word embeddings to achieve the SOTA performance.

Without any doubt, in all cases, our ESCRTopic framework consistently outperforms all baselines across all metrics. This is due to ESCR objective, which allows the topic model to learn how to reconstruct the document based on semantic space constraints for the representation of topics and word embedding. As a result, the semantic space of words and topics becomes sparse. ESCRTopic can effectively reduce the generation of invalid and empty topics via the constraints of the clustering size. By training models based on the ESCR, the resulting topic results are reliable, stable and non-repetitive.

5.3   Ablation Study

To demonstrate the capability of ESCR, the core module of the ESCRTopic, we design an ablation experiment here. ESCRTopic w/o ESCR will preserve the objective function of the topic model only for the document reconstruction objective function. And NSTM w Entropy is the NSTM model with entropy regularization. Table 3 presents experimental results comparing the two models.

Table 3. Comparison for different model variants.

Methods	K=10		K=20		K=30	
	TC	TD	TC	TD	TC	TD
NSTM	0.404	0.810	0.404	0.800	0.383	0.799
NSTM w Entropy	0.271	0.750	0.411	0.713	0.299	0.701
ESCRTopic w/o ESCR	0.501	0.710	0.456	0.709	0.391	0.559
ESCRTopic	0.492	0.900	0.481	0.950	0.431	0.925

Table 3 shows that removing the ESCR module from the ESCRTopic model does not significantly affect the TC metrics, but it does cause a significant decrease in the TD metrics. The TD metric expresses the degree of specificity between the topics generated by the model. ESCRTopic without ESCR regularization are unable to distinguish between topics well and suffer from topic collapsing problem. The NSTM and NSTM with Entropy baseline models exhibit significant differences in TD metrics compared to the ESCRTopic model proposed in this paper. This suggests that the ESCR module proposed in this paper effectively addresses the issue of topic collapsing on policy data and substantially enhances the quality of topic generation.

5.4   Topic Heat-Map

To validate the quality of generated topics, here we visualizes the topic results of ESCRTopic, LDA and the BERTopic model. The similarity scores between topics are calculated and presented in the form of heat maps in Fig. 4.

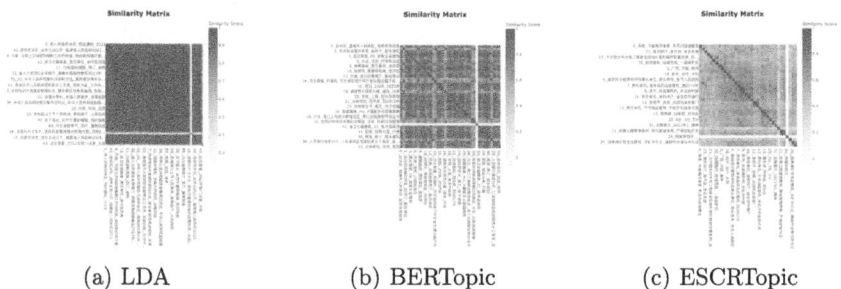

(a) LDA   (b) BERTopic   (c) ESCRTopic

Fig. 4. Visualization of the heat-map of topic similarity generated by LDA, BERTopic and ESCRTopic model.

Figure 4(a) and 4(b) display the heat maps of topic similarity for the LDA and BERTopic models, respectively. The overall topic blocks appear darker in color, and the color differences between topics are smaller, indicating higher generated topic similarity. Figure 4(c) displays the experimental results of the ESCRTopic proposed in this paper. The overall color between the topic blocks is lighter compared to both Fig. 4(a) and Fig. 4(b), indicating a lower similarity between the topics. Our proposed ESCRTopic model, based on vector soft clustering regularization, can generate clearer, more distinguishable, and diverse topic results. It can effectively analyze the topics of massive policy texts and generate reliable results.

6   Conclusion

In this work, we analyze the reason of why existing topic models do not perform well on policy data: topic collapsing caused by the long-tailed problem of policy data. We further present the ESCRTopic model that regularization topic modeling to make sparse allocation between topics and words from the perspective of clustering, which significantly mitigating the topic collapsing problem.

References

1. Abdelrazek, A., Eid, Y., Gawish, E., Medhat, W., Hassan, A.: Topic modeling algorithms and applications: a survey. Inf. Syst. **112**, 102131 (2023)
2. Blei, D.M., Ng, A.Y., Jordan, M.I.: Latent dirichlet allocation. J. Mach. Learn. Res. **3**(Jan), 993–1022 (2003)

3. Cao, Z., Li, S., Liu, Y., Li, W., Ji: A novel neural topic model and its supervised extension. In: Proceedings of the AAAI Conference on Artificial Intelligence, vol. 29 (2015)
4. Cuturi, M.: Sinkhorn distances: lightspeed computation of optimal transport. In: Advances in Neural Information Processing Systems, vol. 26 (2013)
5. Dumais, S.T.: Latent semantic analysis. Ann. Rev. Inf. Sci. Technol. (ARIST) **38**, 189–230 (2004)
6. Grootendorst, M.: BERTopic: neural topic modeling with a class-based TF-IDF procedure. arXiv preprint arXiv:2203.05794 (2022)
7. Jelodar, H., et al.: Latent dirichlet allocation (LDA) and topic modeling: models, applications, a survey. Multimed. Tools Appl. **78**, 15169–15211 (2019)
8. Miao, Y., Yu, L., Blunsom, P.: Neural variational inference for text processing. In: International Conference on Machine Learning, pp. 1727–1736. PMLR (2016)
9. Nguyen, T., Luu, A.T.: Contrastive learning for neural topic model. In: Advances in Neural Information Processing Systems, vol. 34, pp. 11974–11986 (2021)
10. Niu, L., Dai, X., Zhang, J., Chen, J.: Topic2Vec: learning distributed representations of topics. In: 2015 International conference on Asian language processing (IALP), pp. 193–196. IEEE (2015)
11. Wang, C., Zhang, Y., Yao, S.: Text-independent speaker verification based on mutual information disentanglement. In: Yuan, L., Yang, S., Li, R., Kanoulas, E., Zhao, X. (eds.) WISA 2023. LNCS, vol. 14094, pp. 309–318. Springer, Singapore (2023). https://doi.org/10.1007/978-981-99-6222-8_26
12. Wang, D., et al.: Representing mixtures of word embeddings with mixtures of topic embeddings. arXiv preprint arXiv:2203.01570 (2022)
13. Xu, W., Desai, J., Sengamedu, S., Jiang, X., Iannacci, F.: S2vNTM: semi-supervised VMF neural topic modeling. arXiv preprint arXiv:2307.04804 (2023)
14. Yan, X., Guo, J., Lan, Y., Cheng, X.: A biterm topic model for short texts. In: Proceedings of the 22nd International Conference on World Wide Web, pp. 1445–1456 (2013)
15. Zhao, H., Phung, D., Huynh, V., Le, T., Buntine, W.: Neural topic model via optimal transport. arXiv preprint arXiv:2008.13537 (2020)
16. Zipf, G.K.: Human behavior and the principle of least effort: an introduction to human ecology. Ravenio Books (2016)

Attention-Based Spatial-Temporal Fusion Networks for Traffic Flow Prediction

Jiaying Wang[1,2] (ID), Heng Yang[1,2], Jing Shan[1,2(✉)] (ID), Xiaoxu Song[1,2], and Junyi Jiang[3,4]

[1] Shenyang University of Technology, Shenyang 110870, China
mavis0129@126.com
[2] Shenyang Key Laboratory of Intelligent Technology of Advanced Industrial Equipment Manufacturing, Shenyang 110870, China
[3] Liaoning Technical University, Fuxin 123000, China
[4] Liaoning Economic Vocational and Technical College, Shenyang 110122, China

Abstract. Traffic flow prediction plays a crucial role in the development of intelligent transportation systems and smart cities. Recent advancements in deep network models have improved the accuracy of traffic flow prediction. However, existing methods often focus on the close proximity of distance and time, neglecting the dynamic characteristics of traffic flow influenced by various factors such as periodicity, tendency, functional similarity, etc. To address this issue and effectively capture the dynamic spatial-temporal correlation, we propose a novel approach called Attention-based Spatial-Temporal Fusion Networks (ASTFN) for modeling the fusion of dynamic temporal and spatial features. ASTFN utilizes a multi-layer encoder architecture and consists of three main modules: the traffic data embedding module, the spatial-temporal attention module, and spacial-temporal fusion module. To evaluate the performance of ASTFN, we conducted extensive experimental studies on four real-world traffic datasets. The results demonstrate the superiority of ASTFN over other state-of-the-art methods.

Keywords: Traffic flow prediction · Spatial-Temporal · Attention

1 Introduction

In recent years, with the rapid development of intelligent transportation system, traffic flow prediction plays a crucial role in alleviating the congestion problem of fast roads and high-flow urban roads. Traffic flow prediction represents a typical spatial-temporal data prediction challenge. Traffic data is collected through sensors at fixed locations and specific time intervals. Adjacent position data are not independent but dynamically correlated with each other.

Traffic flow has two particular characteristics. First, due to functional similarity, the traffic flow of two long-distance locations could be highly similar, which indicates that spatial dependency can extend over long distance. Take two main roads near two universities for example, although far apart from each other,

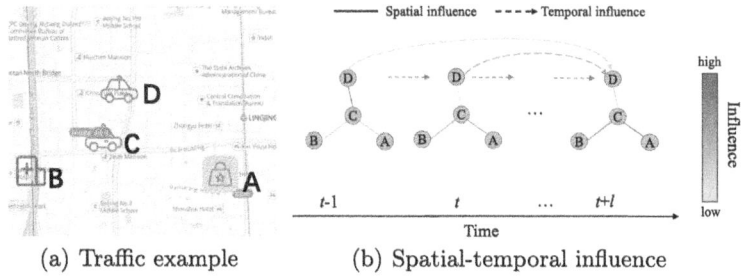

(a) Traffic example   (b) Spatial-temporal influence

Fig. 1. The spatial-temporal correlation diagram of traffic flow

they may still have similar traffic conditions during rush hours. Second, traffic influence may change over time. Taking Fig. 1 as an example, the traffic flow at points A, B and D dynamically affect traffic flow at point C. Effectively capturing dynamic correlations in spatial-temporal features is crucial for accurately predicting traffic flow, which is a non-trivial problem.

To address existing challenges, we propose a novel deep network model, i.e., Attention-based Spatial-Temporal Fusion Networks (ASTFN) for traffic flow prediction. The main contributions of this paper are summarized as follows:

- We designed a new spatial-temporal embedding module to extract time proximity, periodicity and tendency from traffic data.
- We proposed a new spatial-temporal encoder to adaptively capture the important spatial and temporal information.
- We conducted extensive experimental studies on four real-world datasets. The results demonstrate the effectiveness of our approach.

2   Related Work

Traffic flow prediction, as a major component of smart cities, has been widely studied by scholars in the past decades. In the early stages, traffic prediction primarily relied on regression models based on mathematical statistics. For example, ARIMA and VAR are two models to fit the nonlinearity in time to represent more complex dependencies, but these models are not good at capturing sudden traffic flow changes.

Machine learning methods such as Support Vector Regression (SVR), k-Nearest Neighbors (KNN), and Support Vector Machines (SVM) have shown promise in modeling traffic flow data. However, to improve the performance and effectiveness of these models, careful feature engineering is necessary.

Just as deep learning has made breakthroughs in many fields such as cross-modal generation [8], natural language processing [9] and link prediction [15], currently, the state-of-the-art methods for traffic flow prediction are deep learning-based approaches that can be divided into two categories.

GNN-Based Methods. Graph neural networks (GNNs) have gained significant popularity in traffic prediction due to their ability to capture both temporal and spatial dependencies using graph structures. DCRNN [4] modelled traffic flow as a diffusion process on directed graph. STGCN [14] proposed a convolutional structure, which brings faster training with fewer parameters. MTGNN [11] proposed a joint framework for modeling multivariate time series data and learning graph structures, which could handle multivariate time series. STFGNN [3] proposed a data-driven approach to generate "time graphs" to compensate for correlations that may not be reflected in spatial graphs.

Attention Based Methods. The attention mechanism has demonstrated remarkable success in various domains such as natural language processing, image recognition, and time series analysis [7]. GMAN [16] proposed an encoder-decoder structure comprising multiple spatial-temporal attention blocks that simulate the impact of spatial-temporal factors on traffic conditions. TFormer [12] proposed multiple attention mechanisms to extract spatial-temporal features dynamically. ASTGNN [1] proposed an attention-based spatial-temporal graph convolutional network to address traffic flow prediction challenges. PDFormer [2] incorporated a spatial self-attention module to capture dynamic spatial dependencies and a traffic delay sensing module to simulate traffic characteristics.

3   Preliminaries

In this paper, we define the traffic road network as an undirected graph $\mathcal{G} = (V, E)$, which consists of vertices and edges, where V is a finite set of N nodes, in which $N = |V|$. $E \subseteq V \times V$ is a set of edges between nodes. We use A to represent the adjacency matrix of graph $\mathcal{G}$.

We use $X_t \in \mathbb{R}^{N \times C}$ to represent the traffic flow in the road network at time t of N nodes. We use traffic flow tensor $\mathcal{X} = (X_1, X_2, \cdots, X_T) \in \mathbb{R}^{T \times N \times C}$ to represent the traffic flow of all nodes on T time slices.

Traffic flow prediction aims to predict the traffic flow at a future time given the historical road flow data. Given the traffic flow feature tensor $\mathcal{X}$, and various historical measurements of all nodes on the past T time slices, the target is to learn the mapping function f from the observed data to predict the future T' steps' traffic flow:

$$[X_{(t-T+1)}, \cdots, X_t; \mathcal{G}] \xrightarrow{f} [X_{(t+1)}, \cdots, X_{(t+T')}]. \tag{1}$$

4   Methods

Our framework consists of one data embedding layer, several spatial-temporal encoders and one output layer. Figure 2 illustrates the overall framework.

4.1   Data Embedding Layer

The data embedding layer converts the input traffic flow data into a high-dimension representation. We introduce a new spatial-temporal embedding mechanism to extract data features more comprehensively. Firstly, the data $\mathcal{X}$

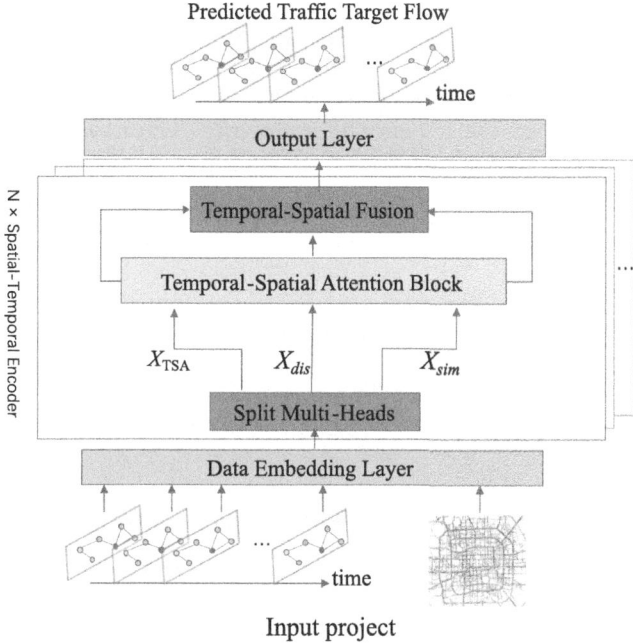

Fig. 2. The Overall Framework of ASTFN

is converted into high-dimension data $\mathcal{X}_{data} \in R^{T \times N \times d}$ through the fully connected layer, where d is the embedding dimension. After that, the spatial graph Laplacian embedding is used to encode the traffic graph structure data. In order to better capture the relationship between nodes, we use the Laplacian eigenvectors for feature decomposition, which can preserve the similarity and connection pattern between nodes. For an adjacency matrix A of graph $\mathcal{G}$, we calculate the symmetric normalization of the Laplacian matrix $L = I - D^{-\frac{1}{2}} A D^{-\frac{1}{2}}$, where D is the degree matrix and I is the identity matrix. We then perform the eigenvalue decomposition $L = U^\top \Lambda U$ to obtain the eigenvalue matrix Λ and the eigenvector matrix U. By assigning k minimal non-trivial eigenvectors and using them for linear projection, we could get a new vector $X_{sp} \in \mathbb{R}^{N \times d}$.

In order to enable the model to understand the order and position of each element in the input sequence, time-position encoding was introduced into the input data using the original Transformer model [7]. The position information is represented by $X_{pe} \in \mathbb{R}^{T \times d}$.

Besides, urban traffic flow has a explicit periodicity like morning and evening rush hours. Therefore, we introduce three embedding vectors to describe the period: T_{near} represents the recent impact, T_{period} represents the periodicity and T_{trend} to reflect the trend. In Fig. 3, time proximity $\mathcal{X}_{near} \in \mathbb{R}^{N \times d}$ covers the historical data in close proximity to the predicted value (blue), time period $\mathcal{X}_{period} \in \mathbb{R}^{N \times d}$ considers the last few days of the cycle (green), and time trend $\mathcal{X}_{trend} \in \mathbb{R}^{N \times d}$ considers the long-term trend characteristics (yellow). Finally,

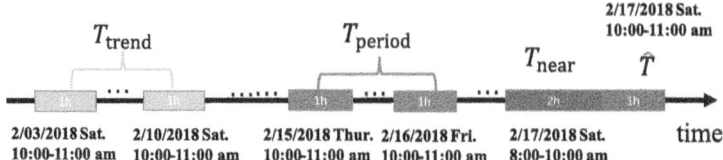

Fig. 3. Three types of the time segments (Color figure online)

we add the above embedding vectors to get the output of the data embedding layer as depicted in Eq. 2.

$$\mathcal{X} = \mathcal{X}_{data} + \mathcal{X}_{near} + \mathcal{X}_{period} + \mathcal{X}_{trend} + X_{sp} + X_{pe}. \tag{2}$$

4.2 Spatial-Temporal Encoder Layer

We propose a special encoder layer based on self-attention mechanism to simulate the spatial-temporal dependencies of complex dynamic traffic networks, which consists of temporal self-attention (TSA) and spatial self-attention (SSA). The TSA is used to capture dynamic and remote temporal patterns, and the SSA module is capable of capturing short-range and similar dynamic spatial dependencies, as shown in Fig. 4.

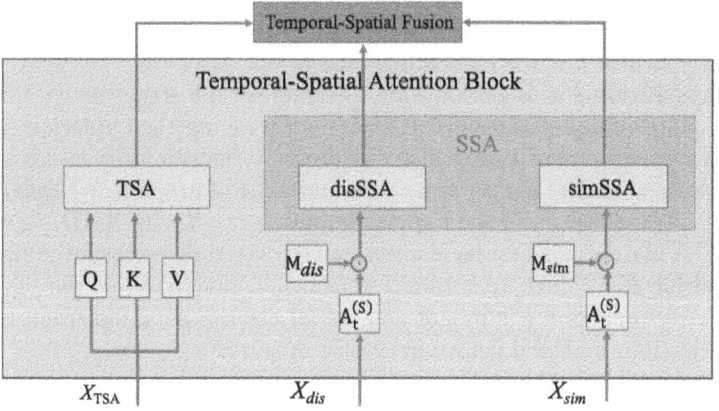

Fig. 4. Spatial-Temporal Encoder

Temporal Self-attention. In the time dimension, traffic flow is affected by different time slices such as proximity, periodicity, and so on. Therefore, we use the temporal self-attention module to capture different importances adaptively. Formally, for a node n, we firstly calculate the query, key and value matrices as:

$$Q_n^T = X_n W_Q^T, K_n^T = X_n W_K^T, V_n^T = X_n W_V^T, \tag{3}$$

where $W_Q^T, W_K^T, W_V^T \in \mathbb{R}^{d \times d'}$ are learnable parameters, and X_n is the slice on the sequence node. We then apply the self-attention operation on the time dimension and get the time dependency between all time slices of node n as:

$$A_n^T = \frac{(Q_n^T)(K_n^T)^\top}{\sqrt{d'}}, \tag{4}$$

in which d' is the dimension of the query, key and value matrices of the attention mechanism. Finally, we can get the output of the time self-attention module as follows:

$$TSA(Q_n^T, K_n^T, V_n^T) = softmax(A_n^T)V_n^T. \tag{5}$$

Spatial Self-attention. We design a new spatial-temporal attention module for capturing dynamic spatial dependencies in traffic networks. This module includes spatial attentions $disSSA$ and $simSSA$, which can capture both short-range and similar dynamic spatial dependencies. Formally, at time t, we firstly obtain the query, key and value matrices of the spatial self-attention operation as:

$$Q_t^S = X_t W_Q^S, K_t^S = X_t W_K^S, V_t^S = X_t W_V^S, \tag{6}$$

where $W_Q^S, W_K^S, W_V^S \in \mathbb{R}^{d \times d'}$ are parameters that can be learned from traffic data. Then, using the learned Q_t^S and K_t^S, we calculate the correlations (i.e. attention score) between each element of the two input vectors. This process simulates the interaction between nodes by applying self-attention operations in the spatial dimension, and obtains the spatial dependencies of all nodes in time t, which is defined as:

$$A_t^S = \frac{(Q_t^S)(K_t^S)^\top}{\sqrt{d'}}. \tag{7}$$

in which d' represents the dimension of the query, key, and value matrices of the attention mechanism. Obviously, the spatial dependency $A_t^S \in \mathbb{R}^{N \times N}$ between nodes varies on different time slices. We get the output of the spatial self-attention module as:

$$SSA(Q_t^S, K_t^S, V_t^S) = softmax(A_t^S)V_t^S. \tag{8}$$

We introduce two graph mask matrices M_{dis} and M_{sim} to capture both short-range and similar spatial dependencies in traffic data. From a short-range perspective, we define the binary proximity mask matrix M_{dis} with a weight of 1 only if the distance between two nodes is less than the threshold λ, and 0 otherwise. The distance proximity spatial self-attention (disSSA) is defined as:

$$disSSA = softmax\left(A_t^{(S)} \odot M_{dis}\right) V_t^{(S)}. \tag{9}$$

where $\odot$ represents the Hadamard product.

From a remote view, we use the DTW algorithm to calculate the historical traffic flow similarity between nodes. We select K nodes with the highest similarity of each node as their similar neighbours. Then we build binary similar mask matrix M_{sim} to find distant node pairs that exhibit similar traffic patterns The similar spatial self-attention (simSSA) can be defined as:

$$simSSA = softmax(A_t^{(S)} \odot M_{sim})V_t^{(S)}. \tag{10}$$

4.3 Temporal-Spatial Fusion

We propose a method that integrates temporal characteristics, spatial proximity and spatial similarity to cope with traffic flow changes. The model can integrate spatial and temporal information synchronously through three kinds of attention heads: TSA, disSSA and simSSA. Formally, the temporal-spatial attention block is defined as:

$$TS_f = W_t \odot TSA + W_d \odot disSSA + W_s \odot simSSA, \tag{11}$$

where $W_t, W_d, W_s \in \mathbb{R}^{d \times d}$ is learnable parameter matrices, disSSA, simSSA and TSA have multiple attention heads.

We use a fully connected feed-forward network on TS_f to get the output $\mathcal{X}_o \in \mathbb{R}^{T \times N \times d}$. In order to preserve the original information regardless of depth, we use skip connections consisting of 1×1 convolution, which converts the output $\mathcal{X}_o$ to skipped $\mathcal{X}_s$ after each temporal-spatial encoder layer. The outputs of the skip connection layer are then summed to obtain the final hidden state $\mathcal{X}_h$. After multi-step prediction, the final hidden state $\mathcal{X}_h$ is converted to the final output as:

$$\hat{\mathcal{X}} = Conv2\left(Conv1\left(\mathcal{X}_h\right)\right), \tag{12}$$

where $\hat{\mathcal{X}} \in \mathbb{R}^{T \times N \times d}$ is the prediction result of step T', $Conv1$ and $Conv2$ are 1×1 convolutions.

5   Experiments

5.1   Datasets

To evaluate the performance of the model, we selected four real-world public traffic datasets to conduct comparative experiments, including graph-based highway traffic datasets (i.e., PeMS04, PeMS08 [6]) and grid-based citywide datasets (i.e., NYCTaxi [5] and CHIBike [10]). The details of these datasets are given in Table 1.

Table 1. Details of the datasets

Datasets	Time Range	Sampling Interval	Nodes	Timesteps	Data Volume
PeMS04	01/01/2018–02/28/2018	5 min	307	16992	5216544
PeMS08	07/01/2016–08/31/2016	5 min	170	17856	3035520
NYCTaxi	01/01/2014–12/31/2014	30 min	75	17520	1314000
CHIBike	07/01/2020–09/30/2020	30 min	270	4416	1192320

5.2   Settings

Dataset Preprocessing. We split the graph-based traffic datasets PeMS04 and PeMS08 into training set, verification set and test set with the ratio of 6:2:2. And we used the data of the past 12 steps to make prediction for the next 12 steps. For the grid-based datasets NYCTaxi and CHIBike, the segmentation ratio was 7:1:2. We use the inflow and outflow of the past 6 steps to predict the next inflow and outflow.

Evaluation Metrics. We used two indices in the experiments, mean absolute error (MAE) and root mean square error (RMSE) to measure the difference between the predicted results of the traffic model and the actual observed values. Missing values are excluded during the experiment, the settings are consistent with other state-of-the-art approaches such as [2,13]. We repeated each of the experiment ten times to calculate the average value as the final result.

5.3   Baselines

We compared our model with the two categories of state-of-the-art models:

- graph-based neural network including DCRNN [4], STGCN [14], MTGNN [11] and STFGNN [3].
- self-attention based models including GMAN [16], TFormer [12], ASTGNN [1] and PDFormer [2].

5.4   Comparison and Result Analysis

We compared our model with eight state-of-the-art methods on four datasets as shown in Table 2, in which bold result indicates the best effect, and underline result indicates the second best effect. Based on the comparison, we can observe that self-attention based models, including TFormer, GMAN, ASTGNN and PDFormer, usually obtain better prediction results than GNN-based methods due to the addition of attention mechanisms. Our ASTFN approach achieved the best performance across PeMS04 and PeMS08 datasets. For NYCTaxi and CHIBike datasets, our approach won in 4 test points, PDFormer won in the other 4 test points, in which our method came in second, but the gap was very small. It is worth noting that the data volume of NYCTaxi and CHIBike datasets is relatively small compared to PeMS04 and PeMS08 dataset. The result confirmed that our model has great advantages in capturing spatio-temporal characteristics of traffic data, and our approach is superior on large datasets.

Table 2. Performance comparison with state-of-art methods

Model	PeMS04		PeMS08		NYCTaxi				CHIBike			
					Inflow		Outflow		Inflow		Outflow	
	MAE	RMSE	MAE	RMSE	MAE	RMSE	MAE	RMSE	MAE	RMSE	MAE	RMSE
DCRNN	22.737	36.575	18.185	28.176	14.421	23.876	12.828	20.067	4.236	5.992	4.211	5.824
STGCN	21.758	34.769	17.838	27.122	14.377	23.860	12.547	19.962	4.212	5.954	4.148	5.779
MTGNN	19.076	31.564	15.396	24.934	14.194	23.663	12.272	19.563	4.112	5.807	4.086	5.669
STFGNN	19.830	31.870	16.636	26.206	15.336	26.112	13.178	21.627	4.234	5.933	4.264	5.875
GMAN	19.139	31.601	15.307	24.915	14.267	23.728	12.273	19.594	4.115	5.910	4.090	5.675
TFormer	18.916	31.349	15.192	24.883	13.995	23.487	12.211	19.522	4.071	5.878	4.037	5.638
ASTGNN	18.601	31.028	14.974	24.710	13.844	23.177	12.112	19.201	4.068	5.818	3.981	5.609
PDFormer	18.321	29.965	13.583	23.505	**13.152**	**21.957**	11.575	18.394	3.950	5.559	**3.837**	**5.402**
ASTFN	**17.322**	**28.876**	**12.918**	**22.120**	13.279	22.212	**11.482**	**18.302**	**3.912**	**5.530**	3.850	5.410

5.5   Ablation Study

To further investigate the effectiveness of different components in ASTFN, we conducted ablation experiments and compared them on the PeMS04 and NYC-Taxi datasets. We have designed four variant versions of ASTFN, including:

- ASTFN-noTA: This variant eliminates the TSA block to investigate the usefulness of the temporal self-attention module in model prediction.
- ASTFN-noSSA: This variant removes disSSA and simSSA to investigate the benefits of modeling spatial features of traffic networks.
- ASTFN-noMask: This variant removes the two mask matrices M_{dis} and M_{sim}, meaning that the spatial attention mechanism focuses on all nodes.
- ASTFN-noTE: This variant eliminates three temporal properties of time embedding, including time proximity $\mathcal{X}_{near}$, time period $\mathcal{X}_{period}$ and time trend $\mathcal{X}_{trend}$, to investigate the effectiveness of traffic data time embedding.

Figure 5 shows how these variants were compared on the two datasets. For the NYCTaxi dataset, only the inflow results are reported because the inflow and outflow results are similar.

Based on the results, we can draw conclusions that the three temporal characteristics captured in the embedding layer have advantages in terms of time dependency. The TSA and SSA have a significant improvement in capturing dynamic temporal and spatial dependency. ASTFN brings a significant performance improvement over ASTFN-noMask, highlighting the effectiveness of mask matrices to identify local and global spatial dependencies.

5.6   Case Study

In order to intuitively study the role of attention mechanism in our model, we analyzed the spatial attention weight graph obtained by adding mask matrices to the spatial-temporal attention module of ASTFN model, and conduct a case study on NYCTaxi dataset.

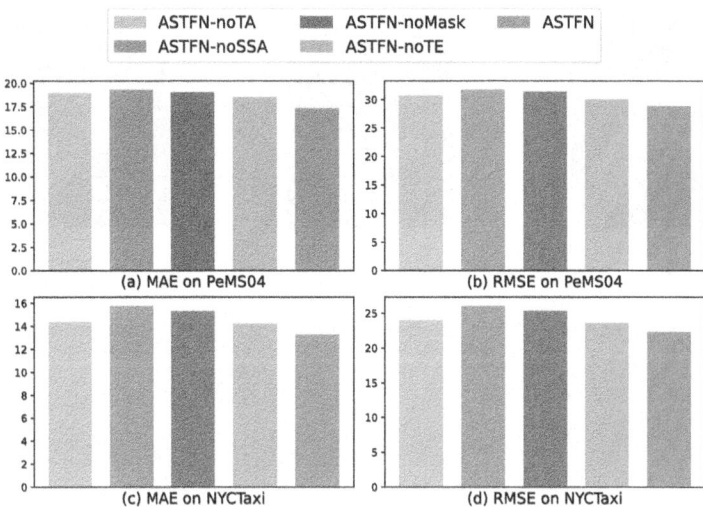

Fig. 5. Ablation Study on PeMS04 and NYCTaxi inflow

There are 75 regions in NYCTaxi dataset. Firstly we visualized the two mask matrices in disSSA and simSSA as shown in Fig. 6(a) and Fig. 6(b). By combining the two matrices, we can calculate the visualized attention matrix, which is shown in Fig. 6(c). The cell at (i, j) in the matrix is the attention weight of region i to region j. It represents the correlation between the two regions. The brighter the cell, the stronger the connection between the two regions.

There is a green box in the attention matrix, which denotes the correlation between region 37 and region 41. Next, we analyzed the specific traffic flow correlation between regions 37 and 41, which is shown in Fig. 6(d). We can see the result is reasonable, as the two regions are close in space and have similar flow patterns. Now let us investigate more examples. We can find that similar flows could happen not only between adjacent regions such as regions 9 and 2 as depicted in Fig. 6(e), but also far away regions such as 9 and 60. While low attention regions such as 9 and 59 show almost no correlation as shown in Fig. 6(g). This confirms the role of the spatial proximity mask matrix M_{dis} and the similar mask matrix M_{sim}.

In Fig. 6(h), we present the attention regions in heatmap. The result can show the distribution of attentions for region 9 more clearly. Finally, we compared the real and predicted values of region 9 as shown in Fig. 6(i), which confirms the effectiveness of our approach.

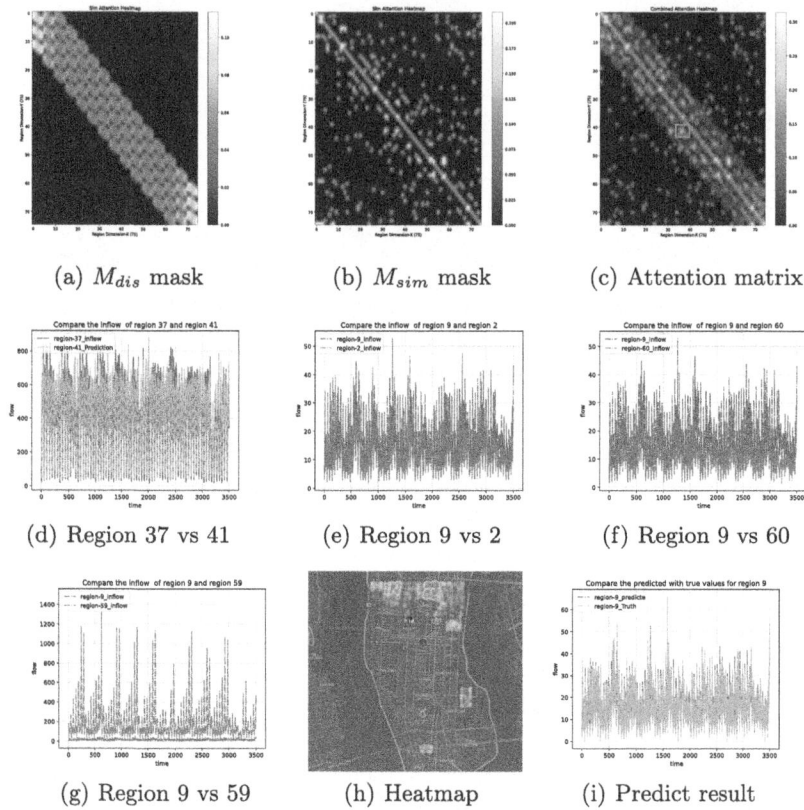

(a) M_{dis} mask   (b) M_{sim} mask   (c) Attention matrix

(d) Region 37 vs 41   (e) Region 9 vs 2   (f) Region 9 vs 60

(g) Region 9 vs 59   (h) Heatmap   (i) Predict result

Fig. 6. Case study of traffic flow

6   Conclusion and Future Work

In this paper, a new attention-based spatial-temporal attention model, ASTFN, was proposed to solve the traffic flow prediction problem. The model combines spatial-temporal embedding mechanism and spatial-temporal attention encoder layer including masking attention in spatial dimension and temporal dimension to captures dynamic spatial-temporal features of traffic data. Experiments on four real-world datasets demonstrate the superiority of ASTFN over other state-of-art methods.

Since ASTFN is a general spatial-temporal prediction framework for graph-structured data and grid data, it can be applied to other tasks, such as estimating arrival time of transportation resource and scheduling. We will explore more potential of the approach in future work.

Acknowledgments. This work is partly supported by the National Natural Science Foundation of China (Nos. 61702346 and 61702345), Basic Scientific Research Project of Liaoning Provincial Department of Education (No. JYTMS20231226), Ministry of

Education industry-university cooperative education project (No. 231002108104009) and China Machinery Industry Education Association 2024 annual project on the integration of industry and education (No. ZJJX24CY008).

References

1. Guo, S., Lin, Y., Wan, H., Li, X., Cong, G.: ASTGNN: learning dynamics and heterogeneity of spatial-temporal graph data for traffic forecasting. IEEE Trans. Knowl. Data Eng. **34**, 5415–5428 (2022)
2. Jiang, J., Han, C., Zhao, W.X., Wang, J.: Pdformer: propagation delay-aware dynamic long-range transformer for traffic flow prediction. In: AAAI Conference on Artificial Intelligence (2023)
3. Li, M., Zhu, Z.: Spatial-temporal fusion graph neural networks for traffic flow forecasting. In: National Conference on Artificial Intelligence (2021)
4. Li, Y., Yu, R., Shahabi, C., Liu, Y.: Diffusion convolutional recurrent neural network: data-driven traffic forecasting. arXiv Learning (2017)
5. Liu, L., Zhen, J., Li, G., Zhan, G., Lin, L.: Dynamic spatial-temporal representation learning for traffic flow prediction. IEEE Trans. Intell. Transp. Syst. **22**, 7169–7183 (2019)
6. Song, C., Lin, Y., Guo, S., Wan, H.: Spatial-temporal synchronous graph convolutional networks: a new framework for spatial-temporal network data forecasting. In: Proceedings of the AAAI Conference on Artificial Intelligence, pp. 914–921 (2020)
7. Vaswani, A., et al.: Attention is all you need. In: Neural Information Processing Systems (2017)
8. Wang, J., Hao, S., Shan, J., Song, X.: Visual language–let the product say what you want. In: Proceedings of the AAAI Conference on Artificial Intelligence, pp. 23841–23843 (2024)
9. Wang, J., Shan, J., Santos, O.E., Bao, J.: High quality error-tolerant phrase mining on text corpus. Expert Syst. Appl. **171**, 114557 (2021)
10. Wang, J., Jiang, J., Jiang, W., Li, C., Zhao, W.X.: Libcity: an open library for traffic prediction. In: Proceedings of the 29th International Conference on Advances in Geographic Information Systems (2021)
11. Wu, Z., Pan, S., Long, G., Jiang, J., Chang, X., Zhang, C.: Connecting the dots: multivariate time series forecasting with graph neural networks. In: Proceedings of the 26th ACM SIGKDD International Conference on Knowledge Discovery & Data Mining (2020)
12. Yan, H., Ma, X.: Learning dynamic and hierarchical traffic spatiotemporal features with transformer. Cornell University - arXiv (2021)
13. Yao, H., et al.: Deep multi-view spatial-temporal network for taxi demand prediction. In: AAAI Conference on Artificial Intelligence (2018)
14. Yu, B., Yin, H., Zhu, Z.: Spatio-temporal graph convolutional networks: a deep learning framework for traffic forecasting. In: Proceedings of the Twenty-Seventh International Joint Conference on Artificial Intelligence (2018)
15. Zhai, H., Cao, X., Sun, P., Shen, D., Nie, T., Kou, Y.: Rule-enhanced evolutional dual graph convolutional network for temporal knowledge graph link prediction. In: Yuan, L., Yang, S., Li, R., Kanoulas, E., Zhao, X. (eds.) WISA 2023. LNCS, vol. 14094, pp. 64–75. Springer, Singapore (2023). https://doi.org/10.1007/978-981-99-6222-8_6
16. Zheng, C., Fan, X., Wang, C., Qi, J.: GMAN: a graph multi-attention network for traffic prediction. In: AAAI 2019, pp. 1234–1241 (2020)

Detection Model of News Distortion Based on Chinese-German Bilingual Knowledge Graphs

Ye Liang(✉)

School of Information Science and Technology, Beijing Foreign Studies University, Beijing, China
liangye@bfsu.edu.cn

Abstract. To reduce the repeated misinterpretation of China's news events in other countries' media, the author constructed a detection model of news distortion based on cross-linguistic knowledge graphs. The research object is news on the well-known portal websites in China and Germany. With the help of the affinity propagation based on bilingual knowledge graphs and the bilingual sentiment dictionaries, the author proposes an automatic discovery mechanism of news distortion in the news transmission across languages. Compared with other traditional methods, the BKG-AP algorithm has the best performance in news clustering, and its F1 score is up to 85.2%. The experiments show that the detection model of news distortion based on Chinese-German bilingual knowledge graphs can discover distorted news more quickly, comprehensively and accurately. Therefore, it can be an effective tool for China to know the public opinion abroad that departs from facts in time and take measures to clarify misunderstandings.

Keywords: Bilingual Knowledge Graphs · News Distortion · Public Opinion Detection

1 Overview

The successful 2022 Olympic Winter Games made Beijing again the focus of the world's attention. The news about the Olympics was also translated into various languages and spread in different regions. At the same time, the Western media made use of the issue and distorted the facts, which had incalculable negative impact on the international image of China. For example, the German sports news agency Sport-Informations-Dienst (SID) took China's targeted rectification work on lean meat extract out of context and reported under the title of "Lean Meat Extract: NADA Warns against Eating Meat in China", which discredited the food safety of Olympic athletes. Against this backdrop, an efficient and accurate cross-linguistic comparative study on big data of news is of great practical significance and application value for the Chinese government and enterprises to know the public opinion that departs from facts in time, take measures to clarify misunderstandings, stop the repeated false assertion, avoid domino effect, and prevent anti-Chinese political incidents [1].

© The Author(s), under exclusive license to Springer Nature Singapore Pte Ltd. 2024
C. Jin et al. (Eds.): WISA 2024, LNCS 14883, pp. 512–523, 2024.
https://doi.org/10.1007/978-981-97-7707-5_42

News distortion means that in the process of news transmission, the communicator adds, subtracts, selects, reshapes or distorts the content or sentiment of the in-formation, resulting in the discrepancy between the content or emotional tendency and the original news [2]. Information loss, distortion and misinterpretation are some specific forms. The detection of news distortion is not about screening the authenticity of each news item, but about targeting and giving feedback on the different interpretations that may arise in the news transmission. A case in point is that the OCOG pays attention to whether there is any emotional bias in its official news when the news is reproduced. Such practice helps maintain its image and clarify the facts in time.

Most of the existing detection algorithms of news distortion are based on the model of news collection, analysis and detection in the same language. However, on today's highly developed network, there are a number of multilingual news communicators. Because of them, news can easily break the language barrier and spread rapidly among different regions. This phenomenon not only opens a door for the general public to obtain information around the world, but also lays a hidden danger for news distortion in transmission. Unlike what they do in news reproduction in the same language, when they spread news in another language, multilingual journalists are easily influenced by their own history, culture, politics and religion background as well as individual knowledge and values. Their cross-linguistic processing may lead to news distortion, resulting in changes in news content, perspectives and sentiment, and thus affecting public opinion in the direction favorable to them. Because of the inadequate number of multilingual personnel, the government departments and enterprises affected find it difficult to track and find the distortion in time, and lack technical support for the positive guidance of public opinion. Fortunately, this detection model of cross-linguistic news distortion can help address this problem.

This paper pays attention to the possible distortion in the cross-linguistic news transmission between Chinese and German media. It takes the advantage that the research team has experts in both Chinese and German languages and uses news from mainstream portal websites in the two countries, such as Tencent, Sina, and Der Spiegel as research objects. This paper focuses on the possible reverse interpretation in news reproduction about the same event between the two languages and builds a classification model of news sentiment based on big data of news in Chinese and German and the bilingual knowledge graphs. It also tracks the information flow of events related to the 2022 Olympics to verify whether this model is valid and applicable.

This paper makes contributions in the following three respects:

Proposing an important but long-neglected research question: detection of cross-linguistic news distortion. Focusing on the difference of expression about the same event in different languages, the detection model quickly makes cross-linguistic horizontal comparisons among the vast amount of news on the Internet to identify news clusters with highly distorted information.

Collecting data of news about the 2022 Olympics, including more than 10,000 news items from mainstream portal websites in Chinese and German. This effort has prepared the experimental data for detection of cross-linguistic information distortion and algorithm verification.

Building a detection model of cross-linguistic news distortion based on knowledge graphs. Compared with the extraction method of cross-linguistic topic clustering based on the word-separated mapping, this model can improve the recall and precision of cross-linguistic news clusters by introducing the priori knowledge of linguistic experts.

2   Related Work

In recent years, the academic community has adopted natural language processing, machine learning, knowledge graphs and other techniques to study the detection of news distortion, and has made achievements.

2.1   Detection of News Distortion

News reporting is how journalists construct the world. In this perspective, journalists filter, interpret and reproduce news facts based on their own cognition and cultural backgrounds. Specifically, news distortion can be divided into three parts: information loss, information distortion and information misinterpretation. Information loss happens in the process of information transmission in which the communicator only intercepts and paraphrases the part that he/she wishes to convey. This is usually called interpretation out of context in Chinese. Information distortion means that the original information is reshaped by the communicator according to his/her personal wishes and becomes fake information in the context of the event. And information misinterpretation is a distorted interpretation of the event.

According to different research ideas about the detection of news distortion, early researchers usually make judgements based on users' comments [3, 4]. Most news websites allow readers to comment. By collecting readers' feedback, researchers can uncover news distortion to some extent. Generally speaking, the impact of news on the viewers is proportional to their number and is also positively related to the number and the quality of comments. News with few comments is not valuable for detection of distortion because its influence is limited, and it will not have significant impact on public opinion. The drawback of this method is that it relies too much on the comments of "informed" people. This method performs well in vertical fields such as travel and food notes, but it has low detection rate of valid comments in news.

Another method to tell the credibility of news is to check the writing style, citation relationships and past behavior characteristics of the journalist. Researchers need to collect the journalist's articles in advance and establish mapping relation-ships between the author and his/her articles, interaction relationships between his/her and other authors, and cross reference relationships between articles. This method highly relies on the size of the dataset and the effectiveness of pre-training.

2.2   Deep Learning

As the deep learning technology matures, researchers can combine features of press releases with technologies, such as Convolutional Neural Networks (CNN) [5], Recurrent Neural Networks [6], Bidirectional Long Short-Term Memory [7], Support Vector

Machines [8] and Gated Recurrent Unit [9]. This method can extract deep semantic meaning from the text, and thus can be used in wider areas and has better effect [21]. However, such methods can only be applied to the detection of chapter-level news distortion. They do not make the cross comparison between information on the same topic and cannot be used in sparse datasets that lack auxiliary information for judgment.

2.3 Natural Language Processing

Thanks to natural language processing, detection techniques of information similarity between articles have been introduced one after another. Such detection approaches are mainly divided into two categories: statistical-based and semantic-based. The main idea of the approach based on statistics is to find out the keywords of each article, group the articles by topics on the basis of keyword matching and word frequency features, and then detect whether there is news distortion among similar articles. Common methods are Longest Common Substring [10], Edit Distance, Jaccard Index, Dice Coefficient and Term Frequency-inverse Document Frequency [11]. The approach based on semantics works by introducing external knowledge. Common methods include the lexicon and the Vector Space Model [12].

2.4 Knowledge Graphs

Knowledge graphs can integrate information and knowledge from different sources. Such graphs are highly suitable for comparison between cross-linguistic information and can achieve more comprehensive detection of news distortion. They take knowledge as the object and show the development and structures of disciplines. They are endowed with the properties and features of both "graph" and "spectrum". In recent years, many knowledge graphs, both general and domain-specific, have played a great role in the research on big data and artificial intelligence, and have become valuable resources in these fields. Some graphs are created to analyze data about e-commerce products, some are applied to study the mentorship between historical celebrities in Song Dynasty, and some are even used to build a big data platform for sci-tech knowledge discovery. In 2012, Google constructed its own knowledge graphs to improve the service quality of its search engines. Subsequently, knowledge graphs have become a hotspot in research and various graphs have emerged, such as CN-DBpedia and Zhixin of Baidu in China and XLore [13], XLORE2 [14], Wiki CiKE [15], Concept Net5.5 [16], CLEQS, DBpedia NIF [17], EventKG [18], Body-MindLanguage [19], CrossOIE [20] in other countries.

3 Detection Model of News Distortion Based on Bilingual Knowledge Graphs

This section introduces problems in the detection of cross-linguistic news distortion, the construction of bilingual knowledge graphs under the guidance of linguists, and the detection model of news distortion based on Chinese-German bilingual knowledge graphs.

3.1 Problems

With the help of natural language processing and knowledge graphs, the detection of information similarity between articles is becoming increasingly sophisticated. By clustering information on the same topic, researchers can discern whether there is a conflict between information in the cluster. However, the current general-purpose technical models, which perform well in the same language, are much less effective in cross-linguistic contexts. Because the simple combination of multiple conventional monolingual knowledge graphs fails to express effectively in cross-cultural contexts.

For example, in Russian-language coverage of the Beijing Winter Olympics, the word Президент (president) was commonly used to refer to the IOC President Thomas Bach. Without knowledge of the Russian language and the context, re-searchers would miss the news related to Президент in the evaluation category of IOC when they produced the news clusters about the Olympics in Russian and Chinese. Similarly, in the news of Tagesspiegel, one of the largest websites in the German-speaking world, in just three months from December 17, 2021 to March 17, 2022, there were up to five references to Russian President Putin (see Table 1).

Table 1. Five references to Russian President Putin.

German	English	Number of occurrences
Wladimir Putin	Vladimir Putin	948
Präsident Putin	President Putin	791
Kriegsverbrecher	War criminal	29
(russischer) Diktator	(Russia)Dictator	30
Zar, Putin	czar, Putin	3

Meanwhile, the service quality of bilingual knowledge graphs depends largely on the quality of knowledge. Conventional monolingual knowledge graphs are mostly constructed by fully automated methods without any human involvement and the data of the graphs mainly come from the Internet. Although the fully automated methods can save time and manpower, they cannot meet the requirements of precision and recall of cross-linguistic clustering of bilingual news. Specifically:

1. Concepts, relationships, and events across languages and cultures are complex and ambiguous.
2. The quality of source data is poor and the information sources from different languages do not have consistent standards when describing events.
3. The detection of news distortion has high requirements for timeliness and relevant context, and thus the prior knowledge of experts in the certain field is important.

3.2 Construction of Multilingual Knowledge Graphs

The precision of the detection of news distortion is closely related to the recall of the clustering of cross-linguistic news on the same topic. Therefore, we need to construct

Chinese-German bilingual knowledge graphs guided by linguists to create clusters of topics more accurately and efficiently, and then explore a wider range of information distortion that crosses languages and cultures. Since linguists are a scarce resource, it is impossible for them to participate in the whole process of the construction of knowledge graphs. Therefore, we must find a low-cost model for linguists to participate. To this end, the author has designed a data model of bilingual knowledge graphs with linguist participation. This model has a data structure that can more accurately describe the relevance of cross-linguistic information.

Data Models. To better describe the construction of bilingual knowledge graphs, the author first defines three concepts: Conceptual Knowledge Graph (CKG), Instance Knowledge Graph (IKG) and Factual Knowledge Graph (FKG). CKG only has concept nodes, IKG contains both entity nodes and event nodes, and FKG is the set of CKG and IKG.

A data model is an abstract model that organizes data elements and standardizes their relationships with each other and their relationships with the attributes of real entities. In the data model of bilingual knowledge graphs, there are five main types: concepts, entities, events, attributes, and relationships. Among them, "concepts", "entities" and "events" are types of nodes, while "relationships" are types of edges. Both nodes and edges have attributes. In the data model, there are three types of nodes: concept nodes, entity nodes, and event nodes. The attributes of a concept node include its unique identifier, source, start time, end time, and text representation. The attributes of an entity node include its unique identifier, source dataset, type, start time, end time, and text representation. And the attributes of an event node include its unique identifier, event type, time, and location (Fig. 1).

Concepts nodes		Entities nodes		Events nodes	
Attribute	**Sample**	**Attribute**	**Sample**	**Attribute**	**Sample**
ID	AUI:A030592	ID	IUI:C007124	ID	EUI:E032118
Source	Tagesspiegel	Source	Tagesspiegel	Type	IUI:C007124
Type	AUI	Type	CUI:C0302960	Time	2022-02-20 20:00:00
Start time	2021-12-11 13:23:42	Start time	2022-02-04 20:00:00	Location	The National Stadium
End time	-1	End time	2022-02-20 21:30:00	Text	Closing ceremony
Text	Президент	Text	Winter Olympics		

Fig. 1. Data models.

Information Processing Framework. For news X and Y in Chinese and German respectively, the detection model of cross-linguistic news distortion detects the differences between the two with rule bases, dictionaries, CKG, and IKG. FKG consists of two parts: CKG and IKG. CKG is constructed on the basis of expert toolsets, open and related knowledge graphs, expert knowledge bases, and analysis criteria. Meanwhile, the construction of IKG is based on expert toolsets, data sources, open datasets, and media data of online news. The process of constructing CKG includes data transformation (from ER to RDF) and data fusion (concept fusion and relationship fusion). And the process of constructing IKG includes structured data transformation (from ER to RDF), unstructured data transformation (extraction of entities and relationships) and data fusion (alignment of entities and relationships) (Fig. 2).

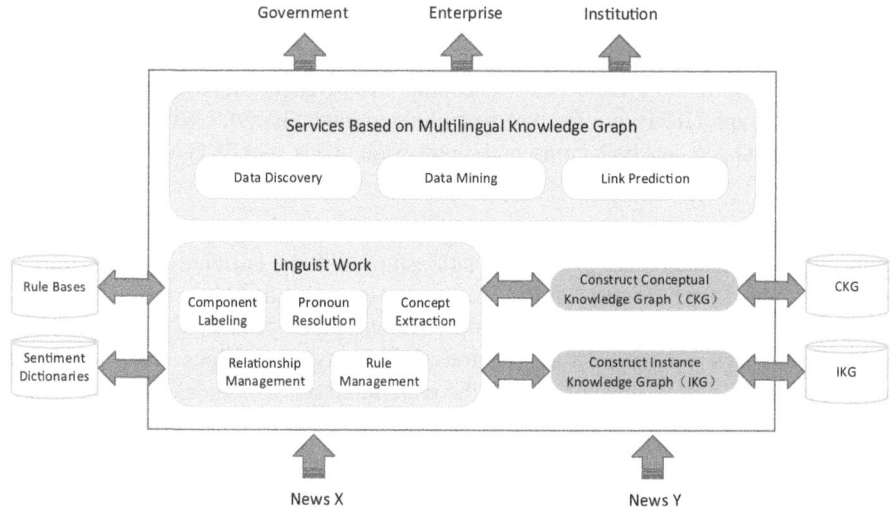

Fig. 2. Information processing framework.

Involvement of Linguistic Experts. In the field of artificial intelligence and big data, linguistic experts play an increasingly important role in the process of constructing knowledge graphs. Although machine learning or deep learning-based solutions are widely used in many tasks due to their efficiency, rule-based approaches are often preferred because they are more interpretable and thus support interactive debugging. There are two kinds of widely used rules: rules created manually by linguistic experts and weakly supervised rules generated automatically by machines. The former requires experts to write tagging heuristics on the basis of their domain-specific knowledge. The knowledge graph contains many concepts, relationships, triples, and rules specific to the flow of information across languages and requires linguistic experts to confirm with prior knowledge.

There is no doubt that the involvement of experts can make topic clustering and the detection of distortion more precise. However, since linguists are a scarce resource, they cannot be required to participate in the entire process of the construction of knowledge graphs. Therefore, a key consideration is when and where to include the work of linguists can save time and minimize the cost. Make sure what the position of linguists and the role they play in the processing framework is. If we build the information processing framework in this way, we can integrate the work of linguists into the process of constructing knowledge graphs.

There are four tools in the cross-linguistic information processing framework: OP-inspection (a tool for discovering new concepts and relationships), OP-annotation (a tool for experts' annotation), OP-synonym (a tool for the construction of synonyms) and OP-rule (a tool for the construction and maintenance of a rule base). With OP-inspection, we can apply algorithms of machine learning to the fusion of concepts and relationships, which helps in the construction of CKG. With OP-annotation, we can build a corpus of entities and relationships, which helps in constructing events, entities and relationships, and thus contributing to the construction of IKG. With OP -synonym, we can add new synonyms, which helps in the construction of entities. OP-rule is based on relationships and entities that can be extracted and on data that can be transformed from ER to RDF/OWL. With OP-rule, we can integrate construction with rule-making. From the perspective of both linguistic experts and machines, the involvement of experts in these four tools will make the construction of bilingual knowledge graphs more efficient.

3.3   The Detection Model of News Distortion Based on Knowledge Graphs

Since FKG contains IKG and CKG, and IKG is based on CKG, we first discuss how to construct CKG and then introduce how to create IKG. The process of producing CKG includes data transformation (from ER to RDF) and data fusion (the fusion of concepts and relationships). The data in CKG is mainly extracted from knowledge bases (data sources and knowledge bases on related topics), relevant standards, open- source knowledge graphs (such as SNOMEDCT and DBpedia), and knowledge and rules summarized by linguists. The process of constructing IKG involves data transformation (from ER to RDF) and data fusion (the fusion of events, entities and relationships). At the same time, data required by IKG are from mainstream online media, public datasets (such as MIMIC-III), and media reviews.

Construction of CKG. There are three types of data sources in the construction of CKG: data from ER in relational databases, data from RDF in graphical databases, and data reviewed or added by linguists. The main work is to get triples from structured databases, from linked data, from semi-structured data and from texts.

Construction of IKG. To construct IKG is to represent instance data with graphs. This process includes the transformation of structured and unstructured data. The transformation functions of the latter include new word discovery, the extraction of entity, relationship and event, and entity alignment.

Contributions of Linguistic Experts. There are four tools in the toolset of linguistic experts to help showing the details on how these tools contribute to the construction of CKG and IKG. As we can see, OP- synonym, OP-annotation and OP-rule help in constructing IKG, and OP-inspection is conductive to the construction of CKG. OP-annotation provides an interface that allows linguistic experts to annotate unstructured data on the basis of instances and relationships in IKG and CKG. Then the annotated results are stored in the corpus of entities or relationships. OP-rule standardizes the mapping rules from ER to RDF and the extraction of concepts, entities, event nodes and relationships. Experts can ensure that knowledge graphs are constructed more precisely. By integrating with ML algorithms, FKG can be comprehensive.

4   Experiments and Analysis

To test whether the Chinese-German bilingual knowledge graphs proposed in this paper are effective in the detection of news distortion, the author compared the Chinese-German knowledge graphs with the traditional clustering algorithms. Also, the author studied precision and recall of the hybrid clustering of the bilingual news with the support of two monolingual knowledge graphs of Chinese and German, and analyzed the effect of the detection of news distortion on the basis of sentiment dictionaries.

4.1   Construction of News Corpus

Firstly, with Python, the author collected news on the Beijing Winter Olympics from December 17, 2021 to March 17, 2022 on China's and Germany's mainstream portals, such as Tencent, Sina and Der Spiegel. The valid news items collected were more than 10,000 in total. Then the author reduced noise through data cleaning, segmented texts in the Chinese news with Jieba, made lemmatization in the German news with CST's Lemmatiser, and removed stop words in the bilingual news respectively with their dictionaries. Finally, the author got a bilingual corpus of 8479 items that could be processed and analyzed.

To effectively evaluate the results of the topic clustering, the author invited a group of experts in Chinese and German languages to manually annotate the content of news items, classify them according to 20 predetermined topics, and rate the sentiment of each news item. Then the author invited another group of experts to review the classification results.

4.2   Comparison of Effects of Cross-Linguistic News Clustering

Table 2. Comparison results of cross-linguistic news clustering.

Method	Precision	Recall	F1 score
K-means	65.5	67.8	66.6
LDA	61.2	64.3	62.7
CNN	68.4	71.1	69.7
MKG-AP	78.7	81.5	80.1
BKG-AP	83.6	86.9	85.2

The author created news clusters with the Chinese-German bilingual corpus through K-means, LDA, CNN, MKG-AP (the affinity propagation based on monolingual knowledge graphs), and BKG-AP (the affinity propagation based on bilingual knowledge graphs). The comparison results are shown in Table 2.

The author established mapping of cross-linguistic word meanings in news texts through Chinese-German dictionaries. Then then author found that the three methods,

K-means, LDA, and CNN, were unsatisfactory in three indexes of news clustering: precision, recall, and F1 score. Supported by knowledge graphs, MKG-AP improved more than 10% in the effect of clustering and could show the classification of news content more accurately and comprehensively. With the help of bilingual knowledge graphs, BKG-AP did very well in clustering and its precision rate exceeded 80%. Since some news had two or more subjects, the experimental results had an error of 10%-20%, which was in a reasonable range.

4.3 Detection of News Distortion Based on Sentiment Dictionaries

To detect news distortion, after creating clusters, the author analyzed the sentiment of news within the same cluster article by article. In this stage, the author did the sentiment calculation with sentiment dictionaries: HowNet of CNKI for Chinese, and General Inquirer (the German version) for German. The calculations revealed that, due to the different national conditions, there were significant differences in cognition and sentiment between some Chinese and German reports on the same type of events. Some of the calculation results are listed in Table 2.

Table 3. Detection results of news distortion based on sentiment dictionaries.

Category	Summary	Attitude	Sentiment value	Average score of sentiment value	
				German	Chinese
COVID -19 epidemic control	Of the 2.5 million COVID-19 tests, nearly 500 were positive	Criticizing the closed-loop management of Beijing Winter Olympics. Athletes must be quarantined after testing positive, including those from Germany. Because of China's "dynamic zero-COVID" policy, spectators cannot be present on site and journalists are not allowed to "go out of the loop"	−7	−3	9
Opening and closing ceremonies	There was a joyful atmosphere at the closing ceremony of the Paralympic Games, but the last-minute review intensified	The Chinese organizers' approach was controversial: the translation of the IPC President's speech omitted the word "peace": "hopes for peace" was translated as "hope to be a family", and "champions for peace" was not translated at all	−4	−2	10
Athletes	Ukraine at the Paralympic Games	"It is a miracle that we are able to participate in the Paralympic Winter Games", said the Ukrainian athletes	3	−1	5

The second example in Table 3 shows that German journalists, influenced by their knowledge background and values, did obvious news processing, which led to news distortion.

5 Conclusion and Prospect

Focusing on the different descriptions of the same event in the news of China and Germany, in this paper the author developed a detection model of news distortion on the basis of Chinese-German bilingual knowledge graphs. This study took the Beijing Winter Olympics as the starting point, took the corpus of the bilingual news as the basis, and compared four common methods for news clustering with the BKG-AP algorithm proposed in this paper through experiments. As a result, BKG-AP had better performance in three indexes: precision, recall and F1 score. With the help of BKG-AP and the sentiment dictionaries, the author found the distorted news reports in a more effective way and achieved the expected goal.

This paper offered a new idea for the detection of news distortion across languages, but there are still limitations and shortcomings. The method proposed in this paper is only applicable to the distortion of all news on a certain topic in two languages, it cannot identify which news is true and which is distorted, and it requires the interpretation of linguists. Therefore, how to further reduce the involvement of experts is to be continued in subsequent research.

Acknowledgement. This work was supported in part by the Fundamental Research Funds for the Central Universities (No. 2024TD001), and the First-class Disciplines Construction Foundation of Beijing Foreign Studies University (No. YY19SSK02).

References

1. Liang, Y., Qin, Y.: Cross-lingual public opinion tracing based on blockchain technology. In: 2020 17th International Conference on Web Information Systems and Applications, pp. 607–617 (2020). https://doi.org/10.1007/978-3-030-60029-7_54
2. Liang, Y., Xu, L., Huang, T.: Sentiment tendency analysis of NPC & CPPCC event in German news. In: 2019 16th International Conference on Web Information Systems and Applications, pp. 298–308 (2019). https://doi.org/10.1145/2976749.2978341
3. Zong, H., Wu, X., Xue, C., Chen, F.: Distribution law of user comments on hot news. In: 2017 International Conference on Progress in Informatics and Computing (PIC), pp. 461–465 (2017). https://doi.org/10.1109/PIC.2017.8359592
4. Alhujaili, R.F., Yafooz, W.M.S.: Sentiment analysis for YouTube videos with user comments: review. In: International Conference on Artificial Intelligence and Smart Systems (ICAIS), pp. 814–820 (2021)https://doi.org/10.1109/ICAIS50930.2021.9396049
5. Xin, R., Zhang, J., Shao, Y.: Complex network classification with convolutional neural network. Tsinghua Sci. Technol. **25**(4), 447–457 (2020). https://doi.org/10.26599/TST.2019.9010055
6. Uçkun, F.A., Özer, H., Nurbaş, E., Onat, E.: Direction finding using convolutional neural networks and convolutional recurrent neural networks. In: 2020 28th Signal Processing and Communications Applications Conference (SIU), pp. 1–4 (2020). https://doi.org/10.1109/SIU49456.2020.9302448

7. Akula, S.P., Kamati, N.: Credibility of social-media content using bidirectional long short-term memory-recurrent neural networks In: International Conference on Emerging Techniques in Computational Intelligence (ICETCI), pp. 170–175 (2021)https://doi.org/10.1109/ICETCI51973.2021.9574061

8. Mohan, L., Pant, J., Suyal, P., Kumar, A.: Support vector machine accuracy improvement with classification. In: 2020 12th International Conference on Computational Intelligence and Communication Networks (CICN), pp. 477–481 (2020). https://doi.org/10.1109/CICN49253.2020.9242572

9. Santur, Y.: Sentiment analysis based on gated recurrent unit. In: International Artificial Intelligence and Data Processing Symposium (IDAP), pp. 1–5 (2019).https://doi.org/10.1109/IDAP.2019.8875985

10. Prozorov, D., Yashina, A.: The extended longest common substring algorithm for spoken document retrieval. In: 2015 9th International Conference on Application of Information and Communication Technologies (AICT), pp. 88–90 (2015). https://doi.org/10.1109/ICAICT.2015.7338523

11. Tongman, S., Wattanakitrungroj, N.: Classifying positive or negative text using features based on opinion words and term frequency - inverse document frequency. In: 2018 5th International Conference on Advanced Informatics: Concept Theory and Applications (ICAICTA), pp. 159–164 (2018). https://doi.org/10.1109/ICAICTA.2018.8541274

12. Premalatha, R.: Srinivasan, S.: Text processing in information retrieval system using vector space model. In: International Conference on Information Communication and Embedded Systems (ICICES2014), pp. 1–6 (2014). https://doi.org/10.1109/ICICES.2014.7033837

13. Wang, Z.G., et al.: XLore: a large-scale English-Chinese bilingual knowledge graph. In: Proceedings of the International Semantic Web Conference, pp. 121–124 (2013)

14. Jin, H.L., et al.: XLORE2: large-scale CrossLingual knowledge graph construction and application. Data Intell. 1(1), 77–98 (2019)

15. Wang, Z.G., et al.: Transfer learning based CrossLingual knowledge extraction for Wikipedia. In: Proceedings of the 51st Annual Meeting of the Association for Computational Linguistics, pp. 641–650 (2013)

16. Zhou, Y.L., Steven, S., Shah, J.: Predicting ConceptNet path quality using crowdsourced assessments of naturalness. In: Proceedings of the Web Conference, pp. 2460–2471 (2019)

17. Bu, Q., et al.: Using microtasks to crowdsource DBpedia entity classification: a study in workflow design. Semantic Web 9(4), 1–18 (2017)

18. Gottschalk, S., Demidova, E.: EventKG: A multilingual EventCentric temporal knowledge graph. In: Proceedings of the 15th Extended Semantic Web Conference, pp. 272–287 (2018)

19. Gromann, D., Hedblom, M.M.: Body-Mind-Language: multilingual knowledge extraction based on embodied cognition. In: Proceedings of the 5th International Workshop on Artificial Intelligence and Cognition, pp. 20–33 (2017)

20. Cabral, B.S., et al.: CrossOIE: cross-lingual classifier for open information extraction. In: Proceedings of the 14th International Conference on Computational Processing of the Portuguese Language, pp. 368–378 (2020)

21. Jiang, J., Xu, L.: Jointly learning structure-augmented semantic representation and logical rules for knowledge graph completion. In: Proceedings of the 20th International Conference on Web Information Systems and Applications, pp. 52–63 (2023)

Knowledge-Aware Self-supervised Educational Resources Recommendation

Jing Chen, Yu Zhang$^{(\boxtimes)}$, Bohan Zhang, Zhenghao Liu, Minghe Yu, Bin Xu, and Ge Yu

School of Computer Science and Engineering, Northeastern University, Shenyang 110167, China
zhangyu@mail.neu.edu.cn

Abstract. In the evolving landscape of educational technology, personalized recommendation systems play a pivotal role in delivering tailored educational content to learners. The knowledge-aware recommendation has emerged as a new trend in this field. However, knowledge graphs often suffer from sparsity and noise, characterized by long-tail entity distributions. We propose a novel framework, Knowledge-Aware Self-Supervised Educational Resources Recommendation (KASERRec), which leverages an integrated approach using knowledge graph embeddings, Light Graph Convolution Network(LightGCN), and cross-view knowledge contrastive learning to address the challenges inherent in educational resources recommendation. Focused on enhancing the discoverability of long-tail educational content, which often contains valuable but overlooked knowledge points, KASERRec introduces innovative modules designed to optimize recommendation accuracy and diversity. Employing a real-world educational knowledge graph, the framework demonstrates significant improvements in personalized learning experiences by effectively recommending educational resources that cater to the specific needs and preferences of learners. Our evaluations on the MOOCCube dataset highlight the framework's superiority over existing methods in terms of recommendation relevance and user satisfaction.

Keywords: Knowledge-aware Recommendation · Self-Supervised Learning · Contrastive Learning

1 Introduction

The digital transformation of education has led to an exponential increase in the amount and variety of educational resources available online [1]. This offers learners unprecedented access to learning materials, but it simultaneously presents significant challenges in navigating this vast landscape effectively. Personalized recommendation systems are crucial in this context, as they can guide learners to the most relevant resources based on their unique preferences, learning objectives, and academic backgrounds. However, traditional recommendation systems often face difficulties in the educational context where the goal is not just to recommend any content, but to guide learners towards resources that are pedagogically valid and aligned with their evolving educational needs and academic goals [2].

Traditional methods like collaborative filtering [3] rely heavily on extensive user-item interaction histories, struggling with new users or recently introduced courses and learning materials. To overcome these limitations, recent advancements have integrated knowledge graphs with recommendation systems [4], providing a structured representation of complex relationships among educational entities such as course prerequisites and topics. This approach not only mitigates the cold start problem by inferring the relevance of new items based on their connections within the graph but also addresses the long-tail problem [5] by highlighting less popular yet educationally valuable materials.

Building on these insights, our framework, KASERRec, harnesses the power of knowledge graphs in conjunction with advanced machine learning techniques to create a knowledge-aware recommendation system. By doing so, KASERRec aims to not only alleviate the common issues faced by traditional models but also enhance the personalization and educational relevance of the recommendations.

2 Preliminaries

This section outlines the symbols and definitions utilized throughout this paper and delineates the Knowledge-aware recommendation task.

User-Item Interactions Graph. In our task, we manage a collection of learners and educational resources, denoted by $\mathcal{U}$ for users (learners) and $\mathcal{V}$ for items (educational resources). For each learner $u \in \mathcal{U}$ and each resource $v \in \mathcal{V}$, we establish a binary graph $\mathcal{G}_u = (u, y_{uv}, v)$, which signifies the interactions between learners and resources. Here, $y_{uv} = 1$ represents a learner u who has engaged with resource v.

Educational Knowledge Graph. Our paper represents the structured knowledge about educational resources using a knowledge graph, defined by the set of triplets $\mathcal{G}_k = (h, r, t)$. In this graph, h and t are nodes in the set of knowledge entities $\mathcal{E}$. The relationship $r \in \mathcal{R}$ represents the semantic connections between these entities, such as (concept, is_part_of, course). It is crucial to note that the set of educational resources $\mathcal{V}$ is a subset of the entity set $\mathcal{E}$.

Task Formulation. Our recommendation task is symbolized as $\mathcal{F}(u, v | \mathcal{G}_u, \mathcal{G}_k, \theta)$, that effectively predicts educational interactions. This model utilizes both the interactions graph $\mathcal{G}_u$ and the educational knowledge graph $\mathcal{G}_k$ to forecast the likelihood that a learner u will engage with a resource v. The learnable parameters θ is fine-tuned to optimize educational content delivery. The output from $\mathcal{F}$ is a score in the range $[0, 1]$, indicating the probability of engagement.

3   Methodology

In this section, we detail the KASERRec framework, specifically designed to enhance educational resource recommendations.

3.1   Knowledge Graph Relation Weighted Aggregator

3.1.1 Attention-Based Relation Weight Function. In the educational domain, our knowledge graph serves as a comprehensive framework, mapping the intricate relationships between diverse educational elements such as videos, concepts, courses, and fields, all structured into meaningful triplets (h, r, t). We then define a novel relation-aware GNN network to capture the contextual information between nodes. Unlike traditional GCN models that primarily focus on direct relationships, our relation-aware GNN is uniquely tailored to navigate the educational landscape, capturing not just direct educational ties but also nuanced semantic relationships and contextual dependencies critical in learning environments. Here, inspired by the graph attention mechanism detailed in HGT [6], we have defined a new relation-aware weighting function α to compute the importance of each triplet, with the specific formula as follows:

$$\alpha(h,r,t) = \frac{e_h W^Q \cdot (e_t W^K \odot e_r W^R)^T}{\sqrt{d}}, \tag{1}$$

where the embedding of head and tail entity is denoted as $e_h \in \mathbb{R}^d$ and $e_t \in \mathbb{R}^d$ respectively, and e_r is the embedding of relation r. Here, W_Q, W_K and W_R are three different attention-based trainable weight matrices with dimension $\mathbb{R}^{d \times d}$. The head entity's information is encoded into a query vector through a transformation using W^Q, which is crucial for subsequent matching with the key vector. W^K transforms the tail entity's embedding into the key space. The query and key mechanism enables the model to accurately assess the relevance of different entity and relationship combinations. W^R is specifically designed to adjust the relationship embeddings to suit the model's requirements. In this formula, the symbol $\odot$ denotes element-wise multiplication, also known as the Hadamard product. By computing the element-wise product of relation embedding vectors and entity embedding vectors [7,8], relational information is preserved during information propagation and aggregation processes.

To ensure comparability among different triplets associated with the same entity, we consider the entity's neighbors $\mathcal{N}_h$ and utilize the softmax function to normalize the weight scores:

$$\beta(h,r,t) = \frac{\exp(\alpha(h,r,t))}{\sum_{(h,r',t' \in \mathcal{N}_h)} \exp(\alpha(h,r',t'))}. \tag{2}$$

3.1.2 Knowledge Graph Relation Weighted Aggregator. Previous research [9] efforts only incorporated relations into the weighting factors during propagation periods. It is important to recognize that combinations of connection relations and entities have different contextual meanings. For example, (zoology, contains, ornithology) and (zoology, precedes, ornithology), where zoology connects to ornithology twice, but each time plays a different role (parent and prerequisite). To effectively navigate these complex educational relationships, our aggregator incorporates context from these connections, distinguishing between different instructional needs, such as foundational knowledge versus advanced exploration. Therefore, we design the knowledge-aware aggregator as follows:

$$e_h^{(l)} = \frac{1}{|\mathcal{N}_h|} \sum_{(h,r,t)\in\mathcal{N}_h} \beta(h,r,t)\mathbf{e}_r \odot e_t^{(l-1)}, \tag{3}$$

where $\mathcal{N}_h$ represents the set of first-order neighbors of entity h in the knowledge graph, and l denotes the number of layers in the aggregator. Since $\mathcal{V} \subset \mathcal{E}$, we generate the embedding representation of item v along the path $e_v \leftarrow e_{t_l} \leftarrow e_{t_{l-1}} \leftarrow ... \leftarrow e_{t_2} \leftarrow e_{t_1}$ after l layers of iteration.

To better capture the collaborative signals between users and items, we consider the roles of users in the interaction graph $\mathcal{G}_u$. This allows us to integrate the knowledge graph $\mathcal{G}_k$ and the user-item interaction graph $\mathcal{G}_u$. We tailor user embeddings by aggregating the embeddings of related educational content, ensuring that each learner's academic profile and needs are reflected in their personalized learning path. The specific formula is as follows:

$$e_u^{(l)} = \frac{1}{|\mathcal{N}_u|} \sum_{i\in\mathcal{N}_u} e_v^{(l-1)}. \tag{4}$$

here, e_u and e_v represent the embeddings of users and items, respectively, with e_v obtained from the aforementioned Eq. 3. Furthermore, stacking multiple layers has been demonstrated to effectively leverage higher-order connectivity to obtain different receptive fields. We further summarize to form the final user and item representations:

$$e_h = \text{Readout}[(e_h^{(0)}, e_h^{(1)}, ..., e_h^{(L)})] = \sum_{l=0}^{L} e_h^{(l)}, \tag{5}$$

$$e_u = \text{Readout}[(e_u^{(0)}, e_u^{(1)}, ..., e_u^{(L)})] = \sum_{l=0}^{L} e_u^{(l)}. \tag{6}$$

where L is the number of aggregation layers, for each user/item, we obtain L embeddings through L layers of GNN, and then obtain the final representation through a readout function.

3.2 Semantic Masking Autoencoder Module

Real-world knowledge graphs are often sparse and thus exhibit a long-tail distribution [10]. In such distributions, a small "head" segment possesses a high volume of exposure, while the "long tail" consists of a large number of segments with values significantly lower than the head group. As illustrated in Fig. 1, our constructed educational knowledge graph also conforms to a long-tail distribution. This reflects a vast array of niche educational topics and resources remain largely unexplored despite their potential value. This is disadvantageous for recommendation systems seeking to discover those potentially overlooked but promising peripheral areas or knowledge concepts.

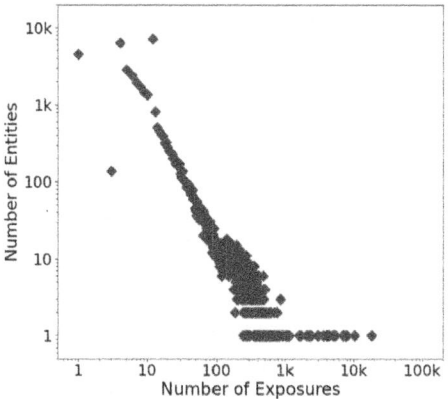

Fig. 1. Long-tail entity distributions in MOOC KG

To effectively address the challenges posed by the long-tail distribution within educational knowledge graphs, and to enhance the discovery and recommendation of underrepresented but potentially valuable educational content, a robust and innovative approach is essential. This underscores the critical role of generative self-supervised learning approach [11]. By selectively masking and reconstructing significant educational relationships in our graph, the autoencoder learns to identify and emphasize the subtler, yet crucial educational connections often missed in traditional approaches. When the autoencoder accurately reconstructs important triples during the masking process, it indicates that the model has acquired the capability to recover information from incomplete data, which may help the model generalize better to less frequently occurring nodes [12].

Based on the knowledge-aware weighting function we designed, as shown in Eq. 2, we can filter out triples that are particularly important for a given entity. However, to ensure that these scores have global comparability, we utilize neighbors of the entity nodes to compute a global weighting score for the triples:

$$\gamma(h, r, t) = |N_h| \cdot \beta(h, r, t) = \frac{|N_h| \cdot \exp(\alpha(h, r, t))}{\sum_{(h, r', t') \in N_h} \exp(\alpha(h, r', t'))}, \qquad (7)$$

And then, we obtain a set of globally highest-scoring masked triples:

$$\mathcal{M}_k = \{(h, r, t) \mid \gamma(h, r, t) \in \mathrm{topk}(\Gamma; k)\}. \qquad (8)$$

here, Γ represents the distribution of scores $\gamma(h, r, t)$, and then we remove the set of high-scoring triples $\mathcal{M}_k$ from the original educational knowledge graph to obtain an enhanced subgraph $\mathcal{G}_m$, denoted as $\mathcal{G}_k \backslash \mathcal{M}_k$.

To adapt to the altered structure of the knowledge graph post-masking, we employ the aggregators proposed above to update the node embeddings within the enhanced subgraph $\mathcal{G}_m$, as described in Eqs. 4 and 5. Since M_k is invisible during the aggregation stage, it can be effectively utilized as self-supervised labels for reconstruction. To ensure that our model effectively learns these key connections that have been masked, we have designed a loss function L_m, specifically tailored to enhance our model's predictive ability for these selected triples:

$$\mathcal{L}_m = \sum_{(h,r,t) \in M_k} -\log \sigma \left(e_h^T \cdot (e_t \odot e_r) \right). \tag{9}$$

where σ represents the sigmoid function. By optimizing this loss function during training, the model is compelled to adjust its embeddings and relational predictions to better match the actual data. This approach enhances the educational value of our knowledge graph, ensuring that even the less prominent educational topics and resources are accurately mapped and recommended, thus broadening the learning horizons for students.

3.3   Cross-View Knowledge Contrastive Learning

After implementing multiple propagation layers within the Educational Knowledge Graph, we utilize the Light Graph Convolutional Network (Light-GCN) [18] to capture nuanced collaborative signals inherent in the interactions between students and educational resources. The selection of Light-GCN is based on its simplicity and efficiency in learning node embeddings without the need for complex transformations or non-linearities.

$$X_u^{(k+1)} = \sum_{v \in \mathcal{N}_u} \frac{1}{\sqrt{|\mathcal{N}_u||\mathcal{N}_v|}} X_v^{(k)}; X_v^{(k+1)} = \sum_{u \in \mathcal{N}_v} \frac{1}{\sqrt{|\mathcal{N}_u||\mathcal{N}_v|}} X_u^{(k)}. \tag{10}$$

where $X_u^{(k+1)}$ represents the embedding of user u at layer K, and $\mathcal{N}u$ denotes the set of neighbors of user u. The normalization factor $\frac{1}{\sqrt{|\mathcal{N}_u||\mathcal{N}_v|}}$ is used to mitigate the influence of node degree on the learning process, which helps in maintaining a balance between the influence of high-degree and low-degree nodes.

To integrate the diverse signals and features captured separately in the collaborative and UI graphs, we need a method that not only aligns these embeddings but also enhances their representational power across different graph structures. Cross-graph contrastive learning provides a mechanism to unify these diverse embeddings, ensuring they are coherent and mutually informative [13].

Considering that item embeddings from the collaborative graph and the UI graph reside in distinct representational spaces, we employ Multi-Layer Perceptrons (MLPs) to map item representations from two different views-the collaborative relational signals and the knowledge graph signals into a unified latent space:

$$z_v^* = \text{MLP}^*(x_v^*); \text{MLP}^*(x) = W_2^*(\sigma(W_1^*x + b_1^*)) + b_2^*. \tag{11}$$

where σ denotes the activation function, W_1^*, W_2^* and b_1^*, b_2^* are the weights and biases of the MLPs.

Subsequently, as inspired by [14], we utilize the contrastive learning objective to ensure the representational efficacy of the aligned embeddings across different views.

$$\mathcal{L}_c = \sum_{v \in \mathcal{V}} -\log \frac{\exp(d(z_v^u, z_v^k)/\tau)}{\sum_{j \in \{v,v',v''\}} \left(\exp(d(z_v^u, z_v^k)/\tau) + \exp(d(z_j^u, z_v^k)/\tau)\right)}. \tag{12}$$

here, v' and v'' represent randomly selected negative samples for item v from each of the two views, respectively. The similarity function $d(\cdot)$ employs cosine similarity, normalized to unit length to stabilize the training dynamics. The temperature parameter τ is finely tuned to modulate the model's sensitivity to variations in educational content similarity.

3.4 Loss Optimization Module

In educational recommendation systems, accurately predicting student learning preferences is critical for delivering personalized educational resources. To achieve this objective, we employ the Bayesian Personalized Ranking (BPR) loss function L_{rec}, a key component of our training framework. Rooted in the principles of collaborative filtering adapted for education, the BPR loss optimizes the prediction of student-resource interactions by contrasting successful learning interactions (positive samples) with non-interactions (negative samples), thus refining our understanding of subtle student learning preferences and needs.

$$\mathcal{L}_{rec} = \sum_{(u,v,j) \in D} -\log \sigma(\hat{y}_{uv} - \hat{y}_{uj}). \tag{13}$$

where D encompasses the dataset of all user-item pairs, u denotes a user, v represents an item with which the user has interacted, and j symbolizes an item with which the user has not interacted. The sigmoid function σ maps the score differences into a probability space, indicating the likelihood of a student benefiting more from one educational resource over another, based on their interaction history and learning outcomes. In the culmination of our training process tailored for educational environments, we define the total loss function as the sum of three distinct components-each reflecting a different aspect of learning dynamics-thus ensuring a well-rounded approach to personalized educational recommendations. This total loss function is expressed as:

$$\mathcal{L} = \mathcal{L}_{rec} + \lambda_1 \mathcal{L}_m + \lambda_2 \mathcal{L}_c. \tag{14}$$

here, the parameters λ_1 and λ_2 are regularization coefficients that control the relative importance of the self-supervised loss L_m and the contrastive loss L_c in the overall optimization process. These coefficients are crucial for balancing the influence of each loss component on the model's training, allowing for fine-tuning the model's focus on unsupervised learning cues or the alignment of embeddings across different graph representations.

4   Evaluation

4.1   Experimental Setup

4.1.1 Datasets. To comprehensively evaluate the efficacy of our proposed method, we employed the public benchmark dataset: MOOCCube.MOOCCube is an open data repository tailored for researchers in the fields of natural language processing, knowledge graphs, data mining, and related areas, focusing on massive open online education [15]. The repository comprises course data and student behavioral data sourced from the real usage environment of XuetangX. To ensure the quality of the dataset, we employ a 10-core setting, which means retaining only those users and items with at least ten interactions. In our experimental setup, we randomly partitioned the dataset into distinct subsets: 70% for training, 10% for validation, and 20% for testing. The specific statistics of the dataset are illustrated in Table 1:

Table 1. Statistics of MOOCCube Dataset.

Statistics	MOOCCube
# Users	143209
# Items	38181
# Interactions	5,969,162
Knowledge Graph	
# Entities	153,474
# Relations	6
# Triplets	764,541

4.1.2 Evaluation Protocols. To ensure a fair comparison of our experimental results, we employed widely used metrics in recommendation systems, specifically Recall@ (R@k) and NDCG@k (N@k), to evaluate model performance. The values of k were selected from the set {5, 10, 20}. We reported the average measures across all users in the test set as the final results.

4.1.3 Baselines. To validate the effectiveness of our proposed model, we compared it with four representative types of baseline models, including regularization-based (CKE) [16], matrix factorization-based (BPR) [17] and GCN-based (LightGCN, KGIN) [8,18] models.

- **CKE**: Integrates TransR-encoded item semantics with denoising autoencoders for knowledge-driven representations.

- **BPR**: Utilizes matrix factorization with pairwise ranking loss to handle implicit feedback.
- **LightGCN**: Simplifies graph convolution by removing non-linearities to directly enhance user and item embeddings.
- **KGIN**: Employs heterogeneous propagation to merge collaborative filtering with knowledge graph embeddings.

4.2   Performance Evaluation

We summarize all the experimental results as shown in Table 2:

Table 2. Performance Comparison on MOOCCube Dataset

Model	Recall			NDCG		
	R@5	R@10	R@20	N@5	N@10	N@20
BPR	0.1392	0.2643	0.3801	0.2748	0.2976	0.3413
CKE	0.1758	0.2941	0.4150	0.3002	0.3247	03652
LightGCN	0.1536	0.2833	0.4037	0.2905	0.3125	0.3543
KGIN	0.1825	0.3089	0.4313	0.3257	0.3450	0.3854
KASER	**0.1963**	**0.3237**	**0.4520**	**0.3387**	**0.3588**	**0.4008**

The BPR model demonstrated the lowest performance, indicating its limitations in handling complex user-item interactions without contextual information. LightGCN surpassed BPR, illustrating the advantages of directly utilizing graph structures in learning user and item embeddings. However, the CKE model, which integrates contextual knowledge, showed superior outcomes, confirming that incorporating additional data sources from a knowledge graph can enhance recommendation quality. KGIN, leveraging the rich semantic relations within the knowledge graph, performed better than the previous models, highlighting the significance of semantic understanding in recommendations. Our proposed model, KASERRec, outperformed all baseline models, achieving the highest metrics. This underscores the effectiveness of integrating a knowledge-aware strategy in uncovering valuable yet underrepresented content.

4.3   Ablation Study

We then investigate the validity of the model, and to do so, we construct three model variants. The specific results are presented in Table 3.

The experimental results indicate that removing the masking autoencoder component (w/o mask autoencoder) had the most significant impact on performance. This suggests that masking and reconstruction are effective strategies that can appropriately utilize highly valuable knowledge triples for recommendations. The reconstruction based on the relation-weighted masking mechanism

Table 3. Ablation Study Results

Ablation Settings	Recall			NDCG		
	R@5	R@10	R@20	N@5	N@10	N@20
KASER	**0.1963**	**0.3237**	**0.4520**	**0.3387**	**0.3588**	**0.4008**
w/o mask autoencoder	0.1922	0.3204	0.4501	0.3352	0.3561	0.3989
w/o Relation-M	0.1948	0.3015	0.4513	0.3371	0.3575	0.4001
w/o CL	0.1944	0.3008	0.4506	0.3376	0.3569	0.3992

(w/o Relation-M) demonstrates the importance of utilizing relational semantics to enhance recommendation precision. Furthermore, excluding the Contrastive Learning component (w/o CL) also leads to a decrease in overall performance metrics.

4.4 Case Study

In this case study, we selected a random user, $U\_8665386$, from the MOOC-Cube dataset to compare the number of long-tail items recommended by our proposed KASERRec model and a baseline recommendation model. We defined entities in the educational knowledge graph with fewer than 10 exposures as long-tail items. We retrieved the top-20 recommendations for user $U\_8665386$ from each model. The results show that the baseline model included an average of 2 long-tail items in its top-20 recommendations for this user. In contrast, KASER-Rec's top-20 recommendations contained 7 long-tail items. This indicates that KASERRec's knowledge-aware approach effectively highlights valuable yet less visible educational content, directly addressing the long-tail issue.

5 Related Work

Knowledge-Aware Recommendation. Incorporating external knowledge into recommendation systems has emerged as a potent solution to enhance the recommendation quality beyond the traditional collaborative filtering techniques. Knowledge-aware recommendation systems leverage structured knowledge from various sources, such as knowledge graphs, to provide context-rich recommendations [19]. Based on how information from knowledge graphs is represented and utilized, existing methods integrating knowledge graphs into recommendations can be categorized into three types: 1)Path-based recommendation methods extract meta-path information from knowledge graphs according to specific rules and use this information to predict user-item interactions. Zhao et al. proposed a method [20] that integrates multiple meta-paths into a meta-graph representation, effectively capturing user interests; 2)Embedding-based recommendation methods predict user-item interactions by constructing embedding representations of users and items. Zhang et al.'s CKE model [16] is recognized as one of the earliest embedding-based methods that incorporate knowledge graphs, enhancing latent representations with structured knowledge. 3)Network propagation-based recommendation methods focus on learning the weights of edges on the knowledge graph automatically during the recommendation process, using the weights to identify the most likely paths for recommendation explanations. The KGAT [9] model employs a graph attention network on the collaborative knowledge graph to learn embeddings of entities and their semantic relationships. The KGCL [21] and MCCLK [13] models introduce contrastive learning to improve knowledge-aware recommendations by better aligning representations from different views.

6   Conclusion

In this paper, we present KASERRec, a novel recommendation framework integrating knowledge graphs and advanced machine learning techniques, enhances educational resource recommendations by addressing key challenges such as the long-tail and the cold start problem. Specifically, the framework utilizes relation-weighted GNN to intricately model the complex relationships within educational data, employs semantic masking autoencoders to enhance the representation of long-tail educational content, and integrates cross-view knowledge contrastive learning to align and refine user and item embeddings. The experimental results on the MOOCCube dataset have demonstrated the superiority of our proposed model. Future work will focus on dynamically updating the knowledge graph and refining educational outcomes to further optimize personalized learning paths.

Acknowledgement. Research was sponsored in part by National Natural Science Foundation of China (62137001, 62272093) and in part by Fundamental Research Funds for the Central Universities (N2116007) and Project of the Association of Fundamental Computing Education in Chinese Universities (2023-AFCEC-184).

References

1. Hill, J.R., Hannafin, M.J.: Teaching and learning in digital environments: the resurgence of resource-based learning. ETR&D. **49**, 37–52 (2001). https://doi.org/10.1007/bf02504914
2. Urdaneta-Ponte, M.C., Mendez-Zorrilla, A., Oleagordia-Ruiz, I.: Recommendation systems for education: systematic review. Electronics **10**, 1611 (2021)
3. He, X., Liao, L., Zhang, H., Nie, L., Hu, X., Chua, T.-S.: Neural collaborative filtering. In: Proceedings of the 26th International Conference on World Wide Web, pp. 173–182 (2017)
4. Abu-Salih, B., Alotaibi, S.: A systematic literature review of knowledge graph construction and application in education. Heliyon **10**, e25383 (2024)
5. Bai, J., Nie, T., Shen, D., Kou, Y., Yu, G.: Bi-directional neighborhood-aware network for entity alignment in knowledge graphs. In: Zhao, X., Yang, S., Wang, X., Li, J. (eds.) Web Information Systems and Applications, pp. 64–76. Springer, Cham (2022). https://doi.org/10.1007/978-3-031-20309-16

6. Hu, Z., Dong, Y., Wang, K., Sun, Y.: Heterogeneous graph transformer. In: Proceedings of the Web Conference 2020, pp. 2704–2710 (2020)
7. Sun, Z., Deng, Z.-H., Nie, J.-Y., Tang, J.: RotatE: knowledge graph embedding by relational rotation in complex space. arXiv preprint arXiv:1902.10197 (2019)
8. Wang, X., et al.: Learning intents behind interactions with knowledge graph for recommendation. In: Proceedings of the Web Conference 2021, pp. 878–887 (2021)
9. Wang, X., He, X., Cao, Y., Liu, M., Chua, T.-S.: KGAT: knowledge graph attention network for recommendation. In: Proceedings of the 25th ACM SIGKDD International Conference on Knowledge Discovery & Data Mining, pp. 950–958 (2019)
10. Peng, C., Xia, F., Naseriparsa, M., Osborne, F.: Knowledge graphs: opportunities and challenges. Artif. Intell. Rev. **56**, 13071–13102 (2023)
11. Yang, Y., Huang, C., Xia, L., Huang, C.: Knowledge graph self-supervised rationalization for recommendation. In: Proceedings of the 29th ACM SIGKDD Conference on Knowledge Discovery and Data Mining, pp. 3046–3056 (2023)
12. Xia, L., Huang, C., Huang, C., Lin, K., Yu, T., Kao, B.: Automated self-supervised learning for recommendation. In: Proceedings of the ACM Web Conference 2023, pp. 992–1002 (2023)
13. Zou, D., et al.: Multi-level cross-view contrastive learning for knowledge-aware recommender system. In: Proceedings of the 45th International ACM SIGIR Conference on Research and Development in Information Retrieval, pp. 1358–1368 (2022)
14. Wang, K., Liu, Y., Sheng, Q.Z.: Swift and sure: hardness-aware contrastive learning for low-dimensional knowledge graph embeddings. In: Proceedings of the ACM Web Conference 2022, pp. 838–849 (2022)
15. Yu, J., et al.: MOOCCube: a large-scale data repository for NLP applications in MOOCs. In: Proceedings of the 58th Annual Meeting of the Association for Computational Linguistics, pp. 3135–3142. Association for Computational Linguistics, Online (2020)
16. Zhang, F., Yuan, N.J., Lian, D., Xie, X., Ma, W.-Y.: Collaborative knowledge base embedding for recommender systems. In: Proceedings of the 22nd ACM SIGKDD International Conference on Knowledge Discovery and Data Mining, pp. 353–362 (2016)
17. Rendle, S., Freudenthaler, C., Gantner, Z., Schmidt-Thieme, L.: BPR: bayesian personalized ranking from implicit feedback. arXiv preprint arXiv:1205.2618 (2009)
18. He, X., Deng, K., Wang, X., Li, Y., Zhang, Y., Wang, M.: LightGCN: simplifying and powering graph convolution network for recommendation. In: Proceedings of the 43rd International ACM SIGIR Conference on Research and Development in Information Retrieval, pp. 639–648 (2020)
19. Guo, Q., et al.: A survey on knowledge graph-based recommender systems. IEEE Trans. Knowl. Data Eng. **34**, 3549–3568 (2022)
20. Zhao, H., Yao, Q., Li, J., Song, Y., Lee, D.L.: Meta-graph based recommendation fusion over heterogeneous information networks. In: Proceedings of the 23rd ACM SIGKDD International Conference on Knowledge Discovery and Data Mining, pp. 635–644 (2017)
21. Yang, Y., Huang, C., Xia, L., Li, C.: Knowledge graph contrastive learning for recommendation. In: Proceedings of the 45th International ACM SIGIR Conference on Research and Development in Information Retrieval, pp. 1434–1443 (2022)

JEAPC: A Joint Extraction Model of Action Sequence from Chinese Instructions for Home Service Robot

Bin Wang[4], Haoyu Wang[1], Xianshan Li[1,3] ⓘ, and Fenda Zhao[1,2,3](✉) ⓘ

[1] School of Information Science and Engineering, Yanshan University,
Qinhuangdao 066004, China
zfd@ysu.edu.cn
[2] School of Information Science and Engineering, Xinjiang University of Science
and Technology, Korla 841000, China
[3] Key Laboratory for Software Engineering of Hebei Province, Yanshan University,
Qinhuangdao 066004, China
[4] Hebei Telecom Co., Ltd., Shijiazhuang 050035, China

Abstract. The language interaction between humans and robots is one of the critical issues in the field of home service robots. In particular, irrelevant information in feature vectors interferes with the extraction task during Chinese instruction parsing. Moreover, the relations between feature vectors of different time steps affect the accuracy of action sequence extraction. In this paper, overlapping action entities in Chinese instructions are labeled through span-based mode, and a Joint Extraction Model of Action Sequences with Partition Coding(JEAPC) is proposed for Chinese instructions. The JEAPC is divided into four modules: BERT, partition encoder, and two decoders. BERT is utilized to obtain the feature vector of each Chinese character in the instructions. The partition encoder is composed of an entity gate, and an action gate, in which the features in the vector are classified, and the irrelevant features are filtered through multi-dimensional vector operations. Furthermore, adversarial training is employed to improve the robustness of JEAPC. Extensive experiments are conducted on a self-built Chinese instructions dataset(FCI) and three entity and relation extraction datasets (CoNLL04, ADE, and SciERC). The experimental results show that the JEAPC can accurately generate action sequences from Chinese instructions and obtain optimal results compared to several competitive approaches.

Keywords: Human-robot interaction · Home service robots · Chinese instruction parsing · Adversarial training

1 Introduction

Home service robots do not have the ability to interact naturally with humans through language, and using natural language to give instructions to robots

C. Jin et al. (Eds.): WISA 2024, LNCS 14883, pp. 536–548, 2024.
https://doi.org/10.1007/978-981-97-7707-5_44

remains a challenging problem [1, 2]. Therefore, it is a critical problem in human-robot interaction that enables robots to understand human instructions correctly. The goal of instruction parsing for home service robots is extracting useful information from natural language instructions and converting the corresponding service tasks into robot action sequences.

The current instruction parsing method combines deep learning to establish a mapping relation between instructions and action sequences of home service robots [3]. The sequence labeling method only assigns one label to each word in the instructions. If there is an action entity corresponding to multiple action relations, the sequence labeling method is not capable of completely labeling every action relation [4]. Previous methods classified entities and relations by sharing features directly [5], which is unreasonable since the information about entities and relations may be conflicting. Chinese instructions have no special inter-word delimiters and no specific fixed grammar to follow [6]. Thus, the pre-trained language model is not able to guarantee the robustness of the joint extraction model.

In this study, a joint extraction model of action sequences with partition coding is proposed for Chinese instructions. Span-based mode labels entities and relations of self-built Chinese instructions dataset FCI(Follow Chinese Instructions). The corresponding fixed-dimensional feature vector of each Chinese character is obtained by BERT. Employing the partition encoder, which contains an entity gate and an action gate, the features are divided into three categories: features that are only useful for classifying action entities, features that are only useful for classifying action relations, and features that are useful for both classification tasks. Hence, action sequences are generated by the action entity classifier and the action relation classifier. Furthermore, adversarial training is utilized to perturb the model and improve its robustness.

In summary, this study's contributions are three-fold: 1) A joint extraction model of action sequences with partition coding for Chinese instructions is suggested, which labels overlapping action entities using span-based mode to parse Chinese instructions containing complex action relations effectively; 2) During model training, the robustness of the proposed model is improved by adding perturbations to the encoded feature vectors. The experimental results on CoNLL04, ADE, and SciERC show that after introducing adversarial training, F1 is improved by 2.75%, 1.37%, and 0.96%, respectively; 3) The suggested model obtains accurate action sequence extraction results on FCI, and the F1 value achieves 97.89%. On the CoNLL04, ADE, and SciERC datasets, the proposed model achieves distinguished results with F1 values of respectively 72.82%, 80.56%, and 48.41% on relation extraction compared to other competitive models. Our self-build dataset FCI will be released.

2   Related Work

Early methods for extracting critical information from natural language instructions mainly used dependency parsing and rule-based methods [7]. The overall intention and core information of a task in a natural language instruction is often contained in various vital verbs. Based on the dependency path of verbs, a critical verb extraction method based on dependency analysis is proposed by Zhang et al. [8]; this method constructed features combining key verbs with their context information and achieved a satisfactory parsing result in English instructions. Mensio et al. [9] used an external knowledge base to enhance the understanding ability of the robot's instructions. Zhao et al. The study by Sharma et al. [10] allowed robots to learn to sequence actions in natural language instructions. Chen et al. [3] enabled a robot to automatically fill in information missing from the instruction using a commonsense reasoning approach. However, all the works mentioned above were parsing English instructions. On the other hand, extracting critical information from Chinese texts is more challenging than from English ones [11]. Hence, Chinese text instructions in the home environment are more colloquial and contain richer parts of speech, so the corresponding grammatical structure is more complex, which increases the difficulty of extracting information from Chinese instructions. The instruction above parsing method for English instructions is not capable of achieving satisfactory performance in parsing complex Chinese instructions.

To reduce the restrictions on Chinese instructions and enable home service robots to better understand the intentions of users, deep learning method was used to establish the mapping relation between instructions and the robot action sequence. Whether it is to encode the specific features to the extraction task in order so that the features do not affect each other [5], or to encode the common features derived by the encoder through independent sub-modules to encode the specific features to the extraction task in [12], they both lack in terms of simulating the effects between extraction tasks. In previous studies [13], inter-task influence was simulated by sharing the same features.

As a pre-trained language model, BERT [15] described character-level, word-level, and sentential features in combination with the context of the text. In this study, BERT is used as the encoder to obtain the feature vector corresponding to each Chinese character in the instruction. Sequence labeling mode can only assign one single label to each word in the instructions. Subsequently, it is not capable of handling the case of overlapping entities, while the span-based mode is able to search the whole statement and cover overlapping entities [16]. Recently, span-based models have been used to jointly extract named entities and relations in [17]. [18] implements Chinese nested named entity recognition based on boundary prompt. In order to lable the complex relations between entities in the instructions as much as possible, span-based mode is employed to lable FCI.

3   The Proposed Model

In this section, the joint extraction model of action sequences with partition coding for Chinese instructions is introduced in detail, as illustrated in Fig. 1. First, each word in the Chinese text instructions is encoded using BERT to obtain a vector representation of fixed dimensions. Next, a partition encoder

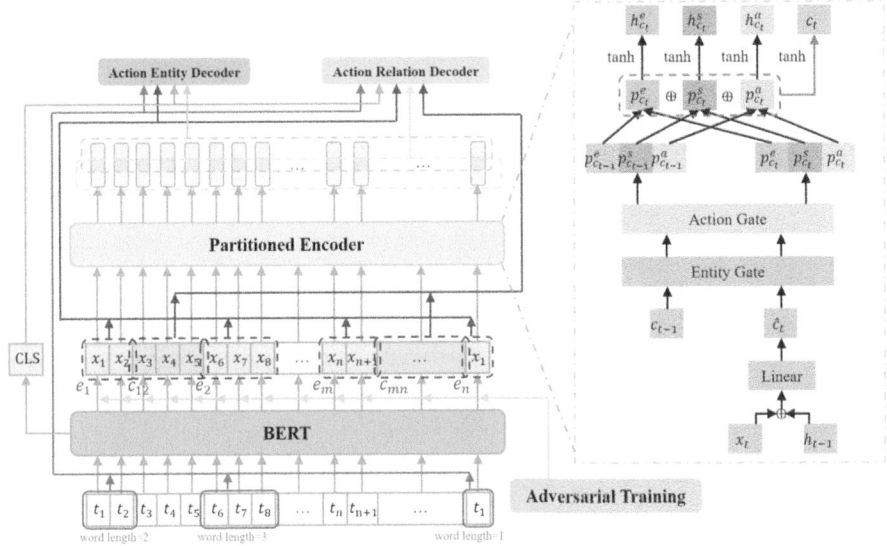

Fig. 1. Overall flowchart of JEAPC.

is utilized to model task interaction and acquire task-specific feature vectors. Finally, a linear classifier and the softmax activation function are applied to obtain action entities, and a linear classifier and the sigmoid activation function are used to obtain action relations. In the process of model training, adversarial training is introduced to improve the robustness of the model.

3.1   Partition Encoder

The partition encoder is a recurrent feature encoder that partitions the feature vectors encoded by BERT. At each time step, every neuron is divided into an entity partition, an action partition, and a shared partition. Afterward, the three partitions are combined according to different task characteristics to generate task-specific features. The functions for the segmented candidate neuron $\hat{c}_t$ are as follows:

$$
\begin{aligned}
h_{t-1} &= \tanh(c_{t-1}) \\
\hat{c}_t &= \tanh(\text{Linear}(x_t \circ h_{t-1}))
\end{aligned}
\tag{1}
$$

where c_{t-1} is the neuron that stores the information of the previous time step, h_{t-1} represents the hidden state of c_{t-1}, x_t represents the input of the partition encoder at the current time step, and $\circ$ denotes concatenation.

The cummax activation function is used by the entity gate e_{c_t} and the action gate a_{c_t} to determine two demarcation points, which are able to divide the neuron into three parts, as illustrated in (2):

$$e_{c_t} = \text{cummax}(\text{Linear}(x_t \circ h_{t-1}))$$
$$a_{c_t} = 1 - \text{cummax}(\text{Linear}(x_t \circ h_{t-1})) \tag{2}$$

The three partitions of the neuron c_{t-1} from the previous time step $p^s_{c_{t-1}}$, $p^e_{c_{t-1}}$ and $p^a_{c_{t-1}}$ are calculated as (3):

$$p^s_{c_{t-1}} = e_{c_{t-1}} \cdot a_{c_{t-1}}$$
$$p^e_{c_{t-1}} = e_{c_{t-1}} - p^s_{c_{t-1}} \tag{3}$$
$$p^a_{c_{t-1}} = a_{c_{t-1}} - p^s_{c_{t-1}}$$

$\hat{c}_t$'s three partitions $p^s_{\hat{c}_t}$, $p^e_{\hat{c}_t}$ and $p^a_{\hat{c}_t}$ are calculated as (4):

$$p^s_{\hat{c}_t} = e_{\hat{c}_t} \cdot a_{\hat{c}_t}$$
$$p^e_{\hat{c}_t} = e_{\hat{c}_t} - p^s_{\hat{c}_t} \tag{4}$$
$$p^a_{\hat{c}_t} = a_{\hat{c}_t} - p^s_{\hat{c}_t}$$

Combined with the partition information of c_{t-1} and $\hat{c}_t$, the entity partition $p^e_{c_t}$, action partition $p^a_{c_t}$ and shared partition $p^s_{c_t}$ of c_t are calculated as follows:

$$p^e_{c_t} = p^e_{c_{t-1}} \cdot c_{t-1} + p^e_{\hat{c}_t} \cdot \hat{c}_t$$
$$p^a_{c_t} = p^a_{c_{t-1}} \cdot c_{t-1} + p^a_{\hat{c}_t} \cdot \hat{c}_t \tag{5}$$
$$p^s_{c_t} = p^s_{c_{t-1}} \cdot c_{t-1} + p^s_{\hat{c}_t} \cdot \hat{c}_t$$

Input the information in the three partitions into the activation function to obtain entity feature h_e, action feature h_a and shared feature h_s:

$$h^e_{c_t} = \tanh(p^e_{c_t} + p^s_{c_t})$$
$$h^a_{c_t} = \tanh(p^a_{c_t} + p^s_{c_t}) \tag{6}$$
$$h^s_{c_t} = \tanh(p^s_{c_t})$$

c_t is calculated as follows:

$$c_t = \text{Linear}(p^e_{c_t} \circ p^s_{c_t} \circ p^a_{c_t}) \tag{7}$$

where c_t is the neuron that stores the information of the current time step, $\circ$ denotes concatenation.

3.2 Action Entity Decoder

Note that each action parameter in the action sequence is an entity. $E = \{E_1, E_2, \ldots, E_{len}\}$ represents the set of all candidate entity samples, where E_i denotes the set of all entities with length i and len is the maximum entity length. The input of the action entity classifier consists of the following features:

- The span of each candidate entity store the start and end locations of that candidate entity in a sentence. Perform a max pooling over the span of each entity to get $f(E)$.

- The word length of each entity in the current candidate pair w_i, where $i = 1, 2, \ldots, len$.
- BERT generates a CLS token as a semantic representation of the entire instruction. Get the CLS vector c without max pooling.
- The entity feature h_e and shared feature h_s generated in the partition encoder.

Concatenate the above-mentioned features as the input of the entity classifier to obtain the final entity classification y^{Entity}, which is made up of a linear classifier and the Softmax activation function:

$$y^{Entity} = \text{softmax}(\text{Linear}(f(E) \circ w_i \circ c \circ h_e \circ h_s)) \qquad (8)$$

3.3   Action Relation Decoder

Among all candidate entity samples E, an action relation may exist between every two entity samples. The input of the action relation classifier consists of the following features:

- Each of the two entity samples act as a candidate pair for (e_i, e_j). Span of each entity on the candidate pair obtained by max pooling $f(E)$.
- The word length of each entity in the current candidate pair w_i, where $i = 1, 2, \ldots, len$.
- The text between each candidate pair serves as the context feature to classify action relation.
- The CLS vector c generated by BERT without max pooling.
- The action feature h_a and shared feature h_s generated in the partition encoder.

All of the aforementioned features are concatenated as the input of the action classifier, containing a linear classifier and the Sigmoid function. The final action classification y^{Action} is:

$$y^{Action} = \sigma(\text{Linear}(f(E) \circ w_i \circ c(e_i, e_j) \circ c \circ h_a \circ h_s)) \qquad (9)$$

Given a confidence threshold α, any action related to a score superior to α is considered valid. After the parsed action is obtained, the final action sequence is acquired by combining the entities parsed in the previous step.

3.4   Adversarial Training

During adversarial training, small perturbations are added to the samples. To cope with disturbances, the robustness of the proposed model is improved, so its ability to parse Chinese instructions is ameliorated. The idea of the FGM (Fast Gradient Method) [22] method is to add disturbance δ along the direction of gradient boosting and use L_2 normalization. FGM makes two gradient updates at each time step. Firstly, the disturbance that maximizes the loss function is

Table 1. Main results on CONNL04, ADE and SCIERC datasets.

Dataset	Model	Entity			Relation		
		P	R	F1	P	R	F1
CoNLL04	Multi-turn QA [4]	89.00	86.60	87.80	69.20	68.20	68.90
	SpERT [5]	88.25	89.64	88.94	73.04	70.00	71.47
	HfGCN [19]	88.40	89.51	88.95	74.87	69.91	72.30
	JEAPC (ours)	89.19	88.69	88.94	78.05	68.25	**72.82**
ADE	Relation-Metric [20]	86.16	88.08	87.11	77.36	77.25	77.29
	SpERT	88.69	89.20	88.95	77.77	79.96	78.84
	PFN [14]	-	-	89.6	-	-	80.00
	HfGCN	88.55	90.86	89.69	78.54	82.43	80.44
	JEAPC (ours)	88.74	91.00	89.85	78.16	83.12	**80.56**
SciERC	DyGIE [21]	-	-	65.20	-	-	41.60
	SpERT	68.53	66.73	67.62	49.79	43.53	46.44
	JEAPC (ours)	65.76	67.36	66.55	50.00	46.92	**48.41**

obtained through gradient ascent. Secondly, the model parameter that minimizes the loss function is obtained through gradient descent. The perturbation calculation formula using FGM algorithm is as follows:

$$r_{adv} = \frac{\epsilon \cdot g}{\|g\|_2} \tag{10}$$

4 Experiments

4.1 Experiments on Public Datasets

We compare JEAPC with other joint entity/relation extraction models on three publicly available English datasets CoNLL04 [23], ADE [24] and SciERC [25]. CoNLL04 was extracted from news articles. ADE is extracted from medical reports. SciERC was generated from 500 abstracts of AI papers.

The experimental results of the previous models and JEAPC on the English datasets mentioned above are shown in Table 1. On CoNLL04, the proposed model obtained 72.82 F1. Compared with Multi-trun QA, SpERT and HfGCN, our model improves by 3.92%, 1.35% and 0.52%, respectively. On the ADE dataset, JEAPC obtains 80.56 F1. Compared to the results in Relation-Metric, SpERT, PFN, and HfGCN, our model achieves 3.27%, 1.72%, 0.56% and 0.12% improve- ment in F1, respectively. On the SciERC dataset, JEAPC achieved 48.41 F1. Compared with DyGIE and SpERT, the model improves by 6.81% and 1.97% on F1.

Table 2. Influence of the partition encoder and adversarial training on CONNL04, ADE and SCIERC datasets.

Dataset	Partition Encoder	Adversarial Training	Entity			Relation		
			P	R	F1	P	R	F1
CONNL04	✗	✗	85.54	88.79	87.13	67.36	69.43	68.38
	✓	✗	84.76	88.14	86.42	70.73	68.72	69.71
	✗	✓	86.87	88.88	87.86	77.97	65.40	71.13
ADE	✗	✗	87.39	90.37	88.85	74.91	81.89	78.22
	✓	✗	87.59	90.33	88.94	75.72	82.05	78.77
	✗	✓	87.96	90.69	89.30	76.60	82.83	79.58
SCIERC	✗	✗	66.69	68.66	67.66	48.42	45.48	46.90
	✓	✗	64.72	66.94	65.81	48.70	46.30	47.47
	✗	✓	67.00	67.83	67.41	50.63	45.38	47.86

4.2   Ablation Study

In order to verify the influence of the partition encoder module and adversarial training module on experimental performance, comparative experiments are conducted on CoNLL04, ADE and SciERC. The experimental results in Table 2 show that the action classification results on CoNLL04, ADE and SciERC are respectively improved by 1.33%, 0.55% and 0.57% when using the partition encoder. After using adversarial training, the relational classification results of CoNLL04, ADE and SciERC are improved by 2.75%, 1.36% and 0.96%, respectively. It is proved that the introduction of adversarial training can greatly improve the ability of the model to correctly extract key information.

4.3   Experiments on FCI

Data Processing. In this study, oral instructions in daily life are collected from the family environment. The categories of entity in these instructions are divided into five categories: "物品(Item)", "人(Person)", " 机器人(Robot)", "智能设备(Smart Device)" and "地点(Place) ". The categories of action relations between entities are divided into eight categories: "抓取(Grasp)", "放置(Put)", "移动(Move)", "打开(Open)", "关闭(Close)", "执行程序(Execute Program)", "优先执行(Priority)" and "其他(Other)".

Each spoken instruction is annotated according to the defined categories of entities and actions, indicating the starting and ending position of each entity in the sentence. The head and tail entities of each action are also indicated. Table 3 describes the way of using span-base mode to annotate the Chinese instruction "汤姆, 从餐桌拿个苹果(Take an apple from the table, Tom)".

Table 3. An example of chinese instruction annotation.

"tokens": 汤姆从餐桌拿个苹果 (Tom, take an apple from the table.)
"entities": ["type": "机器人(Robot)", "start": 0, "end": 2, "type": "物品(Item)", "start": 3, "end": 5, "type": "物品(Item)", "start": 7, "end": 9]
"actions": ["type": "移动(Move)", "head": 0, "tail": 1, "type": "抓取(Grab)", "head": 1, "tail": 2]

Table 4. Results on FCI.

Model	F1
SpERT	97.54
JEAPC (ours)	97.89

Evaluation Criteria. The loss function $\mathcal{L}^{Entity}$ for evaluating entity types uses cross-entropy is as follows:

$$\mathcal{L}^{Entity} = -\sum_{j=1}^{m} \sum_{i=1}^{n_{Entity}} \left(\hat{y}_j^{i,Entity} \log(y_j^{i,Entity}) + (1 - \hat{y}_j^{i,Entity}) \log(1 - y_j^{i,Entity}) \right) \tag{11}$$

where n_{Entity} is the number of entity samples in the current instruction, and $\hat{y}_j^{i,Entity}$ is the correct entity type label. The loss function $\mathcal{L}^{Action}$ for evaluating action types uses bivariate cross-entropy:

$$\mathcal{L}^{Action} = -\sum_{j=1}^{m} \sum_{i=1}^{n_{Action}} \left(\hat{y}_j^{i,Action} \log(y_j^{i,Action}) + (1 - \hat{y}_j^{i,Action}) \log(1 - y_j^{i,Action}) \right) \tag{12}$$

where n_{Action} is the number of actions that exist between entities in the current instruction sequence, and $\hat{y}_j^{i,Action}$ is the correct action type label. Joint loss function:

$$\mathcal{L} = \mathcal{L}^{Entity} + \mathcal{L}^{Action} \tag{13}$$

Extraction Results on FCI. The comparison results of SpERT and JEAPC on FCI are shown in Table 4. Compared with SpERT, JEAPC improves F1 by 0.35%. For home service robots, there is a special phenomenon in which the same verb corresponds to different execution results in Chinese instructions. Using span-based mode to label the overlapping action entities in the Chinese instructions. Taking "关电脑"(turn off the computer) and "关门"(close the door) as examples, the action in "关电脑"(turn off the computer) corresponds to the execution of a set of programs in the robot system, while the action in "关门"(close the door) requires the robot to move in the scene. Based on the span-based mode, similar situations are distinguished to extract the action sequence that is more suitable for the actual application scenario.

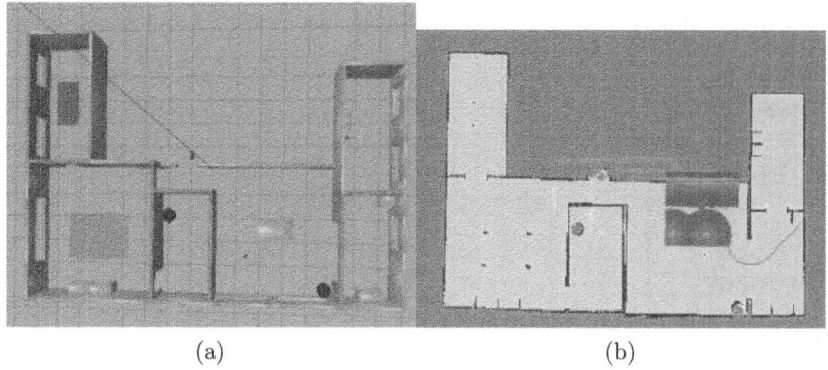

(a)   (b)

Fig. 2. (a) Simulation of a home environment. (b) A moment in the navigation.

4.4 Simulation Experiment

In this paper, Gazebo simulation software is used to build a common home simulation environment, and TurtleBot is used to simulate the robot's execution process of the user's instructions. The environment includes a living room, kitchen, storage, toilet, and bedroom, as shown in Fig. 2(a).

We use gmapping algorithm to draw 2D raster map according to mobile robot mileage data and lidar data, and use Rviz visualization software to view the map. After starting the voice recognition node, the user can post the command "汤姆去箱子里拿个苹果", and the recognition result of the voice command can be viewed in the iat_publish. The robot parses the Chinese instruction, judges the target position that needs to be moved at present, and generates the text that feeds back to the user. After obtaining the target position coordinates, the robot avoids obstacles on the way and navigates to the box's position, as shown in Fig. 2(b).

5 Conclusions

In this paper, a joint extraction model of action sequences with partition coding is proposed for Chinese instructions. First, a partitioned encoder is employed to simulate the influence between the action relation extraction task and the action entity extraction task. Due to the partition encoder, the feature vectors conducive to extracting action sequences from Chinese instructions are obtained. Second, adversarial training is utilized to improve the robustness of the proposed model. Finally, the experimental results on FCI show that JEAPC accurately extracts the robot action sequence from the Chinese instructions. JEAPC is evaluated on three public datasets, including CoNLL04, ADE, and SciERC, to verify its generalization. The results show that the suggested model performs better than previous baseline ones regarding named entity recognition and relation classification.

Acknowledgement. This work was supported in part by the Natural Science Foundation of Xinjiang Uygur Autonomous Region, China Grant No. 2022D01A59, National Natural Science Foundation of China under Grand No. U20A20167, Key Research Foundation of Integration of Industry and Education and the Development of New Business Studies Research Center, Innovation Capability Improvement Plan Project of Hebei Province under Grand No. 22567637H, Hebei Province Central Leading Local Science and Technology Development Project, 246Z1817G.

References

1. Martins, P.H., Custódio, L., Ventura, R.: A deep learning approach for understanding natural language commands for mobile service robots. arXiv preprint arXiv:1807.03053 (2018)
2. Ishikawa, S., Sugiura, K.: Target-dependent UNITER: a transformer-based multimodal language comprehension model for domestic service robots. IEEE Robot. Autom. Lett. **6**(4), 8401–8408 (2021)
3. Chen, H., Tan, H., Kuntz, A., Bansal, M., Alterovitz, R.: Enabling robots to understand incomplete natural language instructions using commonsense reasoning. In: 2020 IEEE International Conference on Robotics and Automation (ICRA), Paris, France, pp. 1963–1969. IEEE (2020)
4. Li, X., et al.: Entity-relation extraction as multi-turn question answering. In: Proceedings of the 57th Annual Meeting of the Association for Computational Linguistics, Florence, Italy, pp. 1340–1350. Association for Computational Linguistics (2019)
5. Eberts, M., Ulges, A.: Span-based joint entity and relation extraction with transformer pre-training. In: Proceedings of the 24th European Conference on Artificial Intelligence (ECAI) - Including 10th Conference on Prestigious Applications of Artificial Intelligence (PAIS), vol. 325, pp. 2006–2013. Santiago de Compostela, Spain (2020)
6. Zhao, S., Cai, Z., Chen, H., Wang, Y., Liu, F., Liu, A.: Adversarial training based lattice LSTM for Chinese clinical named entity recognition. J. Biomed. Inform. **99**, 103290 (2019)
7. Misra, D.K., Sung, J., Lee, K., Saxena, A.: Tell me dave: context-sensitive grounding of natural language to manipulation instructions. Int. J. Robot. Res. **35**(1–3), 281–300 (2016)
8. Zhang, S., Jiang, J., He, Z., Zhao, X., Fang, J.: A novel slot-gated model combined with a key verb context feature for task request understanding by service robots. IEEE Access **7**, 105937–105947 (2019)
9. Mensio, M., Bastianelli, E., Tiddi, I., Rizzo, G.: Mitigating bias in deep nets with knowledge bases: the case of natural language understanding for robots. In: Proceedings of the AAAI 2020 Spring Symposium on Combining Machine Learning and Knowledge Engineering in Practice (AAAI-MAKE), Palo Alto, CA, USA, vol. 2600, p. 20 (2020)
10. Sharma, S., Gupta, J., Tuli, S., Paul, R.: Goalnet: inferring conjunctive goal predicates from human plan demonstrations for robot instruction following. arXiv preprint arXiv:2205.07081 (2022)
11. Zhao, S., Hu, M., Cai, Z., Chen, H., Liu, F.: Dynamic modeling cross-and self-lattice attention network for Chinese NER. In: Proceedings of the AAAI Conference on Artificial Intelligence, vol. 35, no. 16, pp. 14515–14523. AAAI Press (2021)

12. Wei, Z., Su, J., Wang, Y., Tian, Y., Chang, Y.: A novel cascade binary tagging framework for relational triple extraction. In: Proceedings of the 58th Annual Meeting of the Association for Computational Linguistics, pp. 1476–1488. Association for Computational Linguistics, Online (2020)

13. Wang, Y., Yu, B., Zhang, Y., Liu, T., Zhu, H., Sun, L.: Tplinker: single-stage joint extraction of entities and relations through token pair linking. In: Proceedings of the 28th International Conference on Computational Linguistics (COLING), Barcelona, Spain (Online), pp. 1572–1582. International Committee on Computational Linguistics (2020)

14. Yan, Z., Zhang, C., Fu, J., Zhang, Q., Wei, Z.: A partition filter network for joint entity and relation extraction. In: Proceedings of the 2021 Conference on Empirical Methods in Natural Language Processing (EMNLP), Online and Punta Cana, Dominican Republic, pp. 185–197. Association for Computational Linguistics (2021)

15. Devlin, J., Chang, M., Lee, K., Toutanova, K.: BERT: pre-training of deep bidirectional transformers for language understanding. In: Proceedings of the 2019 Conference of the North American Chapter of the Association for Computational Linguistics: Human Language Technologies (NAACL-HLT 2019), Volume 1 (Long and Short Papers), Minneapolis, MN, USA, pp. 4171–4186. Association for Computational Linguistics (2019)

16. Joshi, M., Chen, D., Liu, Y., Weld, D.S., Zettlemoyer, L., Levy, O.: Spanbert: improving pre-training by representing and predicting spans. Trans. Assoc. Comput. Linguist. **8**, 64–77 (2020)

17. Li, Y., Wang, C., Lin, Y., Lin, Y., Chang, L.: Span-based relational graph transformer network for aspect-opinion pair extraction. Knowl. Inf. Syst. **64**(5), 1305–1322 (2022)

18. Li, Z., Song, M., Zhu, Y., Zhang, L.: Chinese nested named entity recognition based on boundary prompt. In: Yuan, L., Yang, S., Li, R., Kanoulas, E., Zhao, X. (eds.) WISA 2023. LNCS, vol. 14094, pp. 331–343. Springer, Singapore (2023). https://doi.org/10.1007/978-981-99-6222-8_28

19. Nong, W., Zhang, T., Yang, S., Hu, N., He, X.: HfGCN: hierarchical fused GCN for joint entity and relation extraction. In: 2021 IEEE International Conference on Big Knowledge (ICBK), Auckland, New Zealand, pp. 307–314. IEEE (2021)

20. Tran, T., Kavuluru, R.: Neural metric learning for fast end-to-end relation extraction. arXiv preprint arXiv:1905.07458 (2019)

21. Luan, Y., Wadden, D., He, L., Shah, A., Ostendorf, M., Hajishirzi, H.: A general framework for information extraction using dynamic span graphs. In: Proceedings of the 2019 Conference of the North American Chapter of the Association for Computational Linguistics: Human Language Technologies, Volume 1 (Long and Short Papers), Minneapolis, Minnesota, pp. 3036–3046. Association for Computational Linguistics (2019)

22. Miyato, T., Dai, A.M., Goodfellow, I.J.: Adversarial training methods for semi-supervised text classification. In: 5th International Conference on Learning Representations (ICLR), Toulon, France. OpenReview.net (2017)

23. Roth, D., Yih, W.: A linear programming formulation for global inference in natural language tasks. In: Proceedings of the Eighth Conference on Computational Natural Language Learning (CoNLL-2004) at HLT-NAACL 2004, Boston, Massachusetts, USA, pp. 1–8. Association for Computational Linguistics (2004)

24. Gurulingappa, H., Rajput, A.M., Roberts, A., Fluck, J., Hofmann-Apitius, M., Toldo, L.: Development of a benchmark corpus to support the automatic extraction of drug-related adverse effects from medical case reports. J. Biomed. Inform. **45**(5), 885–892 (2012)
25. Luan, Y., He, L., Ostendorf, M., Hajishirzi, H.: Multi-task identification of entities, relations, and coreference for scientific knowledge graph construction. In: Proceedings of the 2018 Conference on Empirical Methods in Natural Language Processing (EMNLP), Brussels, Belgium, pp. 3219–3232. Association for Computational Linguistics (2018)

Spatio-Temporal Motion Topology Aware Graph Convolutional Network for Skeleton-Based Action Recognition

Ji Ma, Wei Liu, Linlin Ding, and Hao Luo$^{(\boxtimes)}$

School of Information, Liaoning University, Shenyang 110036, China
{maji,dinglinlin,luohao}@lnu.edu.cn

Abstract. Graph Convolutional Networks (GCNs) have gained significant attention and application in skeleton-based action recognition tasks due to their superior ability to process the intrinsic topological information of skeletons. In GCNs-based methods, extracting discriminative features from skeletal topology is crucial for improving recognition accuracy. However, most works alternately extract spatial and temporal features separately, neglecting the complex spatio-temporal parallel features during motion. To address this issue, a novel Spatio-Temporal Motion Topology Aware Graph Convolutional Network (STMTA-GCN) for learning rich spatio-temporal motion information flow is proposed in our work. The core of this network is the Spatio-Temporal Motion Feature Extraction (STMFE) block that globally considers the learning of temporal, spatial, and spatio-temporal information flow between human joints, which includes two modules: Spatio-Temporal Interaction Enhanced Graph Convolution (STIE-GC) and Multi-scale Temporal Information Extraction (MTIE). STIE-GC can directly analyze the interrelations between joints and their spatio-temporal neighbors in the spatio-temporal graph, and capture channel-level spatio-temporal parallel features. MTIE is used to enhance the model's perception ability of temporal dynamic features. Extensive experiments on three public datasets, NTU RGB+D 60, NTU RGB+D 120, and NW-UCLA, demonstrate that our method achieves better or comparable performance compared to state-of-the-art methods.

Keywords: Skeleton-based action recognition · Graph convolutional networks · Spatio-temporal parallel features

1 Introduction

In recent years, human action recognition has become a hot topic in video understanding tasks due to its wide application in fields such as intelligent surveillance, motion analysis, virtual reality, and so on [1,6,22]. So far, various modalities of data have been explored for human action recognition tasks, including RGB

J. Ma and W. Liu—These authors contribute equally to this work.

© The Author(s), under exclusive license to Springer Nature Singapore Pte Ltd. 2024
C. Jin et al. (Eds.): WISA 2024, LNCS 14883, pp. 549–560, 2024.
https://doi.org/10.1007/978-981-97-7707-5_45

frames, depth, and skeleton data. Compared to other modalities, skeleton data is more practical as it can represent human actions with compact information and inherently adapt to changes in background, lighting, and perspective. Therefore, skeleton-based action recognition has attracted much attention and research. Considering that the human skeleton can naturally be represented as a graph, Graph Convolutional Networks (GCNs) are used to process skeleton data to capture the intrinsic relationships between joints. As shown in Fig. 1(a), most existing GCNs architectures first use 2-D spatial convolution to extract local spatial features between physically connected joints in each frame, and then use 1-D convolution with fixed kernels to extract simple temporal features of the same joint. However, in fact, actions are always represented by the simultaneous movement of joints in time and space. The work in [14] believes that complex spatio-temporal joint-dependent dependencies in human motion patterns have a positive impact on human behavior recognition. In this paper, we introduce a new learning strategy. As shown in Fig. 1(b), based on the spatial and temporal clues, we add a spatio-temporal clue as a supplement to enhance the extraction performance of the spatio-temporal graph. By sequentially extracting temporal, spatial, and spatio-temporal parallel features, we construct the Spatio-Temporal Motion Feature Extraction (STMFE) block. The full architecture is shown in Fig. 2(b). By stacking multiple STMFE blocks, our Spatio-Temporal Motion Topology Aware Graph Convolutional Network (STMTA-GCN) is established. We evaluated our method on several widely used skeleton-based action recognition datasets, namely NTU RGB+D 60, NTU RGB + D 120, and NW-UCLA, to demonstrate the effectiveness of the proposed method. The main contributions of this work include the following aspects:

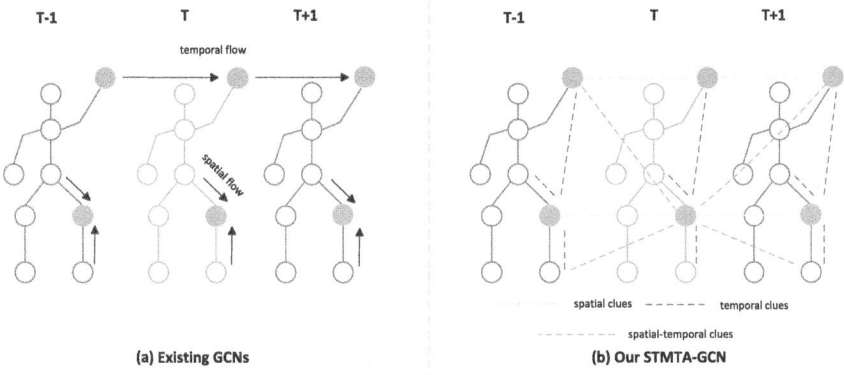

Fig. 1. (a) Illustration of the feature extraction schema used by existing GCNs. (b) Our proposed STMTA-GCN can capture spatial, temporal, and spatio-temporal parallel features through three different clues.

1) We redefine a graph convolution method to extract spatio-temporal parallel features of body nodes during motion from the skeletal graph sequence. Additionally, we introduce a channel-level dynamic awareness strategy for spatial topological information among human body nodes, which together form our STIE-GC module, to enhance the interaction of high-level feature information in both spatial and spatio-temporal dimensions.

2) We propose a MTIE module, which can learn temporal features in a multiscale manner, avoiding the loss of key frames beneficial for action recognition, and combined with frame index semantic information, further providing the ability to capture temporal context features.

3) By integrating STIE-GC and MTIE, a STMTA-GCN with stronger spatiotemporal graph modeling capability for skeleton-based action recognition is obtained. Extensive experimental results have demonstrated the benefits of the proposed method. Our STMTA-GCN performs better than or comparable to state-of-the-art methods on three widely used skeleton-based action recognition.

2   Related Work

2.1   Skeleton-Based Action Recognition

In previous works, GCNs have been applied to skeleton-based action recognition and have attracted widespread attention due to their outstanding performance in this field. Yan et al. [20] proposed ST-GCN, which, for the first time, utilized GCN to model the relationships between joints. Following this work, due to the importance of topology in graph convolution, subsequent researchers have proposed various innovative methods to learn action-related connections between joints. Shi et al. [16] adaptively established joint dependencies based on data-driven approaches as a supplement to priori topology. Li et al. [8] adaptively learned the intrinsic higher-order connection relationships between physically separated joints, thus avoiding the limitations brought by using a fixed human natural connection topology. Ye et al. [21] utilized Dynamic GCN to automatically learn the topological structure combined with global contextual features. Chen et al. [2] proposed CTR-GCN to dynamically learn different topological structures and effectively aggregate joint features from different channels.

2.2   Exploration of Spatio-Temporal Feature

Based on current research on spatio-temporal graph, strategies for learning spatio-temporal graph features can be roughly divided into two categories: (2+1)-D graph convolution and 3-D graph convolution. Specifically, traditional GCNs-based methods extract spatial information and model temporal relationships separately by alternately applying 2-D spatial convolutions and 1-D temporal convolutions. Meanwhile, some concurrent works directly employ 3-D graph convolutions to model spatial and temporal relationships between joints to deeply explore the complex spatio-temporal dependencies. For example, the MS-G3D designed by Liu et al. [14] constructs a large matrix by fusing the adjacency matrices of multiple consecutive frames, capturing cross-frame interaction information between nodes to obtain features that express the flow of spatio-temporal information in the spatio-temporal graph. We believe that the spatio-temporal parallel features and the combined spatial and temporal features each reveal the motion properties of actions from different aspects, and they should be complementary rather than replaceable. Based on this perspective, in our work, we extract spatio-temporal parallel features while preserving the separated spatial and temporal features, employing all three types of features together for action recognition task.

3   Method

3.1   Principles of Skeleton Sequence Processing

Formally, in previous methods [10, 16, 20], the skeleton sequence is constructed as a spatio-temporal graph $G(V, E)$ with N joints and T frames, where $V = \{v_i^t \mid i = 1, 2, \ldots, N; t = 1, 2, \ldots, T\}$ represents the set of all joint nodes in the sequence, and E represents the set of edges connecting these joint pairs. The adjacency matrix $A \in R^{N \times N}$ is then used to represent the strength of connections between nodes. The node feature vector $X \in R^{C \times T \times N}$ is utilized to identify a specific action. In the spatial dimension, the layer-wise update operation of node feature at time t can be formulated as:

$$X_t^{l+1} = \sigma(\ D^{-\frac{1}{2}}\overline{A}D^{-\frac{1}{2}}X_t^l W^l) \tag{1}$$

where X_t^l is the feature of all nodes in the l-th layer at time t, $\overline{A} = A + I$ is the adjacency matrix with added self-loops to maintain identity information, D is the degree matrix of $\overline{A}$, W^l denotes a learnable weight matrix at layer l and $\sigma(\cdot)$ is an activation function.

In the temporal dimension, since only connections between the same joints in consecutive frames need to be considered, $\Gamma \times 1$ classic convolution operation can be directly employed for temporal feature extraction, where Γ is the size of the convolution kernel.

3.2   Spatio-Temporal Motion Topology Aware Graph Convolutional Network

The full architecture of our proposed STMTA-GCN is illustrated in Fig. 2(a). It utilizes seven STMFE blocks to extract temporal, spatial, and spatio-temporal parallel features among joints. Each STMFE block consists of three modules: Spatio-Temporal Interaction Enhanced Graph Convolution (STIE-GC) module,

Multi-scale Temporal Information Extraction (MTIE) module, and basic Spatial Graph Convolutional (S-GC) module. Additionally, to leverage temporal contextual information, we introduce semantic information of frame index. The STIE-GC module and the MTIE module will be described in detail below.

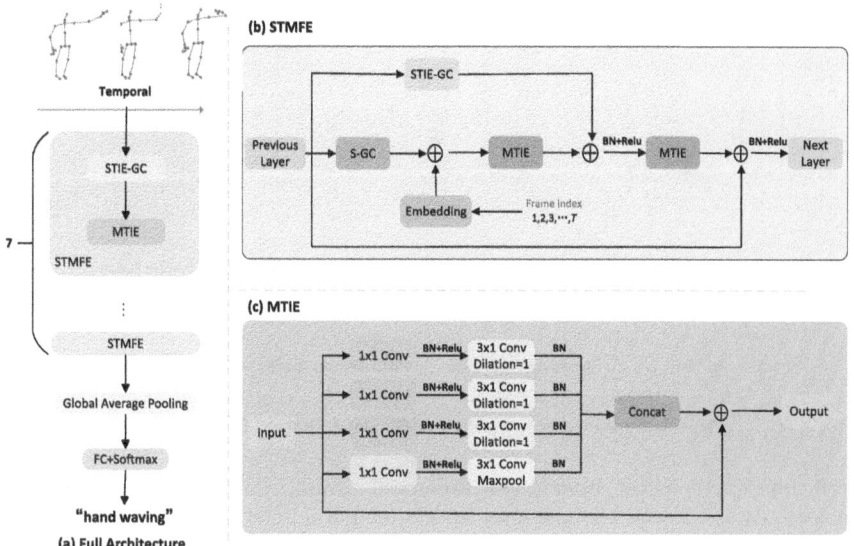

Fig. 2. (a) Illustration of the full architecture of the proposed Spatio-Temporal Motion Topology Aware Graph Convolutional Network (STMTA-GCN). (b) The Spatio-Temporal Motion Feature Extraction (STMFE) block employs two pathways to simultaneously capture channel-level spatio-temporal parallel information and separated spatial and temporal information in the action sequence. (c) The Multi-scale Temporal Information Extraction (MTIE) consists of multiple parallel branches, and the output of each branch is concatenated to obtain the final temporal features.

3.3 Spatio-Temporal Interaction Enhanced Graph Convolution

The framework of the STIE-GC module is shown in Fig. 3. It can be seen that the entire STIE-GC is divided into three parts: (1) Channel-level Topology Aware (yellow area), (2) Spatio-temporal Feature Extraction (blue area) and (3) Spatio-temporal Feature Fusion (orange area). These three parts are described in detail in the following.

Channel-Level Topology Aware. For the input human skeletal joint features $X \in R^{C \times T \times N}$, they are first fed into a 1×1 convolution function ϕ to reduce the channel dimension for computational efficiency. Then, the temporal dimension is compressed via temporal average pooling, and the channel-level topological

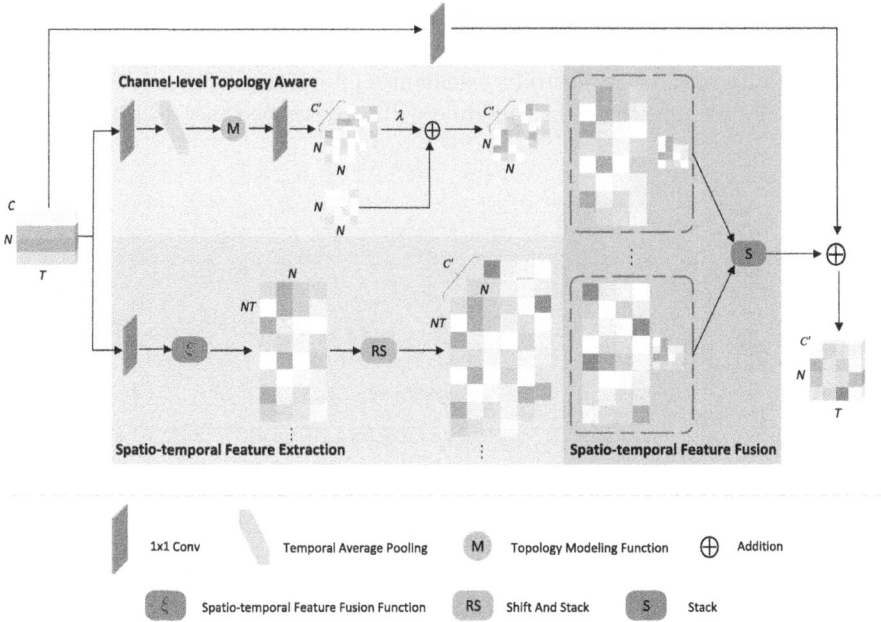

Fig. 3. Spatio-Temporal Interaction Enhanced Graph Convolution: Channel-level Topology Aware learns channel-level spatio-temporal relationships between joints. Spatio-temporal Feature Extraction converts the input features into spatio-temporal interaction features. Spatio-temporal Feature Fusion aggregates information in groups following the principles of skeletal sequence processing and ultimately obtain motion-oriented channel-level spatio-temporal parallel features. (Color figure online)

modeling function $\mathcal{M}(\cdot)$ [2] is employed to model the channel-level relationships between joints, obtaining adjacency matrix groups A' that reflect channel-specific topological relationships. We also utilize a predefined shared topology A reflecting the physical connectivity of the skeleton to refine A' and obtain the final channel-level topological adjacency matrix groups $A^* \in R^{C' \times N \times N}$.

Spatio-Temporal Feature Extraction. The input features $X \in R^{C \times T \times N}$ are first fed into a 1×1 convolution function φ to adjust the channel dimension, during which the number of channels increases from the original C to N. The entire spatio-temporal feature extraction process can be defined as follows:

$$X' = S(\varphi(X)) \tag{2}$$

where X' are the advanced temporal-spatial interaction features. At this moment, the spatial dimension in vector space is expanded from the original N to NT, thus extending from the spatial dimension information of a single frame to the spatio-temporal information of all frames in the entire sequence. Finally, the two-dimensional feature matrix X' is duplicated C' times, shuffled, and then stacked along the channel dimension to obtain the spatio-temporal interaction feature groups $\widetilde{X} \in R^{C' \times NT \times N}$.

Spatio-Temporal Feature Fusion. This part aggregates the spatio-temporal interaction feature groups and the channel-level topological adjacency matrix groups according to Eq. 1 to obtain motion-oriented channel-level spatio-temporal parallel features. Specifically, $\widetilde{X}$ and A^* are first divided into C' groups along the channel dimension, and matrix multiplication is performed within each group to obtain the spatio-temporal interaction features for each channel. Then, the features of each group are aggregated by stacking them together in the channel dimension. Finally, the aggregated advanced features are adjusted to match the dimensions of the original skeleton sequence. The spatio-temporal feature fusion part can be represented as follows:

$$Y = \theta(\psi(\xi((\widetilde{X}_1 A_1^* \,||\, \widetilde{X}_2 A_2^* \,||\, \cdots \,||\, \widetilde{X}_{C'} A_{C'}^*)))) \tag{3}$$

where $Y \in R^{C \times T \times N}$ is the motion-oriented channel-level spatio-temporal parallel features passed into the next layer.

3.4 Multi-scale Temporal Information Extraction

As illustrated in Fig. 2(c), the module consists of four parallel branches. The first three branches capture multi-scale temporal information with different dilation rates, while the fourth branch performs max pooling. To accelerate the training of the network and increase non-linearity, we adopt a BN layer and a ReLU activation layer at the end of each branch. Finally, the temporal features extracted by these four branches are aggregated through a concatenation.

4 Experiments

4.1 Datasets

NTU RGB+D 60. The NTU RGB+D 60 [15] dataset is a large-scale action recognition dataset containing a total of 56,880 skeletal sequences in 60 action categories. Each action is performed by 40 different volunteers and recorded by cameras with three different view angles. There are two standard evaluation benchmarks for this dataset: Cross-Subject (X-Sub) and Cross-View (X-View). In the Cross-Subject setting, 40,320 samples from 20 subjects are used for training, and the rests are used for testing. In the Cross-View setting, 37,920 samples from views 2 and 3 are used for training, while samples from view 1 are used for testing.

NTU RGB+D 120. The NTU RGB+D 120 [12] dataset is an extended version of the NTU RGB+D dataset. The samples were collected from 106 volunteers. The dataset contains 32 setups with varying positions and backgrounds. There are also two standard evaluation benchmarks for this dataset: Cross-Subject (X-Sub), where 106 subjects are divided into training and testing sets, each consisting of 53 subjects; and Cross-Setup (X-Set), where samples are divided into training and testing sets based on on the odd or even setting ID.

NW-UCLA. The NW-UCLA [18] dataset is captured simultaneously by three Kinect cameras from multiple view angles. The dataset covers 10 different action categories with a total of 1494 video clips. Each action performed by 10 different performers. We follow the same evaluation protocol proposed in [18]: data from the first two cameras is used for training, while data captured by the third camera is used for testing.

Table 1. Comparison with state-of-the-art methods on NTU RGB+D 60 and NTU RGB+D 120 datasets. The best value is marked in **bold**, and the second is underlined.

Categories	Methods	NTU-RGB+D 60		NTU-RGB+D 120	
		X-Sub (%)	X-View(%)	X-Sub (%)	X-Set (%)
Non-GCNs	ST-LSTM [13]	69.2	77.7	55.7	57.9
	Ind-RNN [11]	81.8	88.8	-	-
	HCN [9]	86.5	91.1	-	-
	ST-TR [8]	89.9	96.2	85.1	87.1
	Ta-CNN+ [19]	90.7	95.7	85.7	87.3
GCNs	ST-GCN [20]	81.5	88.3	70.7	73.2
	AS-GCN [10]	86.8	94.2	77.9	78.5
	2s-AGCN [16]	88.5	95.1	82.5	84.2
	Shift-GCN [3]	90.7	96.5	85.9	87.6
	Dynamic GCN [21]	91.5	96.0	87.3	88.6
	MS-G3D [14]	91.5	96.2	86.9	88.4
	CTR-GCN [2]	92.4	96.8	88.9	90.6
	InfoGCN [4]	<u>93.0</u>	<u>97.1</u>	**89.8**	**91.2**
Our	STMTA-GCN	**93.4**	**97.3**	<u>89.6</u>	<u>90.8</u>

4.2   Implementation

All experiments were conducted on an RTX 3090 GPU using the PyTorch deep learning framework. The training was conducted using the SGD optimizer with a momentum of 0.9 and weight decay of 0.0004. The training epoch was set to 65. The initial learning rate was set to 0.1 and decayed by a factor of 0.1 at the 35th and 55th epochs. For the three datasets NTU RGB+D 60, NTU RGB+D 120, and NW-UCLA, the batch sizes were set to 64, 64, and 16, respectively.

4.3   Comparison with the State-of-the-Art

We compared our model with state-of-the-art methods (Non-GCNs and GCNs) on the NTU RGB+D, NTU RGB+D 120, and NW-UCLA datasets. The results are shown in Tables 1 and 2 respectively, and we have the following observations:

Our method achieves impressive results on all three datasets, outperforming most existing methods. Compared to the excellent GCNs-based method MS-G3D, which also focuses on spatio-temporal dependencies of joints, our method achieves better performance. Specifically, for the NTU RGB+D 60 dataset, our method improves the classification accuracy by 1.9% and 1.1% on X-Sub evaluation and X-Set evaluation separately. And for the NTU RGB+D 120 dataset, our method shows an improvement of 2.7% and 2.4% in classification accuracy on X-Sub evaluation and X-Set evaluation respectively. On the NW-UCLA dataset, our method improves the classification accuracy by 4.2% compared to the outstanding Non-GCNs method AGC-LSTM. Compared to the excellent GCNs-based method InfoGCN [4], our method has 0.5% higher classification accuracy. The results indicate that our STMTA-GCN can achieve the best performance with a smaller dataset.

Although the performance of our method on the NTU RGB+D 120 dataset is slightly inferior to the current best GCNs-based method, InfoGCN, which utilizes multi-modal representation to improve action recognition performance, its six-stream ensemble mode is overly complex. Overall, our STMTA-GCN is an efficient method for skeleton-based action recognition.

Table 2. Comparison with state-of-the-art methods on the NW-UCLA dataset. The best value is marked in **bold**, and the second is underlined.

Categories	Methods	NW-UCLA Top-1 (%)
Non-GCNs	HBRNN-L [5]	78.5
	Ensemble TS-LSTM [7]	89.2
	AGC-LSTM [17]	93.3
GCNs	SGN [23]	92.5
	Shift-GCN [3]	94.6
	CTR-GCN [2]	96.5
	InfoGCN [4]	<u>97.0</u>
Our	STMTA-GCN	**97.5**

4.4   Ablation Study

We investigated the contribution of key designs in STMTA-GCN to the entire model performance. For simplicity, we only used the X-Sub benchmark on the NTU RGB+D 60 dataset to evaluate the performance of our multiple model variants that remove components from our STMTA-GCN. The experimental results are shown in Table 3. In the table, the "STMTA-GCN wo/STIE-GC" refers to using the adaptive graph convolutional network in [16] instead of the

Table 3. Comparison of the model accuracy when removing different components from STMTA-GCN. The best value is marked in **bold**.

Model variants	X-Sub(%)
STMTA-GCN wo/STIE-GC	89.7
STMTA-GCN wo/MTIE	92.5
STMTA-GCN wo/Semantics	93.0
STMTA-GCN	**93.4**

STIE-GC module. The "STMTA-GCN wo/MTIE" refers to not using MTIE module and only utilizing the convolution operation with fixed kernels described in Sect. 3.1 to extract temporal features. It is worth noting that "STMTA-GCN wo/Semantics" refers to only removing semantic information, and our STIE-GC and MTIE are both retained. We observe that the performance of "STMTA-GCN wo/STIE-GC" drops sharply compared to STMTA-GCN, with a 3.7% decrease in accuracy. This demonstrates that the STIE-GC module is very useful, and the spatio-temporal parallel features between human joints captured can significantly improve model performance. Comparing "STMTA-GCN wo/MTIE" and "STMTA-GCN wo/Semantics", it is found that when MTIE is used alone without Semantics, the model accuracy decreases by 0.5%. This is because the implicit semantic information of frame index can promote STIE to learn temporal features, which is beneficial to action recognition. However, MTIE module improves performance more significantly than semantic information. The main reason is that compared with convolution operations with fixed convolution kernels, the multi-scale approach can extract key frames for action recognition.

5   Conclusion

In this work, a novel STMTA-GCN model is proposed for efficient processing of skeletal spatio-temporal graphs. It mainly consists of two well-designed modules: STIE-GC and MTIE. The STIE-GC module is used to extract spatio-temporal parallel features at the channel level of human joints, while the MTIE module can enhance the model's temporal dynamic feature perception capabilities. Additionally, frame index semantic information is introduced to facilitate the model to learn temporal context information. Comprehensive experiments on three public datasets NTU RGB+D 60, NTU RGB+D 120, and NW-UCLA, demonstrate the effectiveness of each design in the proposed STMTA-GCN and its superiority over existing methods.

Acknowledgements. This work was supported by the Basic Scientific Research Project of Liaoning Provincial Department of Education (LJKFZ20220174), the National Natural Science Foundation of China (No. 62072220, 61502215), the Central Government Guides Local Science and Technology Development Foundation Project of Liaoning Province (No. 2022JH6/100100032), the Natural Science Foundation of Liaoning Province (2022-KF-13-06), China.

References

1. Cai, J., Cai, J.: Brain-machine based rehabilitation motor interface and design evaluation for stroke patients. In: Yuan, L., Yang, S., Li, R., Kanoulas, E., Zhao, X. (eds.) WISA 2023. LNCS, vol. 14094, pp. 625–635. Springer, Cham (2023). https://doi.org/10.1007/978-981-99-6222-8_52

2. Chen, Y., Zhang, Z., Yuan, C., Li, B., Deng, Y., Hu, W.: Channel-wise topology refinement graph convolution for skeleton-based action recognition. In: 2021 IEEE/CVF International Conference on Computer Vision (ICCV), pp. 13339–13348 (2021). https://doi.org/10.1109/ICCV48922.2021.01311

3. Cheng, K., Zhang, Y., He, X., Chen, W., Cheng, J., Lu, H.: Skeleton-based action recognition with shift graph convolutional network. In: 2020 IEEE/CVF Conference on Computer Vision and Pattern Recognition (CVPR), pp. 180–189 (2020). https://doi.org/10.1109/CVPR42600.2020.00026

4. Chi, H.G., Ha, M.H., Chi, S., Lee, S.W., Huang, Q., Ramani, K.: Infogcn: representation learning for human skeleton-based action recognition. In: 2022 IEEE/CVF Conference on Computer Vision and Pattern Recognition (CVPR), pp. 20154–20164 (2022). https://doi.org/10.1109/CVPR52688.2022.01955

5. Du, Y., Wang, W., Wang, L.: Hierarchical recurrent neural network for skeleton based action recognition. In: Proceedings of the IEEE Conference on Computer Vision and Pattern Recognition, pp. 1110–1118 (2015). https://doi.org/10.1109/CVPR.2015.7298714

6. Gaur, U., Zhu, Y., Song, B., Roy-Chowdhury, A.: A "string of feature graphs" model for recognition of complex activities in natural videos. In: 2011 International Conference on Computer Vision, pp. 2595–2602 (2011). https://doi.org/10.1109/ICCV.2011.6126548

7. Lee, I., Kim, D., Kang, S., Lee, S.: Ensemble deep learning for skeleton-based action recognition using temporal sliding LSTM networks. In: 2017 IEEE International Conference on Computer Vision (ICCV), pp. 1012–1020 (2017). https://doi.org/10.1109/ICCV.2017.115

8. Li, B., Li, X., Zhang, Z., Wu, F.: Spatio-temporal graph routing for skeleton-based action recognition. In: Proceedings of the AAAI Conference on Artificial Intelligence, vol. 33, no. 01, pp. 8561–8568 (2019). https://doi.org/10.1609/aaai.v33i01.33018561

9. Li, C., Zhong, Q., Xie, D., Pu, S.: Co-occurrence feature learning from skeleton data for action recognition and detection with hierarchical aggregation. arXiv preprint arXiv:1804.06055 (2018)

10. Li, M., Chen, S., Chen, X., Zhang, Y., Wang, Y., Tian, Q.: Actional-structural graph convolutional networks for skeleton-based action recognition, pp. 3590–3598 (2019). https://doi.org/10.1109/CVPR.2019.00371

11. Li, S., Li, W., Cook, C., Zhu, C., Gao, Y.: Independently recurrent neural network (IndRNN): building a longer and deeper RNN. In: 2018 IEEE/CVF Conference on Computer Vision and Pattern Recognition, pp. 5457–5466 (2018). https://doi.org/10.1109/CVPR.2018.00572

12. Liu, J., Shahroudy, A., Perez, M., Wang, G., Duan, L.Y., Kot, A.C.: NTU RGB+D 120: a large-scale benchmark for 3D human activity understanding. IEEE Trans. Pattern Anal. Mach. Intell. **42**(10), 2684–2701 (2020). https://doi.org/10.1109/TPAMI.2019.2916873

13. Liu, J., Shahroudy, A., Xu, D., Wang, G.: Spatio-temporal LSTM with trust gates for 3D human action recognition. In: Leibe, B., Matas, J., Sebe, N., Welling, M. (eds.) ECCV 2016. LNCS, vol. 9907, pp. 816–833. Springer, Cham (2016). https://doi.org/10.1007/978-3-319-46487-9_50

14. Liu, Z., Zhang, H., Chen, Z., Wang, Z., Ouyang, W.: Disentangling and unifying graph convolutions for skeleton-based action recognition. In: 2020 IEEE/CVF Conference on Computer Vision and Pattern Recognition (CVPR), Los Alamitos, CA, USA, pp. 140–149. IEEE Computer Society (2020). https://doi.org/10.1109/CVPR42600.2020.00022. https://doi.ieeecomputersociety.org/10.1109/CVPR42600.2020.00022

15. Shahroudy, A., Liu, J., Ng, T., Wang, G.: NTU RGB+D: a large scale dataset for 3D human activity analysis. In: 2016 IEEE Conference on Computer Vision and Pattern Recognition (CVPR), Los Alamitos, CA, USA, pp. 1010–1019. IEEE Computer Society (2016). https://doi.org/10.1109/CVPR.2016.115. https://doi.ieeecomputersociety.org/10.1109/CVPR.2016.115

16. Shi, L., Zhang, Y., Cheng, J., Lu, H.: Two-stream adaptive graph convolutional networks for skeleton-based action recognition. In: Proceedings of the IEEE/CVF Conference on Computer Vision and Pattern Recognition, pp. 12026–12035 (2019). https://doi.org/10.1109/CVPR.2019.01230

17. Si, C., Chen, W., Wang, W., Wang, L., Tan, T.: An attention enhanced graph convolutional LSTM network for skeleton-based action recognition. In: 2019 IEEE/CVF Conference on Computer Vision and Pattern Recognition (CVPR), pp. 1227–1236 (2019). https://doi.org/10.1109/CVPR.2019.00132

18. Wang, J., Nie, X., Xia, Y., Wu, Y., Zhu, S.C.: Cross-view action modeling, learning, and recognition. In: 2014 IEEE Conference on Computer Vision and Pattern Recognition, pp. 2649–2656 (2014). https://doi.org/10.1109/CVPR.2014.339

19. Xu, K., Ye, F., Zhong, Q., Xie, D.: Topology-aware convolutional neural network for efficient skeleton-based action recognition. In: Proceedings of the AAAI Conference on Artificial Intelligence, vol. 36, no. 3, pp. 2866–2874 (2022). https://doi.org/10.1609/aaai.v36i3.20191

20. Yan, S., Xiong, Y., Lin, D.: Spatial temporal graph convolutional networks for skeleton-based action recognition. In: Proceedings of the AAAI Conference on Artificial Intelligence, vol. 32, no. 1 (2018). https://doi.org/10.1609/aaai.v32i1.12328. https://ojs.aaai.org/index.php/AAAI/article/view/12328

21. Ye, F., Pu, S., Zhong, Q., Li, C., Xie, D., Tang, H.: Dynamic GCN: context-enriched topology learning for skeleton-based action recognition. In: Proceedings of the 28th ACM International Conference on Multimedia, MM 2020, pp. 55–63. Association for Computing Machinery, New York (2020). https://doi.org/10.1145/3394171.3413941

22. Yu, Z., et al.: Searching multi-rate and multi-modal temporal enhanced networks for gesture recognition. IEEE Trans. Image Process. **30**, 5626–5640 (2021). https://doi.org/10.1109/TIP.2021.3087348

23. Zhang, P., Lan, C., Zeng, W., Xing, J., Xue, J., Zheng, N.: Semantics-guided neural networks for efficient skeleton-based human action recognition. In: Proceedings of the IEEE/CVF Conference on Computer Vision and Pattern Recognition, pp. 1112–1121 (2020)

Breast Mass Classification in Mammograms Based on the Fusion of Traditional and Deep Features

Hongyu Zhang, Zhili Chen$^{(\boxtimes)}$, and Adamu Abubakar Abba

School of Computer Science and Engineering, Shenyang Jianzhu University, Shenyang, China
zzc@sjzu.edu.cn

Abstract. Breast cancer is a disease that affects the life and health of women worldwide. Early detection and treatment of the disease can effectively prolong patients' life and improve their quality of life. Mammography is one of the most commonly used examination methods for breast cancer screening and early diagnosis. Radiologists can differentiate breast masses as benign or malignant according to their shape and appearance. However, the large variation in X-ray appearances of breast masses brings great diagnostic difficulties to doctors. Therefore, this paper proposes a breast mass classification model that fuses traditional and deep features to assist doctors in classifying and diagnosing breast masses. A variety of handcrafted traditional features and the deep features learned from the improved SE-DenseNet model based on DenseNet are integrated. An improved genetic algorithm based on mutual information is used to select the most valuable subset of features in the feature selection stage. The benign and malignant classification of breast masses is finally accomplished through the fusion of multi-classifier decisions. The experimental results show that the classification accuracy of the proposed method for feature fusion has better classification performance compared to using traditional features or deep features separately, indicating that the two types of features contain complementary information, and the fusion of the two can achieve the complementary advantages to a certain extent.

Keywords: Mammographic Images · Deep Learning · Breast Mass Classification · Feature Fusion

1 Introduction

Breast cancer is currently the cancer with the highest number of diagnosed cases and one of the high-risk cancer types, and it is the leading cause of death among women. It was estimated that 684,996 people will die from breast cancer in 2020, and by 2040, the number of cases will increase by nearly 50 percent [1]. During cancer diagnosis, breast masses can be classified into benign and malignant; benign masses do not spread throughout the body and do not reappear after being surgically removed. In contrast, cancer cells from malignant masses spread and invade other body organs, which is one of the many complications that can lead to a patient's death. Breast cancer, if detected at an

© The Author(s), under exclusive license to Springer Nature Singapore Pte Ltd. 2024
C. Jin et al. (Eds.): WISA 2024, LNCS 14883, pp. 561–572, 2024.
https://doi.org/10.1007/978-981-97-7707-5_46

early stage through a combination of surgical removal, radiation therapy, immunotherapy and chemotherapy, can prevent the progression of the disease and eradicate it with a survival rate of 90% or higher.

There are usually four stages required to complete the classification of benign and malignant breast masses: preprocessing, feature extraction, feature fusion and classification, and the key lies in the two parts of feature extraction and feature fusion, Yang et al. [2]. Scholars such as Yang et al. obtained significant results by utilizing the traditional edge features of the mass to perform detailed segmentation and then using the fractal dimension to discriminate the benign and malignant nature of the mass. However, traditional feature extraction methods rely on hand-designed features to characterize image content, while deep learning techniques can automatically extract high-quality features, Rahaman [3] proposes a hybrid deep feature fusion technique (DeepCervix) based on deep learning to classify cervical cells by introducing four types of augmented structured CNN: VGG16, VGG19, XceptionNet, and ResNet50 in order to extract complementary features from different depths of the network thus obtaining better classification results. To summarize, shallow network models with traditional features can achieve faster convergence but tend to underperform in terms of accuracy, while deep networks, although providing higher classification accuracy, have a large number of parameters to be learned, resulting in slow convergence. Therefore, fusing traditional and deep features for mammogram image mass classification has more advantages than using either feature type alone, combining the ability of deep learning to automatically extract complex and high-level features with the accuracy of traditional features in capturing image details. Hu et al. [4] used a convolutional neural network to extract deep features while fusing local binary pattern texture features with rotational invariance to make up for the lack of rotational adaptation of CNN deep features, but the dimensionality of the feature space after feature fusion becomes too high and contains a large amount of redundant information, which leads to the consumption of a large amount of wasted computational resources and does not result in a higher efficiency improvement.

To address the above problems, this paper proposes a classification model based on the fusion of traditional features and deep features, and builds a SE-DenseNet model by combining the SE attention mechanism to the DenseNet network in the feature extraction stage, which is able to strengthen the useful features and suppress the irrelevant ones by learning the important weights of the different channels, and significantly improving the sensitivity and selectivity of the network. A genetic algorithm based on mutual information improvement is used in the feature fusion stage to screen out the most valuable feature subsets, and finally the benign and malignant classification of breast masses is completed by multi-classifier decision fusion, which improves the classification accuracy and enhances the model's ability to differentiate between different types of breast masses, and also improves the model's ability to generalize and reduces the dependence on large-scale training datasets. In addition, it enhances the interpretability of the model in mammographic image analysis, thereby promoting trust and allowing better clinical decision making.

2 Methodology

The mammographic image mass classification model designed in this paper based on the fusion of conventional and deep features is shown in Fig. 1. Firstly, it performs a histogram equalization on the mammogram image in the preprocessing stage, it then removes the noise present in the image at different luminance levels by using median filtering, and finally separates the foreground and the background of the patch of the mammogram image obtained after cropping by using Otsu threshold image segmentation algorithm, making the contour of the mass more visible. In the feature extraction stage, traditional features are extracted as represented by LBP [5] based texture features, HOG [6] and shape features of shape descriptors, and dimensionality reduction is performed using the PCA technique to reduce feature space redundancy. Next, the improved SE-DenseNet is utilized as a deep feature extractor to extract the deep features. The third part of the feature fusion stage is the key stage of the whole model. In this paper, the most valuable features are filtered based on mutual information and an improved genetic algorithm to improve the classification performance. In the fourth stage the decision fusion model constructed by input to XGBoost, AdaBoost and SVM classifiers is used to complete the final classification.

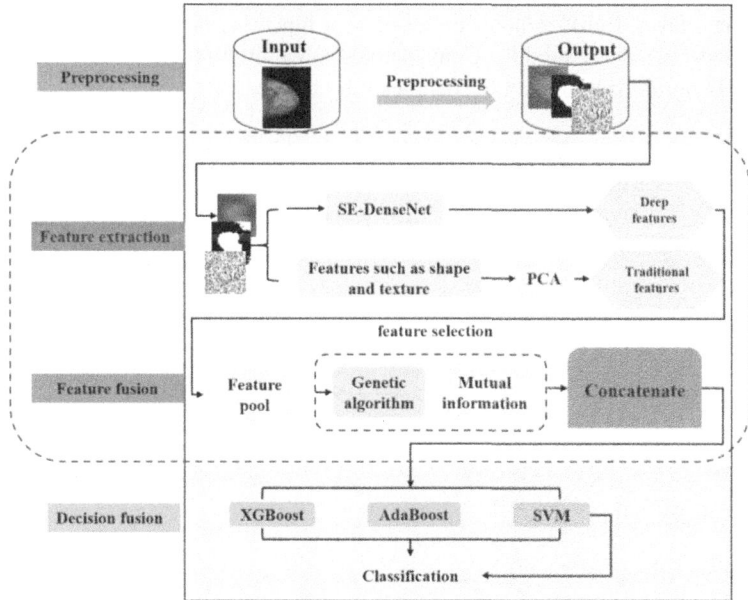

Fig. 1. Flowchart of the traditional and deep feature fusion model.

2.1 The Improved DenseNet Network Mechanism Based on Attention

SE Module. SE (Squeeze-and-Excitation) module is a channel attention module proposed by Hu et al. in 2017, which is a new architectural unit that can be flexibly integrated into different positions of a convolutional neural network and mainly consists of two parts, Squeeze and Excitation. W and H denote the width and height of the feature maps. C denotes the number of channels, and the size of the input feature maps is $W \times H \times C$. The first step is the Squeeze operation (compression), after which the feature map is compressed into a $1 \times 1 \times C$ vector. The next step is Excitation, which consists of two fully connected layers, where SERatio is a scaling parameter, the purpose of this parameter is to reduce the number of channels and thus reduce the computational effort. Liang [7] use medical knowledge graphs for multi-hop reasoning to guide a self-attention-based transformer model for disease prediction.

SE-DenseNet. The structure of the DenseNet [8] is such that each layer is connected to all subsequent layers in a densely connected manner, and the idea behind the SE-DenseNet network improvement is to add SE modules after each 3×3 convolutional layer of DenseNet block, as shown in Fig. 2. In this paper, the DenseNet121 network is used as the base convolutional neural network, which is divided into two structures: dense module and transition layer, and each dense module consists of different numbers of dense layers, and the structure of each dense layer is BN-ReLU-Conv(1×1)-BN-ReLU-Conv(3×3). The SE module is integrated into the dense module to form a new SE-DenseNet structure, corresponding to BN-ReLU-Conv(1×1)-BN-ReLU-Conv(3×3) -SE. Where Conv($m \times m$) denotes the size of the convolutional kernel in the convolutional layer, ReLU denotes the excitation function, and N denotes the number of dense layers in a dense module. Conv denotes convolution, Pool denotes pooling, BN denotes bulk normalization, and FC denotes fully connected.

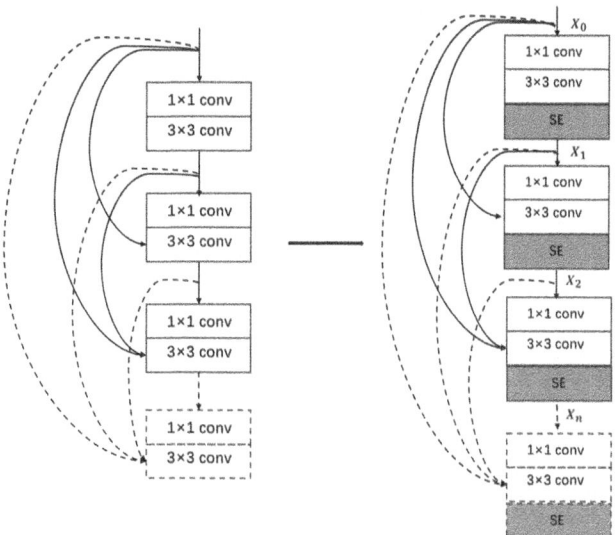

Fig. 2. Structure of SE-DenseNet improved based on DenseNet.

With this improvement, the network not only realizes lossless transmission of the original input information but also automatically adjusts the weights of each channel by learning the network's global information. Subsequently, the useful features are enhanced, and the useless features are suppressed according to their importance, thus realizing the adaptive calibration of the feature channels. The complete SE-DenseNet network framework is shown in Fig. 3, which mainly consists of four SE-Denseblocks and three transition layers, where the number of dense layers in the four SE-Denseblocks is 6, 12, 24 and 16, respectively. After going through the above structure, a global average pooling layer is used to average all pixel values of each feature map, followed by global spatial information summation, and finally a fully-connected layer is used to map the learned features to 2-dimensional vectors in the sample labeling space. The He initialization function to the Pytorch framework is used in the experiments to initialize the parameters of the added SE module, followed by training the model with the training sample set. The experimental procedure uses cross-entropy to calculate the loss value and the Adam method to train the network. The iterative epoch is set to 100, and since there are 1094 images in the training set, a single batch of 64 samples is used as model input, totalling 17 batches. The test set consisted of 483 images with a single batch of 22 samples as model input for a total of 21 batches. The learning rate is continuously adjusted with the depth of training. The initial learning rate is set to 0.01, the momentum is 0.09, the step size of the learning rate decay is 5, and the multiplier factor of the learning rate decay is 0.2, to obtain the final classification results.

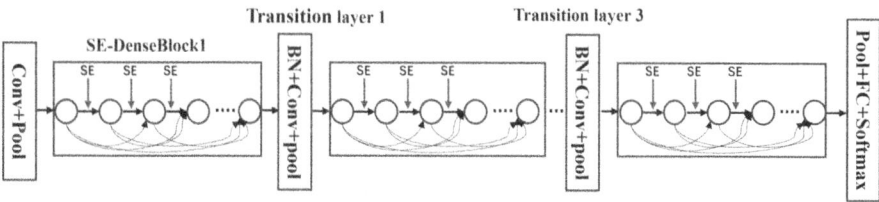

Fig. 3. SE-DenseNet classification structure diagram.

2.2 Based on Improved Feature Selection Methods

Genetic Algorithms and Mutual Information. The key to feature selection is to reduce computational complexity by reducing the number of features to be processed. In addition, it is to improve the system's performance by selecting the best features and eliminating the features that may lead to misclassification. The feature selection of the genetic algorithm [9] is inspired by natural evolutionary mechanisms and employs a global stochastic search strategy, which mimics the processes of reproduction, crossover and mutation in nature. Starting from an initial population, more environmentally adapted offspring are produced through these random processes. Through multiple iterations, the population gradually evolves towards a more optimal region of the optimized search space. In this way, a population gradually evolves to the superior region of search. This successive evolution leads to the aggregation of a group of individuals best adapted to the environment in which they live, thus finding a superior solution.

Mutual information (MI) is calculated based on Shannon Entropy, which measures the dependence or mutual information between two random variables [10]. It determines how much can be learned about a variable by considering another variable. In machine learning, mutual information measures the information about a feature (i.e., its presence or absence), which helps to a great extent in making correct predictions about Y. These two variables present non-negative values of mutual information. MI is zero if both random variables are independent; when this value increases, it means that their dependence increases.

Genetic Algorithm Based on Mutual Information Improvement. In the designed system, Genetic Algorithm Improved Based on Mutual Information (GA-MI) is used to judge the fitness score by embedding the mutual information algorithm into the genetic algorithm and evaluating it as a fitness function, which combines the global search ability of the genetic algorithm and the ability of mutual information to identify the most valuable features, and the improved framework is shown in Fig. 4.

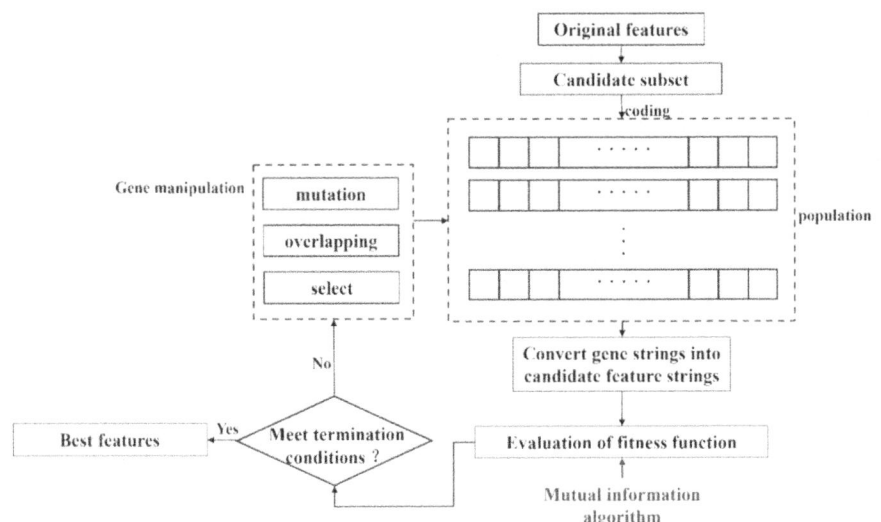

Fig. 4. Schematic diagram of genetic algorithm based on mutual information improvement.

The experiment was executed as follows. A random population of 300 chromosomes is first created, each of which consists of 2,260 individual features, and each of which represents a subset of possible features. These individual features are binary encoded, describing whether a feature is selected or not, with 1 representing selection and 0 for non-selection.

Fitness Function. Mutual information is used as a fitness function to evaluate the relevance of a subset of features to the target variable, in the experiments, by calculating the average value of mutual information between all selected features and the target variable as the fitness.

Selection Mechanism. Adopting Roulette Wheel Selection (RWS), the probability of each chromosome being selected is proportional to its fitness. Adaptations below the average may be eliminated, or probably still be selected in other to maintain the diversity of the population.

Genetic Manipulation. Crossover and mutation of the features screened by the selection mechanism to generate a new population, eliminating the lowest performing individuals; the crossover ratio was exchanged at 50% in this experiment, and a 5% mutation rate was set to change the feature coding in the chromosome.

Termination. The abort condition in this experiment is set to a threshold of 50 epochs or an improvement in fitness of less than 0.005% in 5 consecutive generations.

Evaluation. The accuracy of the decision tree classifier in estimating the selected features is chosen as the evaluation metric. Decision trees are capable of handling both numerical and categorical data without complex preprocessing.

Output. Return the final subset of features that have the highest fitness values after a series of epoch of feature selection. The experiment achieves an accuracy improvement of less than 0.005% when compared in 5 consecutive epochs, and it finally reaches the abort condition after 35 epochs, obtaining the optimal subset of features consisting of 47 traditional handmade features and 72 deep features.

3   Experiment

3.1   Dataset

The DDSM (The Digital Database for Screening Mammography) [11] dataset is the largest digitized film dataset ever compiled. The dataset is divided into four volumes according to the severity of the cases: normal, benign, malignant, and benign with no recall. It includes a total of 2,620 cases, with each case containing two views of the

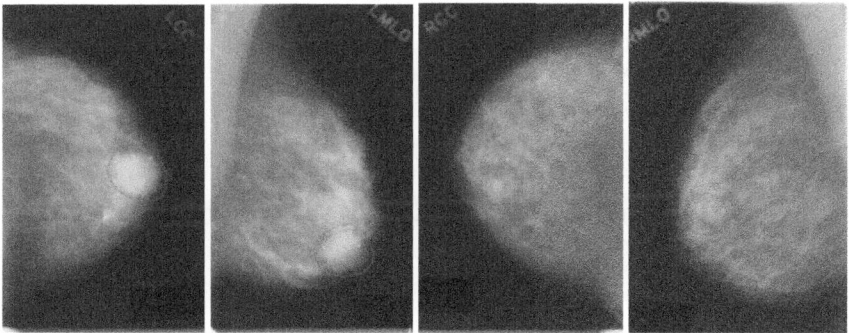

Fig. 5. Schematic diagram of bilateral breast mammographic images of breast masses.

left and right breasts, as shown in Fig. 5. Each case contains two views of the right and left breasts. Corresponding to four images in total and the dataset comes with detailed information about the patient's breast density class and coordinates of the region of interest, along with the patient's age, date of the study as well as the date of digitization, the category of dense tissue, the scanner used for digitization, and the resolution of each image.

Based on the original mammogram images and the txt files containing the image information, in this paper, we pinpoint the regions-of-interest (ROI) by locating the mass coordinates precisely labeled by breast oncologists to the corresponding locations. Subsequently, we further frame the square boundaries of these regions of interest in order to display and analyze the mass features more intuitively. This step is crucial for subsequent image processing, feature extraction, and construction of classification models. Upon using the Matlab software to crop the images, 2417 mass image samples were finally obtained, of which 1367 malignant mass image samples and 1050 benign mass image samples were classified. Table 1 shows the statistics of the data used in the experiment.

Table 1. Statistical table of data used in the experiment.

Class	Number of cases	Number of cases	Mass image patch
Benign	502	1017	1050
Malignant	628	1277	1367
Total	1130	2294	2417

3.2 Experimental Environment and Evaluation Metrics

Experimental environment: operating system: 64-bit Windows 10; CPU: 2.2 GHz; 4.0 GB of running memory and 8.0 GB of storage memory; development environment: PyCharm2021.1; Python version: Python3.8; main libraries: NumPy, OpenCV, Keras, scikit, etc.

In the result evaluation phase of this study, Accuracy, Precision, Recall, and F1 Score are used to assess the Model's performance. The accuracy value is used to measure the total number of correct predictions for all evaluated elements and is calculated as (1).

$$Accuracy = \frac{TP + TN}{TP + TN + FP + FN} \tag{1}$$

The precision rate is the closeness of the predicted results to the actual values and is calculated as (2).

$$Precision = \frac{TP}{TP + FP} \tag{2}$$

Recall, also known as sensitivity, measures the percentage of positive items that are correctly categorized and is calculated as (3).

$$Recall = \frac{TP}{TP + FN} \tag{3}$$

$F1 - Score$ is the harmonic mean of precision and recall, which gives a good indication of a model's performance, and is calculated as (4).

$$F1 - Score = 2 * \frac{Precision * Recall}{Precision + Recall} \tag{4}$$

3.3 Experimental Results and Analysis

In this chapter, the mammogram image mass classification model based ontraditional features and deep feature fusion mainly performs preprocessing, feature extraction, feature fusion and classification. In the feature fusion stage, an improved feature selection method was utilized to filter 2260 hybrid features and finally 119 hybrid features were retained, followed by a feature level data fusion step to stitch the feature vectors as input data for the classification stage. A decision fusion model was utilized in the classification stage to complete the final classification. The experimental results are shown in Table 2.

Table 2. Classification results.

Typology	Model	Accuracy	Recall	Precision	F1-score
Traditional features	SVM	80.34%	74.76%	81.49%	0.78
	Decision Fusion	82.91%	74.85%	85.26%	0.80
Deep Features	VGG16	81.25%	84.62%	83.53%	0.84
	DenseNet	82.62%	84.25%	81.93%	0.83
	SE-DenseNet	84.77%	85.18%	83.79%	0.84
Traditional features + deep features	SVM	90.43%	88.52%	89.31%	0.88
	Decision Fusion	90.64%	89.15%	87.62%	0.89

Comprehensive experimental data shows that the performance of traditional feature fusion is improved after applying the decision fusion model. Compared with the traditional features, the deep learning model performs better in terms of overall performance, with the improved SE-DenseNet deep learning network achieving an accuracy of 84.77% and an F1-score of 0.84. The classification performance after fusion of traditional and deep features is the best of all models, with the accuracy rate increasing to 90.43%, and in the final classification stage, the decision fusion model constructed by XGBoost, AdaBoost, and SVM classifiers further improves the classification accuracy to 90.64%. This demonstrates the complementary effectiveness of feature fusion and the robustness of decision fusion.

3.4 Comparative Experiments

The research in this chapter aims to analyze and evaluate the impact of extracting different features and different fusion strategies on the performance of breast mass classification through comparative experiments. By exploring the combined application of feature fusion and decision fusion, the proposed method in this chapter is compared with the experimental methods of classification of the current methods and the results are shown in Table 3.

Table 3. Comparison of feature fusion classification results with other lesion classification methods.

Author	Methodologies	Dataset	Accuracy	AUC
Mudigonda [12]	Traditional features	MIAS	82.10%	0.85
Zhang [13]	Traditional features	MIAS	89.44%	-
Arora [14]	Deep features	DDSM	88.01%	0.88
Jones [15]	Traditional features + VGG16	DDSM	-	0.80
model of this paper	Traditional features + SE-DenseNet + SVM	DDSM	90.43%	-
model of this paper	Traditional Features + SE-DenseNet + Decision Fusion	DDSM	90.64%	-

In this paper, the model compares experiments by fusing traditional features and deep features using SVM and decision fusion methods respectively, and the results show that the model outperforms previous studies on the DDSM dataset, showing excellent performance in both accuracy and F1-score values. The comparisons show that the use of deep learning models yields better results compared to traditional feature-based classification and the feature fusion strategy is more effective in improving the performance of mass lesion classification on mammogram images compared to the use of traditional features alone or a single deep learning model. Simultaneously fusing decisions from multiple classifiers at the classification stage can further enhance classification performance.

4 Conclusions

This paper proposes a classification model for breast mass classification based on the fusion of traditional and deep features. The key to the entire classification model lies in improving feature extraction and selection methods. Throughout the classification process, experimental data undergo preprocessing, where shape and texture features are extracted as representatives of traditional handcrafted features during the feature extraction stage. An improved SE-DenseNet deep learning network is utilized as the deep feature extractor. Subsequently, in the feature selection stage, a method that replaces mutual information with the fitness function in a genetic algorithm is proposed. All extracted features are then filtered through GA-MI to select the most valuable feature subset as input for the classification stage. In the classification stage, a decision fusion model constructed by multiple classifiers is added to complete the classification. The experiments demonstrate that the overall feature fusion model exhibits improved accuracy and robustness compared to the previous version. Moreover, when compared with the experimental results of other authors on breast mass classification, the proposed classification model demonstrates superior performance. It achieves more accurate differentiation between benign and malignant masses, showcasing better classification performance. Additionally, the integration of traditional features into the model enhances its interpretability for practical diagnosis, making the decision-making process more transparent and understandable to healthcare professionals.

References

1. Sung, H., et al.: Global cancer statistics 2020: GLOBOCAN estimates of incidence and mortality worldwide for 36 cancers in 185 countries. CA Cancer J. Clin. **71**(3), 209–249 (2021)
2. Yang, S.C., et al.: A computer-aided system for mass detection and classification in digitized mammograms. Biomed. Eng. Appl. Basis Commun. **17**(05), 215–228 (2005)
3. Rahaman, M.M., et al.: DeepCervix: a deep learning-based framework for the classification of cervical cells using hybrid deep feature fusion techniques. Comput. Biol. Med. **136**, 104649 (2021). https://doi.org/10.1016/j.compbiomed.2021.104649
4. Dan, H.U., et al.: Visual tracking based on deep features and LBP texture fusion. Comput. Eng. **42**(9), 220–225 (2016)
5. Chen, H., Ma, M., Liu, G., Wang, Y., Jin, Z., Liu, C.: Breast tumor classification in ultrasound images by fusion of deep convolutional neural network and shallow LBP feature. J. Digit. Imaging **36**(3), 932–946 (2023). https://doi.org/10.1007/s10278-022-00711-x
6. Sajid, U., Khan, R.A., Shah, S.M., Arif, S.: Breast cancer classification using deep learned features boosted with handcrafted features. Biomed. Signal Process. Control **86**, 105353 (2023)
7. Liang, Y., Wang, H., Zhang, W.: A Knowledge-guided method for disease prediction based on attention mechanism. In: Zhao, X., Yang, S., Wang, X., Li, J. (eds.) Web Information Systems and Applications: 19th International Conference, WISA 2022, Dalian, China, September 16–18, 2022, Proceedings, pp. 329–340. Springer International Publishing, Cham (2022). https://doi.org/10.1007/978-3-031-20309-1_29
8. Liao, T., et al.: Classification of asymmetry in mammography via the DenseNet convolutional neural network. Eur. J. Radiol. Open **11**, 100502 (2023)
9. Altarabichi, M.G., Nowaczyk, S., Pashami, S., Sheikholharam Mashhadi, P.: Fast genetic algorithm for feature selection-a qualitative approximation approach. In: Proceedings of the Companion Conference on Genetic and Evolutionary Computation, pp. 11–12 (2023)
10. Belghazi, M.I., et al.: Mutual information neural estimation. In: International Conference on Machine Learning, pp. 531–540. PMLR (2018)
11. Heath M, et al.: The digital database for screening mammography. Springer Netherlands (2001). https://doi.org/10.1007/978-94-011-5318-8_75
12. Mudigonda, N.R., Rangayyan, R., Desautels, J.E.L.: Gradient and texture analysis for the classification of mammographic masses. IEEE Trans. Med. Imaging **19**(10), 1032–1043 (2000)
13. Zhang, Y.D., Wang, S.H., Liu, G., Yang, J.: Computer-aided diagnosis of abnormal breasts in mammogram images by weighted-type fractional Fourier transform. Adv. Mech. Eng. **8**(2), 1687814016634243 (2016)

14. Arora, R., Rai, P.K., Raman, B.: Deep feature-based automatic classification of mammograms. Med. Biol. Eng. Comput. **58**, 1199–1211 (2020)
15. Jones, M.A., Faiz, R., Qiu, Y., Zheng, B.: Improving mammography lesion classification by optimal fusion of handcrafted and deep transfer learning features. Phys. Med. Biol. **67**(5), 054001 (2022)

TRoute: Dynamic Time-Dependent Route Recommendation on Road Networks

Xiaolin Han[1,2], Xiurui Hu[1], Chenhao Ma[3], and Xuequn Shang[1(✉)]

[1] Northwestern Polytechnical University, Xi'an, China
{xiaolinh,shang}@nwpu.edu.cn, huxr@mail.nwpu.edu.cn
[2] Laboratory for Advanced Computing and Intelligence Engineering, Wuxi, China
[3] Chinese University of Hong Kong, Shenzhen, China
machenhao@cuhk.edu.cn

Abstract. Recommending routes for different origin-destination pairs poses a significant challenge in transportation and logistics. Traditional algorithms often overlook time-dependent reachable time, which is influenced by dynamic traffic conditions and road characteristics. However, in congested traffic conditions, the shortest route may take longer to travel than alternative routes, potentially causing delays that disrupt passengers' subsequent schedules and plans. In this paper, we introduce a novel data-driven method called **TRoute**, which focuses on recommending **T**ime-dependent **Route**s adaptable to changing traffic conditions. Our approach employs a deep generative model to automatically infer latent patterns, specifically reachable times under varying traffic conditions and road properties, for these dynamic routes. Through extensive evaluation using two real trajectory datasets, our method exhibits significant performance improvements, achieving 14.35% and 14.02% improvements in precision and recall, respectively, compared to existing methods.

Keywords: Route Recommendation · Time-dependent Route · Dynamic Traffic Condition

1 Introduction

The swift rise in urbanization and vehicle usage has escalated traffic management and congestion issues, adversely affecting environmental sustainability and quality of life. Consequently, route recommendation has become essential, as it assists users in identifying the optimal routes to their destinations.

Route recommendation within road networks is a thoroughly researched area with numerous practical applications. The fundamental issue at hand is determining the optimal route between starting and ending points. Traditional route planning techniques [4,8,9,19,32] depend on static maps that utilize historical average travel times. Nevertheless, these approaches frequently overlook the dynamic characteristics of traffic and the intricate patterns emerging from real-time data.

C. Jin et al. (Eds.): WISA 2024, LNCS 14883, pp. 573–585, 2024.
https://doi.org/10.1007/978-981-97-7707-5_47

In recent years, progress in data collection and communication technologies has enabled the acquisition of large-scale, dynamic traffic data [22,23]. This data encompasses extensive traffic trajectories that reflect the spatial and temporal dynamics of traffic conditions across different routes. Leveraging this information can enhance route recommendation accuracy, thereby improving traffic flow and reducing congestion.

Efforts to address route planning challenges using deep learning models have gained attraction recently. For instance, DeepST [21] employs a deep probabilistic learning framework to identify popular routes from historical taxi trip data. This model integrates the sequential nature of transitions, the influence of destinations, real-time traffic conditions, and representations of destinations. On the other hand, the state-of-the-art method, NeuroMLR [18], introduces a generative model that uses inductive learning to recommend routes aimed at efficiently reaching destination locations. However, NeuroMLR overlooks the critical factor of reachable time, which considers dynamic traffic conditions and specific road properties such as speed limits and the number of lanes. This omission can lead to suboptimal routing recommendations, which could disrupt subsequent schedules and plans of passengers.

Our Contributions: Our primary objective is to comprehensively capture the underlying patterns in dynamic traffic conditions. This task is challenging due to the constantly changing and diverse nature of traffic conditions across different locations and times. Even for the same source and destination points, traffic conditions may fluctuate over time. To address this issue, we introduce a novel generative neural network, named **TRoute**, designed to capture these dynamic traffic conditions. The novelty of our approach lies in our ability to capture dynamic traffic conditions by extracting latent reachable time from real-time traffic condition during journeys, which is non-trivial. Specifically, we infer latent reachable time by maximizing the likelihood of observing trajectories under varying traffic conditions.

Additionally, we study how to improve the effectiveness of our model by incorporating road-specific properties. As reachable time can be influenced by various factors related to individual road properties, such as maximum speed and number of lanes, we propose learning latent patterns within these road properties. By jointly considering these learned patterns and the dynamic traffic conditions mentioned earlier, we infer reachable time, effectively integrating static road properties and dynamic traffic conditions into the process. Subsequently, optimal routes are dynamically recommended based on the inferred reachable time.

We have performed substantial experiments on two real trajectory datasets against existing methods. Our model achieves 14.35% and 14.02% improvements on average w.r.t. precision and recall, respectively.

2   Related Work

2.1   Traditional Methods for Route Recommendation

Some strategies enhance traditional shortest path algorithms by including time-dependent travel times into the cost function. These methods model the travel time on each road segment based on the time of departure, allowing the algorithms to adjust for traffic conditions throughout the day. Examples of such time-dependent shortest path algorithms include the time-dependent Dijkstra's algorithm [4] and the time-dependent A* algorithm [19,32]. Techniques like A* with landmarks [8] and the bidirectional search scheme in A* [9] have also been adapted to accommodate time-dependent scenarios [5,6,29].

Although these methods offer more precise route predictions than static algorithms, they primarily depend on historical traffic data and do not adequately account for real-time spatial and temporal patterns inherent in dynamic traffic data.

2.2   Data-Driven Models for Route Recommendation

These approaches can integrate real-time traffic data to dynamically adjust route predictions, enhancing the accuracy of route recommendations. This capability helps drivers bypass congestion and reduce travel times.

MPR [3] identifies the most frequented route between a source and destination using historical trajectories. It starts by developing a transfer model to calculate the likelihood of transitioning between nodes. The most popular route is then determined by identifying the path that maximizes this probability based on the transfer model. L2R [10] addresses route planning in areas with sparse trajectories by learning routing patterns between frequently traveled regions and applying these patterns to less frequented areas. TALL [36] introduces a deep learning approach to predict user destinations through the understanding of sub-trajectories. However, these models do not accommodate time-sensitive queries. NASF [34] is designed to accurately predict travel times for optimizing route recommendations. Additionally, several sophisticated models [7,22,24,35] focus on estimating travel times from trajectory data. Since networks are prevalent to many data mining tasks [12–17,20,25–28,30], several studies leverage networks or graphs in their research. CSSRNN [33] uses Recurrent Neural Networks to capture trajectory patterns while considering the structural constraints of the road networks. The AyHy-TNet Model [24] introduces an attribute-related hybrid trajectories network to estimate travel times along paths. DEEPST [21] employs a deep probabilistic approach that accounts for previously traveled routes, destinations, and current traffic conditions in generating routes, demonstrating superior performance in producing the most likely routes. NEUROMLR [18] offers an inductive approach for dependable and efficient route recommendations. Furthermore, a context-aware and preference-sensitive routing algorithm [11] has been developed for vehicular routing. Beyond route prediction, numerous innovative models are emerging for general trajectory modeling. Toast [2] integrates traffic patterns with travel semantics for robust representation learning in road networks. MTNet [31] introduces a deep generative framework for broad trajectory modeling and application.

However, these methods focus solely on recommending the shortest route without taking into account the reachable time. It is important to recognize that the shortest path is not always the fastest. In congested traffic conditions, the shortest route may require more time to traverse compared to alternative routes, potentially disrupting passengers' subsequent schedules and plans.

3   Problem Definition

In this section, we first present definitions of the basic concepts. Then, we give a formal problem definition.

Definition 1 (Road Network). *A road network can be described as a directed graph $G = (V, E)$, where V represents the set of vertices denoting intersections or junctions, and E includes directed edges representing road segments that link these intersections.* □

Definition 2 (Trajectory). *A point p_{t_i} is a triplet (t_i, x, y) comprising a timestamp t_i, and the geographic coordinates, latitude x and longitude y, of its position at that time. A trajectory T is a chronologically ordered sequence of points $\langle p_{t_1}, \cdots, p_{t_i}, \cdots, p_{t_n} \rangle$ where $t_1 < \ldots < t_i < \ldots < t_n$.* □

Definition 3 (Route). *A route $R(s, d) = \{v_1, \cdots, v_k\}$ defines a simple path from the source node $s = v_1$ to the destination node $d = v_k$ in the road network G, which is devoid of cycles.* □

Definition 4 (Time-dependent Route). *A time-dependent route $R(s, d, t_j) = \{v_1, \cdots, v_k\}$ specifies a path from the source node $s = v_1$ to the destination node $d = v_k$, conditioned on the departure time t_j and the traffic conditions in the road network G, and it is also free from cycles.* □

Problem 1 (Time-dependent Route Recommendation). Given a road network G, a historical database D of trajectories, and a tuple (s, d, t_j), where s and d are the source and destination nodes respectively, and t_j is the departure time, the objective is to deduce the route $R(s, d, t_j)$ that best fits the traffic patterns in D. □

4   Proposed Model

4.1   Framework

Figure 1 illustrates the overall framework of our proposed model, TRoute. Initially, it infers the latent reachable time from both dynamic traffic conditions and road properties (refer to Sect. 4.2). Subsequently, it utilizes the learned latent reachable time to dynamically recommend the next path (see Sect. 4.3).

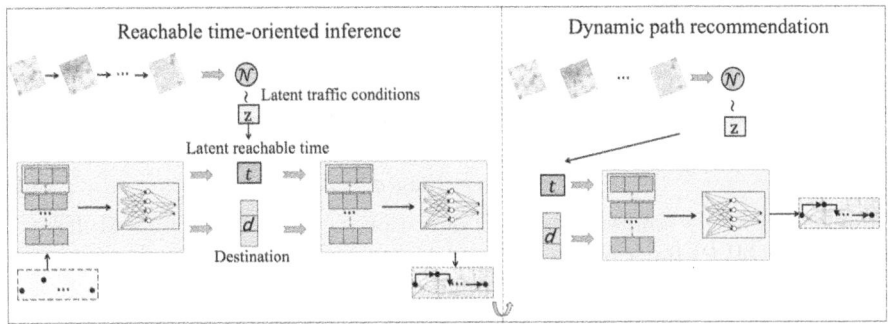

Fig. 1. Overview of our model. The notation $\mathcal{N}$ denotes the Gaussian distribution.

4.2 Reachable Time Inference

In this section, we introduce the method of inferring the expected reachable time by leveraging real-time traffic data and road properties to dynamically optimize route planning.

Firstly, we construct a real-time traffic speed matrix Z, which encapsulates the current traffic conditions and is derived from extensive urban trajectory data, including trajectories from private cars, taxis, and others. Specifically, we partition the city map into grids and compute the average speed of traffic within each grid for each time period. To distill spatio-temporal traffic features from these real-time traffic speed matrices, we employ Convolutional Neural Network (CNN) as represented by the equation:

$$f_1(Z) = \text{CNN}(Z), \tag{1}$$

where the function $f_1(\cdot)$ is structured using a CNN. The CNN model operates on each temporal snapshot Z_{t_i}.

Following this, the latent variable z, representing potential traffic conditions, is inferred from the traffic patterns. This variable z is sampled from a Gaussian distribution $\mathcal{N}$, expressed as:

$$z \sim \mathcal{N}(\mu_Z, \text{diag}(\sigma_Z^2)), \tag{2}$$

where the mean μ_Z and the standard deviation σ_Z are learned by a Multilayer Perceptron (MLP) function $g_1(f_1(Z))$ during the training phase.

Road properties, such as maximum speed limit, number of lanes, road type, and one-way flag, can be converted into contextual embeddings. We employ one-hot encoding to transform them into vectors: the maximum speed limit vector f_{ms}, the number of lanes vector f_{nl}, road type vector f_{rt}, and one-way flag vector f_{ow}. These vectors are then concatenated to form $f_c \in \mathbb{R}^{|E| \times n_f}$, as expressed by the equation:

$$f_c = f_{ms} || f_{nl} || f_{rt} || f_{ow}, \tag{3}$$

where nf represents the dimension of feature embedding for f_c of each edge e, and $\|$ denotes the concatenation operator.

The latent road feature r_i of road i is derived from its own road properties. The generation of r_i is sampled from a Gaussian distribution $\mathcal{N}$:

$$r_i \sim \mathcal{N}(\mu_{f_c}, \mathrm{diag}(\sigma^2_{f_c})) \,, \tag{4}$$

where the mean μ_{f_c} and the standard deviation σ_{f_c} are learned by a Multilayer Perceptron (MLP) function $g_2(f_2(f_c))$ during the training phase.

We combine the latent road feature and latent traffic pattern using a Linear layer:

$$x_i = \mathrm{Linear}(W_1 r_i + W_2 z) \,, \tag{5}$$

Then, the speed of road i can be sampled from a Gaussian distribution:

$$s_i \sim \mathcal{N}(\mu_{x_i}, \mathrm{diag}(\sigma^2_{x_i})) \,, \tag{6}$$

where the mean μ_{x_i} and the standard deviation σ_{x_i} are learned by a Multilayer Perceptron (MLP) function $g_3(f_3(x_i))$ during the training phase.

The non-linear transformation of the embedding representation of speed s_i and road distance l_i yields the reachable time representation a_i:

$$a_i = \sigma(w_1^T s_i + w_2^T l_i) \,, \tag{7}$$

where the $\sigma(\cdot)$ function represents the activation function.

Finally, the reachable time t_i can be sampled from a Gaussian distribution:

$$t_i \sim \mathcal{N}(\mu_{a_i}, \mathrm{diag}(\sigma^2_{a_i})) \,, \tag{8}$$

where the mean μ_{a_i} and the standard deviation σ_{a_i} are learned by a Multilayer Perceptron (MLP) function $g_4(f_4(a_i))$ during the training phase.

The optimization function of the model is:

$$max \log p_\theta(T^{(1)}, T^{(2)}, \cdots, T^{(N)}). \tag{9}$$

The evidence lower bound (ELBO) corresponding to each trajectory is derived as:

$$\log p_\theta(T) \geq \mathcal{L}_{\mathrm{ELBO}}(\theta) = \mathbb{E}_{q_{\gamma,\phi}(z,r,s,t|T)} \left[\log \frac{p_{(\phi,\gamma,\theta)}(z, r, s, t, T)}{q_{\gamma,\phi}(z, r, s, t|T)}\right] . \tag{10}$$

Therefore, the final optimization function is:

$$\max \mathbb{E}_{q_{\gamma,\phi}(z,r,s,t|T)} \left[\log \frac{p_{(\phi,\gamma,\theta)}(z, r, s, t, T)}{q_{\gamma,\phi}(z, r, s, t|T)}\right] . \tag{11}$$

The overall procedure of our proposed algorithm is depicted in Algorithm 1. The fundamental concept involves leveraging convolutional neural networks to extract latent traffic patterns z from traffic data Z. Subsequently, the latent road features r_i are inferred from the road properties. The combination of latent traffic patterns and road features constitutes the speed s of the road. Following this, the latent reachable time t_i can be inferred from the transformation of road length and road speed.

Algorithm 1: Training of our model

Input : The trajectory T
Output: Model parameters
/* Infer latent traffic pattern z */
1 $f_1(Z) \leftarrow \text{CNN}(Z_{t_i})$
2 $\mu_Z, \sigma_Z^2 \leftarrow \text{MLP}(f_1(Z))$
3 $z \sim \mathcal{N}(\mu_Z, \text{diag}(\sigma_Z^2))$
/* Infer latent road featuren r */
4 $f_c \leftarrow f_{ms} \parallel f_{nl} \parallel f_{rt} \parallel f_{ow}$
5 $\mu_{f_c}, \sigma_{f_c}^2 \leftarrow \text{MLP}(f_2(f_c))$
6 $r \sim \mathcal{N}(\mu_{f_c}, \text{diag}(\sigma_{f_c}^2))$
/* Infer the speed s of the road   */
7 $x_i \leftarrow \text{Linear}(W_1 r_i + W_2 z)$
8 $\mu_{x_i}, \sigma_{x_i}^2 \leftarrow \text{MLP}(f_3(x_i))$
9 $s_i \sim \mathcal{N}(\mu_{x_i}, \text{diag}(\sigma_{x_i}^2))$
/* Infer latent reachable time t_i */
10 $a_i \leftarrow \sigma(w_1^T s_i + w_2^T l_i)$
11 $\mu_{a_i}, \sigma_{a_i}^2 \leftarrow \text{MLP}(f_4(a_i))$
12 $t_i \sim \mathcal{N}(\mu_{a_i}, \text{diag}(\sigma_{a_i}^2))$
13 Optimize the Equation 11
14 **return** *Updated model parameters*

4.3 Dynamic Route Recommendation

In this section, we introduce the methodology for dynamically recommending optimal routes based on latent reachable time.

The optimal path optimization based on the generation model is as follows:

$$p_\theta(e_i|e_{<i}, h_i) = \text{Mult}(\text{softmax}(\sigma(\alpha^T h_i + \beta^T d + \gamma^T t_i))), \qquad (12)$$

where e_i denotes the next road segment selected by the model. h_i and d represent the embedded representation of the current road and the destination, respectively, which are obtained through Lipschitz embeddings [1]. Furthermore, t_i denotes the learned reachable time of the road. The softmax($\cdot$) function maps the embedded representation of the trajectory to the road space, and the final route with the highest probability is chosen as the selected route. The optimal path optimization aims to select the next path with the shortest reachable time towards the destination d.

The overall process of dynamic route recommendation algorithm is shown in Algorithm 2. The key idea is to compute the optimal next path by evaluating the probabilities of all possible roads based on their embedded features and current conditions, selecting the route with the highest likelihood of minimizing reachable time.

Algorithm 2: Dynamic route recommendation

    Input   : The trajectory $T_{<i}$, the learned parameters in Algorithm 1
    Output: The recommended route e_i
1 score$_i \leftarrow \sigma(\alpha^T h_i + \beta^T d + \gamma^T t_i)$
    /* Probabilities for each road segment   */
2 $p_i \leftarrow$ softmax(scores)
    /* Select segment with highest probability   */
3 $e_i \leftarrow$ argmax(p_i)
4 **return** e_i

---

5   Experiments

5.1   Experimental Setup

Dataset. We evaluate our approach using two openly accessible datasets, similar to the benchmark study in [18]. The road network details for each city were acquired from OpenStreetMap, and their statistics are outlined in Table 1.

Table 1. Dataset statistics after pre-processing.

Statistics	Harbin(HRB)	Beijing(BJG)
No. of nodes	6, 598	31, 199
No. of edges	16, 292	72, 156
No. of trajectories	1, 133, 548	1, 382, 948
Avg trip length (km)	10.92	7.39
Avg number of edges/trip	56.81	36.08

- Harbin (HRB) [22]: This dataset comprises actual trajectories from 13,000 taxis in Harbin, China. The time gap between two successive trajectory points is approximately 0.5 min.
- Beijing (BJG) [23]: This dataset includes real trajectories from 67,000 taxis in Beijing, China. The sampling interval between consecutive trajectory points is around 3.03 min.

Preprocessing. We follow the preprocessing procedures outlined in [18]. Specifically, we employ map-matching techniques to align the GPS trajectories extracted from the taxi datasets with the road network obtained from Open-StreetMap.

Competitors. We compared our proposed with several advanced methods.

- NEUROMLR [18]: An advanced model for route recommendation that introduces a generative model conditioned on real-time traffic conditions. It comprises two variants: NEUROMLR-Dijkstra (NEUROMLR-D) and NEUROMLR-Greedy (NEUROMLR-G), using Dijkstra and Greedy algorithms, respectively.

- DEEPST [21]: A sophisticated model for spatial transition learning on road networks. It presents a deep probabilistic model considering three factors: the past traveled route, the destination, and real-time traffic conditions.
- CSSRNN [33]: A deep learning-based approach for modeling spatiotemporal data such as trajectory data and traffic flow data. By combining these two types of layers, CSSRNN effectively models and predicts spatiotemporal sequence data.
- QP (Quickest Path): A method utilized to determine the quickest route between two locations. In this algorithm, the weight of each edge on the road network corresponds to the mean travel time of the route, calculated from historical datasets.
- SP (Shortest Path): A method employed to compute the shortest path between the source and destination locations.

Performance Metrics. To comprehensively evaluate the performance of our proposed method and compare it with other baseline approaches, we utilize performance metrics as in [18], namely Precision and Recall. These metrics enable effective comparison between the generated route R^* and the ground truth time-dependent path R. Given the true route R, we generate the most likely route R^*. The metrics are defined as:

$$\text{Precision} = \frac{\sum_{e \in (R \cap R^*)} l(e)}{\sum_{e \in R^*} l(e)} , \quad (13) \qquad \text{Recall} = \frac{\sum_{e \in (R \cap R^*)} l(e)}{\sum_{e \in R} l(e)}, \quad (14)$$

where $l(e)$ denotes the length of edge e.

Hyperparameter Settings. We perform hyperparameter tuning using grid search for all methods on all datasets. The hyperparameter scopes include learning rate from the set {0.001, 0.005, 0.01, 0.05, 0.1}, batch size from {8, 16, 32, 64, 128}, embedding size of trajectory points from {32, 64, 128, 256, 512}, and MLP hidden size {32, 64, 128, 256, 512}.

Table 2. Effectiveness on all datasets.

Algorithm	Precision(%)		Recall(%)	
	HRB	BJG	HRB	BJG
NEUROMLR-D	66.1	77.9	49.6	76.5
NEUROMLR-G	59.6	75.6	48.6	74.5
CSSRNN	49.8	59.5	51.1	68.8
DEEPST	51.9	60.3	27.3	33.2
SP	46.4	59.2	31.3	55.5
QP	40.7	51.4	28.6	50
TRoute	**66.7**	**78.4**	**50.5**	**76.7**

5.2   Effectiveness Evaluation

We evaluate the effectiveness of all methods presented in Table 2. It's worth noting that we provide average results over five runs for all methods. We find:

(1) Our model, TRoute, consistently outperforms other competitors in terms of precision and recall. It achieves the highest precision and recall scores on both datasets, indicating its effectiveness in accurately recommending the next paths. Specifically, TRoute shows an average improvement of 14.35% and 14.02% on two datasets compared to other baselines with respect to precision and recall, respectively. This underscores the effectiveness of considering latent reachable time in our model architecture.

(2) Non-learning-based methods, such as SP and QP, exhibit relatively lower recalls compared to other methods, suggesting potential challenges in recommending time-dependent routes compared with learning-based methods. This limitation arises from their inability to learn both road-related patterns and real-time traffic patterns from dynamic data. However, these patterns prove beneficial for recommending paths, especially under varying traffic conditions.

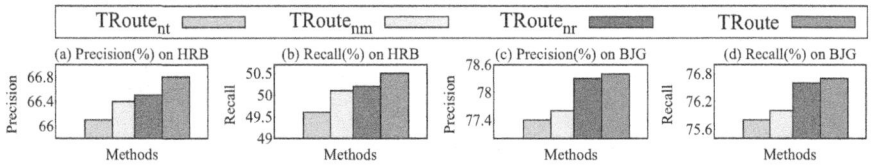

Fig. 2. The ablation study w.r.t. evaluation metrics on two real datasets.

5.3   Ablation Study

We conduct an ablation analysis on the TRoute model to evaluate the impacts of its primary designs, as illustrated in Fig. 2. One variant, denoted as $TRoute_{nt}$(no traffic), ignores dynamic traffic conditions. Another variant, labeled as $TRoute_{nm}$(no maximum speed), omits maximal speed limits from road properties. The third variant, termed $TRoute_{nr}$ (no road), excludes consideration of road properties altogether. From Fig. 2, we find:

(1) TRoute significantly outperforms $TRoute_{nt}$, with the most notable performance gap observed in the BJG dataset. This emphasizes the pivotal role of dynamic traffic conditions in route recommendation.

(2) TRoute surpasses $TRoute_{nm}$ across both datasets, particularly excelling in the BJG dataset. This underscores the benefits of incorporating maximum speed limits when inferring latent patterns for reachable time.

(3) TRoute shows superior performance compared to $TRoute_{nr}$, with considerable enhancements observed in the HRB dataset. This validates the importance of utilizing road properties in inferring reachable time for route recommendation.

6   Conclusions

This study delves into the challenge of dynamic time-dependent route recommendation within road networks. We introduce a novel generative neural network named TRoute, designed to infer latent reachable time from evolving traffic conditions on route. Additionally, TRoute enhances the effectiveness by incorporating static road properties, such as speed limits and the number of lanes. Through evaluation on two real trajectory datasets, our approach demonstrates improvements of 14.35% and 14.02% compared with existing competitors in terms of precision and recall, respectively.

Acknowledgement. Xiaolin han was supported by NSFC under Grant 62302397, the fund of Laboratory for Advanced Computing and Intelligence Engineering (No. 2023-LYJJ-01-021) and Fundamental Research Funds for the Central Universities, China under Grant No. D5000230191. Chenhao Ma was partially supported by NSFC under Grant 62302421, Basic and Applied Basic Research Fund in Guangdong Province under Grant 2023A1515011280, the Guangdong Provincial Key Laboratory of Big Data Computing, The Chinese University of Hong Kong, Shenzhen.

References

1. Bourgain, J.: On lipschitz embedding of finite metric spaces in hilbert space. Israel J. Math. **52**, 46–52 (1985)
2. Chen, Y., et al.: Robust road network representation learning: when traffic patterns meet traveling semantics. In: CIKM, pp. 211–220 (2021)
3. Chen, Z., Shen, H.T., Zhou, X.: Discovering popular routes from trajectories. In: ICDE, pp. 900–911. IEEE (2011)
4. Dehne, F., Omran, M.T., Sack, J.R.: Shortest paths in time-dependent fifo networks. Algorithmica **62**(1–2), 416–435 (2012)
5. Delling, D., Nannicini, G.: Core routing on dynamic time-dependent road networks. INFORMS J. Comput. **24**(2), 187–201 (2012)
6. Demiryurek, U., Banaei-Kashani, F., Shahabi, C., Ranganathan, A.: Online computation of fastest path in time-dependent spatial networks. In: Pfoser, D., Tao, Y., Mouratidis, K., Nascimento, M.A., Mokbel, M., Shekhar, S., Huang, Y. (eds.) SSTD 2011. LNCS, vol. 6849, pp. 92–111. Springer, Heidelberg (2011). https://doi.org/10.1007/978-3-642-22922-0_7
7. Fu, K., Meng, F., Ye, J., Wang, Z.: Compacteta: a fast inference system for travel time prediction. In: SIGKDD, pp. 3337–3345 (2020)
8. Goldberg, A.V., Harrelson, C.: Computing the shortest path: a* search meets graph theory. In: SODA, vol. 5, pp. 156–165 (2005)
9. Goldberg, A.V., Kaplan, H., Werneck, R.F.: Reach for a*: shortest path algorithms with preprocessing. In: The Shortest Path Problem, pp. 93–139. Citeseer (2006)
10. Guo, C., Yang, B., Hu, J., Jensen, C.: Learning to route with sparse trajectory sets. In: ICDE, pp. 1073–1084. IEEE (2018)
11. Guo, C., Yang, B., Hu, J., Jensen, C.S., Chen, L.: Context-aware, preference-based vehicle routing. VLDBJ **29**, 1149–1170 (2020)
12. Han, X.: Traffic incident detection: a deep learning framework. In: MDM, pp. 379–380. IEEE (2019)

13. Han, X., Cheng, R., Grubenmann, T., Maniu, S., Ma, C., Li, X.: Leveraging contextual graphs for stochastic weight completion in sparse road networks. In: SDM. SIAM (2022)

14. Han, X., Cheng, R., Ma, C., Grubenmann, T.: Deeptea: effective and efficient online time-dependent trajectory outlier detection. PVLDB **15**(7), 1493–1505 (2022)

15. Han, X., Dell'Aglio, D., Grubenmann, T., Cheng, R., Bernstein, A.: A framework for differentially-private knowledge graph embeddings. J. Web Semant. **72**, 100696 (2022)

16. Han, X., Grubenmann, T., Cheng, R., Wong, S.C., Li, X., Sun, W.: Traffic incident detection: a trajectory-based approach. In: ICDE, pp. 1866–1869 (2020)

17. Han, X., Grubenmann, T., et al.: FDM: effective and efficient incident detection on sparse trajectory data. Inf. Syst. 102418 (2024)

18. Jain, J., Bagadia, V., Manchanda, S., Ranu, S.: Neuromlr: robust & reliable route recommendation on road networks. NeurIPS **34**, 22070–22082 (2021)

19. Kanoulas, E., Du, Y., Xia, T., Zhang, D.: Finding fastest paths on a road network with speed patterns. In: ICDE, pp. 10–10. IEEE (2006)

20. Li, X., Cheng, R., Najafi, M., Chang, K., Han, X., Cao, H.: M-cypher: a GQL framework supporting motifs. In: CIKM, pp. 3433–3436 (2020)

21. Li, X., Cong, G., Cheng, Y.: Spatial transition learning on road networks with deep probabilistic models. In: ICDE, pp. 349–360. IEEE (2020)

22. Li, X., Cong, G., Sun, A., Cheng, Y.: Learning travel time distributions with deep generative model. In: WWW, pp. 1017–1027 (2019)

23. Lian, J., Zhang, L.: One-month Beijing taxi gps trajectory dataset with taxi ids and vehicle status. In: Proceedings of the First Workshop on Data Acquisition to Analysis, pp. 3–4 (2018)

24. Lin, X., Wang, Y., Xiao, X., Li, Z., Bhowmick, S.S.: Path travel time estimation using attribute-related hybrid trajectories network. In: CIKM, pp. 1973–1982 (2019)

25. Ma, C., Cheng, R., Lakshmanan, L.V., Han, X.: Finding locally densest subgraphs: a convex programming approach. PVLDB **15**(11), 2719–2732 (2022)

26. Ma, C., Fang, Y., Cheng, R., Lakshmanan, L.V., Han, X.: A convex-programming approach for efficient directed densest subgraph discovery. In: SIGMOD (2022)

27. Ma, C., Fang, Y., Cheng, R., Lakshmanan, L.V., Han, X., Li, X.: Accelerating directed densest subgraph queries with software and hardware approaches. VLDBJ **33**(1), 207–230 (2024)

28. Ma, C., Fang, Y., Cheng, R., Lakshmanan, L.V., Zhang, W., Lin, X.: Efficient algorithms for densest subgraph discovery on large directed graphs. In: SIGMOD, pp. 1051–1066 (2020)

29. Nannicini, G., Delling, D., Schultes, D., Liberti, L.: Bidirectional a* search on time-dependent road networks. Networks **59**(2), 240–251 (2012)

30. Wang, H., Liu, J., Peng, C., Sun, H.: Representation learning of multi-layer living circle structure. In: WISA, pp. 125–136 (2023)

31. Wang, Y., Li, G., Li, K., Yuan, H.: A deep generative model for trajectory modeling and utilization. PVLDB **16**(4), 973–985 (2022)

32. Wang, Y., Li, G., Tang, N.: Querying shortest paths on time dependent road networks. PVLDB **12**(11), 1249–1261 (2019)

33. Wu, H., Chen, Z., Sun, W., Zheng, B., Wang, W.: Modeling trajectories with recurrent neural networks. In: IJCAI, vol. 25, pp. 3083–3090

34. Wu, N., Wang, J., Zhao, W.X., Jin, Y.: Learning to effectively estimate the travel time for fastest route recommendation. In: CIKM, pp. 1923–1932 (2019)

35. Yuan, H., Li, G., Bao, Z., Feng, L.: Effective travel time estimation: when historical trajectories over road networks matter. In: SIGMOD, pp. 2135–2149 (2020)
36. Zhao, J., Xu, J., Zhou, R., Zhao, P., Liu, C., Zhu, F.: On prediction of user destination by sub-trajectory understanding: a deep learning based approach. In: CIKM, pp. 1413–1422 (2018)

MGCNet: A Multi-scale Grouped Convolution-Based Seal Detection Method for Painting and Calligraphy Works

Yuzheng Liu, Min Li, and Xueqing Zhao[✉]

School of Computer Science, Xi'an Polytechnic University, Shaanxi Key Laboratory of Clothing Intelligence, Xi'an, China
2207211010@stu.xpu.edu.cn, 2207211080@stu.xpu.deu.cn,
zhaoxueqing@xpu.edu.cn

Abstract. With the advancement of digital culture, there is a continuous integration of deep learning and artwork. This paper proposes a multi-scale grouped convolutional model, abbreviated as MGCNet, to address the time-consuming and laborious issue of traditional seal detection. Firstly, the multi-scale grouping convolution is constructed based on the principles of grouping and multi-scale ideas, allowing for the extraction of features with varying semantic information. Secondly, the convolution module in Bottleneck is replaced within the YOLOv8 feature extraction layer, resulting in a reduction of parameters in the model without compromising detection accuracy. Additionally, Mixed Local Channel Attention (MLCA) is introduced to further optimize the number of parameters in the MGCNet network model. Finally, an analysis is conducted on both the Self-constructed Painting and Calligraphy seal dataset and the PASCAL VOC2012 dataset. The method is compared with existing convolutional modules such as PConv, DWConv, and SCConv focusing on seal region detection accuracy, computational complexity, number of parameters, and model weight file size. The experimental results reveal that compared to YOLOv8, the proposed method achieves an 11.3% reduction in the number of parameters on the PASCAL VOC2012 dataset while enhancing accuracy by 0.7%. This not only reduces storage costs but also facilitates deployment in resource-constrained environments such as mobile devices.

Keywords: Seal Detection · YOLOv8 · Self-Attention

1 Introduction

As works of art, calligraphy, and painting contain rich cultural connotations. Each piece of work reflects the artist's aesthetic interests thoughts and feelings in the social and cultural context of the time. Through appreciating these works, people can learn about the historical development and spiritual connotations of

C. Jin et al. (Eds.): WISA 2024, LNCS 14883, pp. 586–594, 2024.
https://doi.org/10.1007/978-981-97-7707-5_48

different cultures, thus enhancing their understanding of and respect for the diversity of the world. As an important part of calligraphy and painting works, seals occupy an important position in Chinese cultural tradition and are one of the treasures of Chinese culture and art.

Traditional seal detection has traditionally been the domain of experts who observe and assess the author's information on the spot. However, with the development of artificial intelligence and deep learning in image processing research, object detection models are effectively addressing the challenges associated with detecting and classifying image content, and it is a key task in the field of computer vision, which can accurately localize and identify the position and class of an object in an image. In addition, mainstream object detection is categorized into two-stage based on candidate regions, one-stage detection algorithms based on regression, like YOLOv5 [10], I-YOLO [7], SSD [5], and two-stage detection algorithms such as Faster R-CNN [6], DynaMask [4].

In 2020, Ultralytics et al. proposed YOLOv5, which introduces a series of new improvements to YOLOv4, significantly increasing the speed and accuracy. In the model training phase, Mosaic data enhancement, adaptive anchor frame computation, and adaptive image scaling are adopted; in the BackBone network, the Focus structure and the CSP structure are fused to improve the performance; in the Neck network, the FPN and PAN structures are introduced to enhance the feature characterization capability. These improvements make YOLOv5 a superior object detection model. In the same year, Wang et al. proposed yolov7 [9], which suggests "scaling" and "composite scaling" methods for real-time target detectors, which can effectively utilize parameters and computations; In 2023, YOLOv8 is the latest model released by Ultralytics. Compared with the previous YOLO series, YOLOv8 not only performs well in tasks such as image classification and object detection but also shows efficient performance in areas such as instance segmentation. In the context of paintings and calligraphy, object detection algorithms can efficiently replace manual labor by quickly identifying seal categories and locations within images. This capability significantly aids users in distinguishing between seals from different authors within a work, thereby providing valuable evidence for verifying the integrity of painting and calligraphy works. However, there is currently limited research on the detection of models in paintings and calligraphy. This project aims to identify a suitable model for detecting seals in paintings and calligraphy, with a particular focus on improving efficiency. While complex detection models may offer high accuracy, they often require a significant number of parameters and result in resource wastage.

On the other hand, lightweight models have lower resource requirements but may sacrifice detection accuracy. The task of detecting seals presents challenges in terms of parameter quantity and computational demands. Therefore, it is important to analyze the advantages of both complex networks and lightweight networks and integrate their respective strengths into a unified model. For the task of detecting seals in paintings and calligraphy, the shape of the image for detection is uniform, resulting in a relatively light burden on network computation and low difficulty in image detection. This paper aims to analyze the advantages of both complex networks and lightweight networks, incorporating their respective strengths. By considering related work on optimizing object detection models to achieve lightweight designs and taking into account the characteristics of the current task, we propose an efficient convolutional module that optimizes computational parameters within neural networks to achieve a lightweight model, we aim to develop a new multiscale grouped convolutional network (MGCNet) as a more efficient solution for seal detection in paintings and calligraphy works. The contributions and innovations of this paper are as follows:

1. New multi-scale grouping-based convolutional modules are proposed to reduce the number of model parameters while minimizing the loss of semantic information.
2. The addition of the MLCA hybrid local channel attention mechanism to the GMPC module continues to optimize the number of network model parameters.
3. Aiming at the problem that the dataset of seals of calligraphy and painting works is small, a new dataset of seals of calligraphy and painting works is constructed.

The rest of this paper is organized as follows: the proposed method is described in Sect. 2; Simulation experiments are presented in Sect. 3; and finally, the conclusions of this paper are given in Sect. 4.

2   The Proposed Method

In this section, we provide a detailed enumeration of the working principle module of the multi-scale group convolution(MGC) module in the MGCNet network. the MGCNet network's Overall architecture is shown in Fig. 1.

2.1   MGC

A novel multi-scale group convolution (MGC) is constructed with two structures: Group convolution(GC) and Partial convolution(PC).In contrast to conventional convolution, GC convolution operates by partitioning the input channel into four equidistant segments and alternately executing 3 × 3 and 5 × 5 convolutions. Subsequently, a 1×1 convolutional layer is incorporated to facilitate the fusion of features. This innovative approach reduces the parameter count compared to traditional convolution, while simultaneously enabling the handling of diverse scales of semantic information through convolutions of varying sizes. The operational principle of PC convolution is derived from the PC convolution utilized in the GSConv [2] and FasterNet network. The input channels are evenly partitioned into two halves, with one half further divided equally into two halves and processed using 3×3 and 5×5 convolutions, respectively. The remaining half is left unprocessed, after which the processed channels and the uninvolved channels

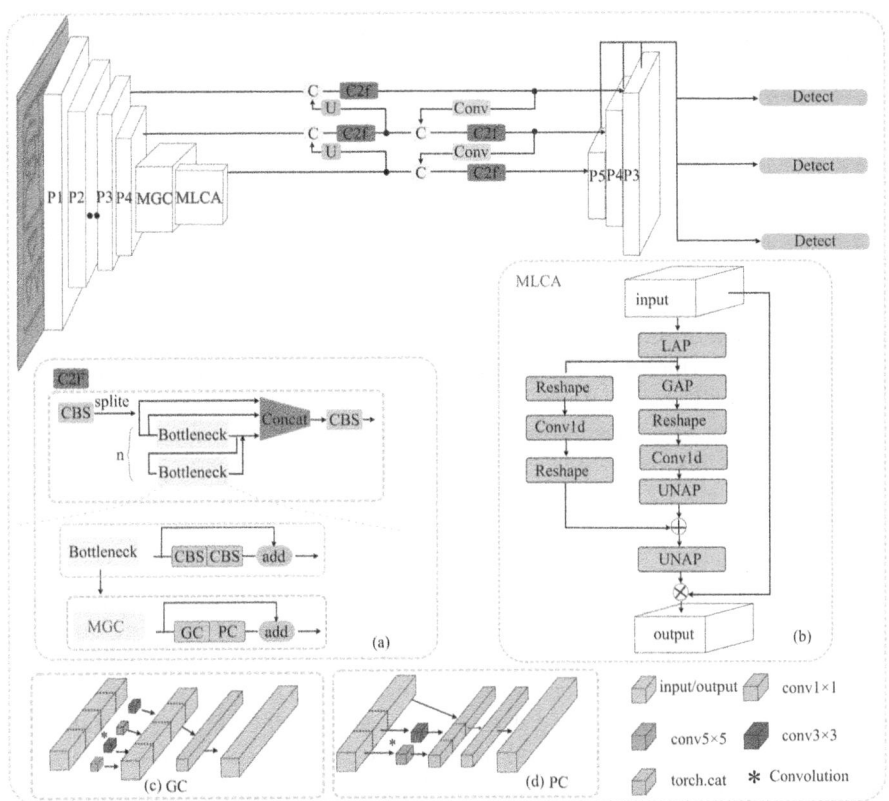

Fig. 1. MGCNet network Overall architecture, (a) is the MGC module, which consists of two convolutions, GC and PC, and is used to replace the Bottleneck module in the C2f structure of the YOLOv8 network, (b) is the module of the MLCA attention mechanism, (c) is the modular diagram of the GC convolution, and (d) is the modular diagram of the PC convolution.

undergo a splicing operation. Ultimately, a 1×1 convolution is employed for feature fusion, effectively reducing parameter count while extracting diverse semantic information. These two convolutions are amalgamated to form the MGC convolution. The advantages of multiple convolutions are scrutinized in terms of model overhead by comparing parameter counts between normal convolution and GC and PC convolution. The combination of these two convolutions forms MGC convolution. The advantages of these convolutions are analyzed in terms of model overhead by comparing parameter counts between ordinary convolutions and GC/PC convolutions. For input $I \in R^{m \times h \times w}$, n convolutional kernels $W \in R^{k \times k}$ are applied to compute the output as $O \in R^{n \times h' \times w'}$. Here I is the input image m is the input channel, h is the height of the input image, w is the width of the input image, n is the output channel, k is the size of the convolution kernel, h is the height of the output image, and w is the width of the output image.

The parametric quantities of the three convolutions are expressed as

$$Conv = m \times n \times k^2 \tag{1}$$

$$PC = (m/4) \times k \times k \times (n/4) + (m/4) \times k \times k \times (n/4) + m \times n \times 1 \times 1 \tag{2}$$

$$GC = 2 \times [(m/4) \times k \times k \times (n/4) + (m/4) \times k \times k \times (n/4)] + m \times n \times 1 \times 1 \tag{3}$$

where input channel m and output channel n, the convolution kernel size is k. Assume that the input convolution kernel size is 3 × 3. Then the number of parameters for normal convolution is 9 × m × n, for PC convolution is 3.125 × m × n, and for GC convolution is 5.25 × m × n. It can be seen that the number of parameters for GC convolution and PC convolution is smaller than that for normal convolution.

2.2   MGCNet

MLCA, Mixed Local Channel Attention, is a novel attention mechanism, to mitigate the issue of accuracy loss due to excessive replacement of MGC modules [8], which combines local and global features, as well as channel and spatial features, resulting in a significant enhancement in accuracy with minimal increase in parametric quantities. Therefore, we designed the MGCNet with the integration of MLCA, which can replace more C2F layers while reducing the number of parameters without compromising accuracy. Simultaneously, the proposed MGCNet can enhance the network's ability to extract valuable features from the data. In conjunction with the information concentration characteristics within the seal area, we did the following when designing the network: substitution of global average pooling with global maximum pooling, during feature map computation, emphasis is placed on extracting key information, and global maximum pooling serves to accentuate the most significant features. To enhance the expressiveness of the MGCNet, it highlights the strongest activation positions within the feature map and enables the model to prioritize essential features. Furthermore, the global maximum pooling exhibits a certain degree of robustness to noise as it focuses primarily on maximum values while disregarding potential noise present in average values. Finally, a Relu activation function can be incorporated before the final output to enhance the model's nonlinearity fitting capability. By introducing an attention mechanism and substituting more MGC modules, we can reduce parameter count without significantly compromising model performance accuracy.

3   Simulation Experiments

In this section, we have employed two datasets to substantiate the generalization capability of our work. The model pre-training weights of YOLOv8n.pt were employed for this purpose. Specifically, the seal detection dataset in paintings and calligraphy comprises a total of 2286 images and encompasses ten author classifications. Additionally, we made use of the public dataset PASCAL VOC2012, which includes a staggering 17125 images across twenty different classifications. By replacing the feature extraction layer of YOLOv8 with the module proposed in this paper, the new network MGCNet is compared and experimented with classical target detection models such as YOLOv8, YOLOv7, YOLOv5, and SSD. The analysis is conducted in terms of mAP accuracy, GFLOP, Parameter count, and the size of the weight file after model training.

(1) Painting and Calligraphy Seal Dataset
Firstly, the validity of the MGCNet model in the painting and calligraphy seal dataset is verified by conducting experiments comparing different target detection models, as shown in Table 1. Additionally, ablation experiments are performed on the baseline model based on the YOLOv8 model by replacing different convolutional modules (PConv, SCConv [3], DWConv [1]) before and after adding the MLCA attention mechanism, as presented in Table 2.

The dataset of Painting and Calligraphy Seal Dataset in Table 1 shows that, firstly, experiments were conducted using different object detection algorithms. It can be seen from the table that the accuracy of YOLOv8 is lower than that of YOLOv7 and YOLOv5, but its parameter count and computational complexity are the lowest among the five object detection algorithms, at 8.1 and 3.01 respectively. The addition of the MGC convolutional module in the experiment resulted in improved precision compared to YOLOv8, with a reduction of 3.7% in parameter count and 14% in computational complexity. The size of the model's trained weight file is only 5.21MB. While reducing the parameter count and computational complexity by 3.7% and 14% respectively, YOLOv8 also significantly reduces the overall size of the model's trained weight file to only 5.21MB while maintaining stable accuracy, thereby releasing more memory space and reducing model complexity.

Table 1. Experimental data of different object detection models on datasets

Models	mAP@0.5/%	GFLOPs	Parameter (M)	Weight File Size (MB)
YOLOv8(baseline)	92.2%	8.1	3.01	5.97
YOLOv7	93.4%	103.3	37	71.4
YOLOv5	93.9%	16.4	7.1	13.7
SSD	63.9%	30.9	25	95.1
Faster-RCNN	80.0%	250.2	28.5	108
MGCNet(ours)	93.1%	7.8	2.59	5.21

Table 2. Validation of MGC and MLCA modules by replacing different convolutional modules in YOLOv8

Models	MLCA	mAP@0.5/%	GFLOPs	Parameter (M)	Size (MB)
YOLOv8(baseline)		92.2%	8.1	3.01	5.97
DWConv		90.9%	7.9	2.71	5.46
SCConv		92.3%	7.9	2.77	5.53
PConv		93.1%	7.8	2.82	5.61
MGCNet(ours)	×	93.1%	7.8	2.59	5.21
MGCNet+MLCA(ours)	√	93.3%	8.3	2.7	5.43

Table 3. Experiments with YOLOv8 adding different modules on the PASCAL VOC2012 dataset

Models	MLCA	mAP@0.5/%	GFLOPs	Parameter (M)	Size (MB)
YOLOv8		68.4%	8.1	3.01	5.97
DWConv		68.1%	8.0	2.86	5.69
SCConv		67.3%	7.9	2.77	5.52
PConv		68.8%	8.0	2.82	5.61
MGCNet(ours)	×	69.1%	8.0	2.84	5.65
MGCNet+MLCA(ours)	√	68.9%	8.5	2.67	5.34

As shown in Table 2, the three convolutional module comparison experiments demonstrate that PConv achieves the highest accuracy rate of 93.1%, while DWConv and SCConv perform poorly. YOLOv8 introduces the MLCA attention mechanism in the feature extraction layer, resulting in an improved accuracy rate of 93.3% compared to YOLOv8 and MGCNet. Replacing more feature extraction layers may lead to a loss of accuracy, but adding the attention mechanism can effectively address this issue.

(2) Generalization experiment
The validity of the MGC module is verified through experiments on a self-constructed dataset. The generalization ability of the module is further confirmed using a public dataset, and a benchmark model is selected for ablation experiments. The comparison of all experimental data is presented in Table 3, showing the impact of adding the MGC module and MLCA attention mechanism.

As shown in Table 3, the accuracy of YOLOv8 on the VOC2012 dataset is 68.4%. Among the three different convolution modules (DWConv, SCConv, and Pconv), PConv convolution achieves the highest accuracy on YOLOv8, while SCConv convolution has the lowest number of parameters and computation on YOLOv8 but results in a significant decrease in accuracy. The inclusion of the MGC module leads to higher accuracy compared to the benchmark model, with a parameter count and computation of 2.84 and 8.0, respectively. It is evident that there is not a significant reduction in the number of parameters, but the addition of the MLCA attention mechanism effectively reduces it to 2.67.

4   Conclusion

Deep learning has replaced the traditional manual detection of seals in paintings and calligraphy to assist individuals in comprehending the content, background, and other aspects of the entire work. This paper introduces a new multi-scale grouped convolutional neural network for seal detection in paintings and calligraphy, known as MGCNet. The design of MGCNet focuses on reducing the number of parameters and floating-point operations through grouping operations, while also further decreasing the number of parameters by enhancing the MLCA attention mechanism without compromising accuracy. In future research, we aim to integrate the characteristics of painting and calligraphy contents with seals, utilizing an object detection model to recognize textual contents while incorporating seal information to enhance understanding and facilitate efficient appreciation of paintings and calligraphy.

Acknowledgment. This work was supported by the National Social Science Foundation of China Art Project (No. 23EH232), the Key Research and Development Program of Shaanxi Province in 2023 (No. 2023-YBGY-404, No. 2023-ZDLGY-48), and Research Center for Culture & Sci-Tech Integration Innovation, Key Research Base of Humanities and Social Sciences of Hubei Province, the Project of Public Digital Cultural Services (GGSZWHFW2024-003), Natural Science Basic Research Program of Shannxi (No. 2021JQ-694) and Shaanxi Province University Young Outstanding Talents Support Program.

References

1. Chollet, F.: Xception: deep learning with depthwise separable convolutions. In: Proceedings of the IEEE Conference on Computer Vision and Pattern Recognition, pp. 1251–1258 (2017)
2. Li, H., et al.: Slim-neck by GSConv: a better design paradigm of detector architectures for autonomous vehicles. arXiv preprint arXiv:2206.02424 (2022)
3. Li, J., Wen, Y., He, L.: Scconv: spatial and channel reconstruction convolution for feature redundancy. In: Proceedings of the IEEE/CVF Conference on Computer Vision and Pattern Recognition, pp. 6153–6162 (2023)
4. Li, R., et al.: DynaMask: dynamic mask selection for instance segmentation. In: Proceedings of the IEEE/CVF Conference on Computer Vision and Pattern Recognition, pp. 11279–11288 (2023)
5. Liu, W., et al.: SSD: single shot multibox detector. In: Leibe, B., Matas, J., Sebe, N., Welling, M. (eds.) ECCV 2016. LNCS, vol. 9905, pp. 21–37. Springer, Cham (2016). https://doi.org/10.1007/978-3-319-46448-0_2
6. Ren, S., et al.: Faster r-cnn: towards real-time object detection with region proposal networks. Adv. Neural Inf. Process. Syst. **28** (2015)
7. Tong, K., Wu, Y.: I-YOLO: a novel single-stage framework for small object detection. In: The Visual Computer, pp. 1–18 (2024)

8. Wan, D., et al.: Mixed local channel attention for object detection. Eng. Appl. Artif. Intell. **123**, 106442 (2023)

9. Wang, C.Y., Bochkovskiy, A., Liao, H.Y.M.: YOLOv7: trainable bag-of-freebies sets new state-of-the-art for real-time object detectors. In: Proceedings of the IEEE/CVF Conference on Computer Vision and Pattern Recognition, pp. 7464–7475 (2023)

10. Zhu, X., et al.: TPH-YOLOv5: improved YOLOv5 based on transformer prediction head for object detection on drone-captured scenarios. In: Proceedings of the IEEE/CVF International Conference on Computer Vision, pp. 2778—2788 (2021)

The International Academic Map of AI Researches——Situation and Trend Exploration Based on WOS Database

Chang Liu⬤, Ying Qin$^{(\boxtimes)}$ ⬤, and Ye Liang⬤

Beijing Foreign Studies University, Beijing 100089, China
`qinying@bfsu.edu.cn`

Abstract. This paper analyzed 65,280 AI research papers from Web of Science (WOS), focusing on the top 20% (13,056 papers) by citation frequency. Using bibliometric and knowledge graph analysis with CiteSpace software, along with Excel, SPSS, and other tools, it explored global AI research trends. Findings suggest a stable, expansive network of AI research globally, with increasingly practical hot spots and expanding interdisciplinary scope. However, there's a noted trend towards technological monopoly due to high research costs.

Keywords: Artificial Intelligence · Bibliometric Analysis · Knowledge Graph · Visualization Analysis · Technology Monopolizes · CiteSpace

1 Introduction

AI research has made significant progress in recent years, thanks to advancements in computing power, data scale, algorithms, and neural networks. Text, image, and speech are important research topics in the field of AI, driving the widespread application of intelligent technology in areas such as healthcare, finance, and transportation. In order to investigate the status, hot issues, trends, and problems of AI research, we conducted a statistical analysis of relevant literature in the AI field over the past 23 years using bibliometrics.

Bibliometric analysis is an effective method for studying the core research, central structure, and evolutionary patterns of a research field. By quantitatively analyzing indicators such as annual publications, high-productive countries or institutions, core authors, and keywords [1], it can demonstrate the research situation. Wang & Yan [2] revealed the current research status and development trend in AI based on the literature during 1999–2018. The research of Tuba Bircan et al. [3] shows that the integration of AI with big data and computing methods has been widely used in the field of social science. E Romero-Riaño [4] explored the keywords that play a guiding role of keywords. The above work reveals the current state of research and hot issues from different perspectives. Still, it lacks higher-level insights. Based on literature from the Web of Science database from 2000 to 2022, we conducted co-occurrence analysis of authors, institutions, keywords, etc., visualizing the development status of international academic AI over the past 23 years.

C. Jin et al. (Eds.): WISA 2024, LNCS 14883, pp. 595–603, 2024.
https://doi.org/10.1007/978-981-97-7707-5_49

The main contributions of this study include: (1) summarizing the current state of AI development through statistics of keywords, research countries, and institutions; (2) visualizing research hotspots, frontiers, and the distribution of research forces; (3) proposing trends analysis in the field of AI.

2   Dataset and Methodology

2.1   Dataset

Web of Science (WOS) is a crucial database for literature research, containing over 10,000 high-impact academic journals. This study focuses on the theme of "artificial intelligence OR AI" in the SCI-Expanded database within the years 2000–2022, resulting in a dataset of 65,280 documents. To ensure quality and representativeness, the top 20% of most cited papers are selected for further analysis of authors, institutions, and countries. Highly cited literature reflects significant influence in the academic community and helps understand research patterns and collaboration networks. This focused dataset enhances the quality of analysis and captures the distribution of research efforts accurately.

2.2   Methodology

Knowledge graph is an interdisciplinary research method based on graph theory, used to construct and represent structured network models. By analyzing and connecting various entities and their relationships, it enables effective organization, retrieval, and inference of knowledge, and is widely used in fields such as information retrieval, intelligent search, and recommendation systems [5]. Research literature is cleaned and the annual publication quantity is described using SPSS to reveal the trend of quantity changes with publication year. Then, CiteSpace [6] is used to analyze the distribution of countries/regions and institutions, co-citation analysis of literature, and co-occurrence analysis of keywords, to draw a knowledge graph, demonstrate the research development status in the field, and conduct keyword co-occurrence, clustering, and emerging research analysis to discover research progress and frontier hotspots in the field.

3   Research Situation, Hot Spots and Frontier

3.1   Research Situation

Between 2000 and 2022, the number of published papers increased exponentially, following Price's curve [7]. The growth rate was consistent, with an average increase of 36.87 papers per year (6.06%) from 2000 to 2015. From 2016 to 2022, the growth accelerated, with an average annual increase of 3,068.67 papers (24.87%). Notably, the growth rates in 2019, 2021, and 2022 were 77.14%, 48.55%, and 40.90% respectively, driven by advancements in deep learning, machine learning algorithms, data availability, industry investment, and successful applications of AI (Fig. 1).

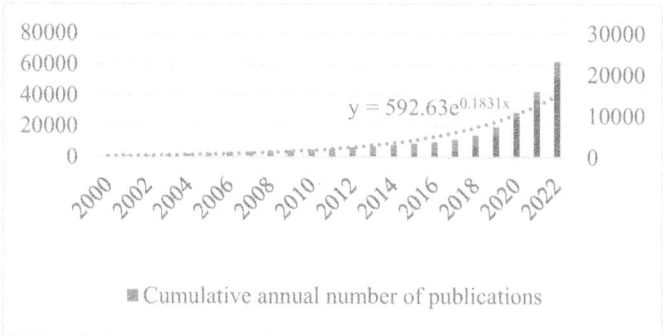

Fig. 1. Number of Published Papers Related to AI from 2000 to 2022

3.2 Research Hot Spots

Word frequency of keywords is crucial for measuring the popularity of research topics [8]. In CiteSpace, a keyword co-occurrence graph was generated with 1326 nodes, 1818 connections, and a network density of 0.0021. The size of nodes indicates keyword occurrence frequency, with machine learning, deep learning, prediction, classification, model, system, neural network, and algorithm identified as research hotspots (Fig. 2).

Fig. 2. Keywords Co-occurrence Map in AI

K-means is used to identify core subgroups in a network by limiting the number of neighbors for each member based on their degree [9]. Core subgroups in keywords were identified by comparing lines with lower connection attribute values and adjusting the number of clusters and similarity threshold. Identified core subgroups include deep learning, artificial neural network, internet of things, swarm intelligence, machine, and electronic skin (Fig. 3).

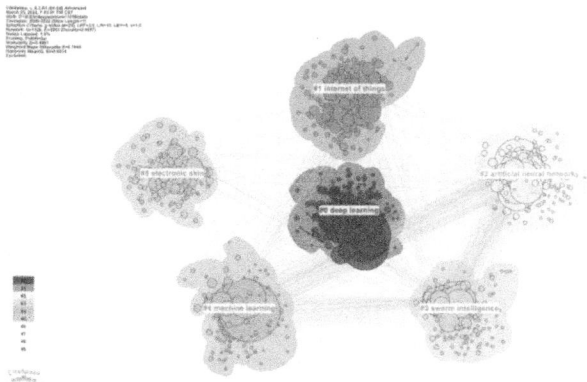

Fig. 3. Keywords Clustering in AI

3.3 Research Frontier

The research frontier encompasses emerging concepts and core issues in development [10]. The emergence detection of keywords in sample literature each year produces a keyword emergence graph. The graph shows the appearance period, with strength indicating the intensity of explosive growth for each keyword in that period.

Keywords	Year	Strength	Begin	End	2000 - 2022
neural networks	2000	65.09	2000	2014	
fuzzy logic	2000	47.97	2000	2016	
identification	2001	10.37	2001	2012	
genetic algorithm	2002	66.98	2002	2017	
knowledge-based systems	2003	3.36	2003	2006	
support vector machines	2004	25.46	2004	2017	
evolutionary computation	2005	12.72	2005	2018	
swarm intelligence	2006	61.44	2006	2017	
computational intelligence	2007	19.01	2007	2018	
model	2008	31.21	2008	2016	
artificial bee colony algorithm	2009	23.14	2009	2018	
artificial neural network	2010	30.79	2010	2018	
artificial bee colony	2011	27.75	2011	2017	
intelligence techniques	2012	6.79	2012	2016	
prediction	2013	9.04	2013	2016	
differential equations	2014	3.94	2014	2015	
electronic skin	2015	8.01	2015	2018	
feature selection	2016	7.01	2016	2017	
activated carbon	2017	4.64	2017	2018	
object recognition	2017	5.81	2017	2018	
plasticity	2018	7.25	2018	2019	
reinforcement learning	2019	7.8	2019	2020	
computational modeling	2020	9.12	2020	2022	
coronavirus	2020	7.51	2020	2022	
energy storage	2020	3.18	2020	2022	
data privacy	2020	3.18	2020	2022	

Fig. 4. Keywords Bursts Map in AI

In summary, from 2000 to 2016, AI research focused on developing technical methods like genetic algorithms and neural networks to enhance AI performance and widen its applications. From 2017 to 2022, the focus shifted towards applying AI in various fields such as medical image processing, environmental research, and energy storage, indicating a move from technical development to practical application (Fig. 4).

The keyword time zone graph in Fig. 5 illustrates the shift in research focus in AI academic research. In the early years, technology and method keywords dominated with a dense distribution trend. In later years, application domain keywords emerged and gradually took over, indicating a change in focus from techniques to practical application. This evolution process validates the analysis and highlights the dynamic nature of AI research frontiers.

This trend underscores the evolution of AI research paradigms and methods. Deep learning and artificial neural networks are becoming central topics, supporting AI development. As technology advances and applications diversify, AI's reach expands across various industries and fields, fostering interdisciplinary collaboration. For example, in AI for medical image analysis, the development of a suitability assessment framework for medical cell images in chromosome analysis has improved the accuracy and efficiency of such analyses [11].

In the future, AI research will focus on seamlessly integrating technology and applications. Advancements in algorithms and technologies will enhance AI performance for diverse scenarios. The growing need for intelligence across industries will drive AI's role in various fields, accelerating the intelligence process. This transformative trend will not only impact AI development but also have profound implications for society.

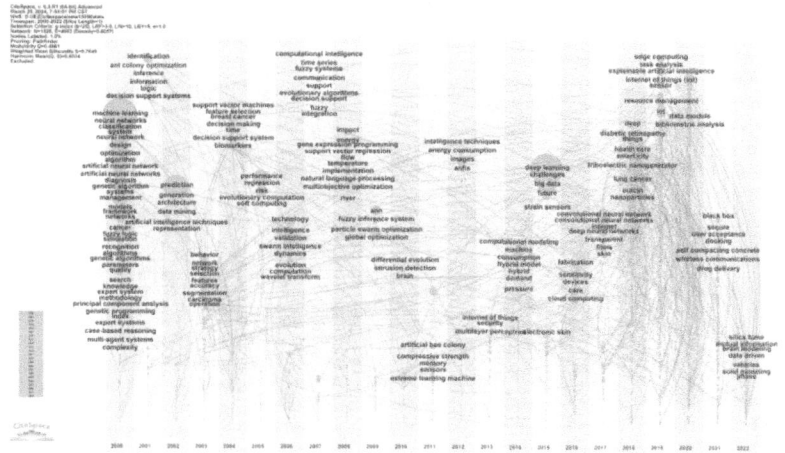

Fig. 5. Keywords Time Zone Map in AI

4 Geographic Distribution

Research strength is reflected in high-impact literature, with 13,056 highly cited papers involving 133 countries and regions. The United States leads in centrality and number of highly cited papers, followed by France, Australia, the UK, and Italy. China ranks sixth in centrality and second in highly cited articles, attributed to efforts in research investment, personnel training, and international cooperation. Chinese researchers' efforts and international collaboration have advanced AI research. However, Chinese AI research still lags behind developed countries like Europe and the US, with low international cooperation.

Research collaboration is limited, with most publications concentrated in high-producing countries. 63 countries produced fewer than 20 publications over 23 years, indicating disparities in research concepts, technical routes, and resource allocation. This suggests potential barriers in international intellectual property rights and data sharing (Fig. 6 and Table 1).

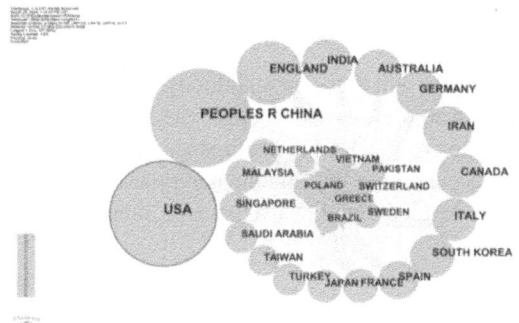

Fig. 6. Country or Region of Author Contribution Map

Table 1. Countries/Regions with More than 400 Highly Cited Papers

	Articles	Centrality	Country/Region		Articles	Centrality	Country/Region
1	3527	0.12	USA	9	426	0.04	Turkey
2	497	0.09	French	10	403	0.04	Taiwan(China)
3	814	0.07	Australian	11	917	0.03	India
4	1453	0.06	U.K	12	783	0.03	Germany
5	674	0.06	Italy	13	428	0.03	Japan
6	3205	0.05	China	14	779	0.02	Canada
7	637	0.05	Spain	15	663	0.01	South Korea
8	780	0.04	Iran				

The University of California System is the most productive institution with 281 highly cited papers. Harvard University ranks first in centrality with 268 papers, followed by the Mayo Clinic and the Chinese Academy of Sciences tying for second place with 242 and 225 papers, respectively.

Highly productive institutions excel in publications but have low centrality, suggesting limited influence in fostering collaboration and exchange. Global AI research remains decentralized and cooperation among institutions is rare (Table 2).

Table 2. Top 10 Institutions for Centrality

	Articles	Centrality	Institutions
1	268	0.23	Harvard University
2	242	0.06	Mayo Clinic
3	225	0.06	Chinese Academy of Sciences
4	211	0.06	University of London
5	191	0.06	University of Texas System
6	152	0.06	Stanford University
7	129	0.06	University System of Ohio
8	125	0.06	University of Toronto
9	121	0.06	Centre National de la Recherche Scientifique
10	121	0.06	Egyptian Knowledge Bank (EKB)

5 Technological Monopoly and Shifting

5.1 Technological Monopoly

Research on AI has become increasingly concentrated in a few countries and institutions in the past decade, as revealed by Niu [12]. By subdividing the period from 2000 to 2022 into five sub-periods, we observe a widening gap in the number of publications between high-producing and low-producing countries/regions. This trend, illustrated in the rose diagram (Fig. 7), reflects a "Matthew effect" where research in AI is highly centralized in a select few countries, leading to a disparity in research output.

2000-2004   2005-2009   2010-2014   2015-2019   2020-2022

Fig. 7. Chart of Country/Region Distribution in Different Periods

A standard statistical test on the number of publications across five time periods revealed that the data does not follow a normal distribution, with rapidly increasing variance. The Shapiro-Wilk (S-W) statistical value is low and has decreased over the years, indicating a significant increase in the unbalanced distribution of literature among countries/regions. This trend suggests the emergence of a technology monopoly (Table 3).

Table 3. Normality Test of the Publications' Number of Countries/Regions

Year	Variance	S-W Value
2000–2004	735.89	0.401
2005–2009	716.18	0.498
2010–2014	1185.44	0.604
2015–2019	24688.67	0.387
2020–2022	72579.82	0.410

5.2 Shifting Landscapes

In the past decade, AI development has been dominated by a few powerful countries and institutions, leading to exclusive "small circle" collaborations and concentrating resources, talent, and funding. This challenges diversity and equity in the global AI research community, increasing competitive pressure and limiting opportunities for other countries. Consequently, this trend restricts the global spread of AI technologies, potentially worsening the digital divide and hindering balanced global economic development.

The shift in AI development patterns is due to high demand for unevenly distributed resources like data, computing power, talent, and funding; the influence of policies and market dynamics attracting resources to certain countries and institutions; and the difficulties faced by regions lacking these resources and policies in making significant AI progress.

6 Conclusion

This paper analyzes AI literature from 2000 to 2022, highlighting the dominance of a few leading institutions and countries in AI research. The reliance on resources and computing power has led to a technology monopoly, hindering diversity and fairness in the global academic research community. This poses challenges to international research cooperation and sharing of AI results.

The authors declare no competing interests related to the content of this article.

Acknowledgment. The work is supported by the National Social Science Foundation (No. 22BYY055).

References

1. Lin, H., Wang, X., Huang, M., et al.: Research hotspots and trends of bone defects based on web of science: a bibliometric analysis. J. Orthopaedic Surg. Res. **15**, 1–15 (2020)
2. Wang, L., Yan, C.: Bibliometric analysis of research on artificial intelligence in information science. Document. Inf. Knowl. 0(1), 53–62 (2020)
3. Bircan, T., Salah, A.A.A.: A bibliometric analysis of the use of artificial intelligence technologies for social sciences. Mathematics **10**(23), 4398 (2022)
4. Romero-Riaño, E., et al.: J. Phys. Conf. Ser. **2046**, 012078 (2021)
5. Zeyuan, L., Yue, C., Haiyan, H., et al.: The Method and Application of Scientific Knowledge Graph. People's Publishing House, Beijing (2008)
6. Yue, C., Chen Chaomei, H., Zhigang, et al.: Principles and Applications of Citation Spatial Analysis: A Practical Guide to CiteSpace. Science Press, Beijing (2014)
7. Price, D.: Little Science. Big Science. University Presses of California, Columbia and Princeton (1965)
8. Zhi, X.: Knowledge graph analysis of library and information science knowledge service. J. Mod. Inf. **33**(2), 166–170 (2013)
9. Yue, H., Liu, S.: Analysis of co-citation structure of management science journals. J. China Soc. Sci. Tech. Inf. **27**(3), 400–406 (2008)
10. Chen, C.: CiteSpaceII: detecting and visualizing emerging trends and transient patterns in science literature. J. Am. Soc. Inf. Sci. Technol. **57**(3), 359–377 (2006)
11. Mo, Z., et al.: A suitability assessment framework for medical cell images in chromosome analysis. In: Yuan, L., Yang, S., Li, R., Kanoulas, E., Zhao, X. (eds) Web Information Systems and Applications. WISA 2023. Lecture Notes in Computer Science, vol. 14094. Springer, Singapore (2023). https://doi.org/10.1007/978-981-99-6222-8_48
12. Niu, J., Tang, W., Xu, F., et al.: Global research on artificial intelligence from 1990–2014: spatially-explicit bibliometric analysis. ISPRS Int. J. Geo-Inf. **5**(5), 66 (2016)

Enhancing Online Education Assessment: A Blockchain-Powered Reliable Behavior Indicator Assessment Framework

Yi Zhang, Peiya Zhang, Yanbin Zhang[✉], Cheqing Jin, Wei Wang, and Bin Su

East China Normal University, Shanghai 200062, China
ybzhang@dase.ecnu.edu.cn

Abstract. The rapid expansion of online education has highlighted the critical need for a reliable and efficient learning evaluation framework. To address this challenge, we propose a blockchain-powered framework for assessing educational behavior indicators. Leveraging blockchain's tamper-proof and traceable characteristics, the framework ensures the security and authenticity of educational data. By integrating multi-source data from the Shuishan Online platform and adopting a multi-dimensional evaluation perspective, the framework comprehensively assesses students' online learning performance. Implementation in a real educational scenario demonstrates the framework's effectiveness in providing credible and comprehensive evaluations of students' learning outcomes, paving the way for a transparent, efficient, and equitable educational environment.

Keywords: Online Learning · Learning Evaluation · Blockchain

1 Introduction

Effective learning evaluation is vital for driving the digital transformation of education. It serves not only as a cornerstone for measuring learning outcomes and ensuring teaching quality but also as a powerful tool for providing personalized learning support and driving educational reform.

In the era of information technology, the education sector faces significant challenges in data security and privacy protection. The emergence of blockchain technology provides a novel solution to these challenges, leveraging its immutability and traceability to offer secure and transparent certification services for student data and meticulously track and record student learning processes. This innovative application model ensures transparent and secure storage and management of student learning data, preventing tampering and misuse, while also employing granular permission management to safeguard privacy and protect sensitive information. By enhancing data reliability and credibility, this blockchain-based Indicators Assessment Framework establishes a solid foundation for subsequent online learning behavior data evaluation.

To comprehensively assess students' learning performance and capabilities, this study has developed a blockchain-based evaluation indicator system that integrates multi-source data. This system, embedded within the online smart education platform Shuishan

C. Jin et al. (Eds.): WISA 2024, LNCS 14883, pp. 604–612, 2024.
https://doi.org/10.1007/978-981-97-7707-5_50

Online [1], adopts a student-centric approach and has collated a substantial set of behavioral indicators, which are stored on the blockchain, encompassing various dimensions of students' online learning behaviors, to construct a five-dimensional index system.

The key contributions of this work are as follows:

(1) By leveraging blockchain technology's decentralized, immutable, and transparent nature, coupled with user-permission access control, the system robustly safeguards the security, credibility, and privacy of educational data. This support for reliable online learning behavior evaluations and educational reforms is thereby ensured.

(2) Through the integration of multi-source learning data and the adoption of a multi-dimensional evaluation perspective, the system comprehensively assesses students' online learning performance, fostering their learning initiative and providing educational decision-makers with a scientific basis for the rational allocation of educational resources.

2 Related Work

Existing online learning platforms predominantly employ centralized databases, rendering them vulnerable to data breaches and privacy compromises due to single point of failure and susceptibility to malicious attacks [2]. In contrast, blockchain technology emerges as a promising alternative due to its decentralized and tamper-proof nature, ensuring data security and authenticity. Wang et al. [3] demonstrated the potential of blockchain by developing a secure data sharing approach for educational resources in IoT environments, utilizing blockchain signatures for user authentication and data encryption to enhance efficiency and security.

Addressing privacy and security concerns, researchers have proposed privacy-preserving data protection mechanisms. Huang Chaoran et al. [4] introduced an approach that combines permission management with Local Differential Privacy (LDP) to enhance data privacy and security while maintaining minimal impact on blockchain performance. Another study by Sun Li [5] explored the transaction processes of consortium blockchains like Fabric and their latest fine-grained data protection mechanisms, such as the privacy dataset, proposing an information protection mechanism for online education resource alliances that was validated in actual blockchain environments. However, existing research primarily focuses on information protection, often neglecting the trustworthiness of information.

The dynamic nature of smart education scenarios necessitates frequent updates and upgrades of smart contracts. However, the immutability of data in blockchain poses a challenge, requiring redeployment of new contract versions and leading to redundant storage and potential security risks. Liu Yunxia et al. [6] proposed a loose coupling model that separates logical operations from data, but this model only supports the update of logical contracts, with all logical contracts still relying on a single data contract. In contrast, the proposed framework employs a fully decoupled contract architecture, further improving upon existing solutions.

3   Credible Educational Behavioral Indicators Assessment Framework

3.1   System Architecture

The learning process data for this study is derived from Shuishan Online, an ECNU-developed online education platform that innovates educational scenarios and blends scalable with personalized education. To cater to educational scenarios, the behavioral indicator system's architecture is built on a layered, trusted framework, cascading into four tiers: Interaction, Functional Service, Contract, and Underlying Chain (as shown in Fig. 1). The key technologies involved are as follows: (1) the construction of a smart contract architecture with separate logic and data to minimize redundant storage and potential security risks during updates; (2) the design of a privacy data protection scheme based on user permission access control to meet different privacy protection needs in educational scenarios; (3) the collaboration with the "Shuishan Online" educational platform to filter and process real-world educational data, offering users trusted services for storing and querying this data.

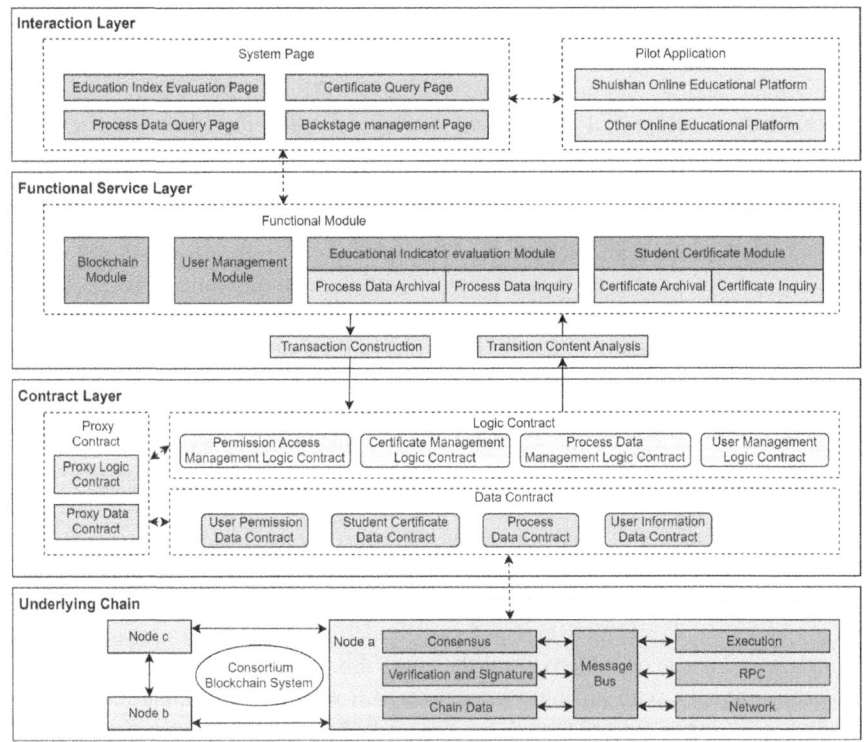

Fig. 1. Overall Architecture of the System

Interaction Layer. The Interaction Layer acts as the interface between user requirements and system functions. When a user initiates a request on the Shuishan Online platform, they are redirected to the system's Interaction Layer page. The Interaction Layer then sends the corresponding operation instructions to the Functional Service Layer. Upon receiving the system's response, the Interaction Layer provides a graphical interface for users to understand the system's feedback.

Functional Service Layer. This layer serves as a middleware between the Interaction Layer and the Contract Layer, featuring four modules: the Blockchain Connection module, which handles communication with the underlying blockchain and encapsulates RPC interfaces for other modules to use; the User Management module, responsible for user registration and permission management; the Student Certificate module, which manages interfaces for storing and retrieving trusted certificates, constructs contract input parameters, and calls lower-level contracts; and the Process Data module, which primarily manages interfaces for storing and retrieving process data.

Contract Layer. The Contract Layer employs a smart contract architecture with separated logic and data (See Sect. 3.2 for details), and upon receiving a transaction constructed by the Functional Service Layer, it reads and writes state data to implement the query and write operations of chain data.

Underlying Chain. The underlying infrastructure utilizes a consortium blockchain system, CITA. CITA micro-service sizes the consensus, validation, chain data, execution, RPC, and network modules of the blockchain, with each node's different modules communicating through a message bus.

3.2 Credible Key Technologies

Smart Contracts with Separation of Logic and Data
The smart contract design in this study employs a fully decoupled contract architecture (as shown in Fig. 2). This architecture, by differentiating between logic operations and data storage, mitigates data redundancy and security risks associated with contract updates, thereby enhancing the security of educational data.

The contracts are classified into four types:

Logic Contracts. These contracts are responsible for implementing specific logic operations and providing interfaces to the Functional Service Layer. They cache call parameters and simple state parameters during transaction execution.

Data Contracts. These contracts are responsible for constructing the formats of different types of educational data and storing the corresponding data as contract state data, providing access and query services for educational data.

Proxy Logic Contracts. These contracts maintain the mapping relationships between logic contract interfaces and addresses, as well as between interfaces and contract version numbers. They encapsulate get and set interfaces for the Functional Service Layer to query and update logic contract addresses.

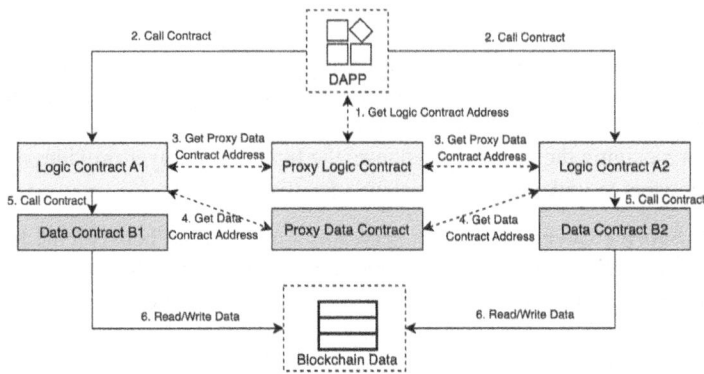

Fig. 2. Architecture of the Smart Contracts

Proxy Data Contracts. These contracts maintain the mapping between data contract interfaces and addresses, as well as between interfaces and contract version numbers. They encapsulate get and set interfaces for logic contracts to call.

Smart Contract-Based Privileged Access Managing

To mitigate the data privacy and security challenges posed by the public and transparent nature of blockchain in smart education, the system incorporates a permission management mechanism based on smart contracts.

Table 1. User classification and access rights

User Category	Read Data	Write Data	Addition and deletion of viewers	Addition and deletion of managers
Trustworthy Institution	√	√	√	√
Data Managers	√	√	×	×
Data Reviewers	√	×	×	×
External Users	×	×	×	×

As shown in Table 1, this mechanism ensures the security and privacy of educational data by assigning different levels of permissions to various user categories within the educational certification scenario.

When faced with the need to certify educational data, the system periodically filters process data from the Shuishan database. The Functional Service Layer then calls the logic contract interface to trigger user permission checks, allowing only users with write data permissions to add new data. Upon successful certification, the Functional Service

Layer receives a transaction receipt from the blockchain, which it parses to retrieve important information such as the transaction hash and contract execution result. The execution result includes the contract address where student learning data is stored, and the system generates a unique QR code for each student's data to bind it to the certified data on the chain. When the QR code is scanned or a data query is initiated, the Functional Service Layer calls the logic contract's query interface to verify the user's read data permissions. If the check passes, the logic contract can proceed to retrieve the corresponding data from the data contract. Upon receiving the transaction receipt, the Functional Service Layer parses it to recover the hexadecimal data stored on the chain into a human-readable format.

3.3 Design of Evaluation Indicator System for Online Learning Behavior

The tag category system encompasses the classification, architecture, and organization of a specific category of objects. In this context, tags are derived from original data and serve as data entities that can be directly leveraged for business purposes and generate value [7]. In the methodology of tag category systems, the object refers to the target that needs to be studied in the real world.

For the online education domain, we have designed a detailed tag category system for both entity objects and relationship objects. The design process of the tag category system focuses on two main themes: data/tag situations and business requirements. The entire design includes five levels of front-end scene categories: entity layer, abstract layer, foundation tags, composite labels, and application scenarios (as shown in Fig. 3), with the overall teaching tag category system for online education becoming increasingly abstract from bottom to top.

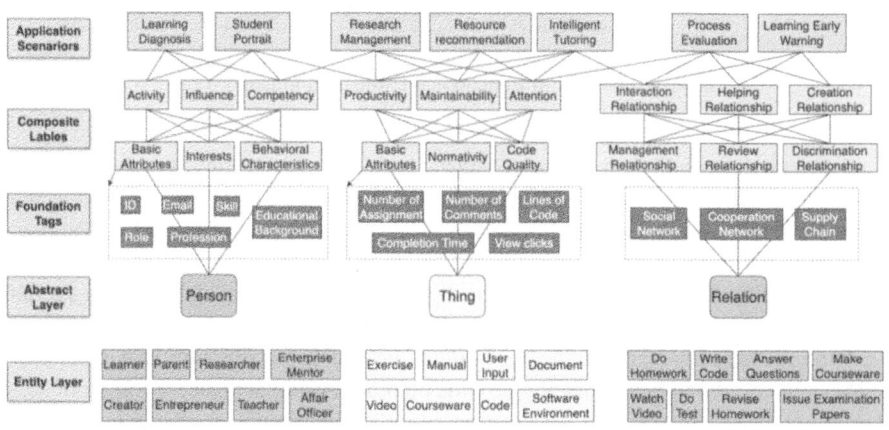

Fig. 3. Design of the Educational Evaluation Indicator Framework

It determines the structure of the original data records and the business-related scenario requirements, and iteratively updates the underlying tag structure based on the real-world situation to fit the updated business needs. This study, based on the actual table structure in the Shuishan Online database, has constructed an indicator system design suitable for the Shuishan platform.

As shown in Fig. 4, a wealth of process data is generated through the platform's various educational phases, including teaching, learning, practicing, testing, creating, and evaluating. Students' process learning behavior data is stored in four different MySQL databases according to the platform's different learning modules: Shuishan Classroom, Shuishan Workspace, Shuishan Judgefield, and Shuishan Codepark. This study selected the "First Lesson for Freshmen" course and collected nearly one million student behavioral records and hundreds of student learning data, and read from the database to certify the student learning behavior data to the blockchain, ensuring the authenticity and reliability of the data.

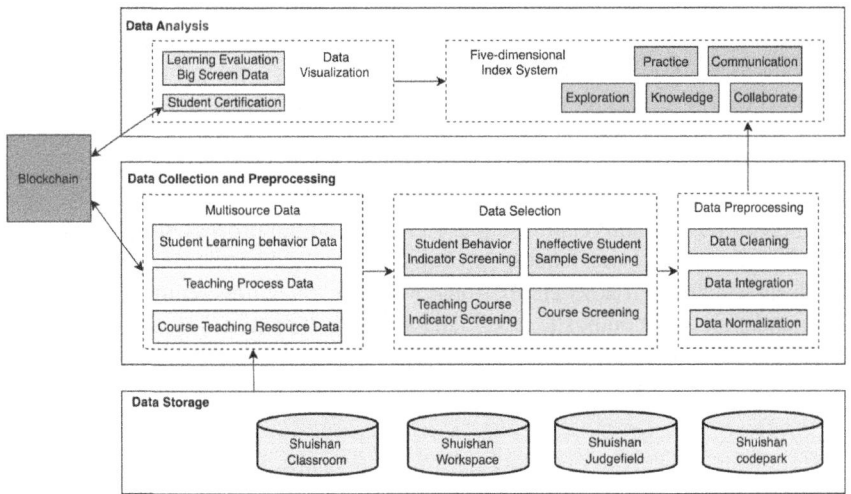

Fig. 4. Processing of Educational Evaluation Indicator Data

Based on these indicators, through data analysis, five dimensions of evaluation indicators were designed to construct a process-oriented evaluation indicator system that integrates multi-source data. Additionally, based on this indicator system, a visualization tool in the form of a digital dashboard has been developed to present and analyze students' online learning behaviors from diverse angles. All this data can be queried through the blockchain, ensuring the security and traceability of educational data.

Through these Evaluation Indicator Systems, educators can precisely understand students' online learning behaviors and provide comprehensive guidance throughout their learning process. This facilitates educational innovation, optimizes curriculum design, and enhances teaching strategies, thereby improving learning effectiveness and educational quality. The comprehensive, intuitive, and diverse dimensional indicator display also helps students to actively identify their learning characteristics, stimulate their learning initiative, and foster autonomous learning habits.

4 Application and Future Prospects

The blockchain-based educational behavior indicator assessment framework proposed in this study ensures the secure storage of educational data while providing authentic and personalized assessments of learning outcomes. This approach establishes the foundation for a transparent, efficient, and equitable educational environment.

To validate the secure storage of educational data at scale, the framework has been successfully implemented in a real educational scenario, the "Computer Science for Freshmen" course on the Shuishan Online Education Platform (as shown in Fig. 5). By leveraging blockchain technology, the framework tracks, records, and evaluates the extensive behavioral data generated by students during their learning journey, offering a more credible, comprehensive, and holistic view of their growth and progress, beyond the traditional focus solely on grades.

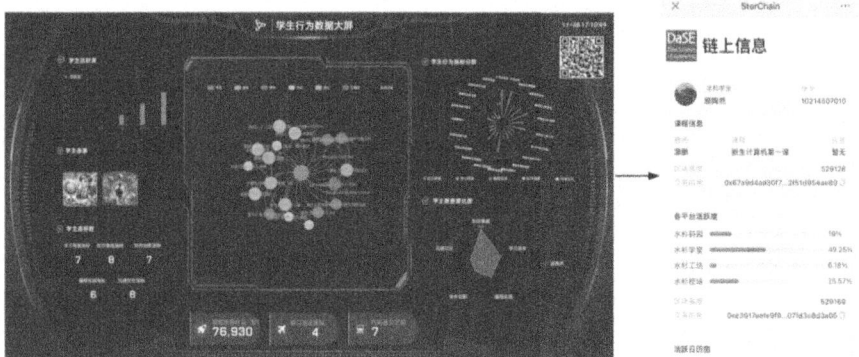

Fig. 5. Student Process Data Tracking and Indicator Evaluation Service

At institutions like East China Normal University, the study has successfully archived over 500,000 educational data records on the blockchain and validated them comprehensively, ensuring their high authenticity and reliability.

This approach is suitable for online course education evaluation and higher education assessment reforms. However, limitations in current cross-chain interoperability and data sharing functionalities present opportunities for future research. Addressing these limitations, specifically through advancements in cross-chain compatibility, integration of diverse educational resources, and customization for individual needs, promises to further propel the digital transformation of education.

References

1. Wang, W., et al.: Shuishan online: constructing and teaching with a data-driven learning platform. In: IEEE International Conference on Engineering, Technology & Education (TALE), pp. 1–8 (2021)

2. Yang, C., Dong, X., Zhang, N., Wen, Z.: Will data sharing scheme based on blockchain and weighted attribute-based encryption. In: International Conference on Web Information Systems and Applications, pp. 391–402. Springer, Singapore (2023)
3. Wang, B.: Data sharing method of Internet of Things environment based on blockchain technology. Wirel. Internet Technol. **20**(16), 158–160+168 (2023)
4. Huang, C., Tong, X., Zhang, Z., Jin, C., Yang, Y., Qin, G.: Research on contract architecture and data privacy for education-oriented blockchain applications. J. East China Normal Univ. (Nat. Sci.) **2022**(5), 61–72 (2022)
5. Sun, L.: Research and application of information protection mechanism for online education resource alliance. Netinfo Secur. **21**(9), 32–39 (2021)
6. Liu, Y.X., Hu, D.S., Jiang, Y.M.: Loose coupling model research for upgrading smart contracts already deployed on blockchain. Appl. Res. Comput. **38**(05), 1309–1313 (2021)
7. Ren, Y.: Label Category System: A Methodology for Business-oriented Data Asset Design, pp. 19–26. Mechanical Industry Press, Beijing (2020)

Author Index

GPSR Compliance

The European Union's (EU) General Product Safety Regulation (GPSR) is a set of rules that requires consumer products to be safe and our obligations to ensure this.

If you have any concerns about our products, you can contact us on ProductSafety@springernature.com

In case Publisher is established outside the EU, the EU authorized representative is:

Springer Nature Customer Service Center GmbH
Europaplatz 3
69115 Heidelberg, Germany

The manufacturer's authorised representative in the EU is Springer
Nature Customer Service Centre GmbH, Europaplatz 3, 69115 Heidelberg,
Germany. If you have any concerns regarding our products, please
contact ProductSafety@springernature.com

Printed and bound by CPI Group (UK) Ltd, Croydon, CR0 4YY
05/05/2026
02102981-0012